Carbohydrate-Based Drugs

Carbohydrate-Based Drugs

Editors

Federica Sodano
Maria Grazia Rimoli

MDPI

Basel • Beijing • Wuhan • Barcelona • Belgrade • Novi Sad • Cluj • Manchester

Editors
Federica Sodano
Department of Pharmacy
University of Naples Federico II
Naples
Italy

Maria Grazia Rimoli
Department of Pharmacy
University of Naples Federico II
Naples
Italy

Editorial Office
MDPI
St. Alban-Anlage 66
4052 Basel, Switzerland

This is a reprint of articles from the Special Issue published online in the open access journal *Pharmaceuticals* (ISSN 1424-8247) (available at: www.mdpi.com/journal/pharmaceuticals/special_issues/carbohydrate_drugs).

For citation purposes, cite each article independently as indicated on the article page online and as indicated below:

Lastname, A.A.; Lastname, B.B. Article Title. *Journal Name* **Year**, *Volume Number*, Page Range.

ISBN 978-3-7258-0466-5 (Hbk)
ISBN 978-3-7258-0465-8 (PDF)
doi.org/10.3390/books978-3-7258-0465-8

Contents

Preface

Dear Colleagues,

Carbohydrates are one of the four major classes of biomolecules. They play important roles in various biological processes, including inflammation and immune response, angiogenesis and metastasis of cancer cells, viral and bacterial infections, and many other cell–cell communications. The tremendous efforts made by glycoscientists in recent decades have allowed carbohydrate-based drugs to pave the way for an anonymous area of the therapeutic world.

Medicinal chemists use carbohydrates, particularly mono- and disaccharides, as valuable carriers or key substructures in order to improve the pharmaceutical phase, pharmacokinetics, and pharmacodynamics of well-known drugs. Resorting to the prodrug strategy or multitarget drugs, the conjugation of a traditional drug to carbohydrates allows many limitations of the parent drug to be overcome, such as lack of site of action specificity, toxicity, and chemical instability. Additionally, carbohydrate-based nanosystems have been shown to better deliver the starting drug or generate selective targeting towards specific body districts, as well as improve the cellular permeability and toxicological profile of the parent drug.

In this Special Issue, "Carbohydrate-Based Drugs", we aim to collect both original articles and up-to-date reviews that focus on the design, synthesis, and biological evaluations of novel carbohydrate-based drugs, with a focus on carbohydrate-based prodrugs and nanosystems.

Federica Sodano and Maria Grazia Rimoli
Editors

pharmaceuticals

Article

Effect of Agave Fructan Bioconjugates on Metabolic Syndrome Parameters in a Murine Model

Eduardo Padilla-Camberos [1], Javier Arrizon [2] and Georgina Sandoval [3,*]

1 Medical and Pharmaceutical Biotechnology Unit, Center for Research and Assistance in Technology and Design of the State of Jalisco, A.C. (CIATEJ), Av. Normalistas No. 800 Col. Colinas de la Normal, Guadalajara C.P. 44270, Jalisco, Mexico
2 Industrial Biotechnology Unit, Center for Research and Assistance in Technology and Design of the State of Jalisco, A.C. (CIATEJ), Camino Arenero 1227, El Bajío del Arenal, Zapopan C.P. 45019, Jalisco, Mexico
3 LIBBA Laboratory, Industrial Biotechnology Unit, Center for Research and Assistance in Technology and Design of the State of Jalisco, A.C. (CIATEJ), Av. Normalistas No. 800 Col. Colinas de la Normal, Guadalajara C.P. 44270, Jalisco, Mexico
* Correspondence: gsandoval@ciatej.mx or gsandoval@confluencia.net; Tel.: +52-333-345-5200

Abstract: Metabolic syndrome is a complex disorder that combines abdominal obesity, dyslipidemia, hypertension, and insulin resistance. Metabolic syndrome affects 25% of the world's population. Agave fructans have shown positive effects on alterations related to metabolic syndrome, so some investigations have focused on their bioconjugation with fatty acids to increase their biological activity. The objective of this work was to evaluate the effect of agave fructan bioconjugates in a rat model with metabolic syndrome. Agave fructans enzymatically bioconjugated (acylated via food-grade lipase catalysis) with propionate or laurate were administered orally for 8 weeks in rats fed a hypercaloric diet. Animals without treatment were used as the control group, as well as animals fed with a standard diet. The data indicate that the group of animals treated with laurate bioconjugates showed a significant decrease in glucose levels, systolic pressure, weight gain, and visceral adipose tissue, as well as a positive effect of pancreatic lipase inhibition. These results allow us to demonstrate the potential of agave bioconjugates, particularly laurate bioconjugates, for the prevention of diseases associated with metabolic syndrome.

Keywords: hypertension; weight gain; oligosaccharides; prebiotics; lipase; acylation

Citation: Padilla-Camberos, E.; Arrizon, J.; Sandoval, G. Effect of Agave Fructan Bioconjugates on Metabolic Syndrome Parameters in a Murine Model. *Pharmaceuticals* **2023**, *16*, 412. https://doi.org/10.3390/ph16030412

Academic Editors: Federica Sodano and Maria Grazia Rimoli

Received: 30 November 2022
Revised: 21 February 2023
Accepted: 2 March 2023
Published: 8 March 2023

1. Introduction

Metabolic syndrome is a cluster of metabolic disorders associated with obesity and type 2 diabetes: insulin resistance, hypertension, hyperlipidemia, hyperglycemia, and cardiovascular disease [1,2]. The prevalence of metabolic syndrome fluctuates worldwide as there is a close association with age, sex, race/ethnicity, and the criteria used for diagnosis [3]. It is estimated that approximately 25% of the world's adult population suffers from metabolic syndrome and that the probability of dying from its complications as well as suffering a stroke increases considerably year by year [4,5]. At present, the main origin of metabolic syndrome has not been established; however, genetic and epigenetic factors, as well as the accelerated lifestyle of individuals and the high caloric intake associated with visceral adiposity, are the main inducers in the development of the syndrome [6,7]. Treatment of people with metabolic syndrome consisting of implementing lifestyle and diet changes, and increasing physical activity can improve the individual components of metabolic syndrome, but reducing cardiovascular risk through treatment of atherogenic dyslipidemia should be addressed directly with medications [8].

To reduce the negative effects of metabolic syndrome on human health, the consumption of bioactive compounds and natural fibers from plants, such as antioxidants and prebiotics, has increased worldwide. In particular, for prebiotics, it is well known that the positive effects of fructans in metabolic syndrome and the biological properties of these polysaccharides depend on the fructosyl linkages inside them, thus they have been used in plant-derived products for the formulation of functional food products against metabolic syndrome.

These plant-derived products have played an important role in maintaining human well-being. For hundreds of years, since ancient times, natural products and their derivatives have been used, mainly in the development of pharmaceuticals for the treatment of human conditions [9]. A recent review by Wang et al. remarked on the potential of several polysaccharides obtained from plants for the treatment of metabolic syndrome, with mechanisms of action associated with the regulation of apoptosis, inflammation, and intestinal microbiota, among others. Such polysaccharides have mainly glycosidic bonds α-(1 $\rightarrow$ 6)-D, α-(1 $\rightarrow$ 4)-D, and β-(1 $\rightarrow$ 4)-D, and their biological activities are closely related to their primary and higher structures [10].

Other well studied groups of plant polysaccharides are inulin and fructans. Fructans from *Agave* genera differ from inulin in the type of linkage. Indeed inulin is lineal, while agave fructans are a complex mixture of fructooligosaccharides containing principally β (2 $\rightarrow$ 1) linkages, and some β (2 $\rightarrow$ 6) and branching moieties, leading to ramified structures [11,12].

Fructans from tequila agave (*Agave tequilana* Weber var. azul) have been shown to have positive effects on metabolic disorders associated with metabolic syndrome [13–16]. For example, studies evaluating the effects of fructans on glucose concentration in animal models have been consistently positive [17]. For instance, Castillo-Andrade et al. [18] evaluated the physiometabolic effects of *Agave salmiana* fructans as a dietary supplement in male Wistar rats. They found that the inclusion of 12.5% *Agave salmiana* fructans in the diet of the animals induced beneficial physiometabolic effects after the seventh day of treatment [18,19].

Functionalization of fructans, such as acylation with fatty acids has also been reported, for instance, for inulin by chemical esterification [20], and by enzymatic bioconjugation of *Agave tequilana* fructans [21]. Some esterified fructans have been tested on metabolic syndrome, but the results were inconclusive, showing an impact on food intake in some cases [20], and no impact on zoometric parameters in others [22].

Animal diet-induced metabolic models are commonly used to study metabolic syndrome because of their simplicity, accuracy, and low cost [23]. The animals most commonly used to develop metabolic syndrome are rats, mainly Wistar and Sprague-Dawley, because they manifest the characteristics of obesity, diabetes, and hypertension. For this study, such a model was chosen and metabolic syndrome was induced with specific diets formulated with fat and carbohydrates that involve a high caloric content [24]. Accordingly, in the present study, we established the metabolic syndrome model in young rats by including a hypercaloric diet and evaluated the effect of agave fructan bioconjugates orally administered on parameters associated with metabolic syndrome.

2. Results and Discussion

2.1. Synthesis of Agave Fructan Bioconjugates

It was previously reported that when enzymatic acylation of agave fructans is performed, only short-chain fructans are acylated [21]; therefore, agave fructans enriched in fructooligosaccharides (FOS) were used. Figure 1 shows the Matrix-Assisted Laser Desorption/Ionization Time-Of-Flight (MALDI-TOF-MS) profile of the FOS used. The m/z distribution ranged from 527.37 to 1826.7, corresponding to FOS with a degree of polymerization (DP) from 3 to 11, respectively (Figure 1). FOS with a DP from 4 to 7 accounted for 53% mol of the total mixture and the most abundant FOS had a DP of 5 (14.8% mol). The mass distribution of this agave FOS mixture used for the acylation reactions is smaller than that of natural agave fructans with DPs ranging from 3 to 30 [10]. This reduction was caused by industrial processing, which favors agave FOS acylation. Indeed, it was previously shown that the immobilized lipase B from *Candida antarctica* Lipozyme™ 435, preferably acylated FOS up to DP 8 [20]. This could be due to the fact that the access of large molecules to the catalytic site of the lipase is hindered due to steric effects [25].

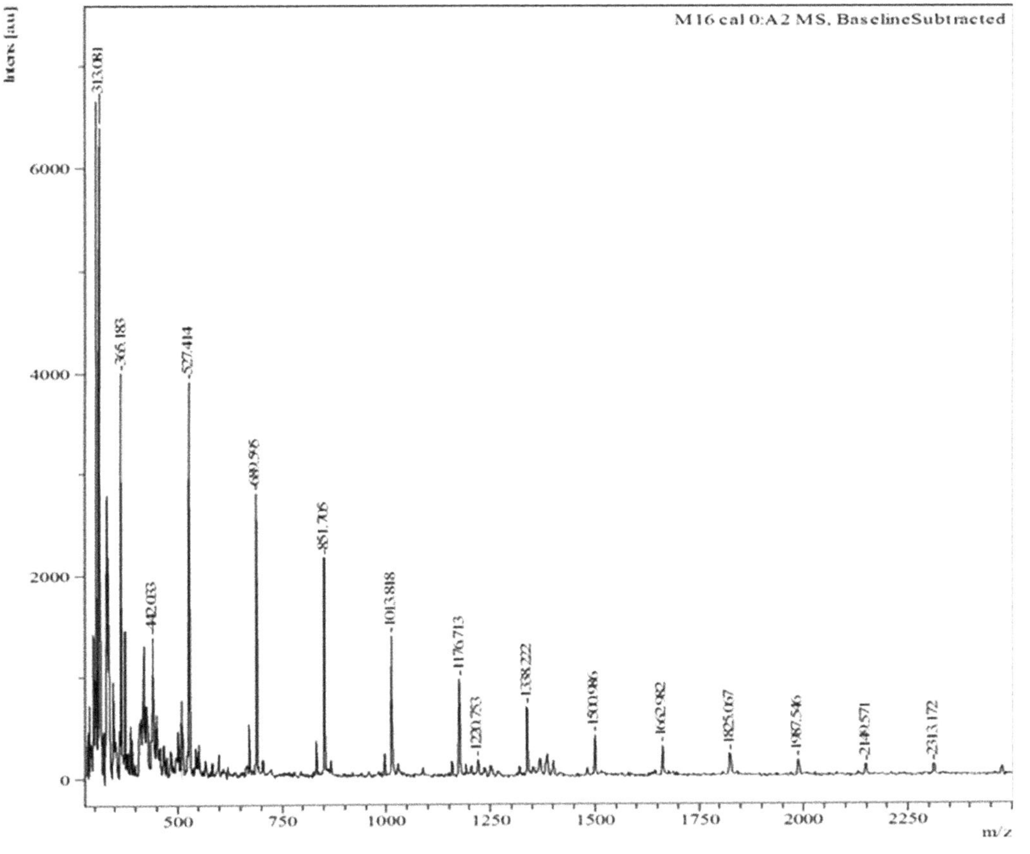

Figure 1. MALDI–TOF–MS profile of the FOS used in this study.

Two kinds of agave fructan bioconjugates were prepared, with short-chain (propionate) and medium-chain (laurate) acyl groups. Both short- and medium-chain fatty acids have been described as beneficial for the colonic gut microbiota [26]. Indeed, the short-chain fatty acids (SCFAs) acetate, propionate, and butyrate are reported as important fuels for intestinal epithelial cells [6]. Regarding propionate, the results of Byrne et al. in nonobese men suggested that colonic propionate may play an important role in human appetitive and reward-based eating behavior [27]. Medium-chain fatty acids and triglycerides ingestion results in ketone body production, provoking a thermogenic response [28]. Lauric acid showed insulinotropic effects in mouse models [29]. Thus, these fatty acids were investigated as acyl donors for agave fructan bioconjugates.

The conversion of these two acyl groups to agave fructan bioconjugates was similar; as an example, Figure 2 shows the HPLC product profile of agave fructan bioconjugates synthesized with laurate. Fructans are not detected by the diode-array, and therefore, the visualized peaks are agave fructan bioconjugate products. Complete consumption of vinyl laurate (acyl donor) was observed. A similar behavior was detected with propionate agave fructan bioconjugates (data not shown).

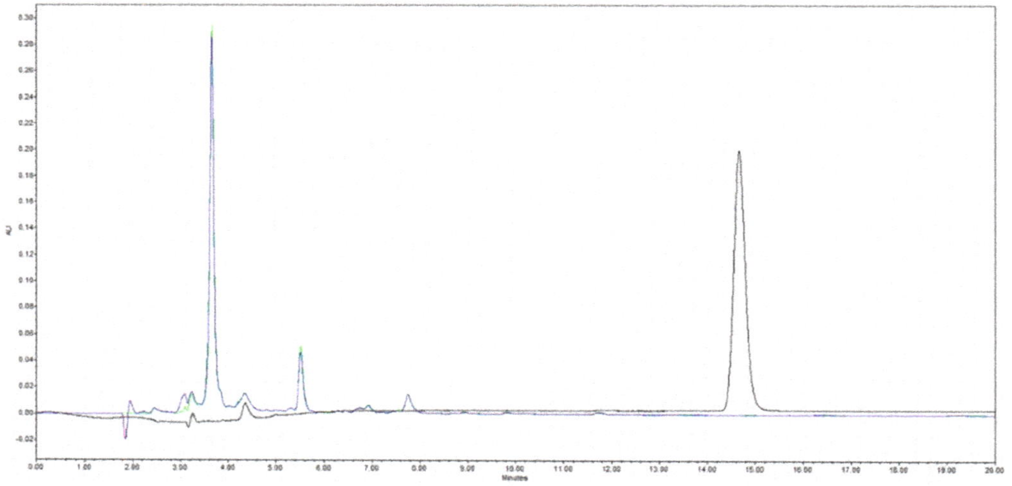

Figure 2. HPLC profile of the agave fructan bioconjugates with vinyl laurate at the beginning (black line) and at the end (green and blue lines) of the reaction.

Additionally, a complex range of agave fructan bioconjugate products is obtained (Figure 3), which is due to the acylation at different positions of the hydroxyl groups of the ramified agave fructans. In the same manner, when acylation was carried out with agave fructans with a higher DP, a complex mixture of agave fructan bioconjugates was also observed [20]. Therefore, the highly complex branched structure of agave fructans could cause acylation in hydroxyl groups at different positions [10,11], which could impart different bioactive properties when compared with linear fructans.

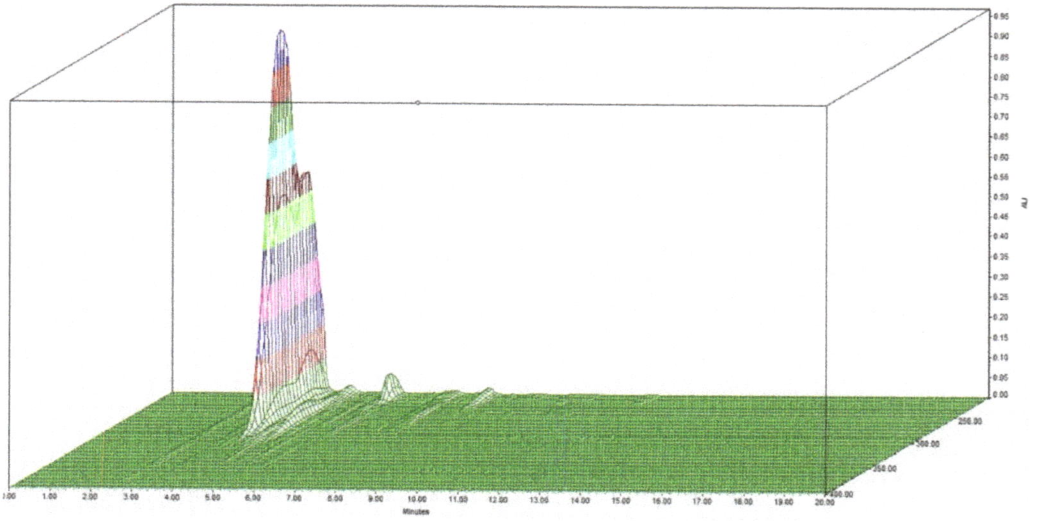

Figure 3. Three−dimensional chromatogram of agave fructan bioconjugates with laurate showing the variety of products due to agave fructan ramification.

2.2. Effects of Acute Supplementation with Agave Fructan Bioconjugates on Metabolic Syndrome Parameters

2.2.1. Induction of Metabolic Syndrome with a Hypercaloric Diet

Most of the parameters that were analyzed showed that the animals that received the hypercaloric diet (HD) developed characteristics related to the metabolic syndrome (Tables 1 and 2). Feed and water consumption in the different experimental groups were as expected according to animal body measures (Table 1). As expected, feed consumption with standard diet (SD) was slightly higher, as HD is more satiating. The use of butter as the main source of fat in the preparation of the hypocaloric diet favored the induction of metabolic syndrome in the animals [24]. Although the scattering in the results is large due to the preparation of the hypercaloric diet in our laboratory, other authors, such as Leonardi et al. [30], also showed that there is considerable variability in animal models of metabolic syndrome. On average, animals treated with the propionate (HDFP) and laurate (HDFL) bioconjugates had lower feed ingestion than those treated with HD and hypercaloric plus fructans (HDF) diets. ST-Onge et al. also showed that medium-chain fatty acid triglyceride consumption by overweight men reduced their food intake [31].

Table 1. Feed and water consumption during the study.

Consumption [1]/Diet [2]	SD	HD	HDF	HDFP	HDFL
Water (mL/d)	23.48	16.74	14.89	13.59	12.66
Feed (g/d)	17.31	15.69	15.68	15.00	15.53

[1] Milliliter per day (mL/d), grams per day (g/d). Values obtained from total water or feed consumed in each group divided by total days of study. [2] Standard diet (SD), hypercaloric diet (HD), hypercaloric diet plus fructans (HDF), hypercaloric diet plus propionate-bioconjugated fructans (HDFP), and hypercaloric diet plus laurate-bioconjugated fructans (HDFL).

Table 2. Biochemical parameters after eight weeks of treatment (mg/dL) [1].

Parameter [2]/Diet [3]	SD	HD	HDF	HDFP	HDFL
HDL	26.74 ± 5.02	28.43 ± 3.53	30.32 ± 5.64	27.69 ± 2.94	26.69 ± 2.45
LDL	22.71 ± 5.16	16.82 ± 7.37	9.02 ± 4.35	16.94 ± 6.00	16.07 ± 5.93
Total cholesterol	57.41 ± 10.17	53.95 ± 5.59	49.51 ± 5.81	56.86 ± 5.35	55.35 ± 5.07
Triglycerides	39.76 ± 13.25	43.45 ± 13.9	50.84 ± 16.07	61.19 ± 19.91	62.96 ± 19.91 *

[1] Asterisks indicate statistically significant differences vs. HD group ($p < 0.05$) after one-way ANOVA and Tukey tests. Values are mean ± standard deviation. [2] HDL, high-density lipoprotein; LDL, low-density lipoprotein. [3] Standard diet (SD), hypercaloric diet (HD), hypercaloric diet plus fructans (HDF), hypercaloric diet plus propionate-bioconjugated fructans (HDFP), and hypercaloric diet plus laurate-bioconjugated fructans (HDFL). * $p < 0.05$.

2.2.2. Metabolic Syndrome Prevention by Agave Fructan Bioconjugates

Table 2 and Figures 4–8 show the biochemical and zoomorphic parameters after eight weeks of standard diet (SD), hypercaloric diet (HD), hypercaloric diet plus fructans (HDF), hypercaloric diet plus propionate-bioconjugated fructans (HDFP), and hypercaloric diet plus laurate-bioconjugated fructans (HDFL).

The animals that received the treatment with bioconjugates of the medium-chain acyl group (HDFL) showed a decrease in glucose levels when compared to the group with a hypercaloric diet (HD), (Figure 4). For the lipid profile parameters (Table 2), no significant differences were observed except for triglycerides where a significant decrease was observed in the HDFL group compared to the HD group; this effect was previously reported in a study with mice induced to obesity with a high-fat diet, which may be due to the proliferation of crypt cells related to lipid metabolism [19,32]. Previous studies have shown that agave fructans lower glucose levels in mice with obesity induced by a high-fat diet [13]. This reduction may be associated with the inhibition of enzymes related to carbohydrate metabolism, such as alpha amylase and alpha glucosidase, because preliminary studies of our working group demonstrated the postprandial hypoglycemic effect of agave fructans. Previous studies showed a significant 15% decrease in postprandial serum glucose values in C57Bl/6J mice fed a high-fat diet with agave fructans [16].

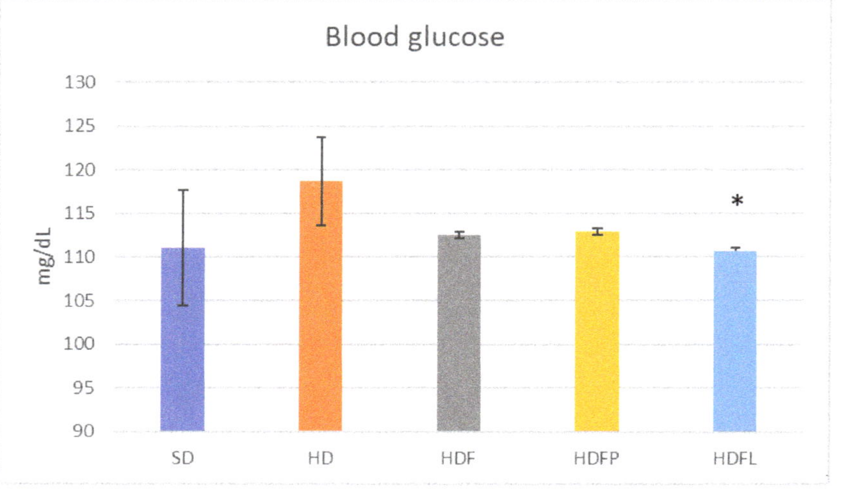

Figure 4. Blood glucose after eight weeks of treatment. Asterisks denote a significant difference vs. HD group ($p < 0.05$) after one-way ANOVA and Tukey tests. Error bars are ± standard deviation. Standard diet (SD), hypercaloric diet (HD), hypercaloric diet plus fructans (HDF), hypercaloric diet plus propionate-bioconjugated fructans (HDFP), and hypercaloric diet plus laurate-bioconjugated fructans (HDFL). * $p < 0.05$.

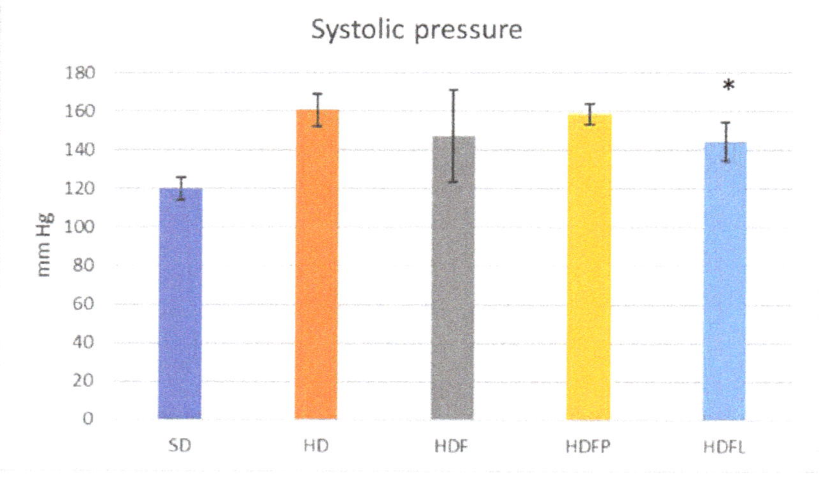

Figure 5. Systolic pressure after eight weeks of treatment. Asterisks denote a significant difference vs. HD group ($p < 0.05$) after one-way ANOVA and Tukey tests. Error bars are ± standard deviation. Treatment acronyms as in Figure 4. * $p < 0.05$.

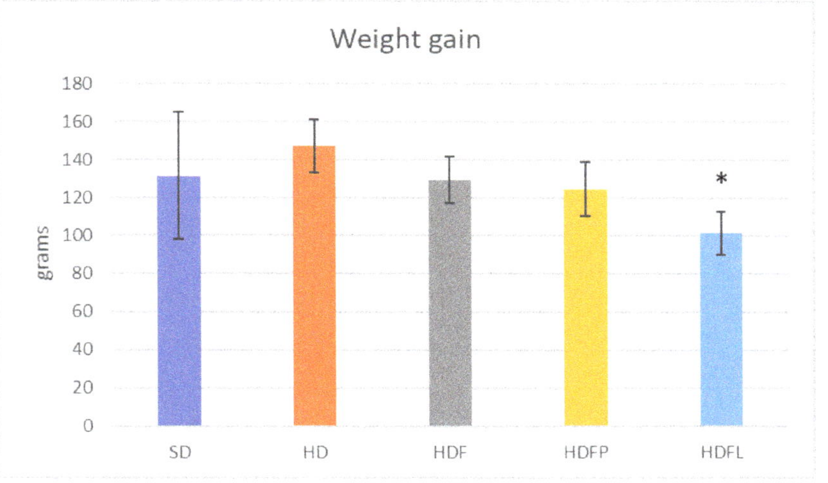

Figure 6. Weight gain after eight weeks of treatment. Asterisks denote a significant difference vs. HD group ($p < 0.05$) after one-way ANOVA and Tukey tests. Error bars are ± standard deviation. Treatment acronyms as in Figure 4. * $p < 0.05$.

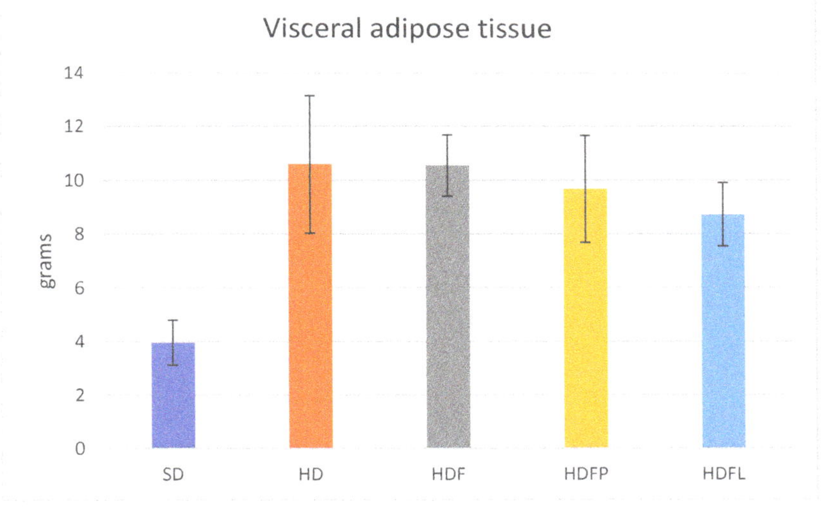

Figure 7. Visceral adipose tissue after eight weeks of treatment. Error bars are ± standard deviation. Treatment acronyms as in Figure 4.

The HDFL sample group of animals showed a significant decrease in blood pressure (systolic) when compared to the HD group (Figure 5). Increased blood pressure is one of the parameters that is affected by high-fat dietary intake, which promotes the appearance of reactive oxygen species mediated by nuclear factor kappa B and proinflammatory cytokines, and activates the renin–angiotensin–aldosterone system, which is widely documented in the etiology of cardiovascular damage [33,34]. It has been reported that some types of soluble dietary fiber, including fructans, produce short-chain fatty acids by microbial fermentation and can regulate blood pressure [35,36]. Indeed, in a study conducted in rats with metabolic syndrome induced with a high fructose diet, it was found that supplementation with inulin-type fructans showed an antihypertensive effect [37].

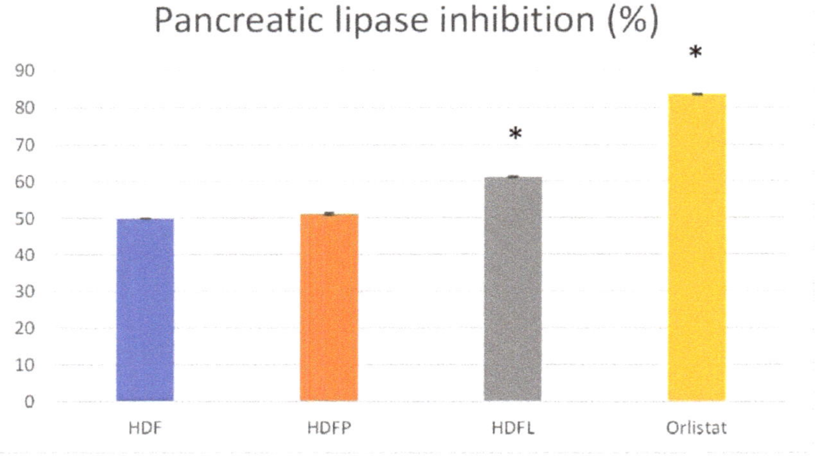

Figure 8. Pancreatic lipase inhibition. Orlistat (6 mg/mL) was used as a positive control. Fructans/agave fructan bioconjugates were evaluated at 1 mg/mL. Asterisks denote a significant difference vs. control ($p < 0.05$) after one-way ANOVA and Tukey tests. Error bars are ± standard deviation. Treatment acronyms as in Figure 4. * $p < 0.05$.

Increases in body weight and body fat are among the main parameters related to metabolic syndrome. Different types of fructans have been studied to evaluate their potential as controllers of these parameters. Additionally, the animals that were treated with the HDFL bioconjugate exhibited significantly decreased weight gain, about 31% less than the HD group. The HDFP and HDF groups also showed a lower weight gain than the HD group; however, this difference was not significant (Figure 6). Visceral adipose tissue (Figure 7) was also about 18% lower in the HDFL group than in the group of animals fed an HD diet and the nonbioconjugated fructan group (HDF). These data show that the HDFL group significantly reduced weight gain and visceral adipose mass in HD-induced rats.

In healthy rats, no difference in body weight gain or adipose tissue weight was observed when fructans were supplemented in the diet [15]; however, in studies performed with mice of the C57Bl/6J strain fed fructans, a decrease in weight gain of the animals was demonstrated, and this effect was attributed to the induction of satietogenic peptides such as glucagon-like peptide-1 (GLP-1) [16]. It has been shown that rats treated with inulin-type fructans have increased serum levels of GLP-1, which, in addition to its satietogenic effect, inhibits macrophage inflammation [38]. Similarly, agave fructans were observed to prevent weight gain and hepatic steatosis in mice with high-fat-diet-induced obesity [13].

In clinical studies, agave fructans showed a beneficial effect on weight control, body fat and triglycerides in obese people during a 12-week treatment [14]. These effects may be due to the prebiotic activity of the fructans that promote the development of a healthy intestinal microbiota and therefore help to improve metabolic diseases [19,39]. Moreover, bioconjugates previously showed higher prebiotic activity than nonbioconjugated fructans by stimulating the growth of selected beneficial probiotic strains of the intestinal microbiota such as *S. boulardii*, *L. lactis*, *L. casei*, and *L. rhamnosus* [40].

Increases in body weight and body fat are among the main parameters related to metabolic syndrome. Different types of fructans have been studied to evaluate their potential as controllers of these parameters.

Unless pancreatic lipase degrades them, dietary fat is not directly absorbed from the intestine. Therefore, pancreatic lipase inhibitors have been studied as a treatment for obesity induced by a high-fat diet [41]. We selected orlistat, a well-studied pharmacological lipase inhibitor, as a positive control, which at 6 mg/mL gave almost 85% lipase inhibition (Figure 8), while the HDFL biconjugate at 1 mg/mL (used at this concentration because of its lower solubility) inhibited the lipase by 60%. This bioconjugate is a promising lipase inhibitor without the disadvantages of using orlistat, which has been associated with several mild-to-moderate gastrointestinal adverse effects [42].

Bioconjugation with propionate (HDFP) had a lesser effect on weight gain, visceral adipose tissue, and pancreatic lipase inhibition, similarly to Chambers et al. [20], who reported only weight and adiposity maintenance in acute supplementation with propionated inulin [20].

These results demonstrated that the bioconjugation of agave fructans has a positive effect on metabolic syndrome prevention, especially in the case of the medium-chain acyl group bioconjugate (HDFL).

3. Materials and Methods

3.1. Synthesis of Agave Fructan Bioconjugates

Agave fructan bioconjugates were enzymatically synthesized as described in Patent MX 358789 [40]. Commercially available organic agave fructans enriched in FOS, and Olifructine^{TM} (Nutriagaves, Guadalajara, Jal., Mexico) were used. The batch of Olifructine^{TM} had 53% mol of FOS of DP 4 to 7 (see Section 2.1 and Figure 1). The biocatalyst was the food-grade immobilized lipase B from *Candida antarctica* Lipozyme^{TM} 435 (Novozymes, Denmark, obtained through the broker Biotecsa, Mexico). Acylants (vinyl propionate and vinyl laurate), as well as solvents, were purchased from Sigma Aldrich (Burlington, MA, USA) at the highest purity available. Confirmation of acylation was performed by HPLC as reported previously [21], using a Luna^{TM} C18 column (Phenomenex, Monterrey, NL, Mexico) in a Waters Acquity^{TM} HPLC, using photodiode-array (PDA) detector at 217 nm; which makes it possible to detect the acylant substrates but not the fructans, with the advantage that the peaks observed once the acylant is consumed, correspond only to products. The mobile phase was methanol–water 90:10 at 0.6 mL/min. FOS are not completely soluble in the reactional system; however, once converted to bioconjugates, their solubility increased as the acylation reaction proceeded, as that kind of compound also has emulsification properties [21,25,43]. At the end of reaction, the immobilized lipase was filtered and the reaction solvent was eliminated in a Buchi^{TM} rotary evaporator. Purification of the bioconjugates was not needed, as the conversion of the limiting substrate (acylant) was almost 100% (see Figure 2), and unreacted fructans were insoluble in the reaction solvent. MALDI-TOF mass spectra analyses were performed in a Microflex LT, Bruker Daltonics system, using 2,5-dihydroxy benzoic acid (DHB) as the ionization matrix. Samples diluted (1:10 for propionate or 1:20 for laurate) were mixed with an equal part (v/v) of the matrix solution (10 mg/mL of DBH in ethanol–water; 1/1; v/v). Then, 1 mL of this mixture was deposited on the target plate and dried at room temperature. The equipment was calibrated from 380 to 3000 m/z with 1-kestose as the standard [43,44].

3.2. Diet-Induced Metabolic Syndrome Model

3.2.1. Preparation of the Special Diet

The hypercaloric feed was formulated based on the AIN-93G diet and according to Cheng et al. [45]

The ingredients: cellulose, carboxymethyl cellulose, L-cystine, choline bitartrate, and butylhydroquinone were purchased from Sigma Aldrich (Burlington, MA, USA). Starch, clarified anhydrous butter, maltodextrin, sucrose, and soybean oil were purchased from a local feed formulating ingredient distributor. Vitamin and mineral premix were purchased from Dyets Inc. (Bethlehem, PA, USA) and casein was purchased from Hilmar Ingredients (Hilmar, CA, USA). The distribution in percentage is shown in Table 3, and macronutrient distribution compared to the standard diet is presented in Table 4. The hypercaloric diet was prepared by mixing all ingredients and baking at 100 °C for 1 h.

Table 3. Distribution of components of the hypercaloric diet.

Component	Percentage
Corn starch	2.448
Anhydrous butter (clarified)	32.00
Casein	23.50
Maltodextrin	16.50
Sucrose	9.00
Mineral premix AIN-93G	4.59
Soybean oil	3.5
Microcrystalline cellulose	3.25
Carboxymethyl cellulose	3.25
AIN-93 vitamin premix	1.31
L-cystine	0.39
Choline bitartrate	0.26
Butylhydroquinone	0.002

Table 4. Macronutrient distribution in standard (SD) and hypercaloric (HD) diets.

Macronutrient	SD [1]	HD
Fat	13.6	58.9
Carbohydrates	60.0	26.25
Proteins	26.0	14.85

[1] The standard diet was acquired from Scientific diets (France), catalog number SAFE A30.

3.2.2. Animals

Thirty male Wistar rats (3–4 weeks, 120 ± 20 g) were randomly housed in acrylic boxes, six animals per box, under standard environmental conditions (12 h artificial light/dark cycle) and water and food ad libitum. The number of animals was determined based on previous studies where 6 animals per group were used [46]. The animals were treated following the guidelines and requirements of the Declaration of Helsinki of the World Medical Association and the recommendations of the Mexican Official Standard for the Production, Care, and Use of Laboratory Animals (Secretariat of Agriculture, Livestock, Rural Development, Fisheries and Food (SAGARPA), Mexican Official Standard NOM-062-ZOO-1999). The protocol was favorably reviewed by the Internal Committee for the Care and Use of Laboratory Animals (CICUAL). Report Code: 2021-002A.

3.2.3. Induction of Metabolic Syndrome and Treatment with Agave Fructans and Bioconjugates

Each box with animals was assigned a group: Group 1 was considered the control group, which continued to receive standard feed (SD) and did not receive treatment. Group 2 was replaced by a hypercaloric diet (own elaboration) and did not receive treatment (HD). Group 3 consisted of a hypercaloric diet and, in addition, a daily dose of 130 mg/kg of nonbioconjugated fructans (HDF) was administered. Groups 4 and 5 were treated with bioconjugates. Group 4 consisted of a hypercaloric diet and was given a daily dose of 130 mg/kg propionate-bioconjugated fructans (HDFP). The fifth experimental group was fed a hypercaloric diet and a daily dose of 130 mg/kg of laurate-bioconjugated fructans (HDFL) was administered.

Samples were administered daily for 8 weeks using an esopharyngeal cannula. The control group was administered phosphate-buffered saline (PBS). Samples were coded to maintain a blinded study.

The following parameters were measured during the study: zoometric data such as weight, height, and body mass index (not shown), as well as water and food consumption, were monitored daily. Food intake was calculated by subtracting the amount of residual food from the amount of supply food. Blood pressure was measured at week 7 by a noninvasive method using an occlusion cuff on the rat's tail (Noninvasive blood pressure system, CODA-M1, Kent Scientific Corporation. Torrington, CT, USA). At the end of 8 weeks, all animals were fasted (12 h) and euthanized by decapitation according to the ethical guidelines described above. Blood was collected in a test tube and centrifuged at 3000× g for 15 min at 4 °C. At the same time, the visceral adipose tissue was excised, washed with 1X PBS solution (Sigma—Aldrich, Burlington, MA, USA), and weighed.

The serum obtained was stored at −80 °C for later use. Finally, glucose, cholesterol, triglycerides, LDL, and HDL parameters were measured in the stored serum using Randox kits (catalog numbers: GL2623, CH200, TR210, CH3841, and CH3811).

3.2.4. In Vitro Study of Pancreatic Lipase Inhibition

Pancreatic lipase participates in the process of absorption of fatty acids from the diet and its inhibition reduces the levels of lipids such as cholesterol in the organism. The activity of porcine pancreatic lipase (type II) was measured by the addition of 1 mM 4-nitrophenyl butyrate as a colorimetric substrate to the fructan samples in 1 mM phosphate-buffered saline (PBS). They were incubated for 30 min at 30 °C and absorbance was measured at 405 nm. Orlistat (6 mg/mL) was used as a positive control based on the method of Kim et al. [41]. Fructans/agave fructan bioconjugates were evaluated at 1 mg/mL.

4. Conclusions

Our findings indicated that bioconjugation of agave fructans enriched in low DP FOS, particularly with medium-chain acyl groups, specifically laurate (HDFL bioconjugate), had a favorable effect on metabolic syndrome prevention. Although metabolic syndrome is a multifactorial condition, and its clinical approach requires integrated management with different pharmacological approaches, HDFL treatment has shown a positive impact on various parameters of metabolic syndrome, and therefore it is a promising dietary adjuvant to prevent metabolic syndrome. However, studies at the cellular and molecular levels are needed to elucidate the mechanisms of action of bioconjugates.

5. Patents

The processes described in our patent MX 358789 were used for the synthesis of agave fructan bioconjugates. A Mexican patent application for the results presented in this manuscript was presented with the number MX/a/2022/012414.

Author Contributions: Conceptualization, G.S.; methodology, J.A., E.P.-C. and G.S.; validation, E.P.-C. and G.S.; formal analysis, E.P.-C. and G.S.; investigation, J.A., E.P.-C. and G.S.; resources, G.S.; data curation, E.P.-C.; writing—original draft preparation, E.P.-C. and G.S.; writing—review and editing, J.A., E.P.-C. and G.S.; supervision, J.A., E.P.-C. and G.S.; project administration, G.S.; funding acquisition, G.S. All authors have read and agreed to the published version of the manuscript.

Funding: This research was funded by the "Consejo Estatal de Ciencia y Tecnología" (COECYTJAL, Jalisco, Mexico), grant number 8975-2020.

Institutional Review Board Statement: The animal study protocol was approved by the Institutional Animal Studies Ethics Committee (Internal Committee for the Care and Use of Laboratory Animals, CICUAL) of the Center for Research and Assistance in Technology and Design of the State of Jalisco, A.C. (CIATEJ), with protocol code 2021-002A and date of approval 13 May 2021.

Informed Consent Statement: Not applicable.

Data Availability Statement: Data is contained within the article.

Acknowledgments: The authors thank Nilda Flores and Carlos González for helping with bioconjugated synthesis and analysis, and Nutriagaves® Group for the kind donation of Olifructine™ (organic agave fructans enriched in FOS).

Conflicts of Interest: The authors declare no conflict of interest. The funders had no role in the design of the study; in the collection, analyses, or interpretation of data; in the writing of the manuscript, or in the decision to publish the results.

References

1. Alberti, K.G.M.M.; Eckel, R.H.; Grundy, S.M.; Zimmet, P.Z.; Cleeman, J.I.; Donato, K.A.; Fruchart, J.-C.; James, W.P.T.; Loria, C.M.; Smith, S.C. Harmonizing the Metabolic Syndrome. *Circulation* **2009**, *120*, 1640–1645. [CrossRef] [PubMed]
2. Wang, X.; Shan, X.; Dun, Y.; Cai, C.; Hao, J.; Li, G.; Cui, K.; Yu, G. Anti-Metabolic Syndrome Effects of Fucoidan from Fucus vesiculosus via Reactive Oxygen Species-Mediated Regulation of JNK, Akt, and AMPK Signaling. *Molecules* **2019**, *24*, 3319. [CrossRef]
3. Rochlani, Y.; Pothineni, N.V.; Kovelamudi, S.; Mehta, J.L. Metabolic syndrome: Pathophysiology, management, and modulation by natural compounds. *Ther. Adv. Cardiovasc. Dis.* **2017**, *11*, 215–225. [CrossRef] [PubMed]
4. González-Álvarez, H.; Bravo-Jiménez, A.; Martínez-Arellanes, M.; Gamboa-Osorio, G.O.; Chávez-Gutiérrez, E.; González-Hernández, L.A.; Gallardo-Ignacio, K.; Quintana-Romero, O.J.; Ariza-Castolo, A.; Guerra-Araiza, C.; et al. In Silico-Based Design and In Vivo Evaluation of an Anthranilic Acid Derivative as a Multitarget Drug in a Diet-Induced Metabolic Syndrome Model. *Pharmaceuticals* **2021**, *14*, 914. [CrossRef] [PubMed]
5. Nilsson, P.M.; Tuomilehto, J.; Rydén, L. The metabolic syndrome—What is it and how should it be managed? *Eur. J. Prev. Cardiol.* **2020**, *26*, 33–46. [CrossRef] [PubMed]
6. Fahed, G.; Aoun, L.; Bou Zerdan, M.; Allam, S.; Bou Zerdan, M.; Bouferraa, Y.; Assi, H.I. Metabolic Syndrome: Updates on Pathophysiology and Management in 2021. *Int. J. Mol. Sci.* **2022**, *23*, 786. [CrossRef] [PubMed]
7. Welty, F.K.; Alfaddagh, A.; Elajami, T.K. Targeting inflammation in metabolic syndrome. *Transl. Res.* **2016**, *167*, 257–280. [CrossRef]
8. Samson, S.L.; Garber, A.J. Metabolic Syndrome. *Endocrinol. Metab. Clin. N. Am.* **2014**, *43*, 1–23. [CrossRef]
9. Waltenberger, B.; Mocan, A.; Šmejkal, K.; Heiss, E.H.; Atanasov, A.G. Natural Products to Counteract the Epidemic of Cardiovascular and Metabolic Disorders. *Molecules* **2016**, *21*, 807. [CrossRef]
10. Wang, X.-F.; Chen, X.; Tang, Y.; Wu, J.-M.; Qin, D.-L.; Yu, L.; Yu, C.-L.; Zhou, X.-G.; Wu, A.-G. The Therapeutic Potential of Plant Polysaccharides in Metabolic Diseases. *Pharmaceuticals* **2022**, *15*, 1329. [CrossRef]
11. Arrizon, J.; Urias-Silvas, J.E.; Sandoval, G.; Mancilla-Margalli, N.A.; Gschaedler, A.C.; Morel, S.; Monsan, P. Production and Bioactivity of Fructan-Type Oligosaccharides. In *Food Oligosaccharides: Production, Analysis and Bioactivity*; Moreno, F.J., Sanz, M.L., Eds.; Wiley-Blackwell: Madrid, Spain, 2014; pp. 184–199.
12. Lopez, M.G.; Mancilla-Margalli, N.A.; Mendoza-Diaz, G. Molecular Structures of Fructans from *Agave tequilana* Weber var. azul. *J. Agric. Food Chem.* **2003**, *51*, 7835–7840. [CrossRef] [PubMed]
13. Márquez-Aguirre, A.L.; Camacho-Ruíz, R.M.; Gutiérrez-Mercado, Y.K.; Padilla-Camberos, E.; González-Ávila, M.; Gálvez-Gastélum, F.J.; Díaz-Martínez, N.E.; Ortuño-Sahagún, D. Fructans from Agave tequilana with a Lower Degree of Polymerization Prevent Weight Gain, Hyperglycemia and Liver Steatosis in High-Fat Diet-Induced Obese Mice. *Plant Foods Hum. Nutr.* **2016**, *71*, 416–421. [CrossRef] [PubMed]
14. Padilla-Camberos, E.; Barragán-Álvarez, C.P.; Diaz-Martinez, N.E.; Rathod, V.; Flores-Fernández, J.M. Effects of Agave fructans (*Agave tequilana* Weber var. azul) on Body Fat and Serum Lipids in Obesity. *Plant Foods Hum. Nutr.* **2018**, *73*, 34–39. [CrossRef]
15. Regalado-Rentería, E.; Aguirre-Rivera, J.R.; Godínez-Hernández, C.I.; García-López, J.C.; Oros-Ovalle, A.C.; Martínez-Gutiérrez, F.; Martinez-Martinez, M.; Ratering, S.; Schnell, S.; Ruíz-Cabrera, M.Á.; et al. Effects of Agave Fructans, Inulin, and Starch on Metabolic Syndrome Aspects in Healthy Wistar Rats. *ACS Omega* **2020**, *5*, 10740–10749. [CrossRef] [PubMed]
16. Urías-Silvas, J.E.; Cani, P.D.; Delmée, E.; Neyrinck, A.; López, M.G.; Delzenne, N.M. Physiological effects of dietary fructans extracted from *Agave tequilana* Gto. and *Dasylirion* spp. *Br. J. Nutr.* **2008**, *99*, 254–261. [CrossRef]
17. Bonsu, N.K.A.; Johnson, C.S.; Mcleod, K.M. Can dietary fructans lower serum glucose? *J. Diabetes* **2011**, *3*, 58–66. [CrossRef]
18. Castillo Andrade, A.I.; Rivera Bautista, C.; Godínez Hernández, C.; Ruiz Cabrera, M.A.; Fuentes Ahumada, C.; García Chávez, E.; Grajales Lagunes, A. Physiometabolic effects of *Agave salmiana* fructans evaluated in Wistar rats. *Int. J. Biol. Macromol.* **2018**, *108*, 1300–1309. [CrossRef]
19. Espinosa-Andrews, H.; Urías-Silvas, J.E.; Morales-Hernández, N. The role of agave fructans in health and food applications: A review. *Trends Food Sci. Technol.* **2021**, *114*, 585–598. [CrossRef]
20. Chambers, E.S.; Viardot, A.; Psichas, A.; Morrison, D.J.; Murphy, K.G.; Zac-Varghese, S.E.K.; MacDougall, K.; Preston, T.; Tedford, C.; Finlayson, G.S.; et al. Effects of targeted delivery of propionate to the human colon on appetite regulation, body weight maintenance and adiposity in overweight adults. *Gut* **2015**, *64*, 1744–1754. [CrossRef]
21. Casas-Godoy, L.; Arrizon, J.; Arrieta-Baez, D.; Plou, F.J.; Sandoval, G. Synthesis and emulsifying properties of carbohydrate fatty acid esters produced from Agave tequilana fructans by enzymatic acylation. *Food Chem.* **2016**, *204*, 437–443. [CrossRef]
22. Hernández-Pérez, S.; Oliart-Ros, R.M.; Casas-Godoy, L.; Sandoval, G.; Guarner-Lans, V.; Castrejón-Téllez, V.; Quevedo-Corona, L.; Peña-Montes, C.; Ramírez-Higuera, A. Beneficial Effects of Fructooligosaccharides Esterified with Lauric Acid in a Metabolic Syndrome Model Induced by a High-Fat and High-Carbohydrate Diet in Wistar Rats. *J. Med. Food* **2022**, *25*, 825–835. [CrossRef]
23. Panchal, S.K.; Brown, L. Rodent Models for Metabolic Syndrome Research. *J. Biomed. Biotechnol.* **2011**, *2011*, 351982. [CrossRef] [PubMed]

24. Rodríguez-Correa, E.; González-Pérez, I.; Clavel-Pérez, P.I.; Contreras-Vargas, Y.; Carvajal, K. Biochemical and nutritional overview of diet-induced metabolic syndrome models in rats: What is the best choice? *Nutr. Diabetes* **2020**, *10*, 24. [CrossRef] [PubMed]
25. van Kempen, S.E.H.J.; Boeriu, C.G.; Schols, H.A.; de Waard, P.; van der Linden, E.; Sagis, L.M.C. Novel surface-active oligofructose fatty acid mono-esters by enzymatic esterification. *Food Chem.* **2013**, *138*, 1884–1891. [CrossRef] [PubMed]
26. Silva, Y.P.; Bernardi, A.; Frozza, R.L. The Role of Short-Chain Fatty Acids from Gut Microbiota in Gut-Brain Communication. *Front. Endocrinol.* **2020**, *11*, 25. [CrossRef]
27. Byrne, C.S.; Chambers, E.S.; Alhabeeb, H.; Chhina, N.; Morrison, D.J.; Preston, T.; Tedford, C.; Fitzpatrick, J.; Irani, C.; Busza, A.; et al. Increased colonic propionate reduces anticipatory reward responses in the human striatum to high-energy foods. *Am. J. Clin. Nutr.* **2016**, *104*, 5–14. [CrossRef] [PubMed]
28. McCarty, M.F.; DiNicolantonio, J.J. Lauric acid-rich medium-chain triglycerides can substitute for other oils in cooking applications and may have limited pathogenicity. *Open Heart* **2016**, *3*, e000467. [CrossRef] [PubMed]
29. Garfinkel, M.; Lee, S.; Opara, E.C.; Akwari, O.E. Insulinotropic potency of lauric acid: A metabolic rationale for medium chain fatty acids (MCF) in TPN formulation. *J. Surg. Res.* **1992**, *52*, 328–333. [CrossRef] [PubMed]
30. Leonardi, B.F.; Gosmann, G.; Zimmer, A.R. Modeling Diet-Induced Metabolic Syndrome in Rodents. *Mol. Nutr. Food Res.* **2020**, *64*, 2000249. [CrossRef]
31. St-Onge, M.P.; Mayrsohn, B.; O'Keeffe, M.; Kissileff, H.R.; Choudhury, A.R.; Laferrère, B. Impact of medium and long chain triglycerides consumption on appetite and food intake in overweight men. *Eur. J. Clin. Nutr.* **2014**, *68*, 1134–1140. [CrossRef]
32. Franco-Robles, E.; López, M.G. Agavins Increase Neurotrophic Factors and Decrease Oxidative Stress in the Brains of High-Fat Diet-Induced Obese Mice. *Molecules* **2016**, *21*, 998. [CrossRef]
33. Furukawa, S.; Fujita, T.; Shimabukuro, M.; Iwaki, M.; Yamada, Y.; Nakajima, Y.; Nakayama, O.; Makishima, M.; Matsuda, M.; Shimomura, I. Increased oxidative stress in obesity and its impact on metabolic syndrome. *J. Clin. Investig.* **2017**, *114*, 1752–1761. [CrossRef]
34. Hsueh, W.A.; Wyne, K. Renin-Angiotensin-Aldosterone System in Diabetes and Hypertension. *J. Clin. Hypertens.* **2011**, *13*, 224–237. [CrossRef]
35. Cronin, P.; Joyce, S.A.; O'Toole, P.W.; O'Connor, E.M. Dietary Fibre Modulates the Gut Microbiota. *Nutrients* **2021**, *13*, 1655. [CrossRef]
36. Nie, Y.; Luo, F. Dietary Fiber: An Opportunity for a Global Control of Hyperlipidemia. *Oxidative Med. Cell. Longev.* **2021**, *2021*, 5542342. [CrossRef]
37. Rault-Nania, M.-H.; Demougeot, C.; Gueux, E.; Berthelot, A.; Dzimira, S.; Rayssiguier, Y.; Rock, E.; Mazur, A. Inulin supplementation prevents high fructose diet-induced hypertension in rats. *Clin. Nutr.* **2008**, *27*, 276–282. [CrossRef]
38. Zhang, Q.; Yu, H.; Xiao, X.; Hu, L.; Xin, F.; Yu, X. Inulin-type fructan improves diabetic phenotype and gut microbiota profiles in rats. *PeerJ* **2018**, *6*, e4446. [CrossRef]
39. Li, H.-Y.; Zhou, D.-D.; Gan, R.-Y.; Huang, S.-Y.; Zhao, C.-N.; Shang, A.; Xu, X.-Y.; Li, H.-B. Effects and Mechanisms of Probiotics, Prebiotics, Synbiotics, and Postbiotics on Metabolic Diseases Targeting Gut Microbiota: A Narrative Review. *Nutrients* **2021**, *13*, 3211. [CrossRef]
40. Sandoval, G.; Arrizon, J.; Gonalez-Avila, M.; Padilla-Camberos, E.; Matinez-Velazquez, M.; Villanueva-Rodriguez, S.; Casas-Godoy, L. Bioconjugate molecules with biological and techno-functional activity, method for the production thereof and use thereof. Patent Number MX 358789, 18 December 2013.
41. Kim, J.; Lee, Y.S.; Kim, C.-S.; Kim, J.S. Betulinic Acid has an Inhibitory Effect on Pancreatic Lipase and Induces Adipocyte Lipolysis. *Phytother. Res.* **2012**, *26*, 1103–1106. [CrossRef]
42. Filippatos, T.D.; Derdemezis, C.S.; Gazi, I.F.; Nakou, E.S.; Mikhailidis, D.P.; Elisaf, M.S. Orlistat-Associated Adverse Effects and Drug Interactions. *Drug Saf.* **2008**, *31*, 53–65. [CrossRef] [PubMed]
43. Campos-Valdez, A.R.; Casas-Godoy, L.; Sandoval, G.; Hernández, L.; Sassaki, G.L.; Alencar de Menezes, L.R.; Campos-Terán, J.; Reyes-Duarte, D.; Arrizon, J. Regioselective synthesis of 6″-O-lauroyl-1-kestose and 6‴-O-lauroylnystose by sequential enzymatic reactions of transfructosylation and acylation. *Biocatal. Biotransform.* **2022**, *40*, 133–143. [CrossRef]
44. Harvey, D.J. Analysis of carbohydrates and glycoconjugates by matrix-assisted laser desorption/ionization mass spectrometry: An update for 2015–2016. *Mass Spectrom. Rev.* **2021**, *40*, 408–565. [CrossRef]
45. Cheng, H.S.; Ton, S.H.; Phang, S.C.W.; Tan, J.B.L.; Abdul Kadir, K. Increased susceptibility of post-weaning rats on high-fat diet to metabolic syndrome. *J. Adv. Res.* **2017**, *8*, 743–752. [CrossRef] [PubMed]
46. Mayer, C.; Côme, M.; Ulmann, L.; Chini Zittelli, G.; Faraloni, C.; Nazih, H.; Ouguerram, K.; Chénais, B.; Mimouni, V. Preventive Effects of the Marine Microalga Phaeodactylum tricornutum, Used as a Food Supplement, on Risk Factors Associated with Metabolic Syndrome in Wistar Rats. *Nutrients* **2019**, *11*, 1069. [CrossRef]

pharmaceuticals

MDPI

Review

Therapeutic Targeting of TLR4 for Inflammation, Infection, and Cancer: A Perspective for Disaccharide Lipid A Mimetics

Holger Heine [1] and Alla Zamyatina [2,*]

1 Research Group Innate Immunity, Research Center Borstel—Leibniz Lung Center, Airway Research Center North (ARCN), German Center for Lung Research (DZL), Parkallee 22, 23845 Borstel, Germany
2 Department of Chemistry, University of Natural Resources and Life Sciences, Muthgasse 18, 1190 Vienna, Austria
* Correspondence: alla.zamyatina@boku.ac.at

Abstract: The Toll-like receptor 4 (TLR4) signaling pathway plays a central role in the prompt defense against infectious challenge and provides immediate response to Gram-negative bacterial infection. The TLR4/MD-2 complex can sense and respond to various pathogen-associated molecular patterns (PAMPs) with bacterial lipopolysaccharide (LPS) being the most potent and the most frequently occurring activator of the TLR4-mediated inflammation. TLR4 is believed to be both a friend and foe since improperly regulated TLR4 signaling can result in the overactivation of immune responses leading to sepsis, acute lung injury, or pathologic chronic inflammation involved in cancer and autoimmune disease. TLR4 is also considered a legitimate target for vaccine adjuvant development since its activation can boost the adaptive immune responses. The dual action of the TLR4 complex justifies the efforts in the development of both TLR4 antagonists as antisepsis drug candidates or remedies for chronic inflammatory diseases and TLR4 agonists as vaccine adjuvants or immunotherapeutics. In this review, we provide a brief overview of the biochemical evidences for possible pharmacologic applications of TLR4 ligands as therapeutics and report our systematic studies on the design, synthesis, and immunobiological evaluation of carbohydrate-based TLR4 antagonists with nanomolar affinity for MD-2 as well as disaccharide-based TLR4 agonists with picomolar affinity for the TLR4/MD-2 complex.

Keywords: immunomodulation; carbohydrates; lipopolysaccharide; lipid A; adjuvant; TLR4 antagonist; TLR4 agonist; innate immunity

Citation: Heine, H.; Zamyatina, A. Therapeutic Targeting of TLR4 for Inflammation, Infection, and Cancer: A Perspective for Disaccharide Lipid A Mimetics. *Pharmaceuticals* **2023**, *16*, 23. https://doi.org/10.3390/ph16010023

Academic Editors: Federica Sodano and Maria Grazia Rimoli

Received: 7 November 2022
Revised: 13 December 2022
Accepted: 14 December 2022
Published: 23 December 2022

1. Introduction

When Pfeiffer in 1892 coined the term "Endotoxin" for a toxin associated with the body of the bacteria and exhibiting the deadly effects of heat-killed cholera bacteria, he could not possibly have known how peculiar and fascinating this molecule would turn out to be [1]. It took more than 50 years to identify this "Endotoxin" as the lipopolysaccharide (LPS) of Gram-negative bacteria [2]. LPS is a bacterial glycan which is composed of three major motifs: a membrane-anchored lipid A, a Kdo- and heptose-containing core oligosaccharide, and a variable O-antigen (Figure 1A). The lipid A portion represents the "endotoxic principle" of LPS [3–6], which was already assumed by Westphal and Lüderitz (1954) and finally proven by Galanos et al. through the generation and subsequent functional analysis of synthetic lipid A structures [7]. The structure of a canonical endotoxic lipid A of *E. coli* relies on a $\beta(1\rightarrow6)$-linked D-glucosamine disaccharide that is hexa-acylated with two long-chain (*R*)-3-alkanoyloxyalkanoyl lipid residues attached at the "distal" GlcN ring and two (*R*)-3-hydroxyalkanoic fatty acids linked to the "proximal" GlcN moiety and carrying two phosphate groups [8–10]. The lipid chains of pro-inflammatory active lipid A variants normally have 12 to 14 carbon atoms. Generally, the endotoxic activity of lipid A is dictated by the following parameters: (i) the chemical structure, length, number, and distribution pattern of lipid chains bound to the disaccharide backbone and (ii) the phosphorylation status of each sugar unit [11].

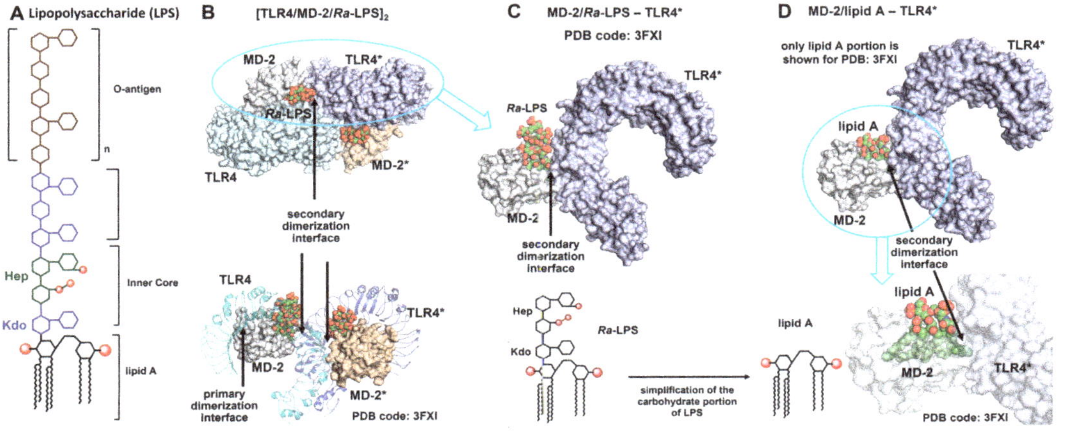

Figure 1. LPS-induced TLR4 complex dimerization. (**A**) Schematic representation of LPS; (**B**) crystal structure of TLR4/MD-2/LPS complex, PDB code: 3FXI; (**C**) formation of a secondary dimerization interface between TLR4/MD-2/LPS and the second TLR4* complex; (**D**) positioning of *E.coli* lipid A in the binding pocket of MD-2 and lipid A participation in the dimerization process. Images were generated with PyMol.

In 1999, the cell surface receptor for LPS had finally been discovered as Toll-like receptor 4 [12,13]. However, the signaling induced by LPS through TLR4 proved to be more complex than with all other TLRs. A sequential event leads to the extraction of LPS molecules by the serum constituent LPS-binding protein (LBP), the transfer to the membrane-tethered GPI-anchored receptor CD14, and the binding to a TLR4 co-receptor, myeloid differentiation 2 (MD-2) [14,15]. MD-2 is a small, secreted molecule that binds to and becomes physically associated with TLR4 [16]. In turn, LPS binding to MD-2 leads to the assembly of a ligand–receptor homodimer formed of two copies of the TLR4/MD-2/LPS complex (Figure 1B) [17–19]. Upon generation of $[TLR4/MD\text{-}2/LPS]_2$ homodimers, the intracellular signaling is initiated through conformational changes in the Toll/IL-1R (TIR) domain of TLR4, enabling the subsequent attachment of signal transduction adapter molecules. Among the TLRs, TLR4 has a special role, as it is the only TLR that uses both signaling adapter molecules, MyD88 (myeloid differentiation primary response 88) [20,21]) and TRIF (TIR (Toll IL-1R)-domain-containing adaptor-inducing interferon-β) [22]. This exclusivity allows TLR4 to drive two different time-shifted sets of immune responses: The first set, starting at the plasma membrane, is MyD88-dependent and leads to a rapid induction of pro-inflammatory cytokines. The second set, however, requires internalization, is TRIF-dependent, and signals from endosomal membranes for the induction of a type I interferon response.

Although the TLR4 complex is a major sensor of Gram-negative bacterial LPS, it can also be activated by various endogenous danger-associated molecules (DAMPs) [23,24] released from lytic cells during host tissue injury or viral infection. Indeed, it has been shown over the last decade that self-encoded molecules derived from damaged or necrotic tissues such as oxidized phospholipids [25], high-mobility group box 1 protein (HMGB1) [26], hyaluronic acid fragments [27], or fibronectins [28] are also able to signal through activation of TLR4. To make LPS-induced cell activation even more complex, it was revealed in 2014 that TLR4 is not the only LPS receptor but that a specific TLR4-independent cytosolic recognition of Gram-negative bacteria and their LPS exists: intracellular caspases (caspase-11 in mouse and caspase-4/5 in humans) were demonstrated to bind LPS and trigger a non-canonical inflammasome activation [29–32]. Inflammasome activation leads to pyroptosis and to the production and release of IL-1β, IL-18, and IL-37—cytokines which are heavily involved in inflammation and inflammatory disorders [33,34].

2. Species-Specific Activity of LPS

Another peculiarity of LPS biology that is not seen to this extent with any other pathogen-associated molecular pattern (PAMP) is its species-specific activity. Much of the in vivo functional analysis of LPS-induced signaling, gene expression, and cytokine production in monocytes and macrophages has been performed using different mouse models. Although the general LPS receptor complex is conserved between humans and mice (and mammals in general), there are some striking differences that impact ligand recognition [35–37].

For example, the tetra-acylated biosynthetic precursor of *E. coli* lipid A, also known as lipid IVa (and its synthetic analogue, compound 406), has endotoxic activity in mice in vitro as well as in vivo [38] but is completely inactive in human cells [39–41]. However, despite being inactive at human (h-)TLR4, the binding to hTLR4 still takes place and therefore this structure has been identified as a very good antagonist in the human system [42]. Similarly, the *N. meningitidis* LpxL1 LPS mutant carrying penta-acylated lipid A activates the murine TLR4/MD-2 complex equally efficiently as a wild-type hexa-acylated LPS, whereas it elicits no or a significantly reduced TLR4-mediated response in native human immune cells and can even act as an antagonist at hTLR4 [43]. The species-specific activity of LPS goes even beyond man and mice: the fivefold-acylated *R. sphaeroides* lipid A is a potent antagonist in human and mouse cells but activates cells of hamster [44] and equine origin [45,46].

These findings started a new research field of the identification and synthetic construction of stable and adjustable LPS/lipid A antagonists in order to explore their therapeutic potential.

3. Therapeutic Potential of TLR4 Antagonists

The first disorder where LPS antagonists come to mind is **sepsis**. In 2017, the estimated number of sepsis-related deaths worldwide was 11 million [47]. Despite a decline in overall incidence, it is still a major public health burden associated in many cases with primary or secondary infections by Gram-negative bacteria and thus their LPS as a virulence factor. Based on promising in vivo data in animal studies and a phase 2 trial, a synthetic antisepsis drug candidate E5564 (Eritoran), an LPS antagonist based on the structure of *R. sphaeroides* lipid A, was tested in a phase 3 trial but failed to improve 28-day mortality [48]. Among the reasons for this failure are significant differences in TLR4 and caspase 4/5/11 signaling and expression in humans and mice. At the time Eritoran was tested in septic patients, the involvement and the importance of caspases in LPS-triggered inflammation was not yet known.

However, the above-mentioned failure of Eritoran in sepsis trials was not the end of its possible use. **Flu pandemics** are another world-wide public health threat with high morbidity and mortality [49]. Although viruses are detected by other PRRs than TLR4 (TLR3, 7, 8, RIG-I), in the more severe outcomes of viral infection such as acute lung injury (ALI), TLR4 signaling is crucially involved through the generation of oxidized phospholipids (OxPAPC) which can also activate TLR4 [25]. It has been proven that administration of Eritoran protects mice from lethal influenza infection by blocking OxPAPC-dependent activation of TLR4 and that the harmful TLR4-mediated signaling can be inhibited also in this case by TLR4 antagonists directly interacting with MD-2 [50].

Another chromatin-associated protein-derived DAMP, HMGB1, which can be secreted from dying cells during **viral infection**, is also sensed by the TLR4 co-receptor MD-2 [51]. The viral-infection-mediated release of HMGB1 triggers pro-inflammatory responses similar to those of LPS and thus supports the significance of the therapeutic potential of TLR4/MD-2 antagonists for symptom release in viral disease. It is assumed that the TLR4 axis is also involved in the detrimental effects of respiratory syncytial virus (RSV) [52], West Nile virus [53], filovirus [54], and influenza virus infections [55,56].

Morbidity and mortality in the course of **acute viral disease** is frequently associated with excessive inflammation triggered by the release of elevated levels of pro-inflammatory cytokines and chemokines which can lead to life-threatening complications such as acute respiratory distress syndrome or ALI, circulatory shock, and myocarditis. Moreover, an upregulation of the TLR4-mediated pro-inflammatory signaling plays an important role in the pathogenesis of **SARS-CoV-2** [57–59]. The innate immune responses to a viral-infection-induced cytokine storm are often systemic and result in sepsis syndrome mimicking the symptoms of Gram-negative bacterial sepsis as reported for SARS-CoV-2 disease [60, 61]. Indeed, TLR4 seems to be a crucial contributor to the pathology and severity of COVID19 [62–64], especially in patients with preexisting cardiovascular comorbidities [65] which can be associated with preceding or inducible endotoxemia [66] and/or the ability of SARS-CoV-2 spike protein to bind and transport bacterial LPS [62].

LPS is a very potent driver of inflammation [67] and thus is clearly involved in many chronic inflammatory disorders such as **atherosclerosis** [68–71]. Here, in particular, the capability of LPS to engage both TLR4 and inflammatory caspases is of crucial importance. A main source of LPS is the microbiome, and dysbiosis leads to impairment of the gut barrier enabling LPS translocation in the blood stream [68]. A number of studies address the role of TLR4 in the regulation of skin defence mechanisms during cutaneous tumorigenic inflammation [72]. Pharmacologic inhibition of the TLR4-driven responses enhanced the skin's photoprotective capacity, attenuating cutaneous stress signaling, and suppressed UV-induced skin tumorigenesis in mice [73].

In rheumatoid **arthritis** patients, the endogenous TLR4 agonist HMGB1 was detected in increased levels in the serum and synovial tissues, providing another link to TLR4 signaling in inflammatory disorders [74]. Similarly, other DAMPs were shown to propel inflammation through the activation of TLR4 in rheumatoid and osteoarthritis, which highlights the inhibition of TLR4-mediated signaling as a potential therapeutic option [75–77].

The contribution of LPS and TLR4 signaling to disease pathology is often not clearly determined and can be positively or negatively associated with a disease as in the case of COPD (chronic obstructive pulmonary disease) [78] and pulmonary fibrosis [79]. In COPD, the expression of TLR4 is elevated, and TLR4, among other TLRs, is also involved in cigarette-smoke-induced inflammation [80]. Polymorphisms that reduce the functionality of TLR4 have been shown to either contribute to COPD development (TLR4-T399I) [81] or reduce COPD severity (TLR4-D299G) [82].

TLR4-mediated production of pro-inflammatory cytokines is also involved in different types of **myocardial inflammation**, and a number of trials evaluated TLR4 as a potential target for improvement of disease outcome and severity [83,84]. Inhibition of TLR4-driven inflammation, which is a major mortality factor in trauma–hemorrhage patients, showed positive results in overcoming hemorrhage-induced cardiac dysfunction in vivo [85].

Epidemiologic studies clearly indicate that early contact with microbes and their LPS decrease the occurrence of **asthma and allergies** in later life [86,87], corroborating the "hygiene hypothesis" introduced by Strachan [88]. However, the role of LPS itself in asthma and allergy development is not that clear because, depending on the timing and the amount of LPS exposure, it can also increase their development [89,90]. To add to this complexity, certain allergens are directly involved in TLR4 signaling: Der p 2, the main allergen of the house dust mite, has not only structural but also functional homology to MD-2 and thus can replace MD-2 in the LPS/MD-2/TLR4 receptor complex [91]. This self-adjuvant effect can play a major role for the allergenicity of other members of the MD-2-like lipid-binding proteins as well [92]. TLR4 was confirmed as a therapeutic target in the management of airway remodeling in house dust mite-induced chronic asthma. Inhibition of TLR4 signaling showed positive effects in a murine chronic asthma model through hampering of the Th2 response and suppression of airway remodeling [93]. In general, the TLR4 signaling pathway was suggested to protect mice against the development of allergic asthma [94]. In contrast to house dust mite allergen, the cat dander protein Fel d 1 enhances TLR4 signaling not by mimicking the TLR4 co-receptor MD-2 or binding to the TLR4/MD2 complex but instead efficiently binding LPS and, therefore, contributing to a lipid transfer mechanism of LPS to MD-2/TLR4 [95]. Similar properties are suggested for a lipocalin-class protein, the dog allergen Can f 6, which can enhance innate immune signaling by LPS and foster airway hypersensitivity in asthma [95].

4. Therapeutic Potential of TLR4 Agonists

In addition to the role of TLR4 in the diverse pathologies mentioned above, there is an increasing number of publications indicating the involvement of LPS and/or TLR4 signaling in **Alzheimer's** disease [96–98]. TLR4 agonists were suggested to have a therapeutic potential for Alzheimer's disease by stimulating the immune system to clear amyloid β. Systemic administration of a weak TLR4 agonist monophosphoryl lipid A (MPLA) led to a substantial reduction in amyloid β load in the brain and to a significant improvement in Alzheimer's-disease-related pathology in mice [99]. Likewise, repeated injections of LPS to transgenic mice to provoke mild but permanent stimulation of the TLR4 axis without inducing pathogenic neuroinflammation led to augmentation of neuronal autophagy, improved cognitive function, and prompted attenuation of the Alzheimer's-disease-related symptoms [100].

Though vaccine **adjuvants** have long been used to either compensate for poor antigen immunogenicity or to reduce the amount of an antigen required for protection, very few have been licensed for human use. The most promising adjuvants that can promote improved innate and adaptive immune responses are those with a preferential Th1 bias, including agonists of certain PRRs and TLR4 in particular [101,102]. Recombinant subunit vaccines supplemented with various formulations (in combination with aluminum salts (AS04) or squalene (AS03)) of the TLR4 agonist MPLA [103] were proven to be more effective than the live attenuated vaccines, thus demonstrating a central role of adjuvants in vaccine development. Currently, MPL® has been approved for use in several human viral vaccines [104]; however, substantial batch-to-batch variations in the content of vaccine-grade MPL® adjuvant (obtained by acid and base hydrolysis of the LPS of *S. enterica* serovar Minnesota Re595) and contaminations with underacylated TLR4-antagonizing lipid A variants bring about serious limitations [105]. It is well known that a moderate activation of TLR4 by exogeneous PAMPs comprises an important mechanism of defense against viral infection in an immunocompetent host by warranting the induction of specific pro-inflammatory signaling pathways [56,106,107]. Therefore, adjuvants targeting TLR4 might provide additional advantages by mounting an intrinsic protective response aimed at the induction of immunological memory. Thus, there is a huge demand for the discovery of more efficient TLR4 agonists other than LPS as non-toxic amplifiers of vaccine efficacy.

It has been proposed that TLR4-driven inflammation could play a dual role in **cancer**: whereas chronic inflammation (e.g., by low doses of LPS released in chronic infection or by endogenous DAMPs produced during infection or sterile inflammation) can facilitate tumor growth and progression [108–114], induced activation of the TLR4 axis improves antitumor responses and is associated with a positive prognosis, suppression of tumor growth, and reduction in cancer metastasis [115–119]. Based on these results, induction and/or amplification of the TLR4 signaling pathway was suggested as a potential treatment strategy in specific cancers [116,120–124].

Indeed, it has been known as early as the 13th century that acute inflammation has the potential to inhibit cancer progression or even eradicate solid tumors: this hypothesis was supported by reports on spontaneous cancer regression following acute bacterial infection [125]. In the early 20th century, a bone surgeon, William B. Coley, demonstrated significant success against advanced-stage bone and soft-tissue sarcoma by applying mixtures of live or killed bacteria to induce topical and systemic acute inflammation [126,127]. The molecular mechanism of this curative effect was not addressed in those times, and the treatments with the preparations of whole bacteria were strongly toxic and often lethal. However, Coley's treatment had a higher success rate (57% complete tumor remission) than contemporary chemo- and radiotherapy [128]. According to current knowledge, the anticancer effect of "Coley vaccine" was related to the activation of TLR2 and TLR4 signaling by its two components, the Gram-positive *S. pyogenes* and Gram-negative *S. marcescens*, respectively.

TLR4 is widely expressed in tumor cells as well as the cells of the tumor microenvironment such as tumor-associated macrophages (TAMs), cytotoxic T cells, myeloid-derived suppressor cells, and natural killer cells. Growing evidences demonstrate that the activation of TLRs in the tumor microenvironment can break the tumor-induced immune tolerance and boost antitumor immunity [129,130]. Furthermore, augmented expression of TLR4 in several immunogenic cancer types predicts positive survival prognosis [131]. In particular, switching TAMs from a pro-tumorigenic and immunosuppressive (M2/repair-type) to an anti-tumorigenic and immunostimulatory (M1/killer-type) phenotype using appropriate TLR agonists is an emerging strategy towards immunity-based approaches for the management of solid cancers. Immunostimulatory M1-type macrophages can stimulate effector cells such as Th1-type cytotoxic T cells and promote tumor regression [132–136].

In fact, TLR4 is increasingly considered a promising therapeutic target for solid cancers, and TLR4 activation by low doses of endotoxic LPS from *E.coli* showed therapeutic promise in inhibiting the progression of osteosarcoma [115], lymphoma [137], subcutaneous glioma [138], and osteosarcoma [115]. TLR4 activation in the microenvironmental cells mediated anticancer effects by stimulating CD8-positive cytotoxic lymphocytes [115] and altering the maturation state of tumor-surrounding dendritic cells [122]. Moreover, preparations of a weak TLR4 agonist MPLA (or its synthetic analogue GLA) restored the intratumoral activity of cytotoxic T cells and elevated pulmonary NK cell cytotoxicity, which led to tumor eradication and a reduction in cancer metastasis [119,139]. Nevertheless, preclinical development of LPS-based cancer treatment turned out to be challenging due to possible adverse effects or clinical inactivity as a result of LPS heterogeneity [120].

Therefore, the development of molecularly defined, homogeneous TLR4 agonists which are safe and well tolerated in humans as potential stand-alone therapeutics for cancer or as supplements in checkpoint inhibitor treatment is highly required. For other tumor types, such as breast or colon cancer, where the overexpression of TLR4 is frequently associated with a negative prognosis, the development of TLR4-inhibiting strategies targeting its co-receptor protein MD-2 could hold therapeutic promise [110,113,140,141].

However, the exact molecular mechanisms driving these different pathologies and the varying importance of LPS- and DAMP-dependent TLR4- or caspase-4/5-induced signaling and cellular activation are often not clear at all. Thus, the design, chemical synthesis, and functional studies of molecularly defined carbohydrate-based LPS mimetics that are capable of either interfering with the TLR4 signaling or enhancing and modulating the TLR4-driven pro-inflammatory responses provide a versatile tool for studying the underlying mechanisms of TLR4-mediated inflammation and has a great therapeutic potential (Figure 2).

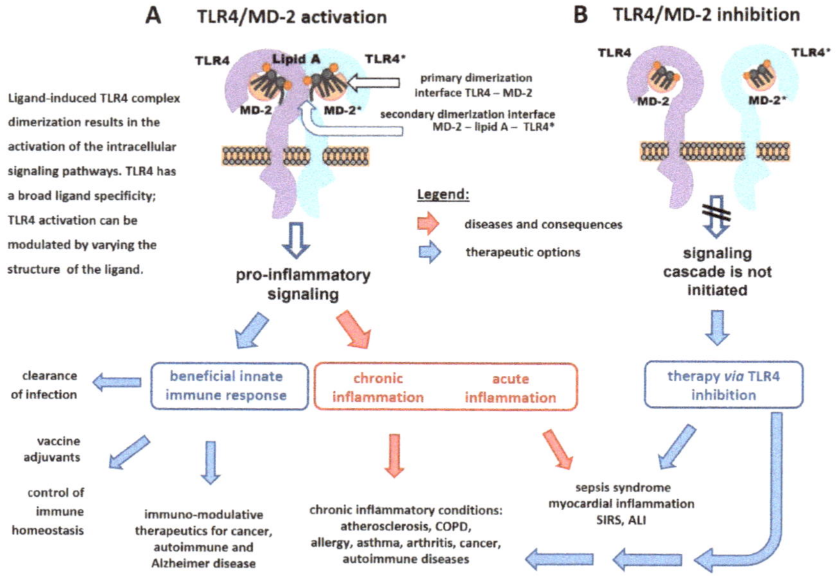

Figure 2. Schematic representation of lipid A-induced activation (**A**) and inhibition (**B**) of the TLR4/MD-2 complex and the options for therapeutic modulation of TLR4-mediated signaling.

5. Challenges in Development of Synthetic TLR4 Ligands

The chemical complexity of LPS and its truncated mutants (*Ra*- and *Re*-LPS) [142], representing the minimal structure needed for full pro-inflammatory activity and proper recognition by the TLR4 complex, renders the synthesis of possible drug candidates a formidable challenge. The complexity and poor predictability of LPS-TLR4 structure–activity relationships [143] as well as evidences for enormous influence of subtle variations in the chemical composition of lipid A (e.g., minor variations in the length of lipid chains) on the expression of TLR4-mediated activity makes rational prediction of biological properties of lipid-A/LPS-derived drugs hardly possible. In addition, the inherent conformational plasticity of LPS-recognizing protein MD-2, which markedly changes its conformation contingent on the structure of a bound ligand [144], makes classical "rational drug design" inefficient in this case. Finally, the species specificity in recognition of the diverse ligands by the TLR4 complex hampers straightforward extrapolation of data obtained in vitro and in animal models for clinical trials. Along these lines, pharmacological exploitation of the LPS/TLR4 complex and rational design of drugs targeting the co-receptor protein MD-2 is far away from being straightforward.

6. Co-crystal-Structure-Based Design of Disaccharide Lipid A Mimetics (DLAMs)

The chemical structure of lipid A is highly conserved within species and relies on a $\beta(1{\rightarrow}6)$ disaccharide backbone (GlcN-$\beta(1{\rightarrow}6)$-GlcN) that is multiply acylated by optically active β-hydroxy fatty acids at equatorially oriented OH groups (positions 2,2′ and 3,3′) (Figure 3). The positions 4 and 6′ in lipid A are free (whereas position 6′ is reserved for linking the core sugars) and the residual OH groups at positions 4′ and 1 are commonly phosphorylated. The secondary 4′-linked phosphate group at the distal GlcN ring is absent in several lipid A modifications which significantly downregulates the TLR4-mediated signaling [9,10]. Similarly, hydrolysis of the intrinsically labile anomeric 1-phosphate group at the proximal GlcN ring leads to attenuation of endotoxic activity which can be applied in biotechnological manipulations of LPS/lipid A structures with the aim to reduce possible adverse effects as in the case of MPLA. The hexa-acylated lipid A of *N. meningitidis* and *E. coli* is the most potent TLR4 agonist known to date, whereas a low degree of acylation is characteristic for lipid A variants with either TLR4 antagonistic properties (as in the case of the tetra-acylated lipid IVa [42] and penta-acylated nonpathogenic lipid A of *R. sphaeroides* [145]) or no TLR4-mediated activity (Figure 3).

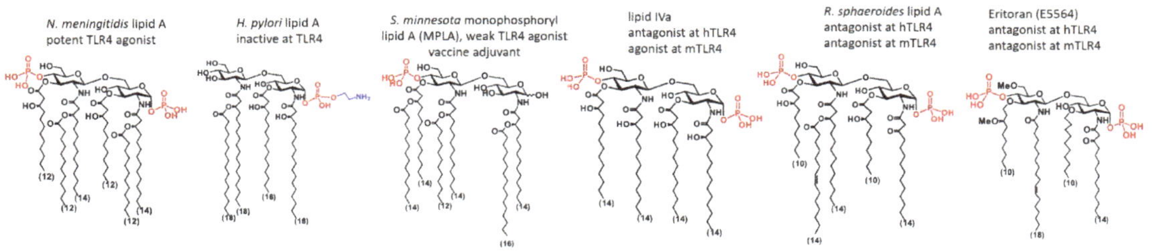

Figure 3. Chemical structure of several lipid A variants.

In line with the published co-crystal structures of *E.coli Ra/Re*-LPS bound to h- or m-TLR4/MD-2 complexes (PDB codes: 3FXI, 3VQ1), respectively, five out of six lipid chains of a typical hexa-acylated lipid A are spaciously incorporated into the hydrophobic binding cavity of the co-receptor protein MD-2 (Figure 4A). The sixth lipid chain (according to the X-ray structure, the 2*N*-linked 3-hydroxyalkanoyl chain attached at the proximal GlcN moiety) is displayed on the surface of MD-2 as it does not fit into the tightly occupied binding pocket [17,146]. This creates a so-called "heterodimerization" or "secondary dimerization interface" which involves a hydrophobic patch on the surface of MD-2 (Phe126, Y131, Leu-rich repeats), an exposed 2*N*-hydroxyalkanoyl chain of lipid A and the Leu-rich repeats (incl. Phe440, Phe463 for hTLR4 and Phe438, Phe461 for mTLR4) of the second TLR4* [17,147]. The formation of a hexameric $[TLR4/MD\text{-}2/LPS]_2$ complex is preceded by significant ligand-driven conformational changes in MD-2 which force the protein in its "closed" conformation able to undergo dimerization (Figure 4B) [144,148]. The phosphate groups of lipid A (1- and 4′-linked) are crucial for binding to hMD-2 where they interact with several positively charged side chains on the rim of the binding groove (mostly Lys and Arg) and are also important for establishing a secondary dimerization interface with the second TLR4* complex [149]. The absence of one of the phosphate groups results in a 100-fold drop in biological activity, partially due to a lack of efficient dimerization, and is manifested by reduced cytokine release [150,151].

The situation is quite different for protein-bound underacylated lipid A variants such as lipid IVa and Eritoran (Figure 3) where all lipid chains are fully incorporated into the binding cleft of hMD-2 (Figure 3C) [42,152]. This type of binding brings hMD-2 into an "open" conformation with the Phe126 loop exposed to solvent which makes dimerization with the second hTLR4* complex impossible and thus renders these lipid A variants potent hTLR4 antagonists (Figure 4D). Accordingly, TLR4 antagonists can tightly bind to the hydrophobic pocket of MD-2 without initiating pro-inflammatory signaling and compete with the natural ligand LPS in occupying the same site on the co-receptor protein.

The chemotype-specific TLR4 activation by variably acylated lipid A variants in different species (human, mice, hamster, horse) is still not a fully explained phenomenon, even though the solved co-crystal structures, mutagenesis studies, and molecular dynamics simulation analyses have provided profound insight into the molecular mechanisms of the species-specific action of lipid IVa on h- and mTLR4/MD-2. Lipid IVa, a tetra-acylated biosynthetic precursor of *E. coli* lipid A, can activate mTLR4/MD-2 by binding to the murine receptor complex in a manner similar to hexa-acylated *E. coli* lipid A, i.e., by exposing a large portion of its 2-*N*-linked β-hydroxyacyl chain on the surface of mMD-2 which shifts the Phe126 residue inwards and enables hydrophobic interactions with the second TLR4* ectodomain leading to heterodimerization [17]. Significant disparities in the binding mode of lipid IVa to hMD-2 (PDB code: 2E59), where the diglucosamine backbone is rotated by

180° and all acyl chains are fully buried in the binding groove, result in the conformational rearrangement in MD-2 and the relocation of Phe126 outwards which brings MD-2 into an "open" conformation unable to undergo heterodimerization [42]. This mode of binding is akin to that of the synthetic TLR4 antagonist Eritoran (PDB code: 2Z65) (Figure 4C,D). Hence, the acylation pattern of lipid A alone does not provide a sufficient basis for reliable and predictable structure–activity relationships in terms of species-specific recognition.

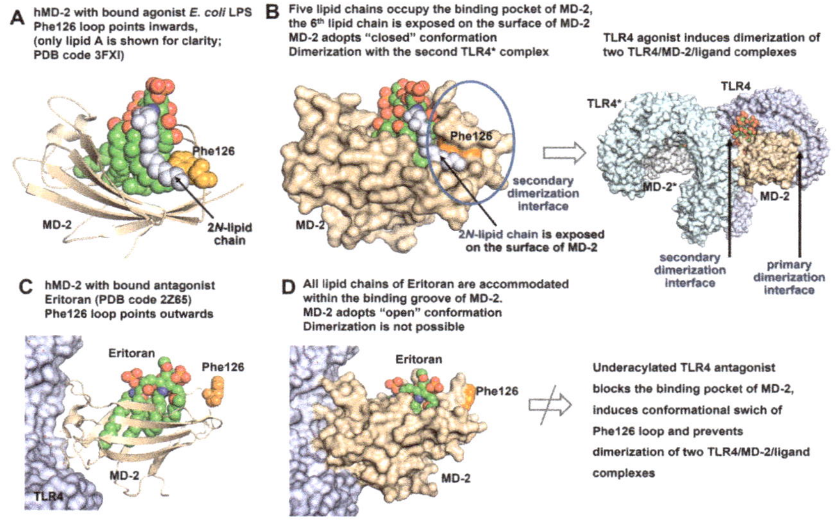

Figure 4. Comparison of co-crystal structures of TLR4/MD-2-bound agonist and antagonist LPS/lipid A variants. (**A**) Co-crystal structure of hTLR4/MD-2/*Ra*-LPS (*E. coli*) complex (PDB code: 3FXI, only hMD-2 with bound lipid A portion of LPS is shown for clarity). The exterior positioning of the 2*N*-acyl chain (colored blue) of hTLR4/MD-2-bound lipid A is supported by interaction with Phe126 (colored orange) which points inwards due to ligand-driven conformational rearrangement in hMD-2. (**B**) Five lipid chains of hexa-acylated lipid A are accommodated in the binding pocket of hMD-2. The 6th 2*N*-linked hydroxyacyl residue is excluded from the binding pocket and, together with a side chain of Phe126, contributes to the formation of a secondary dimerization interface. The LPS binding induces dimerization of two TLR4/MD-2/LPS complexes. (**C**) All lipid chains of a tetra-lipidated TLR4 antagonist Eritoran are intercalated into the binding groove of hMD-2 (PDB code: 2Z65). Phe126 (colored orange) is exposed to solvent. (**D**) hMD-2 with bound antagonist ligand adopts an "open" conformation that results in prevention of dimerization. Images were generated with PyMol.

Biophysical studies have confirmed that the long-chain (*R*)-3-alkanoyloxyalkynoyl residues of lipid A stick snugly together to comprise a voluminous hydrophobic cluster, whereas the β(1→6)-linked diglucosamine skeleton of lipid A possesses a substantial degree of flexibility owing to the possibility of unobstructed rotation around β(1→6) glycosidic and exocyclic oxymethyl (-CH_2O-) linkages (Figure 5A) [153,154]. Thus, protein binding – induced amendment of dihedrals (or torsion angles) around the three-bond β(1→6) glycosidic linkage in the common βGlcN(1→6)GlcN backbone of endotoxic hexa-acylated lipid A results in a repositioning of the "proximal" GlcN ring into a skewed relative orientation (Figure 5B) that might facilitate the exposure of a notorious 2*N*-acyl lipid chain on the surface of MD-2, which, in turn, prompts the dimerization of two ternary TLR4/MD-2/LPS complexes and the following initiation of the intracellular signaling cascade (Figure 4A,B) [155,156]. According to co-crystal structures (PDB: 3FXI/3VQ2/3VQ1), the two GlcN rings of the disaccharide backbone of a protein-bound agonist lipid A (*E. coli* lipid A bound to h- or mTLR4/MD-2 and lipid IVa bound to mTLR4/MD-2) adopt a skewed relative orientation (Figure 5B), whereas the two GlcN rings of an antagonist lipid A (PDB: 2E59, 2Z65) [42,152] are found to accept a co-planar arrangement (Figure 5C) [155].

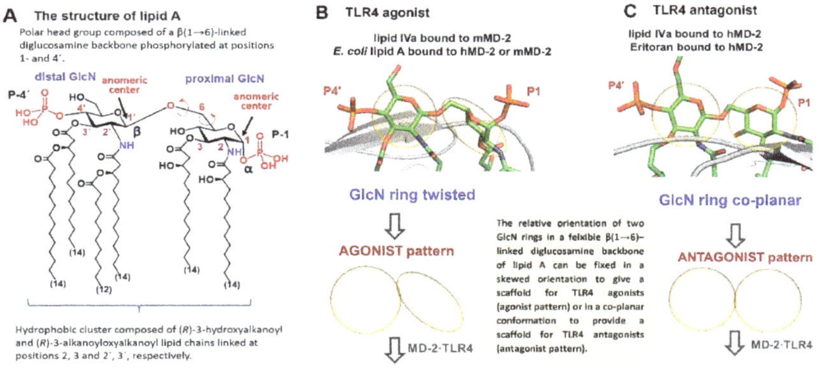

Figure 5. The chemical structure of *E. coli* lipid A and the 3D-tertiary structure of MD-2-bound lipid A in relation to biological activity. (**A**) Chemical structure of *E. coli* lipid A; (**B**) tertiary structure of the diglucosamine backbone of TLR4 agonist lipid A variants bound to MD-2/TLR4 complex; (**C**) tertiary structure of the diglucosamine backbone of TLR4 antagonist lipid A variants bound to MD-2; crystal-structure-based design of synthetic TLR4 ligands.

We assumed that the poor predictability of biological activity of variably acylated lipid A variants (the prediction is commonly based on the number, length, and distribution pattern of lipid chains) can be bypassed by manipulating the inherent flexibility of the carbohydrate backbone of lipid A, which we deemed responsible for the species-specific activity at TLR4. To create versatile TLR4-modulating glycolipids able to either efficiently inhibit the TLR4-driven inflammation through binding to MD-2 or to potently induce the TLR4/MD-2-mediated signaling, both in a species-independent fashion, we designed novel TLR4 ligands based on the disaccharide scaffolds with a fixed rigidified molecular shape corresponding either to an agonist (pyranose rings in a skewed arrangement) [157,158] or antagonist pattern (pyranose rings in a co-planar arrangement) [155,159] (Figure 5). Worth mentioning is that all known TLR4 ligands developed so far had either an ordinary β(1→6) diglucosamine backbone or more bendable skeletons where one or both GlcN rings were exchanged with linear aglycons [160–162]. Moreover, a TLR4 agonist structurally unsimilar to lipid A was developed; however, this small molecule was exclusively mouse-TLR4-specific [163].

As stated above, the rotation about the β(1→6) glycosidic linkage (which is a distinctive feature of the βGlcN(1→6)GlcN backbone of lipid A) brings the proximal GlcN ring in a skewed orientation seen in all agonistic TLR4 co-crystal structures, i.e., for hexa-acylated *E. coli* lipid A bound to hMD-2/TLR4 (PDB code: 3FXI) and mMD-2/TLR4 (PDB code: 3VQ2), as well as for tetra-acylated lipid IVa in a complex with mTLR4 (PDB code: 3VQ1) (Figure 5B). In our approach, the flexible GlcNβ(1→6)GlcN backbone of natural lipid A was exchanged for a rigid nonreducing disaccharide scaffold imitating the 3D-molecular shape of the diglucosamine skeleton of a protein-bound lipid A, which is strikingly different for TLR4 agonist and antagonist ligands (Figure 6A,B). Along these lines, the β,α-1,1′-linked diglucosamine disaccharide with co-planar-arranged sugar rings mimicking the molecular shape of the diglucosamine backbone of hMD-2-bound antagonists lipid IVa and Eritoran (PDB codes: 2E59, 2Z65, respectively) was considered an excellent carbohydrate scaffold for synthetic TLR4 antagonists [155,159] (Figure 6A,C), whereas the α,α-1,1′-linked or β,β-1,1′-linked nonreducing disaccharide scaffolds reflecting the 3D-tertiary structure of the diglucosamine backbone of endotoxic lipid A found in the active [TLR4/MD-2/(*E.coli* lipid A)]$_2$ complexes (PDB codes: 3FXI, 3VQ1, 3VQ2) (Figure 6B) showed a skewed relative orientation of two GlcN rings and thus were believed to provide a perfect scaffold for carbohydrate-based TLR4 agonists [157,164] (Figure 6D,E).

Accordingly, we performed the synthesis of a library of TLR4-modulating glycolipids, disaccharide lipid A mimetics (DLAMs), by replacing the flexible three-bond β(1→6)-diglucosamine frame of lipid A with a conformationally rigid two-bond (1↔1′)-linked disaccharide imitating the molecular shape of TLR4/MD-2-bound lipid A variants (co-planar arrangement of two sugar rings for TLR4 antagonists and twisted relative orientation of two sugar rings for TLR4 agonists) (Figure 6). The compounds are abbreviated DLAMs for "Disaccharide Lipid A Mimetics" with a prefix indicating the anomeric configuration of the 1,1′-glycosidic linkage (βα-DLAMs, αα-DLAMs, and ββ-DLAMs); the latter is decisive for the molecular shape of the scaffold.

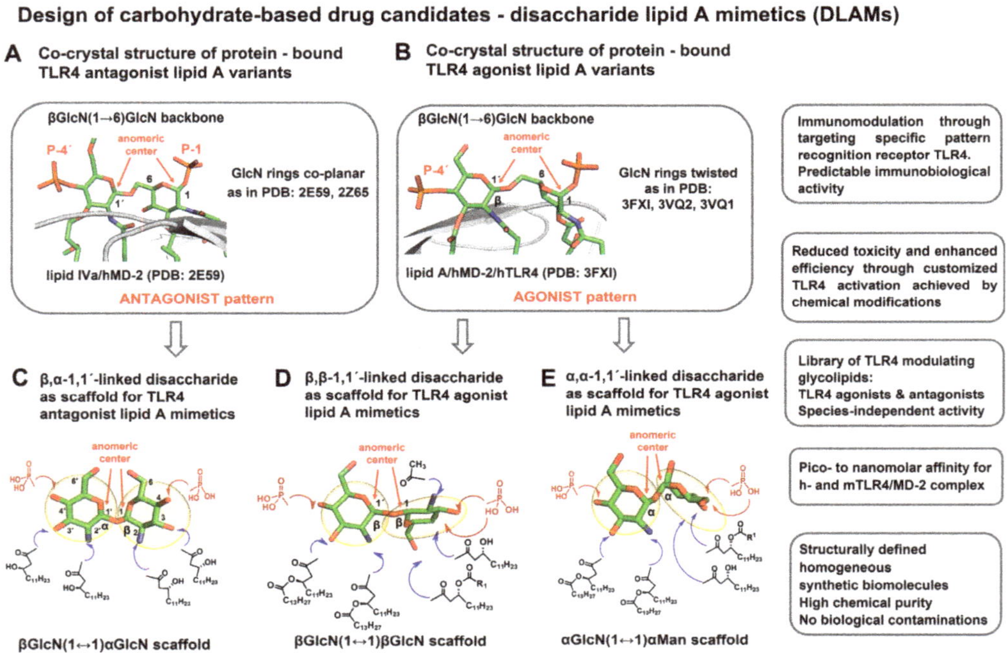

Figure 6. Co-crystal-structure-based design of disaccharide lipid A mimetics (DLAMs). (**A**) The co-crystal structures of hMD-2-bound lipid IVa and Eritoran (PDB codes: 2E59 and 2Z65, respectively) disclose a co-planar relative orientation of two GlcN rings comprising the β(1→6)-linked diglucosamine backbone of lipid A. This arrangement is achieved through rotation about flexible β(1→6) linkage. A specific co-planar relative orientation of the GlcN rings in a βGlcN(1→6)GlcN backbone in a protein-bound lipid A is designated an "antagonist pattern". (**B**) The co-crystal structures of hMD-2/TLR4-bound *E. coli* lipid A (PDB code: 3FXI), mMD-2/TLR4-bound *E. coli* lipid A (PDB code: 3VQ2), and mMD-2/TLR4-bound lipid IVa (PDB code: 3VQ1) reveal a skewed relative orientation of two GlcN rings in the β(1→6)-linked diglucosamine backbone of protein-bound lipid A. This relative orientation of GlcN rings achieved through rotation about β(1→6) linkage is designated an "agonist pattern". (**C**) Design of TLR4 antagonist lipid A mimetics. The co-planar molecular shape of β,α-1,1′-linked diglucosamine mirrors the 3D-topology of the βGlcN(1→6)GlcN backbone of hMD-2-bound TLR4 antagonists shown in (**A**). The sites of attachment of the lipid chains and phosphate groups for the synthesis of lipid A mimetics are represented by blue and red arrows, respectively. (**D**) Design of TLR4 agonist lipid A mimetics based on a β,β-1,1′-diglucosamine scaffold.

The 3D-structure of a synthetic β,β-(1↔1′)-linked diglucosamine scaffold with two GlcN rings in a twisted relative orientation. The sites of possible chemical attachment of functional groups are indicated by arrows: the acylation/phosphorylation pattern of the distal GlcN ring corresponds to that of *E. coli* LPS, whereas the proximal GlcN ring has a variable acylation and phosphorylation pattern. (**E**) Design of TLR4 agonist lipid A mimetics based on an α,α -1,1′-linked disaccharide scaffold. The skewed molecular shape of an α,α-1,1′-linked diglucosamine mirrors the 3D-topology of a βGlcN(1→6)GlcN backbone of MD-2-bound TLR4 agonists shown in (**B**). The acylation and phosphorylation pattern of the distal GlcN ring corresponds to that of *E. coli* lipid A, whereas the proximal pyranose ring (Man) entails variable acylation and phosphorylation pattern. The sites of possible attachment of lipid chains are shown by blue arrows; the sites proposed for chemical phosphorylation are shown by red arrows. Images were generated with PyMol.

7. TLR4 Antagonists Based on a βGlcN(1↔1)αGlcN Scaffold: Nanomolar Potency and Species-Independent Activity of βα-DLAMs

Chemically synthesized, fully orthogonally protected, 1,1′-linked disaccharides were used as the starting point for the assembly of variably acylated and phosphorylated DLAMs. For the synthesis of TLR4 antagonists based on a β,α-1,1′-diglucosamine scaffold, the 2,2′-*N*-Troc, 3,3′-*O*-TBDMS, 4,6-, 4′,6′-*O*-benzylidene acetal-protected disaccharide was assembled by chemical glycosylation (Figure 7) [155]. One phosphate group was attached to each sugar unit, and the acylation pattern followed that of lipid IVa: each GlcN ring was 2-*N*-, 3-*O*-acylated by two long-chain fatty acids of different chemical structures and lengths (tetra-acylated compounds). The βGlcN(1↔1)αGlcN-based DLAMs were designed to match the geometry of the binding pocket of MD-2 (all lipid chains are fitted into the binding cleft of MD-2 analogous to lipid IVa and Eritoran in a complex with hMD-2).

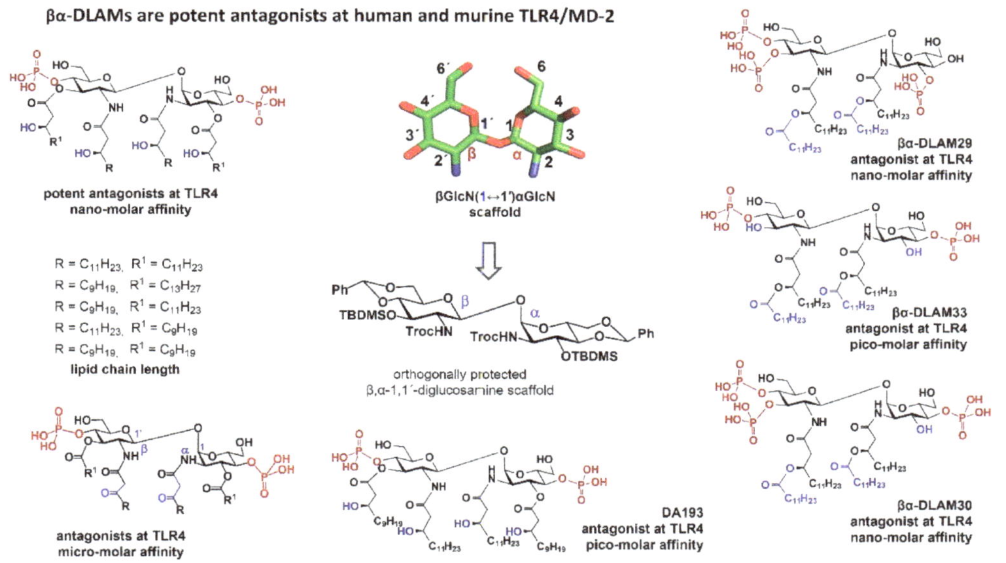

Figure 7. Synthetic disaccharide lipid A mimetics derived from βGlcN(1↔1)αGlcN disaccharide (βα-DLAMs).

In contrast to lipid IVa that acts as an antagonist at hTLR4 but as an agonist at mTLR4, the tetra-acylated βα-DLAMs could not induce the pro-inflammatory signaling in mouse cells owing to the co-planar shape and rigidity of their artificial disaccharide backbone [155,159]. Conversely, all conformationally confined βα-DLAMs performed as potent antagonists at both human and murine TLR4 [165,166]. Indeed, according to molecular dynamics simulations, the tetra-acylated TLR4 antagonist βα-DLAM DA193 was estimated to bind to the co-receptor protein hMD-2 ca. 20-fold more effectively than TLR4 agonist *E. coli* lipid A and 3-fold more strongly than TLR4 antagonist lipid IVa. NMR conformational studies revealed that the distance between the two 4, 4′ phosphate groups in βα-DLAM DA193 roughly corresponds to the distance P1–P4′ found in the native protein-bound lipid IVa.

The NOESY NMR experiments (performed for three variably acylated βα-DLAMs) revealed synΦα/synΦβ geometry around the βα-1,1-glycosidic linkage, which was in line with the double exo-anomeric conformation at both glycosidic torsions [159]. The co-planar conformation was not influenced by the length of multiple lipid chains. In this way, it was proved that the relative orientation of two GlcN rings in the βGlcN(1↔1)αGlcN backbone of the βα-DLAMs has a co-planar arrangement and thus mirrors the 3D-molecular shape of the native β(1→6) diglucosamine backbone of hMD-2-bound antagonists found in the co-crystal structures.

Therefore, fixing the tertiary structure of the carbohydrate backbone of lipid A in an 'antagonist' conformation (co-planar-arranged pyranose rings) through the replacement of a flexible βGlcN(1→6)GlcN backbone of native lipid A with a conformationally constrained β,α(1↔1)-linked diglucosamine scaffold surmounted the species-specific activity and gave rise to synthetic h- and mTLR4-antagonistic glycolipids of nanomolar affinity [165]. However, we observed some lipid chain length-dependent differences in the inhibition of the TLR4-mediated cellular responses by the βα-DLAMs in human and mouse cells, i.e., lipid A mimetic equipped with two C_{14} and two C_{12} lipid chains (DA193) showed the best antagonist potential at hTLR4, whereas the shorter-chain counterparts (i.e., $2xC_{12}$; $2xC_{10}$ lipid chains, DA255) demonstrated a higher affinity for mTLR4 [165] (Figure 7).

To further improve the affinity for MD-2 and to reduce the lipid chain length-driven species specificity in the binding of differently lipidated βα-DLAMs, we turned our attention to the unique structure of the lipid chains of a natural TLR4 antagonist, *R. sphaeroides* lipid A, which has one β-ketoacyl lipid chain (C_{14}) attached at the proximal GlcN ring of the lipid A backbone (Figure 3). The β-keto acids and the β-keto amides commonly exist in a thermodynamic equilibrium with the corresponding enol forms. Although the equilibrium is heavily shifted in favor of the keto form, this ratio could also depend on the nature of the solvent and the protein surrounding of the binding pocket of MD-2. To address a possible influence of keto–enol tautomerism of the β-ketoacyl lipid chains within lipid A molecules, we synthesized a series of βα-DLAMs equipped with two 2*N*-linked β-ketoacyl chains of variable length and investigated their TLR4-inhibiting potential (Figure 7) [166].

Since the binding affinity of DLAMs was supposed to depend both on the lipid composition and on the shape of the hydrophobic region of the molecule, we again modified the acylation pattern and synthesized a series of potent TLR4 antagonists based on a βGlcN(1↔1)αGlcN scaffold acylated at positions 2-*N*, 2′-*N* by the long-chain-branched (*R*)-3-alkanoyloxyalkanoyl fatty acids, whereas hydroxyl groups at positions 3 and 3′were let unsubstituted. Although a typical lipid A molecule has up to two phosphate groups, artificial hyperphosphorylation through appending an extra phosphate at the free OH-3 could increase ionic attraction to the positively charged Lys/Arg side chains on the rim of the binding pocket of MD-2. For that reason, a variable phosphorylation pattern (two, three, and four phosphate groups differently distributed along positions 3 and 4 on both GlcN units) was also included in this study (Figure 7). The antagonist potential of bis-phosphorylated and tri-phosphorylated molecules could be substantially enhanced and fluctuated from pico- to nanomolar concentrations.

Notably, a well-known antisepsis drug candidate E5564 (Eritoran) failed to improve survival in phase III clinical trials, despite an excellent antagonistic activity in vitro and in vivo [48], on account of several immunobiological drawbacks [167,168] as well as owing to the inherent instability of the labile glycosidic phosphate group P-1 turning the compound into inactive dephosphorylated metabolite [169]. Moreover, a possible neutralization by HDL was suggested as a conceivable explanation for the moderate therapeutic efficiency [169]. Therefore, the structure of the disaccharide skeleton of βα-DLAMs, where an unstable glycosidic phosphate group P-1 of lipid A/Eritoran is replaced by the far more durable secondary phosphate P-4, offers an unambiguous advantage for the creation of hydrolytically stable TLR4 antagonists.

8. TLR4 Agonists Based on αGlcN(1↔1)αMan and βGlcN(1↔1)βGlcN Scaffolds: Tailored Modulation of TLR4-Mediated Pro-inflammatory Signaling by αα-DLAMs and ββ-DLAMs

To design potent carbohydrate-based TLR4 agonists, we replaced the flexible β(1→6) glycosidic and exocyclic oxymethyl linkages present in the native βGlcN(1→6)GlcN frame of lipid A with an exceptionally stiff, nonreducing glycosidic linkage of either α,α or β,β configuration (Figure 6D,E). The unique rigidity of the 1,1-glycosidic linkages in the α,α- and β,β-linked nonreducing disaccharides that were supported by the carbohydrate-specific electronic effects was decisive for the immunobiological properties of the DLAMs based thereon.

The skewed relative orientation of the two pyranose rings in α,α- and β,β-1,1′-linked disaccharides is governed by the values of the torsion angles about the (1↔1′) glycosidic linkage which depends on the anomeric configuration and is influenced primarily by anomeric and exo-anomeric effects [170,171]. It has also been shown that the nature of functional/protecting groups and the sites of their attachment has no or little influence on the values of dihedrals around double glycosidic linkages [172]. Specific conformation of the αα-linked disaccharides has been confirmed by molecular dynamics simulations [172, 173], and a predominant gauche–gauche conformation for substituted αα-trehalose [174] and its αGlc(1↔1)αMan analogue [175] was corroborated by X-ray structural studies and conformational analysis.

Similarly, it has been demonstrated that the dihedrals about the β,β-(1↔1′) linkage conferring a skewed topology to the β,β-1,1′-linked disaccharides (Figure 6D) are energetically favored and conserved in differently substituted derivatives [176–179]. A skewed molecular shape as a result of the staggered arrangement of the two β,β-1,1′-linked pyranoses (the staggered low-energy orientation in the positive direction (ca. 50°) places the C-1 *gauche* at the O-5 of the opposite ring) [171] would impose specific spatial orientation to functional groups connected to the twistedly arranged sugar rings (Figure 6). This would allow a presentation of specific long-chain (*R*)-3-hydroxyacyl residues and phosphate groups attached at the proximal sugar ring facing the secondary dimerization interface onto the surface of MD-2 which should make an efficient crosslinking of the second TLR4* receptor complex possible.

In contrast to the natural βGlcN(1→6)GlcN backbone of lipid A, which can spontaneously adjust its 3D-molecular shape (i.e., the relative orientation of GlcN rings) upon interaction with the MD-2/TLR4 complex, the relative stiffness of the α,α-(1↔1′) and ββ-(1↔1′) linkages confers exceptional rigidity of the disaccharide frames of DLAMs based thereon. The skewed relative orientation of the two pyranose rings in α,α-1,1′- and ββ-1,1′-linked disaccharides resembles the 3D tertiary structure of the MD-2-bound β(1→6)-linked diglucosamine backbone of endotoxically active lipid A (Figure 6B). Thus, the α,α-1,1′- and ββ-1,1′-linked disaccharide backbones of DLAMs also retain their specific molecular shape when bound by the MD-2/TLR4 complex.

The sites of attachment of the phosphate groups and branched β-alkanoyloxyalkanoyl chains at the distal GlcN ring reflect the acylation and phosphorylation pattern of *E. coli* lipid A (acyloxyacyl residues at 2′-N and 3′-O, and a phosphate group at position 4′), whereas the proximal sugar moiety (Man for αα-DLAMs and GlcN for ββ-DLAMs) entails two β-hydroxyalkanoyl lipid chains and a variable number of phosphate groups that can be linked at positions 4 or 6 or 3/4 (Figures 8 and 9). To estimate the relative contribution of ionic and hydrophobic interactions supplied by the phosphate groups and acyl chains, respectively, we synthesized four different groups of αα-DLAMs and two series of ββ-DLAMs.

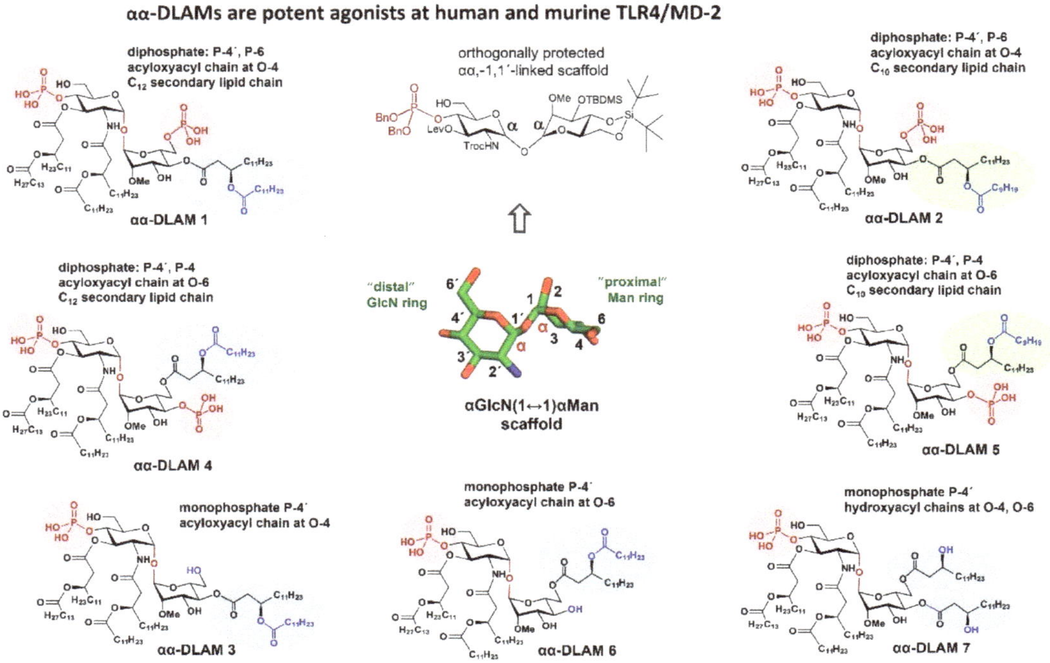

Figure 8. A library of synthetic disaccharide lipid A mimetics based on αGlcN(1↔1)αMan scaffold (αα-DLAMs).

Starting from a synthetically prepared, fully orthogonally protected αGlcN(1↔1)αMan disaccharide [156], three series of hexa-acylated, bisphosphorylated αα-DLAMs were synthesized (Figure 8). The compounds differ in the positioning of the phosphate group and acyl chains attached at the proximal mannose ring. Among diphosphate DLAMs where a second phosphate group is attached either at position 6 of the Man moiety (αα-DLAMs **1** and **2**) or at position 4 of the Man ring (αα-DLAMs **4** and **5**), the length of the secondary acyl chain in a (*R*)-3-acyloxyacyl residue linked to position 4 or 6 varies between C_{12} and C_{10} [157], respectively. The monophosphate αα-DLAMs **3**, **6**, and **7** lack a phosphate group at the Man moiety but differ in the sites of attachment of the branched β-acyloxyacyl chain Accordingly, αα-DLAM **3** lipidated at position 4 is a monophosphate counterpart of αα-DLAM **1**, whereas αα-DLAM **6** with the acyloxyacyl chain at position 6 is a monophosphate counterpart of αα-DLAM **4** (Figure 8).

Upon interaction with the TLR4 complex, the four lipid chains linked at the distal GlcN ring were assumed to tightly bind to the hydrophobic binding pocket of MD-2, whereas the lipid chains and phosphate groups variably linked at the Man moiety in the αα-DLAM **1–7** were expected to face the secondary dimerization interface and to establish persistent ionic and hydrophobic interactions with the second TLR4* complex [156]. The alteration in the phosphorylation pattern (position 4 vs. position 6) at the Man moiety as well as variation in the acylation pattern (position 6 vs. position 4) and the length of the secondary acyl chains (C_{12} vs. C_{10}) affected the efficiency and tightness of the TLR4 complex dimerization, which was subsequently translated in diverging downstream pro-inflammatory signaling [158].

The immunobiological properties of DLAMs were analyzed in a variety of cell types of human and mouse origin such as h- and mTLR4 transfected HEK293 cells, human mononuclear cells (MNC), dendritic cells (DCs), human acute monocytic leukemia cells (THP-1), cultured airway epithelial cell lines (Calu-3 and BEAS-2B), mouse macrophages etc. The α,α-DLAMs and ββ-DLAMs could efficiently trigger the intracellular innate immune signaling that could be readily modulated by specific modification (i.e., the sites of attachment of the functional groups) at the proximal sugar ring of DLAMs. Picomolar concentrations of α,α- and ββ-DLAMs were already sufficient to induce the release of high levels of IL-6, IL-8, TNF-α, IL-1β, and MCP1 in the respective cell types. In contrast to native lipid A and lipid A analogues built on the flexible carbohydrate-containing or aglycon-based skeletons which often demonstrate species-specific activities, the TLR4-mediated pro-inflammatory signaling induced by DLAMs could be gradually and predictably modulated at both human and mouse TLR4 in a species-independent fashion.

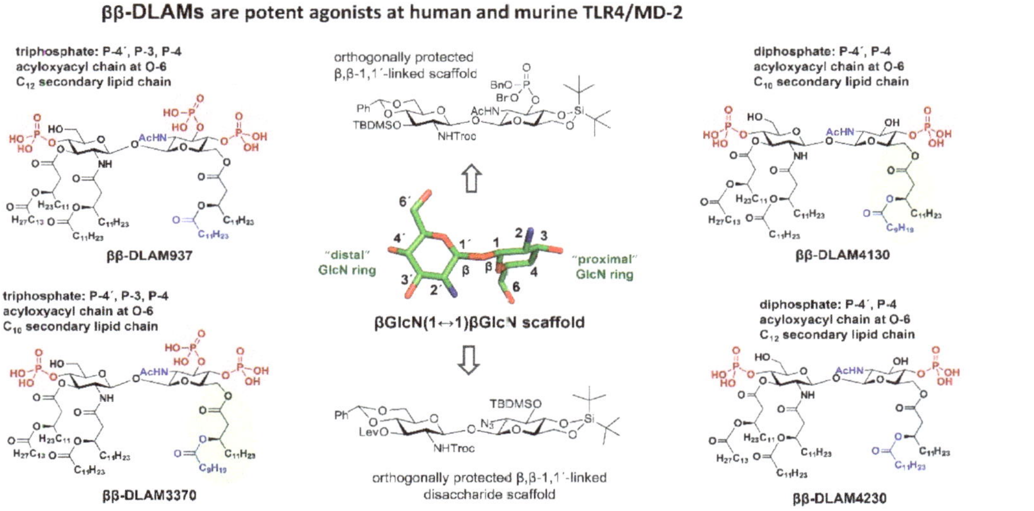

Figure 9. Potent disaccharide-based TLR4 agonists derived from a βGlcN(1↔1)βGlcN scaffold (ββ-DLAMs).

The activation of the NF-κB-regulated signal transduction pathway by αα-DLAMs was analyzed over a broad concentration range through an assessment of the induction of secreted embryonic alkaline phosphatase (SEAP) in hTLR4/MD-2/CD-14-transfected HEK293 cells. The EC_{50} values for the αα-DLAM diphosphates **1, 2** and **4, 5** were in the picomolar range (70 to 400 pM), whereas the TLR4 activation potential of monophosphorylated αα-DLAMs **3, 6, 7** reflected their chemical structure lacking a phosphate group at the Man unit and dropped from a pico- to nanomolar concentration range (EC_{50} 3–60 nM) [158]. The decline in the TLR4-stimulating activity of the αα-DLAM monophosphates can be explained by the absence of a negatively charged phosphate group at the heterodimerization interface leading to feeble and inefficient dimerization [150] as well as by inferior recognition and transfer of molecules through CD14 [180].

The mTLR4 activation profile of αα-DLAMs in mTLR4/mMD-2-transfected HEK293 cells and in mouse macrophages (mBMDM) could be similarly regulated by altering the sites of acylation and phosphorylation at the proximal sugar ring, and the responsiveness of native murine immune cells to αα-DLAM challenge was again in the picomolar range for αα-DLAM diphosphates **1, 2** and **4, 5** (EC_{50} 90–320 pM for release of TNF-α) and, following the previously established structure–activity relationships, fell to nanomolar range in the case of αα-DLAM monophosphates **3, 6, 7** (EC_{50} 7–200 nM).

Picomolar doses of conformationally confined lipid A mimetics αα-DLAMs **1, 2** and **4, 5** also induced a concentration-dependent production of cytokines IL-6 and TNF-α in primary human immune cells (i.e., mononuclear cells (MNCs); EC_{50} 20–140 pM for release of IL-6 and 140–350 pM for release of TNF-α) [158]. The pro-inflammatory responses could be "programmed" and gradually modulated by adjusting the chemical structure of the carbohydrate-based molecules, i.e., by altering the length and position of acyl chains and the sites of attachment of phosphate groups at the proximal Man moiety. Along these lines, the relative potency of the structurally different series of αα-DLAMs to induce the release of pro-inflammatory cytokines in primary murine immune cells adequately correlated with the data obtained in human cells, thus providing proof for species-independent structure–activity relationships.

Our efforts in the simplification of the complex chemical structure of a native TLR4-stimulating ligand *Ra/Re*-LPS upon retaining full biological activity proved to be successful: the agonist DLAMs showed equal or higher potency in inducing the TLR4-specific signal transduction pathway as *Re*-LPS (Kdo_2-lipid A) [158]. Although the composition of DLAMs is as minimal as that of lipid A (two pyranoses, up to six lipid chains, and one/two phosphate groups), the αα-DLAM diphosphates were 5- to 10-fold (depending on the chemical structure) more efficient in inducing the TLR4-mediated pro-inflammatory responses compared with *E. coli* lipid A and αα-DLAM monophosphates were more potent TLR4 activators compared with MPLA in both human and murine immune cells [158].

Further advanced modulation of TLR4-mediated pro-inflammatory responses in a variety of cell types was achieved with the synthesis of ββ-DLAMs—the TLR4-activating glycolipids based on a ββ-1,1′-linked diglucosamine scaffold differing in its 3D-topology from the 1,1′-linked αα- counterpart (Figure 9). Moreover, the primary chemical structure of the βGlcN(1↔1)βGlcN scaffold—with an equatorial 2-*N*-acetamide group instead of an axial 2-*O*-Me group in position 2 of the proximal sugar ring as well as a varying number of phosphate groups attached at positions 3/4 of the proximal GlcN residue—conferred specific immunobiological properties to the ββ-DLAMs based thereon [164]. As changes in the phosphorylation status of the carbohydrate backbone of native lipid A can markedly enhance or reduce the predisposition of LPS to interact with the TLR4 complex, we set out to investigate the influence of increased electronegativity of the sugar head group of DLAMs on the TLR4 activation and synthesized hyperphosphorylated ββ-DLAMs by attaching a third phosphate group at position 3 of the distal GlcN ring (along with 4, 4′ phosphate groups) (Figure 9). As expected, the characteristic geometry of the βGlcN(1↔1)βGlcN disaccharide conferred specific immunobiological behavior to the glycolipid immunomodulator ββ-DLAMs based thereon, and the TLR4-mediated cell activation induced by ββ-DLAMs could be fine-tuned by adding/removing a P-3 phosphate and/or shortening the length of a secondary acyl chain [164].

To summarize, the TLR4-mediated immunobiological activity of agonist DLAMs is governed by the tertiary structure of their conformationally confined 1,1′-linked disaccharide backbones and can be predictably modulated by facile chemical modifications, i.e., by adjustment of the length and distribution pattern of lipid chains on one sugar unit as well as the number and sites of attachment of phosphate groups. The TLR4 stimulatory potential of α,α- and ββ-DLAMs ranges from picomolar concentrations for multiply phosphorylated compounds to nanomolar concentrations for monophosphorylated DLAMs and surpasses the TLR4 activity of the most potent natural TLR4 agonist, *E. coli* lipid A. A further advantage of DLAMs is their improved hydrolytic stability achieved through the replacement of a labile glycosidic phosphate P-1 of lipid A for a stable secondary or primary phosphate group.

9. Conclusions

Given that PRRs including TLRs and their signaling pathways are highly conserved throughout evolution, the pharmacological exploitation of the TLR4 complex can be considered promising in the context of (i) chronic inflammatory disorders (atherosclerosis, arthritis, etc.); (ii) acute inflammation, i.e., SIRS and sepsis; (iii) airway inflammation of both chronic (asthma and allergy) and acute etiology (ALI); (iv) vaccine adjuvant development;

and (v) oncological pathologies where both the activation and inhibition of TLR4 signaling could be of use as a stand-alone or adjuvant therapy.

Drawing on a crystal-structure-inspired design, we created a novel class of carbohydrate-based TLR4 agonists for tailored and controllable modulation of the TLR4-mediated signaling and disaccharide-derived TLR4 antagonists for inhibition of the TLR4-driven pro-inflammatory responses. The immunobiological activity of DLAMs is controlled by the configuration of the glycosidic linkage connecting the anomeric centers of constituting sugars which, in turn, determines the tertiary structure and the molecular shape of the nonreducing disaccharides applied as scaffolds for the assembly of glycophospholipid DLAMs. The relative spatial arrangement of two GlcN rings in the β,α-1,1′-linked diglucosamine was proved to be coplanar, which turned this disaccharide into a perfect scaffold for TLR4 antagonists, whereas α,α-1,1′- and β,β-1,1′-linked disaccharides with their skewed relative orientation of two sugar rings were exploited as scaffolds for TLR4 agonists. The 3D molecular shape of the disaccharide scaffolds (co-planar or tilted) served as the starting point for the design of lipid A mimetics, and the decision on the sites of attachment of the branched lipid chains and phosphate groups was made in agreement with the structural information provided by the co-crystal structures of the TLR4/MD-2/LPS (lipid A) complexes.

One of the major achievements of our investigation relates to surmounting the species specificity in ligand recognition by the TLR4 complex, which provides novel perspectives for a straightforward extrapolation of in vitro data obtained in human cellular models and human/murine immune cells to in vivo studies. To sum up, we generated a versatile library of fully synthetic disaccharide lipid A mimetics (DLAMs), embracing potent TLR4 antagonists with nanomolar affinity for MD-2 and powerful TLR4 agonists with picomolar affinity for TLR4/MD-2 as potential anti-inflammatory drug candidates for sepsis, acute and chronic inflammation as well as unique TLR4-activating immunomodulators with broad therapeutic potential for cancer and prophylactic capacity as vaccine adjuvants.

Author Contributions: Conceptualization, writing—original draft preparation, review and editing A.Z. and H.H. All authors have read and agreed to the published version of the manuscript.

Funding: This research was funded by the Austrian Science Fund (FWF): grant P32397 (A.Z.) and by the German Center for Lung Research (DZL), Airway Research Center North (ARCN) (H.H.).

Institutional Review Board Statement: Not applicable.

Informed Consent Statement: Not applicable.

Data Availability Statement: Data sharing not applicable.

Acknowledgments: Open Access Funding by the Austrian Science Fund (FWF).

Conflicts of Interest: The authors declare no conflict of interest.

Abbreviations

ALI	acute lung injury
BEAS-2B	human bronchial epithelial cell line
Calu-3	human lung adenocarcinoma cell line
CD14	cluster of differentiation 14
COPD	chronic obstructive pulmonary disease
DAMP	danger-associated molecular pattern
DCs	dendritic cells
DLAM	disaccharide lipid A mimetic
GPI	glycosyl phosphatidyl inositol
HEK293	human embryonic kidney 293 cells
HMGB1	high-mobility group box 1 protein
IFN-β	interferon-β
IL-1β	interleukin-1β
IRF3	interferon regulatory factor 3

Kdo	3-deoxy-D-*manno*-octulosonic acid
LBP	LPS-binding protein
LPS	lipopolysaccharide
LRR	leucine-rich repeat
MCP-1	monocyte chemoattractant protein-1
MD-2	myeloid differentiation factor 2
MNC	mononuclear cells
MPLA	monophosphoryl lipid A
MPL®	clinical-grade MPL® adjuvant (GlaxoSmithKline)
MyD88	myeloid differentiation primary response 88
NF-kB	nuclear factor kappa B
PAMP	pathogen-associated molecular pattern
PDB	Protein Data Bank
PRR	pattern recognition receptor
SEAP	secreted embryonic alkaline phosphatase
SIRS	severe inflammatory response syndrome
THP-1	human acute monocytic leukemia cells
TIR	Toll-interleukin-1 receptor
TIRAP	TIR-domain-containing adaptor protein
TLR4	Toll-like receptor 4
TNF-α	tumor necrosis factor-α
TRAF	TNF-receptor-associated factor
TRAM	TRIF-related adaptor molecule
TRIF	TIR-domain-containing adapter-inducing interferon-β

References

1. Pfeiffer, R. Untersuchungen über das Choleragift. *Z. Für Hyg. Und Infekt.* **1892**, *11*, 393–412. [CrossRef]
2. Beutler, B.; Rietschel, E.T. Innate immune sensing and its roots: The story of endotoxin. *Nat. Rev. Immunol.* **2003**, *3*, 169–176 [CrossRef]
3. Rietschel, E.T.; Kirikae, T.; Schade, F.U.; Mamat, U.; Schmidt, G.; Loppnow, H.; Ulmer, A.J.; Zähringer, U.; Seydel, U.; Di Padova, F Bacterial endotoxin: Molecular relationships of structure to activity and function. *FASEB J.* **1994**, *8*, 217–225. [CrossRef]
4. Seydel, U.; Schromm, A.B.; Blunck, R.; Brandenburg, K. Chemical structure, molecular conformation, and bioactivity of endotoxins *Chem. Immunol.* **2000**, *74*, 5–24.
5. Holst, O. Structure of the Lipopolysaccharide Core Region. In *Bacterial Lipopolysaccharides*; Knirel, Y.A., Valvano, M.A., Eds. Springer: Vienna, Austria, 2011; pp. 21–39.
6. Knirel, Y. Structure of O-Antigens. In *Bacterial Lipopolysaccharides*; Knirel, Y.A., Valvano, M.A., Eds.; Springer: Vienna Austria, 2011; pp. 41–115.
7. Galanos, C.; Lüderitz, O.; Rietschel, E.T.; Westphal, O.; Brade, H.; Brade, L.; Freudenberg, M.A.; Schade, U.F.; Imoto, M. Yoshimura, H.; et al. Synthetic and *Escheria coli* free lipid A express identical endotoxic activities. *Eur. J. Biochem.* **1985**, *148*, 1–5 [CrossRef]
8. Schletter, J.; Heine, H.; Ulmer, A.J.; Rietschel, E.T. Molecular mechanisms of endotoxin activity. *Arch. Microbiol.* **1995**, *164*, 383–389 [CrossRef]
9. Trent, M.S.; Stead, C.M.; Tran, A.X.; Hankins, J.V. Diversity of endotoxin and its impact on pathogenesis. *J. Endotoxin Res.* **2006**, *12*, 205–223. [CrossRef]
10. Raetz, C.R.H.; Reynolds, C.M.; Trent, M.S.; Bishop, R.E. Lipid A modification systems in Gram-negative bacteria. *Annu. Rev. Biochem.* **2007**, *76*, 295–329. [CrossRef]
11. Kirikae, T.; Schade, F.U.; Zähringer, U.; Kirikae, F.; Brade, H.; Kusumoto, S.; Kusama, T.; Rietschel, E.T. The significance of the hydrophilic backbone and the hydrophobic fatty acid regions of lipid a for macrophage binding and cytokine induction. *FEMS Immunol. Med. Microbiol.* **1994**, *8*, 13–26. [CrossRef]
12. Kawai, T.; Akira, S. The role of pattern-recognition receptors in innate immunity: Update on Toll-like receptors. *Nat. Immunol.* **2010**, *11*, 373–384. [CrossRef]
13. Kobayashi, M.; Saitoh, S.I.; Tanimura, N.; Takahashi, K.; Kawasaki, K.; Nishijima, M.; Fujimoto, Y.; Fukase, K.; Akashi-Takamura, S.; Miyake, K. Regulatory roles for MD-2 and TLR4 in ligand-induced receptor clustering. *J. Immunol.* **2006**, *176*, 6211–6218 [CrossRef]
14. Ryu, J.K.; Kim, S.J.; Rah, S.H.; Kang, J.I.; Jung, H.E.; Lee, D.; Lee, H.K.; Lee, J.O.; Park, B.S.; Yoon, T.Y.; et al. Reconstruction of LPS transfer cascade reveals structural determinants within LBP, CD14 and TLR4-MD2 for efficient LPS recognition and transfer *Immunity* **2017**, *46*, 38–50. [CrossRef]
15. Delude, R.L.; Savedra, R.; Zhao, H.; Thieringer, R.; Yamamoto, S.; Fenton, M.J.; Golenbock, D.T. CD14 enhances cellular responses to endotoxin without imparting ligand-specific recognition. *Proc. Natl. Acad. Sci. USA* **1995**, *92*, 9288–9292. [CrossRef]

16. Shimazu, R.; Akashi, S.; Ogata, H.; Nagai, Y.; Fukudome, K.; Miyake, K.; Kimoto, M. MD-2, a molecule that confers lipopolysaccharide responsiveness on Toll-like receptor 4. *J. Exp. Med.* **1999**, *189*, 1777–1782. [CrossRef]
17. Ohto, U.; Fukase, K.; Miyake, K.; Shimizu, T. Structural basis of species-specific endotoxin sensing by innate immune receptor TLR4/MD-2. *Proc. Natl. Acad. Sci. USA* **2012**, *109*, 7421–7426. [CrossRef]
18. Resman, N.; Vasl, J.; Oblak, A.; Pristovsek, P.; Gioannini, T.L.; Weiss, J.P.; Jerala, R. Essential roles of hydrophobic residues in both MD-2 and Toll-like receptor 4 in activation by endotoxin. *J. Biol. Chem.* **2009**, *284*, 15052–15060. [CrossRef]
19. Krüger, C.L.; Zeuner, M.-T.; Cottrell, G.S.; Widera, D.; Heilemann, M. Quantitative single-molecule imaging of TLR4 reveals ligand-specific receptor dimerization. *Sci. Signal.* **2017**, *10*, eaan1308. [CrossRef]
20. Kawai, T.; Adachi, O.; Ogawa, T.; Takeda, K.; Akira, S. Unresponsiveness of MyD88-deficient mice to endotoxin. *Immunity* **1999**, *11*, 115–122. [CrossRef]
21. Muzio, M.; Ni, J.; Feng, P.; Dixit, V.M. IRAK (Pelle) Family Member IRAK-2 and MyD88 as Proximal Mediators of IL-1 Signaling. *Science* **1997**, *278*, 1612–1615. [CrossRef]
22. Yamamoto, M.; Sato, S.; Hemmi, H.; Hoshino, K.; Kaisho, T.; Sanjo, H.; Takeuchi, O.; Sugiyama, M.; Okabe, M.; Takeda, K.; et al. Role of adaptor TRIF in the MyD88-independent toll-like receptor signaling pathway. *Science* **2003**, *301*, 640–643. [CrossRef]
23. Seong, S.-Y.; Matzinger, P. Hydrophobicity: An ancient damage-associated molecular pattern that initiates innate immune responses. *Nat. Rev. Immunol.* **2004**, *4*, 469–478. [CrossRef] [PubMed]
24. Kono, H.; Rock, K.L. How dying cells alert the immune system to danger. *Nat. Rev. Immunol.* **2008**, *8*, 279–289. [CrossRef] [PubMed]
25. Imai, Y.; Kuba, K.; Neely, G.G.; Yaghubian-Malhami, R.; Perkmann, T.; van Loo, G.; Ermolaeva, M.; Veldhuizen, R.; Leung, Y.H.C.; Wang, H.; et al. Identification of oxidative stress and Toll-like Receptor 4 signaling as a key pathway of acute lung injury. *Cell* **2008**, *133*, 235–249. [CrossRef]
26. Yang, H.; Wang, H.; Ju, Z.; Ragab, A.A.; Lundbäck, P.; Long, W.; Valdes-Ferrer, S.I.; He, M.; Pribis, J.P.; Li, J.; et al. MD-2 is required for disulfide HMGB1–dependent TLR4 signaling. *J. Exp. Med.* **2015**, *212*, 5–14. [CrossRef] [PubMed]
27. Taylor, K.R.; Yamasaki, K.; Radek, K.A.; Nardo, A.D.; Goodarzi, H.; Golenbock, D.; Beutler, B.; Gallo, R.L. Recognition of Hyaluronan Released in Sterile Injury Involves a Unique Receptor Complex Dependent on Toll-like Receptor 4, CD44 and MD-2. *J. Biol. Chem.* **2007**, *282*, 18265–18275. [CrossRef] [PubMed]
28. Okamura, Y.; Watari, M.; Jerud, E.S.; Young, D.W.; Ishizaka, S.T.; Rose, J.; Chow, J.C.; Strauss, J.F. The Extra Domain A of Fibronectin Activates Toll-like Receptor 4. *J. Biol. Chem.* **2001**, *276*, 10229–10233. [CrossRef] [PubMed]
29. Shi, J.; Zhao, Y.; Wang, Y.; Gao, W.; Ding, J.; Li, P.; Hu, L.; Shao, F. Inflammatory caspases are innate immune receptors for intracellular LPS. *Nature* **2014**, *514*, 187–192. [CrossRef]
30. Lee, B.L.; Stowe, I.B.; Gupta, A.; Kornfeld, O.S.; Roose-Girma, M.; Anderson, K.; Warming, S.; Zhang, J.; Lee, W.P.; Kayagaki, N. Caspase-11 auto-proteolysis is crucial for noncanonical inflammasome activation. *J. Exp. Med.* **2018**, *215*, 2279–2288. [CrossRef]
31. Casson, C.N.; Yu, J.; Reyes, V.M.; Taschuk, F.O.; Yadav, A.; Copenhaver, A.M.; Nguyen, H.T.; Collman, R.G.; Shin, S. Human caspase-4 mediates noncanonical inflammasome activation against gram-negative bacterial pathogens. *Proc. Natl. Acad. Sci. USA* **2015**, *112*, 6688–6693. [CrossRef]
32. Zamyatina, A.; Heine, H. Lipopolysaccharide Recognition in the Crossroads of TLR4 and Caspase-4/11 Mediated Inflammatory Pathways. *Front. Immunol.* **2020**, *11*, 585146. [CrossRef]
33. Hagar, J.A.; Powell, D.A.; Aachoui, Y.; Ernst, R.K.; Miao, E.A. Cytoplasmic LPS activates caspase-11: Implications in TLR4-independent endotoxic shock. *Science* **2013**, *341*, 1250–1253. [CrossRef] [PubMed]
34. Kayagaki, N.; Wong, M.T.; Stowe, I.B.; Ramani, S.R.; Gonzalez, L.C.; Akashi-Takamura, S.; Miyake, K.; Zhang, J.; Lee, W.P.; Muszýnski, A.; et al. Noncanonical inflammasome activation by intracellular LPS independent of TLR4. *Science* **2013**, *341*, 1246–1249. [CrossRef] [PubMed]
35. Vaure, C.; Liu, Y. A Comparative Review of Toll-Like Receptor 4 Expression and Functionality in Different Animal Species. *Front. Immunol.* **2014**, *5*, 316. [CrossRef] [PubMed]
36. Akashi, S.; Nagai, Y.; Ogata, H.; Oikawa, M.; Fukase, K.; Kusumoto, S.; Kawasaki, K.; Nishijima, M.; Hayashi, S.; Kimoto, M.; et al. Human MD-2 confers on mouse Toll-like receptor 4 species-specific lipopolysaccharide recognition. *Int. Immunol.* **2001**, *13*, 1595–1599. [CrossRef] [PubMed]
37. Meng, J.; Drolet, J.R.; Monks, B.G.; Golenbock, D.T. MD-2 residues Tyrosine 42, Arginine 69, Aspartic acid 122, and Leucine 125 provide species specificity for lipid IVa. *J. Biol. Chem.* **2010**, *285*, 27935–27943. [CrossRef]
38. Galanos, C.; Lehmann, V.; Lüderitz, O.; Rietschel, E.T.; Westphal, O.; Brade, H.; Brade, L.; Freudenberg, M.A.; Hansen-Hagge, T.; Lüderitz, T.; et al. Endotoxic properties of chemically synthesized lipid A part structures. Comparison of synthetic lipid A precursor and synthetic analogues with biosynthetic lipid A precursor and free lipid A. *Eur. J. Biochem.* **1984**, *140*, 221–227. [CrossRef]
39. Feist, W.; Ulmer, A.J.; Musehold, J.; Brade, H.; Kusumoto, S.; Flad, H.D. Induction of tumor necrosis factor-alpha release by lipopolysaccharide and defined lipopolysaccharide partial structures. *Immunobiology* **1989**, *179*, 293–307. [CrossRef]
40. Loppnow, H.; Brade, L.; Brade, H.; Rietschel, E.T.; Kusumoto, S.; Shiba, T.; Flad, H.D. Induction of human interleukin 1 by bacterial and synthetic lipid A. *Eur. J. Immunol.* **1986**, *16*, 1263–1267. [CrossRef]
41. Kovach, N.L.; Yee, E.; Munford, R.S.; Raetz, C.R.; Harlan, J.M. Lipid IVA inhibits synthesis and release of tumor necrosis factor induced by lipopolysaccharide in human whole blood ex vivo. *J. Exp. Med.* **1990**, *172*, 77–84. [CrossRef]

42. Ohto, U.; Fukase, K.; Miyake, K.; Satow, Y. Crystal structures of human MD-2 and its complex with antiendotoxic lipid IVa. *Science* **2007**, *316*, 1632–1634. [CrossRef]
43. Steeghs, L.; Keestra, A.M.; van Mourik, A.; Uronen-Hansson, H.; van der Ley, P.; Callard, R.; Klein, N.; van Putten, J.P.M. Differential Activation of Human and Mouse Toll-Like Receptor 4 by the Adjuvant Candidate LpxL1 of Neisseria meningitidis. *Infect. Immun.* **2008**, *76*, 3801–3807. [CrossRef]
44. Lien, E.; Means, T.K.; Heine, H.; Yoshimura, A.; Kusumoto, S.; Fukase, K.; Fenton, M.J.; Oikawa, M.; Qureshi, N.; Monks, B.; et al. Toll-like receptor 4 imparts ligand-specific recognition of bacterial lipopolysaccharide. *J. Clin. Investig.* **2000**, *105*, 497–504. [CrossRef]
45. Irvine, K.L.; Gangloff, M.; Walsh, C.M.; Spring, D.R.; Gay, N.J.; Bryant, C.E. Identification of key residues that confer *Rhodobacter sphaeroides* LPS activity at horse TLR4/MD-2. *PLoS ONE* **2014**, *9*, e98776. [CrossRef]
46. Lohmann, K.L.; Vandenplas, M.L.; Barton, M.H.; Bryant, C.E.; Moore, J.N. The equine TLR4/MD-2 complex mediates recognition of lipopolysaccharide from *Rhodobacter sphaeroides* as an agonist. *J. Endotoxin Res.* **2007**, *13*, 235–242. [CrossRef]
47. Rudd, K.E.; Johnson, S.C.; Agesa, K.M.; Shackelford, K.A.; Tsoi, D.; Kievlan, D.R.; Colombara, D.V.; Ikuta, K.S.; Kissoon, N.; Finfer, S.; et al. Global, regional, and national sepsis incidence and mortality, 1990–2017: Analysis for the Global Burden of Disease Study. *Lancet* **2020**, *395*, 200–211. [CrossRef]
48. Opal, S.M.; Laterre, P.-F.; Francois, B.; LaRosa, S.P.; Angus, D.C.; Mira, J.-P.; Wittebole, X.; Dugernier, T.; Perrotin, D.; Tidswell, M.; et al. Effect of Eritoran, an Antagonist of MD2-TLR4, on Mortality in Patients With Severe Sepsis: The ACCESS Randomized Trial. *JAMA* **2013**, *309*, 1154–1162. [CrossRef]
49. Thompson, W.W.; Shay, D.K.; Weintraub, E.; Brammer, L.; Bridges, C.B.; Cox, N.J.; Fukuda, K. Influenza-Associated Hospitalizations in the United States. *JAMA* **2004**, *292*, 1333–1340. [CrossRef]
50. Shirey, K.A.; Lai, W.; Scott, A.J.; Lipsky, M.; Mistry, P.; Pletneva, L.M.; Karp, C.L.; McAlees, J.; Gioannini, T.L.; Weiss, J.; et al. The TLR4 antagonist Eritoran protects mice from lethal influenza infection. *Nature* **2013**, *497*, 498–502. [CrossRef]
51. Shirey, K.A.; Blanco, J.C.G.; Vogel, S.N. Targeting TLR4 Signaling to Blunt Viral-Mediated Acute Lung Injury. *Front. Immunol.* **2021**, *12*, 705080. [CrossRef]
52. Rallabhandi, P.; Phillips, R.L.; Boukhvalova, M.S.; Pletneva, L.M.; Shirey, K.A.; Gioannini, T.L.; Weiss, J.P.; Chow, J.C.; Hawkins, L.D.; Vogel, S.N.; et al. Respiratory Syncytial Virus Fusion Protein-Induced Toll-Like Receptor 4 (TLR4) Signaling Is Inhibited by the TLR4 Antagonists Rhodobacter sphaeroides Lipopolysaccharide and Eritoran (E5564) and Requires Direct Interaction with MD-2. *mBio* **2012**, *3*, e00218-12. [CrossRef]
53. Sabouri, A.H.; Marcondes, M.C.G.; Flynn, C.; Berger, M.; Xiao, N.; Fox, H.S.; Sarvetnick, N.E. TLR signaling controls lethal encephalitis in WNV-infected brain. *Brain Res.* **2014**, *1574*, 84–95. [CrossRef]
54. Younan, P.; Ramanathan, P.; Graber, J.; Gusovsky, F.; Bukreyev, A. The Toll-Like Receptor 4 Antagonist Eritoran Protects Mice from Lethal Filovirus Challenge. *mBio* **2017**, *8*, e00226-17. [CrossRef] [PubMed]
55. Olejnik, J.; Hume, A.J.; Mühlberger, E. Toll-like receptor 4 in acute viral infection: Too much of a good thing. *PLoS Pathog.* **2018**, *14*, e1007390. [CrossRef] [PubMed]
56. Shinya, K.; Ito, M.; Makino, A.; Tanaka, M.; Miyake, K.; Eisfeld, A.J.; Kawaoka, Y. The TLR4-TRIF pathway protects against H5N1 influenza virus infection. *J. Virol.* **2012**, *86*, 19–24. [CrossRef] [PubMed]
57. Aboudounya, M.M.; Heads, R.J. COVID-19 and Toll-Like Receptor 4 (TLR4): SARS-CoV-2 May Bind and Activate TLR4 to Increase ACE2 Expression, Facilitating Entry and Causing Hyperinflammation. *Mediat. Inflamm.* **2021**, *2021*, 8874339. [CrossRef] [PubMed]
58. Choudhury, A.; Mukherjee, S. In silico studies on the comparative characterization of the interactions of SARS-CoV-2 spike glycoprotein with ACE-2 receptor homologs and human TLRs. *J. Med. Virol.* **2020**, *92*, 2105–2113. [CrossRef]
59. Guan, W.-J.; Ni, Z.-Y.; Hu, Y.; Liang, W.-H.; Ou, C.-Q.; He, J.-X.; Liu, L.; Shan, H.; Lei, C.-L.; Hui, D.S.C.; et al. Clinical Characteristics of Coronavirus Disease 2019 in China. *N. Engl. J. Med.* **2020**, *382*, 1708–1720. [CrossRef]
60. Sohn, K.M.; Lee, S.-G.; Kim, H.J.; Cheon, S.; Jeong, H.; Lee, J.; Kim, I.S.; Silwal, P.; Kim, Y.J.; Paik, S.; et al. COVID-19 Patients Upregulate Toll-like Receptor 4-mediated Inflammatory Signaling That Mimics Bacterial Sepsis. *J. Korean Med. Sci.* **2020**, *35*, 1146183. [CrossRef]
61. Li, G.; Fan, Y.; Lai, Y.; Han, T.; Li, Z.; Zhou, P.; Pan, P.; Wang, W.; Hu, D.; Liu, X.; et al. Coronavirus infections and immune responses. *J. Med. Virol.* **2020**, *92*, 424–432. [CrossRef]
62. Petruk, G.; Puthia, M.; Petrlova, J.; Samsudin, F.; Stromdahl, A.C.; Cerps, S.; Uller, L.; Kjellstrom, S.; Bond, P.J.; Schmidtchen, A.A. SARS-CoV-2 spike protein binds to bacterial lipopolysaccharide and boosts proinflammatory activity. *J. Mol. Cell Biol.* **2021**, *12*, 916–932. [CrossRef]
63. Shirato, K.; Kizaki, T. SARS-CoV-2 spike protein S1 subunit induces pro-inflammatory responses via toll-like receptor 4 signaling in murine and human macrophages. *Heliyon* **2021**, *7*, e06187. [CrossRef] [PubMed]
64. Zhao, Y.; Kuang, M.; Li, J.; Zhu, L.; Jia, Z.; Guo, X.; Hu, Y.; Kong, J.; Yin, H.; Wang, X.; et al. SARS-CoV-2 spike protein interacts with and activates TLR4. *Cell Res.* **2021**, *2021*, 818–820. [CrossRef]
65. Brandão, S.C.S.; Ramos, J.d.O.X.; Dompieri, L.T.; Godoi, E.T.A.M.; Figueiredo, J.L.; Sarinho, E.S.C.; Chelvanambi, S.; Aikawa, M. Is Toll-like receptor 4 involved in the severity of COVID-19 pathology in patients with cardiometabolic comorbidities? *Cytokine Growth Factor Rev.* **2021**, *58*, 102–110. [CrossRef] [PubMed]

66. Kruglikov, I.L.; Scherer, P.E. Preexisting and inducible endotoxemia as crucial contributors to the severity of COVID-19 outcomes. *PLoS Pathog.* **2021**, *17*, e1009306. [CrossRef]
67. Rossol, M.; Heine, H.; Meusch, U.; Quandt, D.; Klein, C.; Sweet, M.J.; Hauschildt, S. LPS-induced Cytokine Production in Human Monocytes and Macrophages. *Crit. Rev. Immunol.* **2011**, *31*, 379–446. [CrossRef]
68. Wang, H.; Reddy, S.T.; Fogelman, A.M. The role of gut-derived oxidized lipids and bacterial lipopolysaccharide in systemic inflammation and atherosclerosis. *Curr. Opin. Lipidol.* **2022**, *33*, 277–282. [CrossRef] [PubMed]
69. Michelsen, K.S.; Doherty, T.M.; Shah, P.K.; Arditi, M. TLR Signaling: An Emerging Bridge from Innate Immunity to Atherogenesis. *J. Immunol.* **2004**, *173*, 5901–5907. [CrossRef]
70. Michelsen, K.S.; Wong, M.H.; Shah, P.K.; Zhang, W.; Yano, J.; Doherty, T.M.; Akira, S.; Rajavashisth, T.B.; Arditi, M. Lack of Toll-like receptor 4 or myeloid differentiation factor 88 reduces atherosclerosis and alters plaque phenotype in mice deficient in apolipoprotein E. *Proc. Natl. Acad. Sci. USA* **2004**, *101*, 10679–10684. [CrossRef]
71. Doherty, T.M.; Shah, P.K.; Arditi, M.; Stoll, L.L.; Denning, G.M.; Weintraub, N.L. Lipopolysaccharide, Toll-Like Receptors, and the Immune Contribution to Atherosclerosis. *Arterioscler. Thromb. Vasc. Biol.* **2005**, *25*, e38–e39. [CrossRef]
72. Dickinson, E.S.; Wondrak, T.G. TLR4-directed Molecular Strategies Targeting Skin Photodamage and Carcinogenesis. *Curr. Med. Chem.* **2018**, *25*, 5487–5502. [CrossRef]
73. Blohm-Mangone, K.; Burkett, N.B.; Tahsin, S.; Myrdal, P.B.; Aodah, A.; Ho, B.; Janda, J.; McComas, M.; Saboda, K.; Roe, D.J.; et al. Pharmacological TLR4 Antagonism Using Topical Resatorvid Blocks Solar UV-Induced Skin Tumorigenesis in SKH-1 Mice. *Cancer Prev. Res. (Phila.)* **2018**, *11*, 265–278. [CrossRef] [PubMed]
74. Chen, Y.; Sun, W.; Gao, R.; Su, Y.; Umehara, H.; Dong, L.; Gong, F. The role of high mobility group box chromosomal protein 1 in rheumatoid arthritis. *Rheumatology* **2013**, *52*, 1739–1747. [CrossRef] [PubMed]
75. Kaur, I.; Behl, T.; Bungau, S.; Kumar, A.; Mehta, V.; Setia, D.; Uddin, M.S.; Zengin, G.; Aleya, L.; Arora, S. Exploring the therapeutic promise of targeting HMGB1 in rheumatoid arthritis. *Life Sci.* **2020**, *258*, 118164. [CrossRef] [PubMed]
76. Abdollahi-Roodsaz, S.; Joosten, L.A.B.; Roelofs, M.F.; Radstake, T.R.D.J.; Matera, G.; Popa, C.; Van der Meer, J.W.M.; Netea, M.G.; van den Berg, W.B. Inhibition of Toll-like receptor 4 breaks the inflammatory loop in autoimmune destructive arthritis. *Arthritis Rheum.* **2007**, *56*, 2957–2967. [CrossRef] [PubMed]
77. Gómez, R.; Villalvilla, A.; Largo, R.; Gualillo, O.; Herrero-Beaumont, G. TLR4 signalling in osteoarthritis—Finding targets for candidate DMOADs. *Nat. Rev. Rheumatol.* **2015**, *11*, 159–170. [CrossRef]
78. Zuo, L.; Lucas, K.; Fortuna, C.A.; Chuang, C.C.; Best, T.M. Molecular Regulation of Toll-like Receptors in Asthma and COPD. *Front. Physiol.* **2015**, *6*, 312. [CrossRef]
79. Bolourani, S.; Brenner, M.; Wang, P. The interplay of DAMPs, TLR4, and proinflammatory cytokines in pulmonary fibrosis. *J. Mol. Med.* **2021**, *99*, 1373–1384. [CrossRef]
80. Bezemer, G.F.G.; Sagar, S.; van Bergenhenegouwen, J.; Georgiou, N.A.; Garssen, J.; Kraneveld, A.D.; Folkerts, G. Dual Role of Toll-Like Receptors in Asthma and Chronic Obstructive Pulmonary Disease. *Pharmacol. Rev.* **2012**, *64*, 337–358. [CrossRef]
81. Speletas, M.; Merentiti, V.; Kostikas, K.; Liadaki, K.; Minas, M.; Gourgoulianis, K.; Germenis, A.E. Association of TLR4-T399I Polymorphism with Chronic Obstructive Pulmonary Disease in Smokers. *Clin. Develop. Immunol.* **2009**, *2009*, 260286. [CrossRef]
82. Sabroe, I.; Whyte, M.K.; Wilson, A.G.; Dower, S.K.; Hubbard, R.; Hall, I. Toll-like receptor (TLR) 4 polymorphisms and COPD. *Thorax* **2004**, *59*, 81.
83. Yang, Y.; Lv, J.; Jiang, S.; Ma, Z.; Wang, D.; Hu, W.; Deng, C.; Fan, C.; Di, S.; Sun, Y.; et al. The emerging role of Toll-like receptor 4 in myocardial inflammation. *Cell Death Dis.* **2016**, *7*, e2234. [CrossRef] [PubMed]
84. Timmers, L.; Sluijter, J.P.G.; Keulen, J.K.v.; Hoefer, I.E.; Nederhoff, M.G.J.; Goumans, M.-J.; Doevendans, P.A.; Echteld, C.J.A.v.; Joles, J.A.; Quax, P.H.; et al. Toll-Like Receptor 4 Mediates Maladaptive Left Ventricular Remodeling and Impairs Cardiac Function After Myocardial Infarction. *Circ. Res.* **2008**, *102*, 257–264. [CrossRef] [PubMed]
85. Zhang, X.; Lu, C.; Gao, M.; Cao, X.; Ha, T.; Kalbfleisch, J.H.; Williams, D.L.; Li, C.; Kao, R.L. Toll-Like Receptor 4 plays a central role in cardiac dysfunction during trauma hemorrhage shock. *Shock* **2014**, *42*, 31–37. [CrossRef] [PubMed]
86. Braun-Fahrländer, C.; Riedler, J.; Herz, U.; Eder, W.; Waser, M.; Grize, L.; Maisch, S.; Carr, D.; Gerlach, F.; Bufe, A.; et al. Environmental exposure to endotoxin and its relation to asthma in school-age children. *N. Engl. J. Med.* **2002**, *347*, 869–877. [CrossRef] [PubMed]
87. Riedler, J.; Braun-Fahrländer, C.; Eder, W.; Schreuer, M.; Waser, M.; Maisch, S.; Carr, D.; Schierl, R.; Nowak, D.; von Mutius, E. Exposure to farming in early life and development of asthma and allergy: A cross-sectional survey. *Lancet* **2001**, *358*, 1129–1133. [CrossRef] [PubMed]
88. Strachan, D.P. Hay fever, hygiene, and household size. *Brit. Med. J.* **1989**, *299*, 1259–1260. [CrossRef]
89. Eisenbarth, S.C.; Piggott, D.A.; Huleatt, J.W.; Visintin, I.; Herrick, C.A.; Bottomly, K. Lipopolysaccharide-enhanced, Toll-like Receptor 4–dependent T Helper Cell Type 2 Responses to Inhaled Antigen. *J. Exp. Med.* **2002**, *196*, 1645–1651. [CrossRef]
90. Thorne, P.S. Environmental endotoxin exposure and asthma. *J. Allergy Clin. Immunol.* **2021**, *148*, 61–63. [CrossRef]
91. Trompette, A.; Divanovic, S.; Visintin, A.; Blanchard, C.; Hegde, R.S.; Madan, R.; Thorne, P.S.; Wills-Karp, M.; Gioannini, T.L.; Weiss, J.P.; et al. Allergenicity resulting from functional mimicry of a Toll-like receptor complex protein. *Nature* **2009**, *457*, 585–588. [CrossRef]
92. Thomas, W.R.; Hales, B.J.; Smith, W.-A. Structural biology of allergens. *Curr. Allergy Asthma Rep.* **2005**, *5*, 388–393. [CrossRef]

93. Li, M.; Wang, Z.N.; Yang, L.F.; Yan, Y.; Cai, L.M.; Li, Y.T.; Qiao, Y.K.; Chen, Z.G. TLR4 antagonist suppresses airway remodeling in asthma by inhibiting the T-helper 2 response. *Exp. Ther. Med.* **2017**, *14*, 2911–2916. [CrossRef] [PubMed]
94. Shalaby, K.H.; Al Heialy, S.; Tsuchiya, K.; Farahnak, S.; McGovern, T.K.; Risse, P.-A.; Suh, W.-K.; Qureshi, S.T.; Martin, J.G. The TLR4–TRIF pathway can protect against the development of experimental allergic asthma. *Immunology* **2017**, *152*, 138–149. [CrossRef] [PubMed]
95. Herre, J.; Grönlund, H.; Brooks, H.; Hopkins, L.; Waggoner, L.; Murton, B.; Gangloff, M.; Opaleye, O.; Chilvers, E.R.; Fitzgerald, K.; et al. Allergens as Immunomodulatory Proteins: The Cat Dander Protein Fel d 1 Enhances TLR Activation by Lipid Ligands. *J. Immunol.* **2013**, *191*, 1529–1535. [CrossRef] [PubMed]
96. Calvo-Rodriguez, M.; García-Rodríguez, C.; Villalobos, C.; Núñez, L. Role of Toll Like Receptor 4 in Alzheimer's Disease. *Front. Immunol.* **2020**, *11*, 1588. [CrossRef]
97. Heneka, M.T.; Golenbock, D.T.; Latz, E. Innate immunity in Alzheimer's disease. *Nat. Immunol.* **2015**, *16*, 229–236. [CrossRef] [PubMed]
98. Balistreri, C.R.; Grimaldi, M.P.; Chiappelli, M.; Licastro, F.; Castiglia, L.; Listì, F.; Vasto, S.; Lio, D.; Caruso, C.; Candore, G. Association between the polymorphisms of TLR4 and CD14 genes and Alzheimer's disease. *Curr. Pharm. Des.* **2008**, *14*, 2672–2677. [CrossRef]
99. Michaud, J.P.; Hallé, M.; Lampron, A.; Thériault, P.; Préfontaine, P.; Filali, M.; Tribout-Jover, P.; Lanteigne, A.M.; Jodoin, R.; Cluff, C.; et al. Toll-like receptor 4 stimulation with the detoxified ligand monophosphoryl lipid A improves Alzheimer's disease-related pathology. *Proc. Natl. Acad. Sci. USA* **2013**, *110*, 1941–1946. [CrossRef] [PubMed]
100. Qin, Y.; Liu, Y.; Hao, W.; Decker, Y.; Tomic, I.; Menger, M.D.; Liu, C.; Fassbender, K. Stimulation of TLR4 Attenuates Alzheimer's Disease–Related Symptoms and Pathology in Tau-Transgenic Mice. *J. Immunol.* **2016**, *197*, 3281–3292. [CrossRef]
101. Morefield, G.L.; Hawkins, L.D.; Ishizaka, S.T.; Kissner, T.L.; Ulrich, R.G. Synthetic Toll-Like Receptor-4 Agonist Enhances Vaccine Efficacy in an Experimental Model of Toxic-Shock Syndrome. *Clin. Vaccine Immunol.* **2007**, *14*, 1499–1504. [CrossRef]
102. Iwasaki, A.; Medzhitov, R. Toll-like receptor control of the adaptive immune responses. *Nat. Immunol.* **2004**, *5*, 987–995. [CrossRef]
103. Mata-Haro, V.; Cekic, C.; Martin, M.; Chilton, P.M.; Casella, C.R.; Mitchell, T.C. The vaccine adjuvant monophosphoryl lipid A as a TRIF-biased agonist of TLR4. *Science* **2007**, *316*, 1628–1632. [CrossRef] [PubMed]
104. Cimica, V.; Galarza, J.M. Adjuvant formulations for virus-like particle (VLP) based vaccines. *Clin. Immunol.* **2017**, *183*, 99–108. [CrossRef] [PubMed]
105. Wang, Y.-Q.; Bazin-Lee, H.; Evans, J.T.; Casella, C.R.; Mitchell, T.C. MPL Adjuvant Contains Competitive Antagonists of Human TLR4. *Front. Immunol.* **2020**, *11*, 577823. [CrossRef] [PubMed]
106. Tsai, S.Y.; Segovia, J.A.; Chang, T.H.; Morris, I.R.; Berton, M.T.; Tessier, P.A.; Tardif, M.R.; Cesaro, A.; Bose, S. DAMP molecule S100A9 acts as a molecular pattern to enhance inflammation during influenza A virus infection: Role of DDX21-TRIF-TLR4-MyD88 pathway. *PLoS Pathog* **2014**, *10*, e1003848. [CrossRef] [PubMed]
107. Shinya, K.; Okamura, T.; Sueta, S.; Kasai, N.; Tanaka, M.; Ginting, T.E.; Makino, A.; Eisfeld, A.J.; Kawaoka, Y. Toll-like receptor pre-stimulation protects mice against lethal infection with highly pathogenic influenza viruses. *Virol. J.* **2011**, *8*, 97. [CrossRef] [PubMed]
108. Zeuner, M.-T.; Krüger, C.L.; Volk, K.; Bieback, K.; Cottrell, G.S.; Heilemann, M.; Widera, D. Biased signalling is an essential feature of TLR4 in glioma cells. *Biochim. Biophys. Acta (BBA)* **2016**, *1863*, 3084–3095. [CrossRef]
109. Ou, T.; Lilly, M.; Jiang, W. The Pathologic Role of Toll-Like Receptor 4 in Prostate Cancer. *Front. Immunol.* **2018**, *9*, 1188. [CrossRef]
110. Wu, K.; Zhang, H.; Fu, Y.; Zhu, Y.; Kong, L.; Chen, L.; Zhao, F.; Yu, L.; Chen, X. TLR4/MyD88 signaling determines the metastatic potential of breast cancer cells. *Mol. Med. Rep.* **2018**, *18*, 3411–3420. [CrossRef]
111. Kelsh, R.M.; McKeown-Longo, P.J. Topographical changes in extracellular matrix: Activation of TLR4 signaling and solid tumor progression. *Trends Cancer Res.* **2013**, *9*, 1–13.
112. Ikebe, M.; Kitaura, Y.; Nakamura, M.; Tanaka, H.; Yamasaki, A.; Nagai, S.; Wada, J.; Yanai, K.; Koga, K.; Sato, N.; et al. Lipopolysaccharide (LPS) increases the invasive ability of pancreatic cancer cells through the TLR4/MyD88 signaling pathway. *J. Surg. Oncol.* **2009**, *100*, 725–731. [CrossRef]
113. Yang, H.; Wang, B.; Wang, T.; Xu, L.; He, C.; Wen, H.; Yan, J.; Su, H.; Zhu, X. Toll-Like Receptor 4 Prompts Human Breast Cancer Cells Invasiveness via Lipopolysaccharide Stimulation and Is Overexpressed in Patients with Lymph Node Metastasis. *PLoS ONE* **2014**, *9*, e109980. [CrossRef] [PubMed]
114. Haricharan, S.; Brown, P. TLR4 has a TP53-dependent dual role in regulating breast cancer cell growth. *Proc. Natl. Acad. Sci. USA* **2015**, *112*, E3216–E3225. [CrossRef] [PubMed]
115. Yahiro, K.; Matsumoto, Y.; Yamada, H.; Endo, M.; Setsu, N.; Fujiwara, T.; Nakagawa, M.; Kimura, A.; Shimada, E.; Okada, S.; et al. Activation of TLR4 signaling inhibits progression of osteosarcoma by stimulating CD8-positive cytotoxic lymphocytes. *Cancer Immunol. Immunother.* **2020**, *69*, 745–758. [CrossRef] [PubMed]
116. Yamazaki, T.; Hannani, D.; Poirier-Colame, V.; Ladoire, S.; Locher, C.; Sistigu, A.; Prada, N.; Adjemian, S.; Catani, J.P.; Freudenberg, M.; et al. Defective immunogenic cell death of HMGB1-deficient tumors: Compensatory therapy with TLR4 agonists. *Cell Death Differ.* **2014**, *21*, 69–78. [CrossRef]
117. Fang, H.; Ang, B.; Xu, X.; Huang, X.; Wu, Y.; Sun, Y.; Wang, W.; Li, N.; Cao, X.; Wan, T. TLR4 is essential for dendritic cell activation and anti-tumor T-cell response enhancement by DAMPs released from chemically stressed cancer cells. *Cell. Mol. Immunol.* **2014**, *11*, 150–159. [CrossRef]

118. Feng, Y.; Mu, R.; Wang, Z.; Xing, P.; Zhang, J.; Dong, L.; Wang, C. A toll-like receptor agonist mimicking microbial signal to generate tumor-suppressive macrophages. *Nat. Commun.* **2019**, *10*, 2272. [CrossRef]
119. Matzner, P.; Sorski, L.; Shaashua, L.; Elbaz, E.; Lavon, H.; Melamed, R.; Rosenne, E.; Gotlieb, N.; Benbenishty, A.; Reed, S.G.; et al. Perioperative treatment with the new synthetic TLR4 agonist GLA-SE reduces cancer metastasis without adverse effects. *Int. J. Cancer* **2016**, *138*, 1754–1764. [CrossRef]
120. Shetab Boushehri, M.A.; Lamprecht, A. TLR4-Based Immunotherapeutics in Cancer: A Review of the Achievements and Shortcomings. *Mol. Pharm.* **2018**, *15*, 4777–4800. [CrossRef]
121. Reisser, D.; Jeannin, J.-F. Lipid A in cancer therapies preclinical results. In *Lipid A in Cancer Therapy*; Jeannin, J.-F., Ed.; Springer New York: New York, NY, USA, 2010; pp. 101–110. [CrossRef]
122. Smith, M.; García-Martínez, E.; Pitter, M.R.; Fucikova, J.; Spisek, R.; Zitvogel, L.; Kroemer, G.; Galluzzi, L. Trial Watch: Toll-like receptor agonists in cancer immunotherapy. *OncoImmunology* **2018**, *7*, e1526250. [CrossRef]
123. Andreani, V.; Gatti, G.; Simonella, L.; Rivero, V.; Maccioni, M. Activation of TLR4 on Tumor Cells In vitro Inhibits Subsequent Tumor Growth In vivo. *Cancer Res.* **2007**, *67*, 10519–10527. [CrossRef]
124. Apetoh, L.; Ghiringhelli, F.; Tesniere, A.; Obeid, M.; Ortiz, C.; Criollo, A.; Mignot, G.; Maiuri, M.C.; Ullrich, E.; Saulnier, P.; et al. Toll-like receptor 4–dependent contribution of the immune system to anticancer chemotherapy and radiotherapy. *Nat. Med.* **2007**, *13*, 1050–1059. [CrossRef] [PubMed]
125. Hoption Cann, S.A.; van Netten, J.P.; van Netten, C.; Glover, D.W. Spontaneous regression: A hidden treasure buried in time. *Med. Hypotheses* **2002**, *58*, 115–119. [CrossRef] [PubMed]
126. Starnes, C.O. Coley's toxins in perspective. *Nature* **1992**, *357*, 11–12. [CrossRef] [PubMed]
127. McCarthy, E.F. The toxins of William B. Coley and the treatment of bone and soft-tissue sarcomas. *Iowa Orthop. J.* **2006**, *26*, 154–158.
128. Nauts, H.C.; McLaren, J.R. Coley Toxins—The First Century. In *Consensus on Hyperthermia for the 1990s: Clinical Practice in Cancer Treatment*; Bicher, H.I., McLaren, J.R., Pigliucci, G.M., Eds.; Springer USA: Boston, MA, USA, 1990; pp. 483–500.
129. Huang, L.; Xu, H.; Peng, G. TLR-mediated metabolic reprogramming in the tumor microenvironment: Potential novel strategies for cancer immunotherapy. *Cell Mol. Immunol.* **2018**, *15*, 428–437. [CrossRef]
130. Cen, X.; Liu, S.; Cheng, K. The Role of Toll-Like Receptor in Inflammation and Tumor Immunity. *Front. Pharmacol.* **2018**, *9*, 878. [CrossRef]
131. Pandey, N.; Chauhan, A.; Jain, N. TLR4 Polymorphisms and Expression in Solid Cancers. *Mol. Diagn. Ther.* **2018**, *22*, 683–702. [CrossRef]
132. Nagarsheth, N.; Wicha, M.S.; Zou, W. Chemokines in the cancer microenvironment and their relevance in cancer immunotherapy. *Nat. Rev. Immunol.* **2017**, *17*, 559–572. [CrossRef]
133. Mills, C.D.; Lenz, L.L.; Harris, R.A. A Breakthrough: Macrophage-Directed Cancer Immunotherapy. *Cancer Res.* **2016**, *76*, 513–516. [CrossRef]
134. Mantovani, A.; Marchesi, F.; Malesci, A.; Laghi, L.; Allavena, P. Tumour-associated macrophages as treatment targets in oncology. *Nat. Rev. Clin. Oncol.* **2017**, *14*, 399–416. [CrossRef]
135. Müller, E.; Speth, M.; Christopoulos, P.F.; Lunde, A.; Avdagic, A.; Øynebråten, I.; Corthay, A. Both Type I and Type II Interferons Can Activate Antitumor M1 Macrophages When Combined With TLR Stimulation. *Front. Immunol.* **2018**, *9*, 2520. [CrossRef] [PubMed]
136. Müller, E.; Christopoulos, P.F.; Halder, S.; Lunde, A.; Beraki, K.; Speth, M.; Øynebråten, I.; Corthay, A. Toll-Like Receptor Ligands and Interferon-γ Synergize for Induction of Antitumor M1 Macrophages. *Front. Immunol.* **2017**, *8*, 1383. [CrossRef] [PubMed]
137. Lu, H.; Betancur, A.; Chen, M.; Ter Meulen, J.H. Toll-Like Receptor 4 Expression on Lymphoma Cells Is Critical for Therapeutic Activity of Intratumoral Therapy With Synthetic TLR4 Agonist Glucopyranosyl Lipid, A. *Front. Oncol.* **2020**, *10*, 1483. [CrossRef] [PubMed]
138. Han, S.; Wang, C.; Qin, X.; Xia, J.; Wu, A. LPS alters the immuno-phenotype of glioma and glioma stem-like cells and induces in vivo antitumor immunity via TLR4. *J. Exp. Clin. Cancer Res.* **2017**, *36*, 83. [CrossRef] [PubMed]
139. Albershardt, T.C.; Leleux, J.; Parsons, A.J.; Krull, J.E.; Berglund, P.; ter Meulen, J. Intratumoral immune activation with TLR4 agonist synergizes with effector T cells to eradicate established murine tumors. *NPJ Vaccines* **2020**, *5*, 50. [CrossRef]
140. Roscigno, G.; Cirella, A.; Affinito, A.; Quintavalle, C.; Scognamiglio, I.; Palma, F.; Ingenito, F.; Nuzzo, S.; De Micco, F.; Cuccuru, A.; et al. miR-216a Acts as a Negative Regulator of Breast Cancer by Modulating Stemness Properties and Tumor Microenvironment. *Int. J. Mol. Sci.* **2020**, *21*, 2313. [CrossRef]
141. Rajamanickam, V.; Yan, T.; Xu, S.; Hui, J.; Xu, X.; Ren, L.; Liu, Z.; Liang, G.; Wang, O.; Wang, Y. Selective targeting of the TLR4 co-receptor, MD2, prevents colon cancer growth and lung metastasis. *Int. J. Biol. Sci.* **2020**, *16*, 1288–1301. [CrossRef]
142. Wang, X.; Quinn, P.J.; Yan, A. Kdo2-lipid A: Structural diversity and impact on immunopharmacology. *Biol. Rev.* **2015**, *90*, 408–427. [CrossRef]
143. Needham, B.D.; Carroll, S.M.; Giles, D.K.; Georgiou, G.; Whiteley, M.; Trent, M.S. Modulating the innate immune response by combinatorial engineering of endotoxin. *Proc. Natl. Acad. Sci. USA* **2013**, *110*, 1464–1469. [CrossRef]
144. Paramo, T.; Piggot, T.J.; Bryant, C.E.; Bond, P.J. The structural basis for Endotoxin-induced allosteric regulation of the Toll-Like Receptor 4 (TLR4) innate immune receptor. *J. Biol. Chem.* **2013**, *288*, 36215–36225. [CrossRef]

145. Christ, W.J.; McGuinness, P.D.; Asano, O.; Wang, Y.; Mullarkey, M.A.; Perez, M.; Hawkins, L.D.; Blythe, T.A.; Dubuc, G.R.; Robidoux, A.L. Total synthesis of the proposed structure of *Rhodobacter sphaeroides* lipid A resulting in the synthesis of new potent lipopolysaccharide antagonists. *J. Am. Chem. Soc. Mass. Spec.* **1994**, *116*, 3637–3638. [CrossRef]
146. Park, B.S.; Song, D.H.; Kim, H.M.; Choi, B.S.; Lee, H.; Lee, J.O. The structural basis of lipopolysaccharide recognition by the TLR4-MD-2 complex. *Nature* **2009**, *458*, 1191–1195. [CrossRef] [PubMed]
147. Teghanemt, A.; Re, F.; Prohinar, P.; Widstrom, R.; Gioannini, T.L.; Weiss, J.P. Novel roles in human MD-2 of Phenylalanines 121 and 126 and Tyrosine 131 in activation of Toll-like receptor 4 by endotoxin. *J. Biol. Chem.* **2008**, *283*, 1257–1266. [CrossRef] [PubMed]
148. Gioannini, T.; Teghanemt, A.; Zhang, D.; Esparza, G.; Yu, L.; Weiss, J. Purified monomeric ligand.MD-2 complexes reveal molecular and structural requirements for activation and antagonism of TLR4 by Gram-negative bacterial endotoxins. *Immunol. Res.* **2014**, *59*, 3–11. [CrossRef]
149. Meng, J.; Gong, M.; Björkbacka, H.; Golenbock, D.T. Genome-wide expression profiling and mutagenesis studies reveal that lipopolysaccharide responsiveness appears to be absolutely dependent on TLR4 and MD-2 expression and is dependent upon intermolecular ionic interactions. *J. Immunol.* **2011**, *187*, 3683–3693. [CrossRef]
150. Casella, C.R.; Mitchell, T.C. Inefficient TLR4/MD-2 heterotetramerization by monophosphoryl lipid A. *PLoS ONE* **2013**, *8*, e62622. [CrossRef]
151. Bohannon, J.K.; Hernandez, A.; Enkhbaatar, P.; Adams, W.L.; Sherwood, E.R. The immunobiology of TLR4 agonists From endotoxin tolerance to immunoadjuvants. *Shock* **2013**, *40*, 451–462. [CrossRef]
152. Kim, H.M.; Park, B.S.; Kim, J.I.; Kim, S.E.; Lee, J.; Oh, S.C.; Enkhbayar, P.; Matsushima, N.; Lee, H.; Yoo, O.J.; et al. Crystal structure of the TLR4-MD-2 complex with bound endotoxin antagonist Eritoran. *Cell* **2007**, *130*, 906–917. [CrossRef]
153. Wang, W.; Sass, H.-J.; Zähringer, U.; Grzesiek, S. Structure and dynamics of ^{13}C,^{15}N-labeled lipopolysaccharides in a membrane mimetic. *Angew. Chem. Int. Ed.* **2008**, *47*, 9870–9874. [CrossRef]
154. Gutsmann, T.; Schromm, A.B.; Brandenburg, K. The physicochemistry of endotoxins in relation to bioactivity. *Int. J. Med. Microbiol.* **2007**, *297*, 341–352. [CrossRef]
155. Artner, D.; Oblak, A.; Ittig, S.; Garate, J.A.; Horvat, S.; Arrieumerlou, C.; Hofinger, A.; Oostenbrink, C.; Jerala, R.; Kosma, P.; et al. Conformationally constrained Lipid A mimetics for exploration of structural basis of TLR4/MD-2 activation by lipopolysaccharide. *ACS Chem. Biol.* **2013**, *8*, 2423–2432. [CrossRef] [PubMed]
156. Adanitsch, F.; Ittig, S.; Stöckl, J.; Oblak, A.; Haegman, M.; Jerala, R.; Beyaert, R.; Kosma, P.; Zamyatina, A. Development of αGlcN (1 ↔ 1) αMan-based Lipid A mimetics as a novel class of potent Toll-like Receptor 4 agonists. *J. Med. Chem.* **2014**, *57*, 8056–8071. [CrossRef] [PubMed]
157. Adanitsch, F.; Shi, J.; Shao, F.; Beyaert, R.; Heine, H.; Zamyatina, A. Synthetic glycan-based TLR4 agonists targeting caspase-4/11 for the development of adjuvants and immunotherapeutics. *Chem. Sci.* **2018**, *9*, 3957–3963. [CrossRef] [PubMed]
158. Heine, H.; Adanitsch, F.; Peternelj, T.T.; Haegman, M.; Kasper, C.; Ittig, S.; Beyaert, R.; Jerala, R.; Zamyatina, A. Tailored Modulation of Cellular Pro-inflammatory Responses With Disaccharide Lipid A Mimetics. *Front. Immunol.* **2021**, *12*, 631797. [CrossRef] [PubMed]
159. Garate, J.A.; Stöckl, J.; del Carmen Fernández-Alonso, M.; Artner, D.; Haegman, M.; Oostenbrink, C.; Jimènez-Barbero, J.; Beyaert, R.; Heine, H.; Kosma, P.; et al. Anti-endotoxic activity and structural basis for human MD-2·TLR4 antagonism of tetraacylated lipid A mimetics based on βGlcN (1 ↔ 1) αGlcN scaffold. *Innate Immun.* **2015**, *21*, 490–503. [CrossRef] [PubMed]
160. Baldridge, J.R.; Cluff, C.W.; Evans, J.T.; Lacy, M.J.; Stephens, J.R.; Brookshire, V.G.; Rong, W.; Ward, J.R.; Yorgensen, Y.M.; Persing, D.H.; et al. Immunostimulatory activity of aminoalkyl glucosaminide 4-phosphates (AGPs): Induction of protective innate immune responses by RC-524 and RC-529. *J. Endotoxin Res.* **2002**, *8*, 453–458. [CrossRef]
161. Ishizaka, S.T.; Hawkins, L.D. E6020: A synthetic Toll-like receptor 4 agonist as a vaccine adjuvant. *Expert Rev. Vaccines* **2007**, *6*, 773–784. [CrossRef]
162. Akamatsu, M.; Fujimoto, Y.; Kataoka, M.; Suda, Y.; Kusumoto, S.; Fukase, K. Synthesis of lipid A monosaccharide analogues containing acidic amino acid: Exploring the structural basis for the endotoxic and antagonistic activities. *Bioorg. Med. Chem.* **2006**, *14*, 6759–6777. [CrossRef]
163. Wang, Y.; Su, L.; Morin, M.D.; Jones, B.T.; Whitby, L.R.; Surakattula, M.M.R.P.; Huang, H.; Shi, H.; Choi, J.H.; Wang, K.-w.; et al. TLR4/MD-2 activation by a synthetic agonist with no similarity to LPS. *Proc. Natl. Acad. Sci. USA* **2016**, *113*, E884–E893. [CrossRef]
164. Strobl, S.; Hofbauer, K.; Heine, H.; Zamyatina, A. Lipid A Mimetics Based on Unnatural Disaccharide Scaffold as Potent TLR4 Agonists for Prospective Immunotherapeutics and Adjuvants. *Chem. Eur. J.* **2022**, *28*, e202200547. [CrossRef]
165. Chebrolu, C.; Artner, D.; Sigmund, A.M.; Buer, J.; Zamyatina, A.; Kirschning, C.J. Species and mediator specific TLR4 antagonism in primary human and murine immune cells by βGlcN (1 ↔ 1) αGlcN based lipid A mimetics. *Mol. Immunol.* **2015**, *67*, 636–641. [CrossRef] [PubMed]
166. Borio, A.; Holgado, A.; Garate, J.A.; Beyaert, R.; Heine, H.; Zamyatina, A. Disaccharide-Based Anionic Amphiphiles as Potent Inhibitors of Lipopolysaccharide-Induced Inflammation. *ChemMedChem* **2018**, *13*, 2317–2331. [CrossRef] [PubMed]
167. Cohen, J.; Opal, S.; Calandra, T. Sepsis studies need new direction. *Lancet Infect. Dis.* **2012**, *12*, 503–505. [CrossRef] [PubMed]
168. Hotchkiss, R.S.; Opal, S.M. Immunotherapy for Sepsis: A new approach against an ancient foe. *N. Engl. J. Med.* **2010**, *363*, 87–89. [CrossRef]

169. Fleischer, J.G.; Rossignol, D.; Francis, G.A.; Chan, T.; Lynn, M.; Wasan, K.M. Deactivation of the lipopolysaccharide antagonist eritoran (E5564) by high-density lipoprotein-associated apolipoproteins. *Innate Immun.* **2012**, *18*, 171–178. [CrossRef]
170. Rao, V.S.R.; Qasba, P.K.; Balaji, P.V.; Chandrasekaran, R. Conformation of Disaccharides. In *Conformation of Carbohydrates*; OPA: Overseas Publishers Association, Ed.; Harwood Academic Publishers: Amsterdam, The Netherlands, 1998; pp. 91–130.
171. Dowd, M.K.; Reilly, P.J.; French, A.D. Conformational analysis of trehalose disaccharides and analogues using MM3. *J. Comput. Chem.* **1992**, *13*, 102–114. [CrossRef]
172. French, A.D.; Johnson, G.P.; Kelterer, A.M.; Dowd, M.K.; Cramer, C.J. Quantum mechanics studies of the intrinsic conformation of trehalose. *J. Phys. Chem. A* **2002**, *106*, 4988–4997. [CrossRef]
173. Nunes, S.C.C.; Jesus, A.J.L.; Moreno, M.J.; Eusebio, M.E. Conformational preferences of α,α-trehalose in gas phase and aqueous solution. *Carbohydr. Res.* **2010**, *345*, 2048–2059. [CrossRef]
174. Färnbäck, M.; Eriksson, L.; Widmalm, G. Octa-*O*-acetyl-α,α-trehalose ethanol disolvate. *Acta Crystallogr. Sect. E* **2004**, *60*, 1483–1485. [CrossRef]
175. Bock, K.; Defaye, J.; Driguez, H.; Bar-Guilloux, E. Conformations in solution of α,α-trehalose, α-D-glucopyranosyl α-D-mannopyranoside, and their 1-thioglycosyl analogs, and a tentative correlation of their behaviour with respect to the enzyme trehalase. *Eur. J. Biochem.* **1983**, *132*, 595–600. [CrossRef]
176. Lee, C.K.; Koh, L.L. Structure of β,β-trehalose. *Acta Crystallogr. Sect. C Cryst. Struct. Commun.* **1993**, *49*, 621–624. [CrossRef]
177. Farnback, M.; Eriksson, L.; Widmalm, G. Octa-*O*-acetyl-β,β-thiotrehalose. *Acta Crystallogr. Sect. C Cryst. Struct. Commun.* **2000**, *56*, 700–701. [CrossRef]
178. Stortz, C.A.; French, A.D. Disaccharide conformational maps: Adiabaticity in analogues with variable ring shapes. *Mol. Simul.* **2008**, *34*, 373–389. [CrossRef]
179. Peric-Hassler, L.; Hansen, H.S.; Baron, R.; Hünenberger, P.H. Conformational properties of glucose-based disaccharides investigated using molecular dynamics simulations with local elevation umbrella sampling. *Carbohydr. Res.* **2010**, *345*, 1781–1801. [CrossRef] [PubMed]
180. Huber, R.G.; Berglund, N.A.; Kargas, V.; Marzinek, J.K.; Holdbrook, D.A.; Khalid, S.; Piggot, T.J.; Schmidtchen, A.; Bond, P.J. A Thermodynamic Funnel Drives Bacterial Lipopolysaccharide Transfer in the TLR4 Pathway. *Structure* **2018**, *26*, 1151–1161.e4. [CrossRef] [PubMed]

pharmaceuticals

Article

Mannosylated Polymeric Ligands for Targeted Delivery of Antibacterials and Their Adjuvants to Macrophages for the Enhancement of the Drug Efficiency

Igor D. Zlotnikov [1,*], Alexander A. Ezhov [2], Rostislav A. Petrov [1], Maksim A. Vigovskiy [3,4], Olga A. Grigorieva [3,4], Natalya G. Belogurova [1] and Elena V. Kudryashova [1,*]

1 Faculty of Chemistry, Lomonosov Moscow State University, Leninskie Gory, 1/3, 119991 Moscow, Russia
2 Faculty of Physics, Lomonosov Moscow State University, Leninskie Gory, 1/2, 119991 Moscow, Russia
3 Institute for Regenerative Medicine, Medical Research and Education Center, Lomonosov Moscow State University, 27/10, Lomonosovsky Ave., 119192 Moscow, Russia
4 Faculty of Medicine, Lomonosov Moscow State University, 27/1, Lomonosovsky Ave., 119192 Moscow, Russia
* Correspondence: zlotnikovid@my.msu.ru (I.D.Z.); helenakoudriachova@yandex.ru (E.V.K.)

Abstract: Bacterial infections and especially resistant strains of pathogens localized in macrophages and granulomas are intractable diseases that pose a threat to millions of people. In this paper, the theoretical and experimental foundations for solving this problem are proposed due to two key aspects. The first is the use of a three-component polymer system for delivering fluoroquinolones to macrophages due to high-affinity interaction with mannose receptors (CD206). Cytometry assay determined that 95.5% macrophage-like cells were FITC-positive after adding high-affine to CD206 trimannoside conjugate HPCD-PEI1.8-triMan, and 61.7% were FITC-positive after adding medium-affine ligand with linear mannose label HPCD-PEI1.8-Man. The second aspect is the use of adjuvants, which are synergists for antibiotics. Using FTIR and NMR spectroscopy, it was shown that molecular containers, namely mannosylated polyethyleneimines (PEIs) and cyclodextrins (CDs) load moxifloxacin (MF) with dissociation constants of the order of 10^{-4}–10^{-6} M; moreover, due to prolonged release and adsorption on the cell membrane, they enhance the effect of MF. Using CLSM it was shown that eugenol (EG) increases the penetration of doxorubicin (Dox) into cells by an order of magnitude due to the creation of defects in the bacterial wall and the inhibition of efflux proteins Fluorescence spectroscopy showed that 0.5% EG penetrates into bacteria and inhibits efflux proteins, which makes it possible to increase the maximum concentration of the antibiotic by 60% and maintain it for several hours until the pathogens are completely neutralized. Regulation of efflux is a possible way to overcome multiple drug resistance of both pathogens and cancer cells.

Keywords: drug resistance; efflux inhibitor; CLSM; moxifloxacin; eugenol; macrophage; flow cytometry

Citation: Zlotnikov, I.D.; Ezhov, A.A.; Petrov, R.A.; Vigovskiy, M.A.; Grigorieva, O.A.; Belogurova, N.G.; Kudryashova, E.V. Mannosylated Polymeric Ligands for Targeted Delivery of Antibacterials and Their Adjuvants to Macrophages for the Enhancement of the Drug Efficiency. *Pharmaceuticals* **2022**, *15*, 1172. https://doi.org/10.3390/ph15101172

Academic Editors: Federica Sodano and Maria Grazia Rimoli

Received: 6 September 2022
Accepted: 16 September 2022
Published: 21 September 2022

1. Introduction

To combat dangerous infections, pathogens resistant to many drugs, and dormant bacteria, it is necessary to create new combined drugs [1,2]. The key direction is targeted drug delivery to increase organ bioavailability, active concentration of the drug, and its duration of action [3–8]. Additional strengthening of the antibiotic can be achieved through the use of substances called efflux inhibitors [9–14]. Currently, antibiotic-resistant pathogens [15], especially those localized in macrophages or granulomas [16], are an acute problem. Targeted delivery of effective fluoroquinolones, inhibitors of bacterial DNA gyrase, by means of mannosylated molecular containers discussed above will allow the drug to be localized in the focus of infection.

On the other hand, antibiotic resistance can be defeated by delivering an adjuvant substance. It is known from the literature that the active efflux of both prokaryotic and eukaryotic cells strongly reduces the efficiency of a large number of antibiotics, and efflux greatly determines the effect of multidrug resistance [12,13]. The study of efflux is a rather complex process; the described techniques are based on the reduction of MIC of model efflux inhibitors (i.e., ethidium bromide, chlorpromazine) with the addition of mainly essential extracts (Thai herbs, allylbenzenes, i.e., EG [11,17–19]. In this paper, we present a new technique for studying efflux using CLSM. The development and use of specific, powerful efflux inhibitors will significantly increase the effectiveness of drug formulations and prevent the development of pathogen resistance.

Previously, a significant potential of allylbenzenes, including eugenol (EG), and terpenoids, which are synergists of fluoroquinolones, has been demonstrated [10,17,20–26]. As promising adjuvants, apiol and its analogs dillapiole, myristicin, and nothoapiol should be noted [27,28]. Some monoterpenes, namely (−)-carveol, geraniol, and citronellol, show a synergy effect when used as additives for conventional antimicrobials (chloramphenicol, minocycline, amoxicillin, and ciprofloxacin) [13,14]. These substances are capable of creating defects in the bacterial membrane [17,23] and probably inhibit enzymes and efflux pumps [9–13,29] of pathogens that cause resistance of strains. Thus, the complex preparation of the fluoroquinolone + adjuvant combination with the function of active targeting of mannose receptors of macrophages is a promising method for the treatment of resistant bacterial infections.

Often, resistant bacteria are localized in macrophages or granulomas [16,30–33]. Macrophages play an important role in both innate and acquired (humoral and cellular) immune responses [34]. The effect on macrophages can be classified into three groups according to the mechanism: activation of macrophages from inactive in M1 (pro-inflammatory) or M2 (pro-tumor or anti-inflammatory) [35]; repolarization of M2 in M1 to reduce the rapid response of the immune system [36,37]; and accumulation of antibacterial substances in macrophages against microorganisms "hidden in the reservoir" [30]. In this work, the focus is on the delivery of antibacterial drugs to alveolar macrophages for the destruction of resistant microorganisms (including *M. tuberculosis* localized in granulomas). The proposed strategy is based on high-affinity binding to CD206 mannose receptors overexpressed on M2 macrophages [38–41]. The use of mannose receptors that recognize Man-, Fuc-, and GlcNAc-residues of oligosaccharides of the cell wall of pathogenic microorganisms has a number of advantages in comparison with the strategy of targeting other receptors: exposure only to pathogenic macrophages, reduced risk of resistance and complications, and increased efficiency (https://www.macrophagetx.com; (accessed on 1 September 2022). CD206 is a C-type lectin (175 kDa), which is a transmembrane protein that binds carbohydrates due to a lectin-like region of eight domains [39,42]. However, since CD206 itself is difficult to isolate, model receptor proteins, for example, concanavalin A (ConA), are often used to screen the affinity of mannosylated delivery systems to macrophage mannose receptors [25,43–48]. The relevance of the ConA model was previously demonstrated using molecular dynamics and neural network analysis [43] as well as in vivo and in vitro experiments, which demonstrated that highly ConA-specific delivery systems effectively target macrophages [49–51].

Thus, using a model receptor, the selection of the greatest affinity for the mannose receptor conjugates and optimization of the structure of the drug delivery system to the macrophage were carried out [25]. The effect of enhancing drug penetration due to delivery systems adsorbed on bacterial wall surfaces was demonstrated [25,52]. The literature describes molecular containers based on liposomes [53,54], cyclodextrins [55,56], oligo- and polyamines [25,57], polyglycols [58–60], chitosan [44,61], and mannan [45,60,61]. Synthesis of molecular containers includes two main parts: container synthesis (carrier molecule) and mannosylation (introduction of a mannose label). The main reactions used in the first stage are as follows: imine (Schiff) reaction between carbonyl (i.e., −C=O groups can be obtained by periodate oxidation of carbohydrates) and amino groups with subsequent reduction of the −C=N bond by borohydride [62]; preparation of carboxymethylated derivatives of carbohydrates by exposure to chlorine acetic and hydrochloric acids [63]; activation of hydroxyl groups with TsCl, substitution for an easily leaving group (often iodide), followed by the introduction of a spacer [56,64]; glutaraldehyde, glyoxal, polyamines (spermine, spermidine), carbodiimide, and diisothiocyanates can be used as crosslinking agents [65,66]; and activation of hydroxyl groups with carbonyl diimidazole followed by attachment to the activated carbonyl group of a nucleophilic agent [25]. The easiest way to perform mannosylation is conjugation of the linear form of mannose formed from cyclic in solution with a delivery system with primary amino groups (reductive amination) [25,44,67]. The introduction of the cyclic form of mannose requires either a preliminary modification of the -OH group or the use of commercially available activated forms, for example, with $-N{=}C{=}S$, $-NH_2$, $-N_3$, and $-C{=}O$ groups [68,69]. High-affinity conjugates with synthetic analogues of oligomannosides can be obtained using available mannose on spacers -O-CH_2CH_2S-Man [56] or on spermine/spermidine [25,44] It is much more difficult to introduce oligomannosides directly into the conjugate: a multi-stage synthesis is required, which reduces the yield [70]. However, it is possible to use a carbohydrate with a reducing end and perform effective mannosylation, as in the case of simple mannose [71]; this approach is used in this work.

The aim of this work is to develop a combined formulation based on 4th-generation fluoroquinolone moxifloxacin (MF) loaded into mannosylated containers as a target delivery system and enhanced with adjuvants (EG [10,17–19,22,72–80], menthol, and safrole [23,28,81]) capable of bypassing the resistance of pathogens and fighting dormant infection.

The approach proposed provides a comprehensive, versatile focus on the problem of pathogen resistance. The above-discussed works were aimed either at creating delivery systems (mannosylated particles) or studying the antibacterial activity of antibiotics or substances from the class of terpenoids, which are practically not soluble in aqueous solution, and thus, they have low efficiency themselves. The advantage of the suggested systems that will ensure their therapeutic potential is the combination of several fronts in the fight against resistance, namely the accumulation of drugs in macrophages with simultaneous enhancing of the antibiotic efficiency due to increased permeability of the bacterial cell membrane and inhibition of efflux protein provided by adjuvants.

2. Results and Discussion

2.1. Synthesis and Physico-Chemical Characteristics of Mannosylated Conjugates

2.1.1. FTIR Spectroscopy

The reactions underlying the synthesis of conjugates 1–3 are as follows (Figure 1): reductive amination of aldehyde groups [67–69,71] and activation of hydroxyl groups of cyclodextrin CDI followed by crosslinking with NH_2– and OH– groups [25,65,82,83]. The lightest conjugate 1 based on amCD and triMan was synthesized according to the scheme in Figure 1a. In the FTIR spectra of amCD-triMan, in comparison with initial compounds (Figure 2a), changes in the region of 2800–3000 cm^{-1} were observed, corresponding to the valence vibrations of C-H and N-H bonds; the intensity of the peak of 1420 cm^{-1} increased, corresponding to the deformation vibrations of N-H at the junction of two fragments: trimannoside and cyclodextrin. Bright changes were recorded in the area of 900–1100 cm^{-1}, which corresponds to the most effective splitting of the reducing end of the saccharide and crosslinking with the formation of C-N bonds (Figure 1a). Ligands 2a and 3a (Figure 1b) are polymers with attached trimannoside (triMan) as labels to the mannose receptors of macrophages. triMan-$(GlcNAc)_2$ reacts with PEI amino groups via the reducing end to form a Schiff base PEI-N=CH-triMan, which is reflected in the IR spectra (Figure 2b and Figure S1): the area of peaks corresponding to N-H and C-H bending decreases (1300–1550 cm^{-1}), and the absorption peak of the bond C=N appears. Grafting of polymer chains with cyclodextrins will allow to shield the charge of PEI amino groups due to its modification and the formation of a protective carbohydrate coat and increase the theoretical drug-capacity since CD are one of the lightest but at the same time effective containers for medicines (ligand 2b, 3b). The activation of secondary hydroxyl groups of cyclodextrin with CDI followed by cyclodextrins' modification with the primary amines of spermine or PEI leads to final grafted conjugates 2b and 3b (Figure 1b).

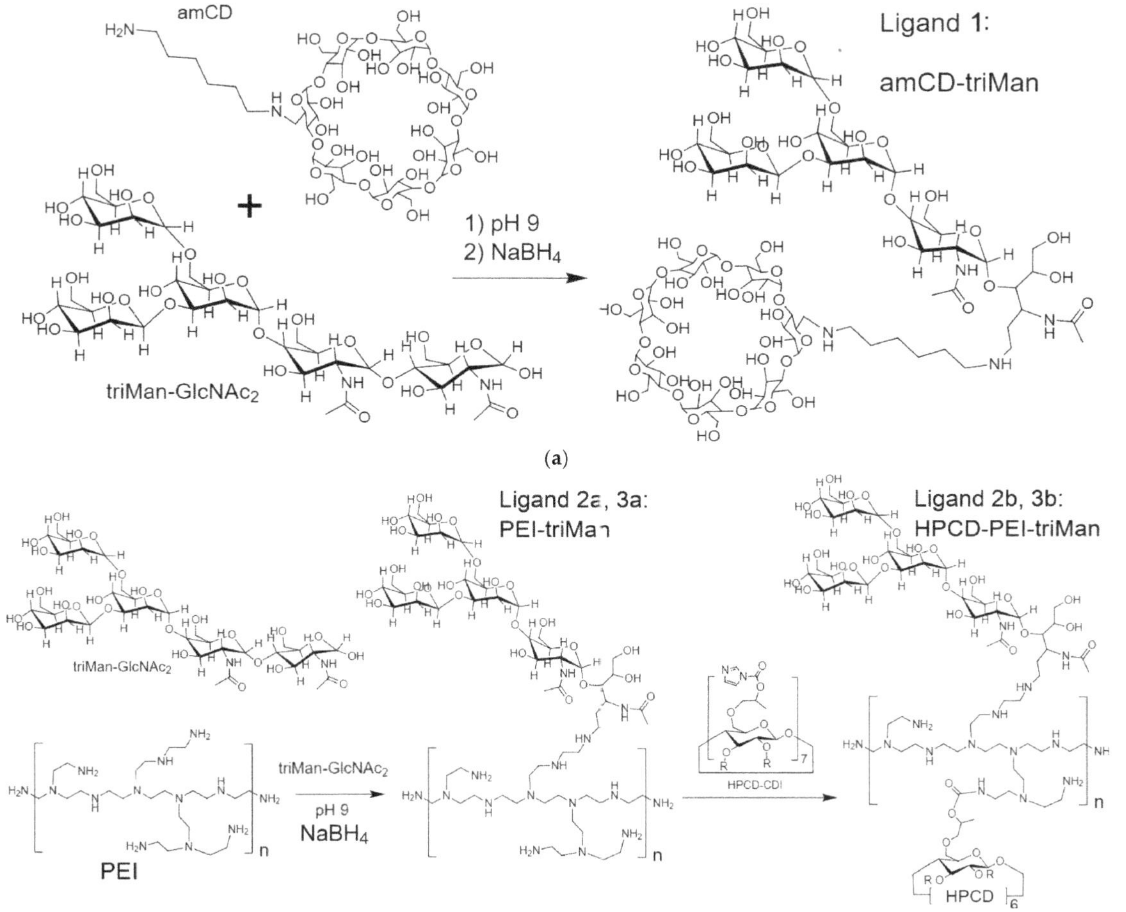

Figure 1. Scheme of synthesis of mannosylated conjugates: (**a**) amCD-triMan; (**b**) PEI1.8-triMan or PEI10-triMan and HPCD-PEI1.8-triMan or HPCD-PEI10-triMan. For PEI 1.8 kDa, n is in the range 3–5; for PEI 10 kDa, n is in the range 20–25.

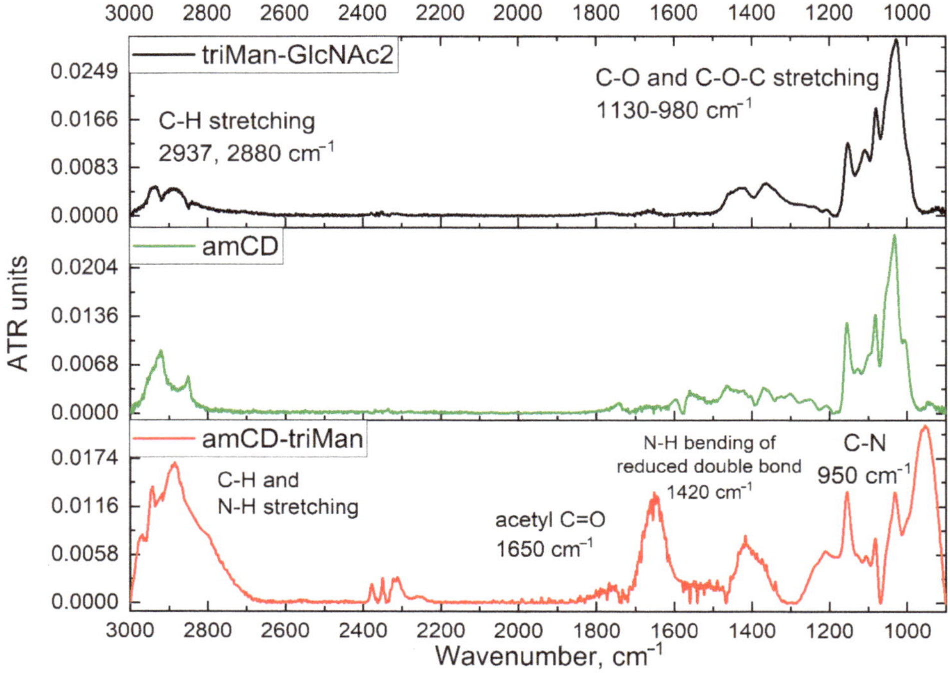

(a)

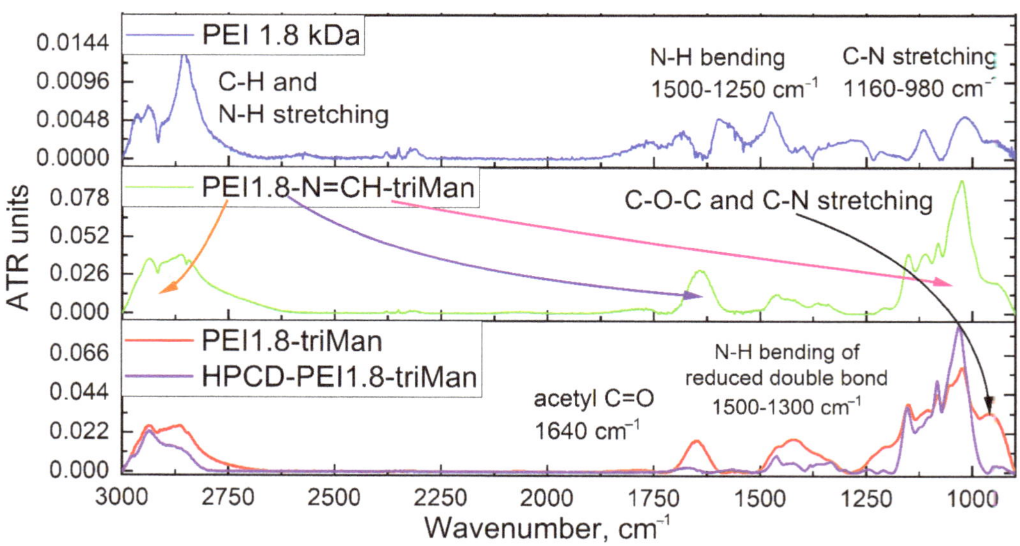

(b)

Figure 2. Fourier transform infrared spectra of: (**a**) triMan-GlcNAc$_2$, amCD, amCD-triMan; (**b**) triMan-GlcNAc$_2$, PEI 1.8 kDa, PEI1.8-N=CH-triMan, PEI1.8-triMan, and HPCD-PEI1.8-triMan. Aqueous solutions, T = 22 °C.

2.1.2. NMR Spectroscopy

The obtained data on the structure of the synthesized conjugates were confirmed by NMR spectroscopy. The ^{1}H NMR spectra of triMan-GlcNAc$_2$, PEI1.8-triMan, and HPCD-PEI1.8-triMan with proton assignment are presented in Figure 3. The signals in the range of 3.9–3.1 ppm and in the range of 5.1–5.3 ppm (Figure 3a) correspond to H2–H5 and H1 of D-mannose or D-glucose (in HPCD) units, respectively [84]. The signals of the anomeric protons H1 lie in a weaker field of 4.6–5.3 ppm [85]. The H2 patterns are shifts of 3.8–3.9 ppm, and for neutral protons, the signals overlap in the region of 3.1–3.8 ppm. The presence of PEI in ligand **2a** (Figure 3b) and the formation of bonds triMan–CH$_2$–NH–PEI in conjugates is confirmed by the signals in the range of 2.5–2.9 ppm according to the literature data for mannosylated chitosan grafted with PEI, PEI copolymers, and CD's derivatives [62,65,86,87]. In the PEI1.8-triMan conjugate, the carbohydrate fragment GlcNAc is in a linear form, while oligosaccharide alone is characterized mainly by a cyclic form of the reducing end; consequently, the graftization of the polymer by trimannoside was carried out according to the scheme (Figure 1b). In the NMR spectrum of PEI1.8-triMan, peaks appear in the region of 4.1–4.0 ppm and 5.0–4.8 ppm, and the peak of H5 shifts from 5.1 to 5.3 ppm (Figure 3a,b) according to the literature data on Man- and GlcNAc-containing sugars [88].

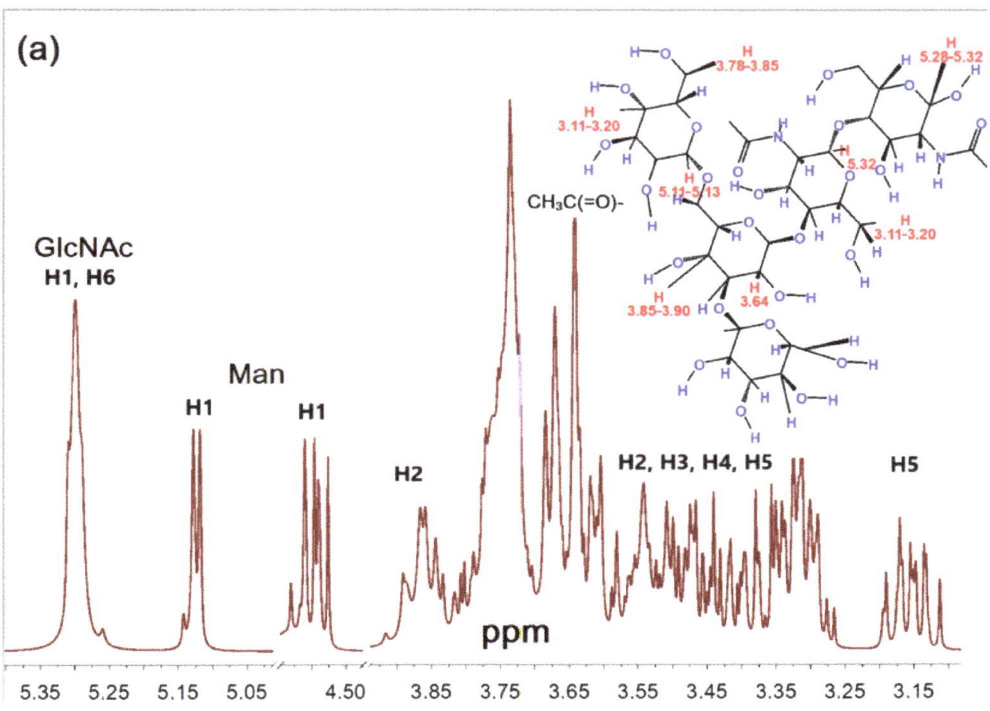

Figure 3. *Cont.*

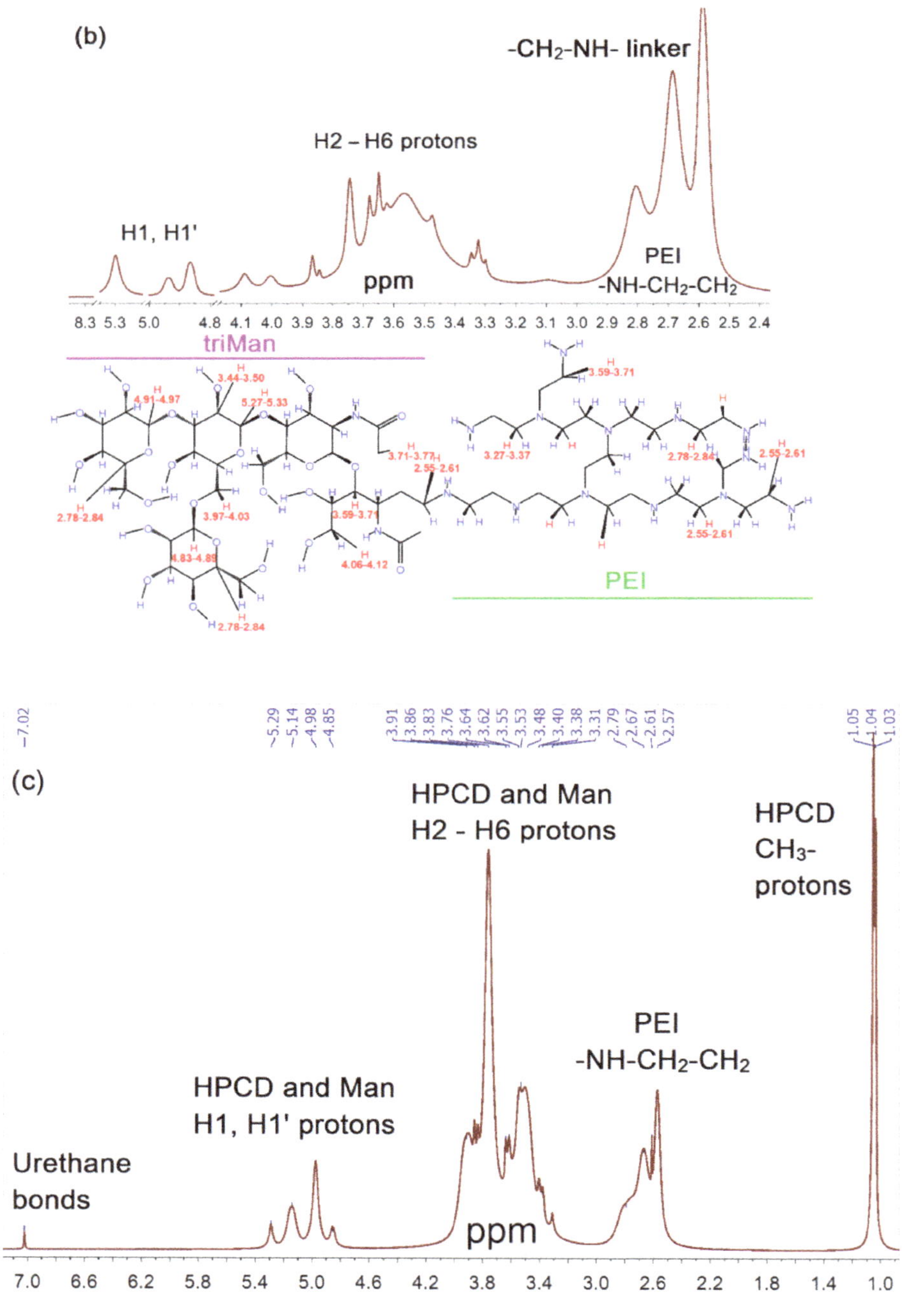

Figure 3. ^{1}H NMR spectra of (**a**) triMan-GlcNAc$_2$, (**b**) PEI1.8-triMan, and (**c**) HPCD-PEI1.8-triMan with proton assignment. D_2O, 400 MHz.

In the ^{1} H NMR spectra of HPCD-grafted conjugate (Figure 3c), both peaks of the initial components and new signals related to urethane bonds -N(H)-C(O) (formed as a result of the interaction of activated HPCD with spermine or PEI) were detected: the chemical shift -NH- at 7.05 ppm and the chemical shift -C(H)-O- at 2.8 ppm characterize the formation of urethane bonds [89,90]. Based on the integral intensity of the characteristic peaks of protons corresponding to triMan, PEI, and HPCD, the molar ratios of the components were calculated (Table 1) correlating with FTIR spectroscopy data.

Table 1. Physico-chemical properties of mannosylated conjugates.

Ligand	Chemical Designation	Molecular Weight, kDa	Molar Ratio *	Hydrodynamic Size **, nm	Zeta Potential **, mV	Polydispersity Index **
1	amCD-triMan	2.1	1:1	<5	−2 ± 1	-
2a	PEI1.8-triMan	15 ± 5	1:15	6 ± 1	−1.1 ± 0.5	0.30
2b	HPCD-PEI1.8-triMan	23 ± 8	5:1:15	17 ± 4	−2.1 ± 0.4	0.28
3a	PEI10-triMan	45 ± 10	1:40	8 ± 1	+20 ± 3	0.5
3b	HPCD-PEI10-triMan	60 ± 7	10:1:40	48 ± 7	+5 ± 2	0.2

* In the specified order. Molar ratio of the components was calculated based on the integral intensity of the characteristic peaks corresponding to triMan, PEI, and HPCD, in the NMR and FTIR spectra. ** Size distribution values by volume, zeta-potential distribution, and polydispersity index were determined by dynamic light scattering method; 0.01 M PBS (pH 7.4).

2.1.3. Physico-Chemical Properties of Mannosylated Ligands

The proposed conjugate's structures and composition (Figure 1) were identified on the basis of conditions and sequence of synthesis reactions, the ratio of components, FTIR and NMR spectral data, and data on the number of mannosylated amino groups by spectrophotometric detection. Physico-chemical properties of synthesized conjugates are summarized in Table 1. Conjugates based on PEI10 have a molecular weight of 45–60 kDa, ligands based on PEI1.8–15–23 kDa, and a "light" 2 kDa amCD-triMan. An important parameter is the hydrodynamic size (Figure S2), which takes into account the hydrate shell of the ligand and the effective particle size. It is believed that the largest particles >200 nm settle well in the lungs and thus enter directly into the alveoli to macrophages through the use of aerosol form of drugs. Smaller particles can be injected intravenously since they do not cause a strong immune response, unlike high-mannose oligosaccharides or mannan. Therefore, a wide range of delivery systems can theoretically provide various options for the treatment of macrophage-associated diseases, considering the specifics of the disease. ζ-potential (Figure S3) of all the final conjugates (1, 2b, 3b) is in the range from −5 to +5 mV.

It is believed that positively or negatively charged (>10 mV) particles interact better with surface proteins or polysaccharides of cell membranes [91,92], which determines effective penetration. However, our strategy is aimed at avoiding non-specific capture of delivery systems, so the charge remains almost neutral (<5 mV). It is worth noting that the high positively or negatively charged (>15 mV) ligands could negatively affect the state of the biomembranes and blood components; for example, they may cause hemolysis of erythrocytes. Therefore, when using PEI as a carrier, it should be modified with neutral agents (such as CDs used in this work) that well-shield the charges groups and protect against aggregation at the same time.

In addition to the size and charges for pharmaceutical preparations, the homogeneity of the obtained ligands is also important. The polydispersity indexes determined by DLS method (PDI, Table 1) of synthesized conjugates show quite acceptable characteristics in terms of homogeneity: it is approximately 0.3 for ligands with a PEI of 1.8 kDa and 0.5 for 10 kDa of polyethylenimine. The relative homogeneity of the samples can also be judged from the hydrodynamic size and the zeta-potential distributions of the particles (Figures S2 and S3), which are which are represented by a single peak.

Purification and homogeneity of samples were analyzed by HPLC. The chromatograms for HPCD-grafted PEI with triMan are shown in Figure S4. For ligands 2b and 3b, retention times are 4.5 and 3.1 min (dead time is 1.7 min). Polymers are eluted from the column by one main peak; however, the peaks are rather blurred to the right due to some polydispersity of the samples. Therefore, for further experiments, fractions were collected that correspond to the maximum peaks on the chromatograms ±0.5 min by retention time.

2.2. Conjugates' Specificity to the Mannose Receptors on the ConA Model

One of the main parameters characterizing the effectiveness of drug delivery systems is affinity for targeted receptors. In this work, high affinity for mannose receptors (CD206) of alveolar macrophages is assumed. ConA was chosen as a model mannose receptor, for which relevance to CD206 was previously proven by methods of molecular dynamics and neural network analysis [43]. According to the literature, the CD206 domain shows a high similarity with mannan binding lectin A, which, in turn, is similar to ConA in terms of carbohydrate binding [93]. This is confirmed by numerous in vivo and in vitro studies [42]. Mannose-containing ligands, pre-selected by ConA assay, show high affinity (by flow cytometry) to CD206+ cells [84,94]. Similarly, mannose-containing ligands, pre-selected by ConA assay, are absorbed by macrophages via a receptor-mediated internalization mechanism [41] and are bio-distributed to lungs due to adsorption by macrophages [50]. All these data support the relevance of using ConA as a mannose receptor model. In this work, we used a previously validated technique [25] for determining the parameters of receptor-ligand binding using FTIR spectroscopy to these multicomponent systems.

Analytically significant absorption bands in the FTIR spectra of ConA are amide I and II (Figures 4a and S5), which are two major bands of the protein infrared spectrum. The amide I band (between 1600 and 1700 cm^{-1}) is mainly associated with the C=O stretching vibration (70–85%) and is directly related to the backbone conformation. Amide II results from the N-H bending vibration (40–60%) and from the C-N stretching vibration (15–40%). The ConA spectrum is also characterized by the amide A absorption band of N-H bonds 2800–3000 cm^{-1} (Figure 4a). At the same time, during complexation with a PEI-triMan ligands, the intensity of the characteristic bands of the protein changes (Figure 4a): the intensity of amide I decreases; in contrast to amide II, the minor peaks of amide III and IV increase and become more noticeable, which indicates the formation of hydrogen bonds, van der Waals, and electrostatic interactions between amino acid residues of the protein and carbohydrate ligand [25,42–45,95–104].

Recently, we showed using molecular dynamics that the "amide" atoms of ConA binding site are involved in complexation: Asn14 (N)–Man (O3), Asp16 (H)–Man (O5), and Arg222 (H)–Man (O2) [43]. This determines the observed changes in the FTIR spectra of the protein when binding to the mannosylated ligand (Figures 4a and S5). At the same time, Arg228, Asn14, Asp208, and Leu99 interact with the terminal mannosyl residue through a network of hydrogen bonds, which explains the appearance of absorption bands in the region of 1380–1480 cm^{-1}. Additionally, the ligand–protein complex is stabilized by stacking interactions (π-CH) of Tyr12 and Tyr100 with mannopyranoside planes, which is reflected in a change in the microenvironment of aromatic tyrosine systems: C–C skeletal deformations (1500 cm^{-1}, overlapped with amide II), C–O stretching vibration of phenol (1270 cm^{-1}), and in-plane C–H deformation vibration (1173 cm^{-1}) [105].

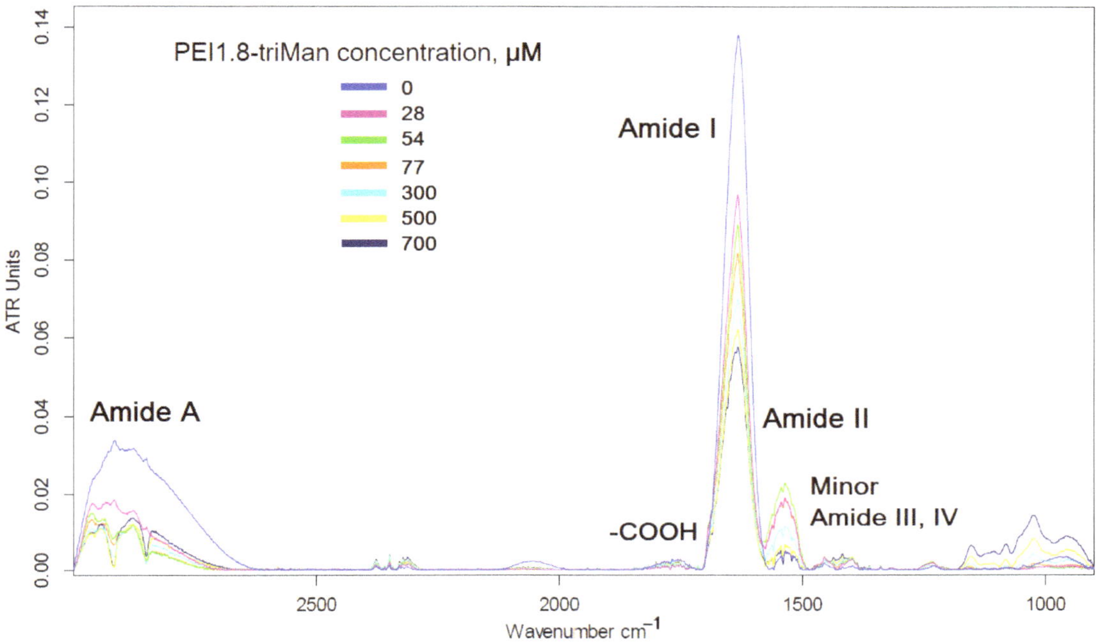

(a)

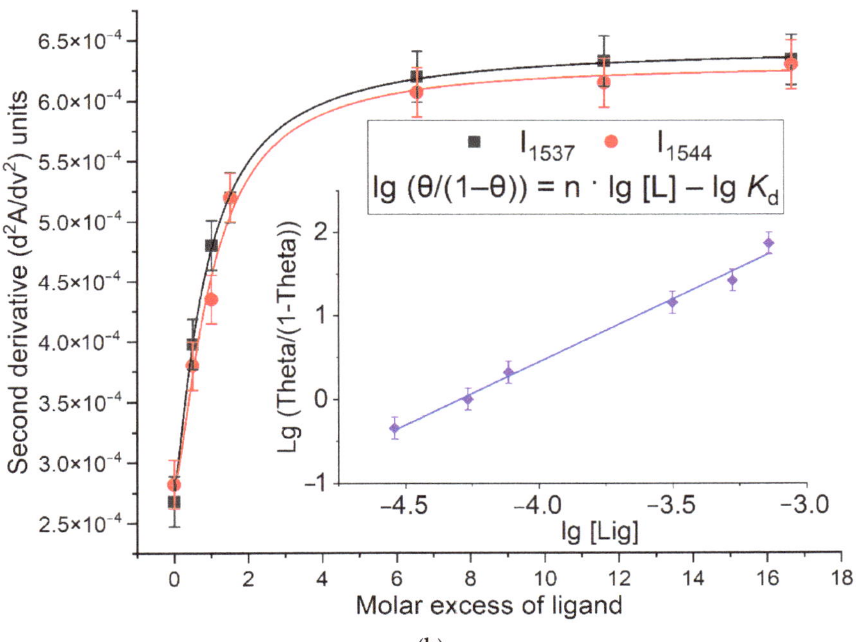

(b)

Figure 4. *Cont.*

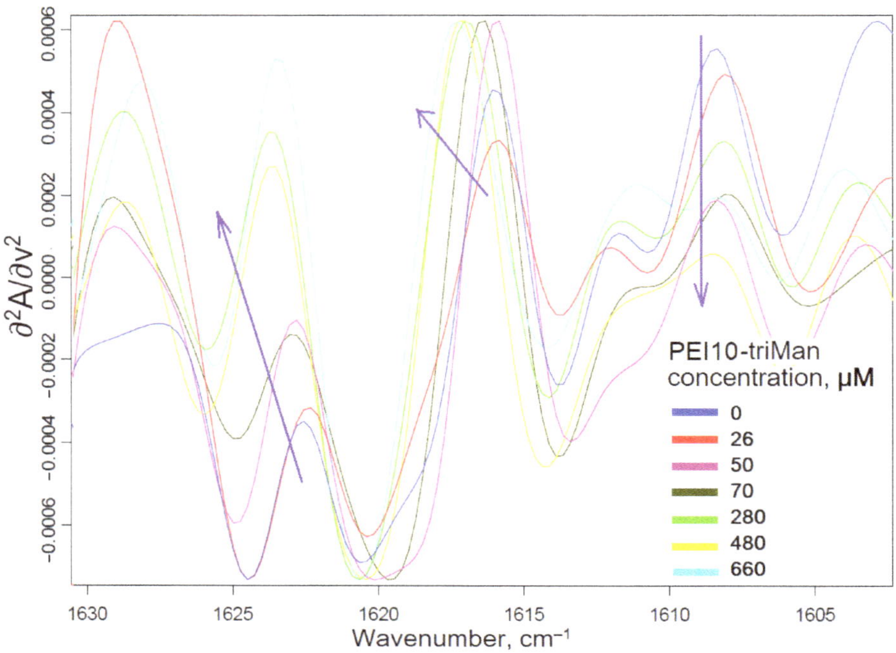

(**c**)

Figure 4. (**a**) Fourier transform infrared spectra of ConA and ConA complexes with PEI1.8-triMan. (**b**) Dependences of the intensity of amide I and amide II peaks in the FTIR spectra from (**a**). On the insert, the Hill linearization plot of the presented curves is given. (**c**) Second derivative (d^2A/dv^2) of FTIR spectra of ConA and ConA complexes with PEI10-triMan. The arrows show the shifts during the complexation of the protein with the increasing concentration of ligand. [L], the equilibrium concentration of the ligand; θ, proportion of bound receptor. C_0(ConA subunit) = 0.1 mM. Natrium-phosphate buffer solution (0.02 M, pH 6.4). $C(Ca^{2+}) = C(Mn^{2+}) = 1$ mM. T = 25 °C.

The dissociation constants were calculated based on the dependence of the amide I and II peak intensity in the second-derivative FTIR spectra (Figure 4b). The complex formation process is described by the scheme ConA + nLigand ⇌ ConA·nLigand and can be linearized in Hill coordinates. The validity of the choice of intensity in the quality of the analytical signal is confirmed by the fact that the energy of valence and deformation vibrations also changes due to the formation of non-covalent protein-ligand bonds, which is demonstrated by short-wave shifts in the differential IR spectra of the second order (Figure 4c). This shift was additionally considered as an analytical signal, and thus, the degree of binding was calculated by several parameters at once (intensity of amides I and II and position of maxima and minima in the differential spectrum).

According to IR spectroscopic analysis, all the ligands considered interact with ConA in a ratio of 1:1 (within the margin of error) except for PEI1.8-triMan, which, being a compact molecule, can interact with a protein in a ratio of 2:1 (Table 2). PEI-triMan ligands are the most affine to mannose receptors among the synthesized ligands (Table 2): $-\lg K_d$ reaches a value of 6.5. PEI-triMan ligands grafted with HPCD are also highly affine. The lightest amCD-triMan ligand demonstrates an affinity for mannose receptors comparable to dimannosides [48], but its advantage is potentially higher loading capacity (LC) due to the presence of CD.

Table 2. ConA-ligand complexation parameters determined by FTIR spectroscopy: complex dissociation constants and stoichiometry. ConA + nLigand ⇌ ConA·nLigand. T = 22 °C. C(Ca^{2+}) = C(Mn^{2+}) = 1 mM. Phosphate buffer solution pH = 6.4.

Ligand	Chemical Designation	n	$-\lg K_d$ (ConA·nLigand)
1	amCD-triMan	1.0 ± 0.2	4.1 ± 0.8
2a	PEI1.8-triMan	1.5 ± 0.1	6.5 ± 0.4
2b	HPCD-PEI1.8-triMan	1.2 ± 0.2	6.0 ± 0.2
3a	PEI10-triMan	1.1 ± 0.2	5.5 ± 0.7
3b	HPCD-PEI10-triMan	0.9 ± 0.2	4.4 ± 0.3

In our opinion, the most optimal and universal system is ligand 2b (HPCD-PEI1.8-triMan, Figure 1b); it has a relatively low molecular weight and therefore is more biodegradable than PEI10. However, at the same time, ligand 2b performs the function of active targeting due to its high affinity to mannose receptors and its LC up to 25% by weight of MF or another antibiotic with an adjuvant.

Earlier, using computer modeling [43], we showed that the binding of triMan and triMan-$GlcNAc_2$ [46,48,95,96,99] with ConA occurs according to an "improved mechanism" in relation to single Man: an average of 15 hydrogen bonds are formed, and 5–9 are additionally mediated by water. In comparison with the mode of binding by proteins of single Man residues, the following amino acid residues are additionally involved: Asp16, Thr15, Pro13, Tyr12, Asp208, Thr226, Glu98, Arg228, and Gly224. This determines the strength of complexes with ConA: $-\lg K_d$ 5 for triMan and 6 for triMan-$GlcNAc_2$. Since the oligosaccharide used in the work was attached by the reducing GlcNAc end (linker analog) to the polymer or CD, it is relevant to compare it with triMan. The affinity of the ligands 2a, 2b, and 3a is 0.5–1.5 orders of magnitude higher than that of the natural ligand trimannoside. Ligands 2a and 2b are comparable and even superior in affinity to mannan ($-\lg K_d$ 6).

2.3. *Moxifloxacin Loading into Molecular Containers*

As discussed above, the fourth-generation fluoroquinolone moxifloxacin was chosen as the main active component due to its wide spectrum of action and applicability for the treatment of respiratory tract diseases. As discussed above, even such strong antibiotics cannot cope with resistant or dormant bacteria. Targeted drug delivery and the use of adjuvant enhancers will help to combat them. In this section, the physicochemical parameters of MF loading into molecular containers are considered.

2.3.1. FTIR Spectroscopy of MF and MF Conjugate

FTIR spectra of MF– and MF–ligand complexes are presented in Figures 5a and S6. Spectra showed aromatic C=C stretching at 1621 (C=O and C=C), 1515, and 1448 cm^{-1}; carboxylic acid C=O stretching at 1715 cm^{-1}; C–N stretching at 1320–1380 cm^{-1}; and C–F stretching at 1188 cm^{-1}. The formation of non-covalent MF complexes with ligands is accompanied by a decrease in the intensity and a change in the shape of these peaks (Figures 5a and S6). Confirmation of the change in the microenvironment of the quinolone structure of MF is the shift of the peak from 1448 to 1454 cm^{-1}. An increase in the resolution of the spectrum can be achieved by differentiating and isolating thin components (Figure 5b). The observed shifts in the position of peaks in the differential spectrum to the high-frequency region correspond to changes in the vibrational energy of C=C, C=O, and C–N bonds in the composition of quinolone and pyrrolidine-piperazine structures following a change in the microenvironment when MF is included in the CD cavity or interaction with the polymer chain of conjugates. Based on the dependences of the intensity or position of the analytical peaks in the FTIR spectra of MF (Figure S6), the parameters of the interaction of MF with ligands were calculated (Table 3). The calculated values of constants in the order of magnitude are 5×10^{-3}–10^{-4} M, which exceeds the previously described values for levofloxacin [25]. The constants for complexes MF–(mannosylated polymers) are comparable to those for the CD–MF complex [89,106–108]. At the same time, polymers grafted with CDs demonstrate values of the dissociation constants of complexes with MF of the order 5×10^{-5}–5×10^{-6} M, while the drug capacity increases by 4–10 molecules for PEI1.8 and PEI10, respectively. Such high-capacity drug carriers can demonstrate the effects of improving pharmacokinetic parameters.

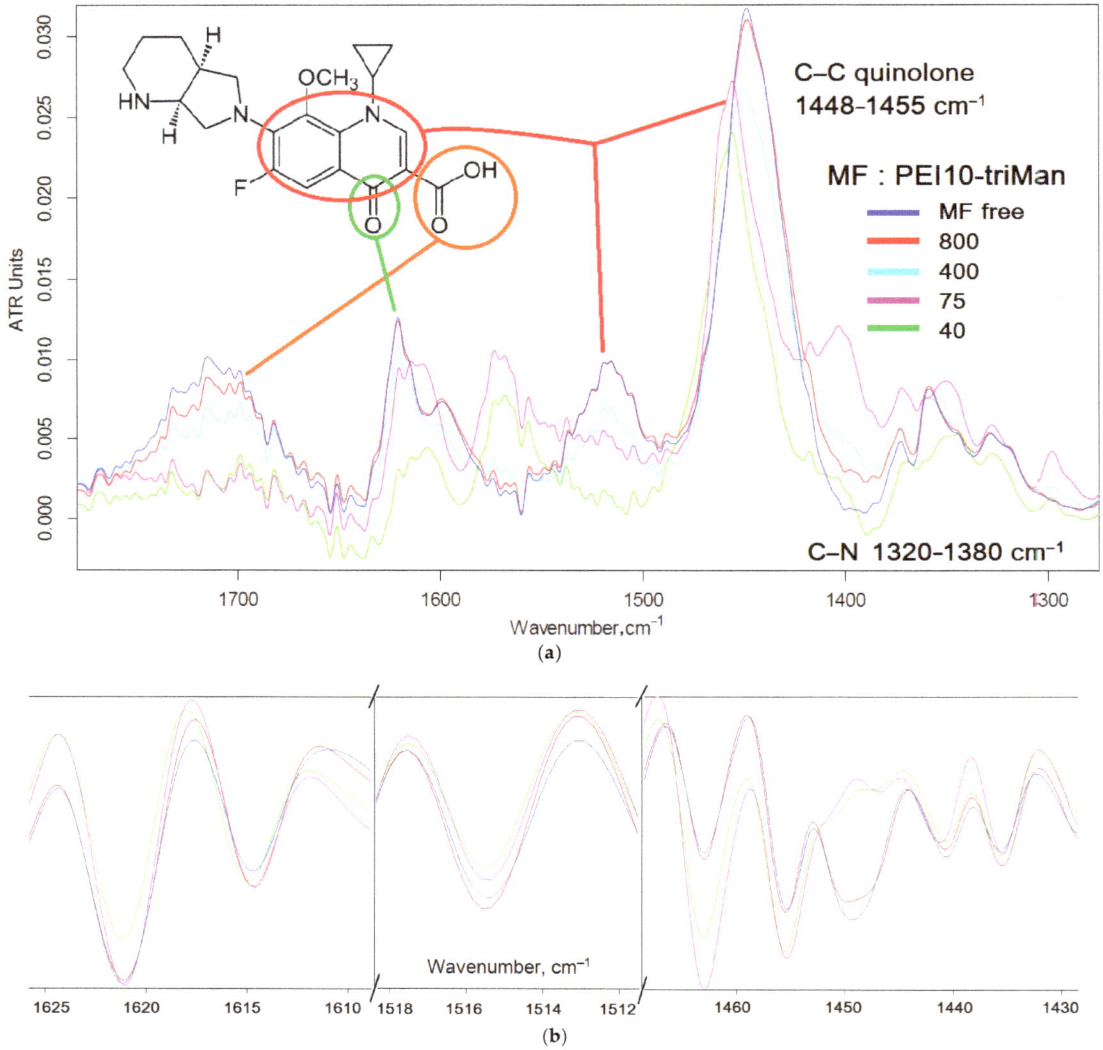

Figure 5. (**a**) FTIR spectra and (**b**) its second derivative (d^2A/dv^2) of MF and MF complexes with PEI10-triMan. Concentrations in (**b**), such as on A. 0.01 M HCl. T = 22 °C.

Table 3. Dissociation constants of MF conjugate complexes determined by FTIR spectroscopy. Number of MF molecules held by one conjugate molecule in CD tori (N_{CD}) and polymer matrix ($N_{polymer}$). Entrapment efficiency (EE) and loading capacity (LC) of MF by molecular containers. C_{ligand_total} = 0.0001 M, C_{MF_total} = 0.001 M. The conditions are given in the Methods section.

Ligand	$-\lg K_d$ (MF–Ligand)	N_{CD} or $N_{polymer}$	EE, %	LC, %
1	4.0 ± 0.2	1–CD	9 ± 1	20 ± 3
2a	3.3 ± 0.3	(4 ± 1)–polymer	24 ± 3	11 ± 2
3a	3.8 ± 0.4	(17 ± 2)–polymer	84 ± 6	15 ± 3

2.3.2. NMR Spectroscopy of MF and MF Conjugate

^{1}H NMR spectra of MF in free form and in the form of non-covalent complexes with ligands 2a and 2b (PEI1.8-triMan and HPCD-PEI1.8-triMan) are presented in Figure 6. The interactions of HPCD and PEI as components of molecular containers with MF can be characterized by induced chemical shifts $\Delta\delta$, equal to the difference of chemical shifts in the complex and single substances (Table 4) [89,109,110]. The interaction of MF with the PEI1.8-triMan polymer chain leads to induced shifts towards a strong field (Table 4), which corresponds to an increase in the hydrophobicity of the microenvironment of drug molecules, which indicates the successful encapsulation of MF into a molecular container. The inclusion of an apolar fragment of the MF molecule into the host hydrophobic cavity of HPCD (Figure 6c) induced a shielding of the inner protons of the glucose units of HPCD, namely H3 and H5, whereas the protons on the exterior of the torus (H1, H2, and H4) were relatively unaffected; this was previously shown in systems where CD formed inclusion complexes with Lev, EG [23], atropine, nitrobenzene, and nicardipine [111–115].

Table 4. ^{1}H-chemical shifts corresponding to MF in free form and in the form of non-covalent complexes with ligands 2a and 2b (PEI1.8-triMan and HPCD-PEI1.8-triMan). D_2O, 400 MHz.

MF Proton Number (Assignment)	^{1}H-Chemical Shifts, ppm		
	MF Free (Figure 6a)	MF + PEI1.8-triMan (Figure 6b)	MF + HPCD-PEI1.8-triMan (Figure 6c)
48 (aromatic)	8.56	8.38	8.51
49 (aromatic)	6.89 and 6.85	7.25 and 7.22	7.30 and 7.27
38 (cyclopropane)	3.34	3.31	3.30
34, 33 (pyrrolidine)	4.08 and 3.76	3.9 and 3.73	3.99 and 3.75 (overlapped)
35, 32 (pyrrolidine)	4.00 and 3.70	3.65 (overlapped) and 3.53	3.88 and 3.51
31 (pyrrolidine–piperidine)	3.89	3.81	3.71
50, 51, 52 (–OCH_3)	3.50	3.39 (overlapped)	3.45
46, 47 (piperidine)	3.03	2.99	3.01
30 (pyrrolidine–piperidine)	2.75	2.74	2.74 and 2.63
36, 37 (piperidine)	1.84–1.87	1.80 and 1.75	1.81
39, 40 (piperidine)	1.75–1.80	NI *	NI
43, 45 (cyclopropane)	0.99–1.05	0.90	0.78
42, 44 (cyclopropane)	0.88 and 1.21	1.07	1.02–1.04

* Not identified.

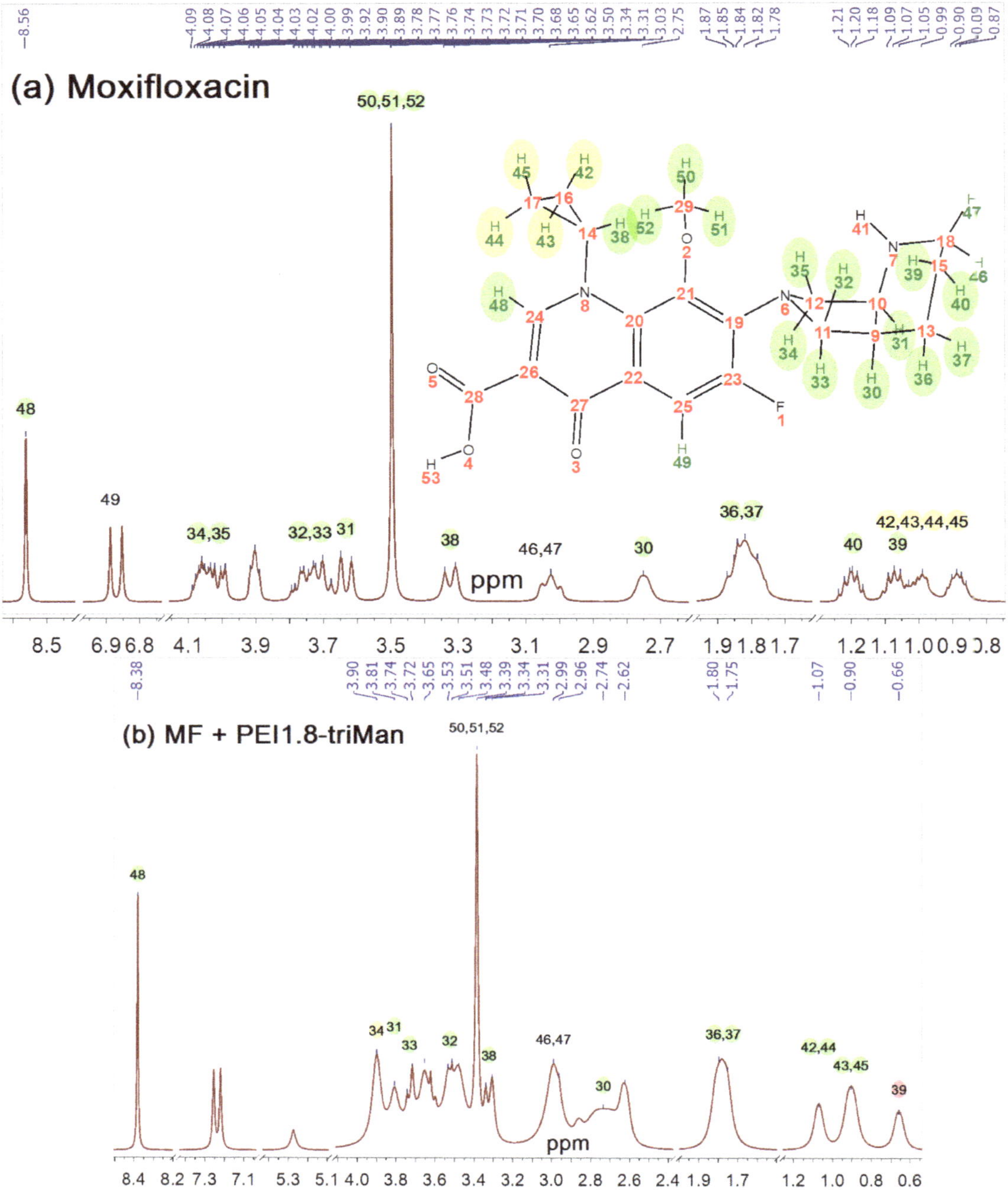

Figure 6. *Cont.*

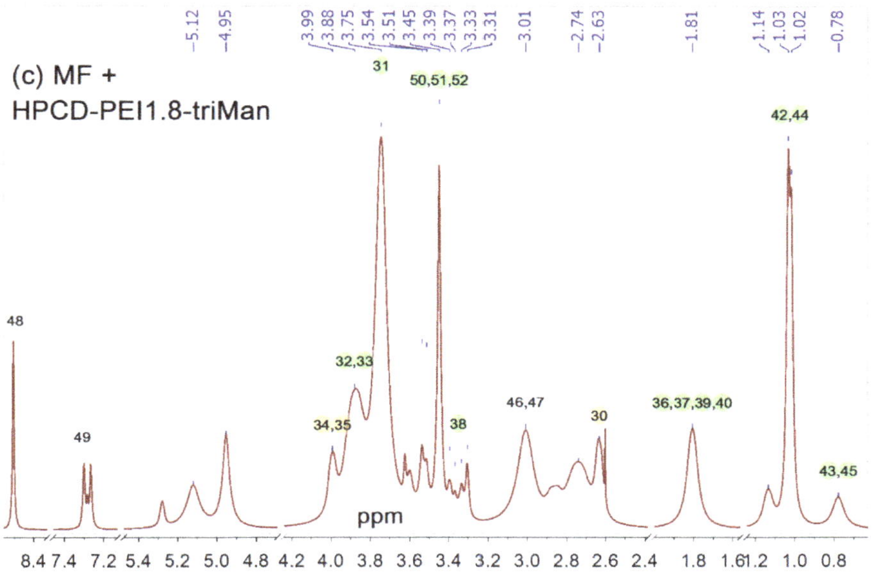

Figure 6. ^{1}H NMR spectra of (**a**) MF, (**b**) MF complexed with PEI1.8-triMan (w:w = 1:1), and (**c**) MF complexed with HPCD-PEI1.8-triMan (w:w = 1:1) with proton assignment. D_2O, 400 MHz.

The inclusion of MF into the HPCD cavity is accompanied by induced chemical shifts of MF protons (Table 4), mainly upfield due to an increase in the hydrophobicity of the environment of the guest molecule. When MF interacts with molecular containers, the induced shift of the proton 49 is knocked out of the common vector, which is probably due to the presence of a neighboring electronegative atom F, which pulls the electron density more efficiently inside the CD cavity. The protons MF 30, 31, and 38 of the piperidine and cycloprapane rings (Figure 6a) are characterized by stronger shifts towards an upfield in the case of a polymer ligand grafted with HPCD in comparison with PEI1.8-triMan, and for the remaining protons, on the contrary, downfield shifts are observed due to the greater donor capacity of PEI nitrogen atoms than OH groups of HPCD. Based on ^{1}H NMR spectra, MF is included in the cyclodextrin by the piperidine end and is buried to the carboxyl group, the location of which is advantageous outside the HPCD cavity and is additionally stabilized by the atoms of the HPCD outer shell [23]. Thus, the structure of the MF inclusion complex in HPCD is similar to those described earlier for levofloxacin [23,111], with the only difference that the piperazine end is completely immersed in the HPCD cavity and not by half, as in the case of piperidine in levofloxacin. A similar conclusion was obtained earlier using computer modeling of fluoroquinolone–cyclodextrin complexation processes [107,116].

2.4. Drugs Interaction with Bacteria

2.4.1. Antibacterial Activity of MF against *B. subtilis*

The use of drug delivery systems as well as adjuvants to drugs allows increasing the therapeutic potential of the drug due to several aspects: (1) due to the formation of non-covalent complexes with a molecular container, the molecules of the therapeutic agent are more stable, and their active concentration is higher; (2) the ligand+adjuvant increases the permeability of the drug into cells; (3) adjuvant inhibits efflux proteins; (4) the delivery system allows one to load a larger amount of the drug and slowly release it while maintaining a given level agent concentrations; and (5) the rate of excretion from the body decreases. The data on item 4 were obtained and analyzed earlier by us: effective action time of levofloxacin (time of 80% release) was increased from 0.5 to 20–70 h due to mannosylated HPCD-PEI conjugates [25]. Point 5 is confirmed by the literature data: due to the conjugation of antibiotics with polymeric (polyelectrolyte) delivery systems, the half-life of fluoroquinolones and their time of effect are increased [5,117].

The role of adjuvants: The authors recently studied the effect of allylbenzenes and terpenoids on Gram-positive and Gram-negative bacteria [23,25]: both individual substances and synergists of antibiotics. It has been shown that among other allylbenzenes and terpenoids studied, EG, menthol and safrole most powerfully enhance the antibiotic. By themselves, these components of essential oils are not capable of efficiently defeating bacterial infection, but they have proven themselves as antibiotic enhancers (Figure 7).

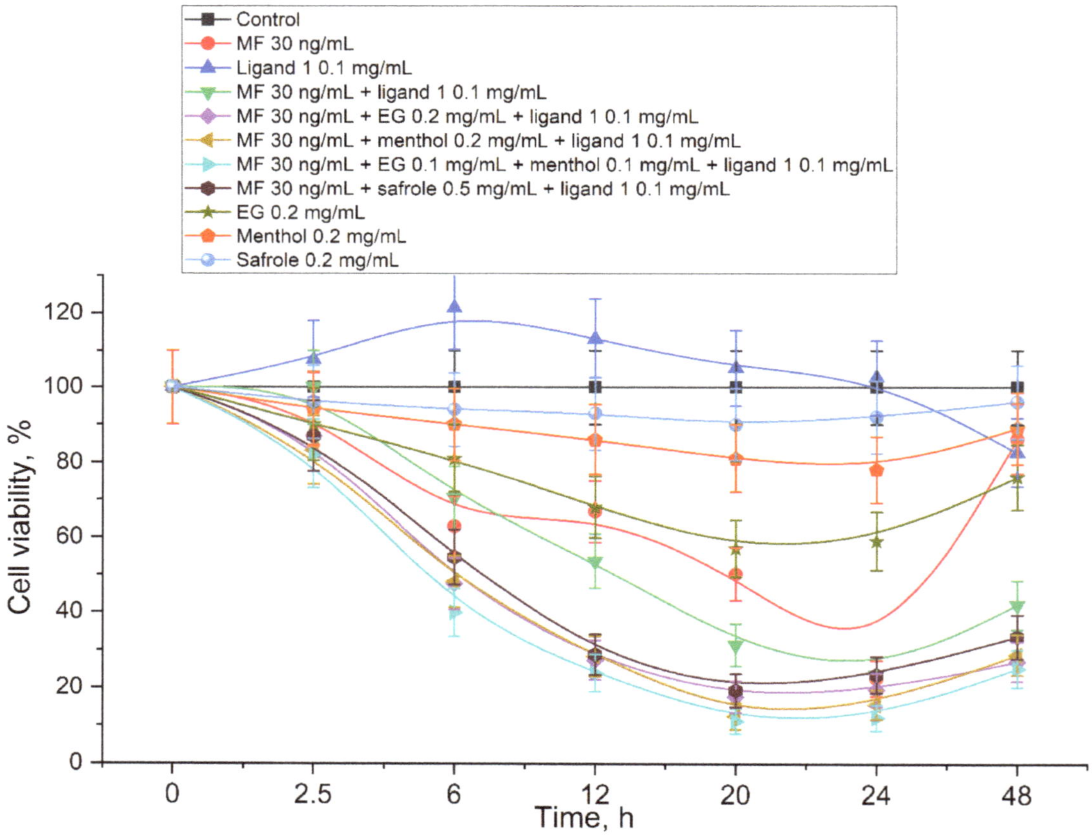

Figure 7. Cell viability of *B. subtilis* as a function of incubation time with drug samples. Major active component is MF; drug loading system—ligand 1 (amCD-triMan); adjuvants—EG, menthol, and safrole. CFU (0 h) = 3 × 10^6. pH 7.4 (0.01 M PBS), 32 °C. The results are presented as the average of three parallel CFU measurements ± SD.

The following is an experiment on points 1–3, which is a model of the action of a combined drug (MF + delivery system + adjuvant) in comparison with control and simple MF in the cell medium. The dependence of the number of surviving cells on the incubation time with MF, MF in the ligand, and MF enhanced with adjuvants EG, safrole, and menthol as well as with EG, safrole, and menthol themselves are presented in Figure 7 and Figure S7. By itself, the drug-free delivery system has practically no effect on cell growth (Figure 7). However, with loaded MF, positive effects of the conjugate on MF action are observed: amCD-triMan does not have a statistically significant effect on MF action up to 24 h; however, with prolonged observation >36–48 h, the effect of MF ends due to MF destruction [106]; on the contrary, MF in a complex form continues to be active. Similar results are described for polymer and 3D matrix structures of cyclodextrins in [104]. «Heavier» mannosylated PEI10 (ligand 3a, Figure S7) makes the antibiotic more active in comparison with the simple form at a concentration of 30 ng/mL, most probably due to the increase in the adsorption degree. The addition of adjuvants (0.1 mg/mL), especially EG and menthol, increases the number of defeated pathogen cells and MF activity by 20–30%, while EG, safrole, and menthol themselves have only a marginal effect on the cells' viability. In addition, the effect of adjuvants is noticeable even with long observation > 24 h, when the active concentration of MF is insufficient to inhibit bacterial growth; however, in systems with adjuvants, MF continues to restrain bacterial reproduction. The effects of both molecular containers and adjuvants as antibiotic enhancers are vividly demonstrated (Figure 7). The triple formulation MF+ligand+adjuvant proved to be a particularly powerful antibacterial agent.

To clarify the mechanisms of action (membrane permeability and efflux inhibition) of the components, CLSM and fluorescent studies of the interaction of drugs with bacteria were carried out.

2.4.2. Confocal Laser Scanning Microscopy (CLSM)

This paragraph shows the role of EG adjuvant and molecular containers and the mechanism of their action together and separately. Namely, we studied the effect of ligand and adjuvant on membrane permeability for an antibiotic and reflux inhibitor; therefore, the purpose of the paragraph is to visualize these processes using CLSM. Dox was taken as the model antibiotic as a widely used drug and fluorescent dye in the visible region, which is necessary for registration on a confocal microscope with laser excitation.

We studied six samples:

(1) Control Dox to study the penetration and accumulation of free matter in cells;
(2) Dox in a polymer ligand to study the effect of polymer on the adsorption efficiency on the cells and penetration of Dox;
(3) Dox in a polymer ligand with grafted CD tori to clarify the role of CD tori in the creation of defects in the membrane and increased Dox (LC) capacity;
(4) Dox with EG as an agent enhancing the membrane permeability and efflux inhibitor;
(5) Dox with preincubated EG only as efflux inhibitor so as to exclude the effect of adjuvant on the membrane permeability due to the desorption of EG;
(6) Combined formulation Dox + ligand with grafted CD tori + EG.

Confocal images of the interaction of Dox (red) and FITC-labeled ligands (green) with *B. subtilis* are shown in Figure 8 (full version—Table S1).

Free Dox accumulates weakly in cells (Figure 8a, Table S1). The maximum concentration among the studied times is reached after 15 min. Over time, Dox is released from the cell by pumping proteins in a process of efflux. Therefore, the drug practically does not accumulate in the cells during the observation period of 90 min.

The effect of adjuvant EG on the accumulation of Dox. When an adjuvant is included in the system simultaneously with Dox, an increase in the degree of Dox accumulation in bacterial cells is observed (Figure 8b,c, Table S1) due to the occurrence of defects in the cell membrane [17] and inhibition of pump proteins [10,11]. The effect of gradual perforation of the cell membrane was confirmed by a decrease in CFU over time during incubation of cells with EG (Figure S8a). When EG acts in two directions (creates defects in the membrane and inhibits pump proteins), the effect of enhancing drug penetration increases over time (Figure 8b, Table S1). However, at the same time, during the observation, EG did not have enough time to block the efflux of proteins because the effluxed Dox was observed in solution.

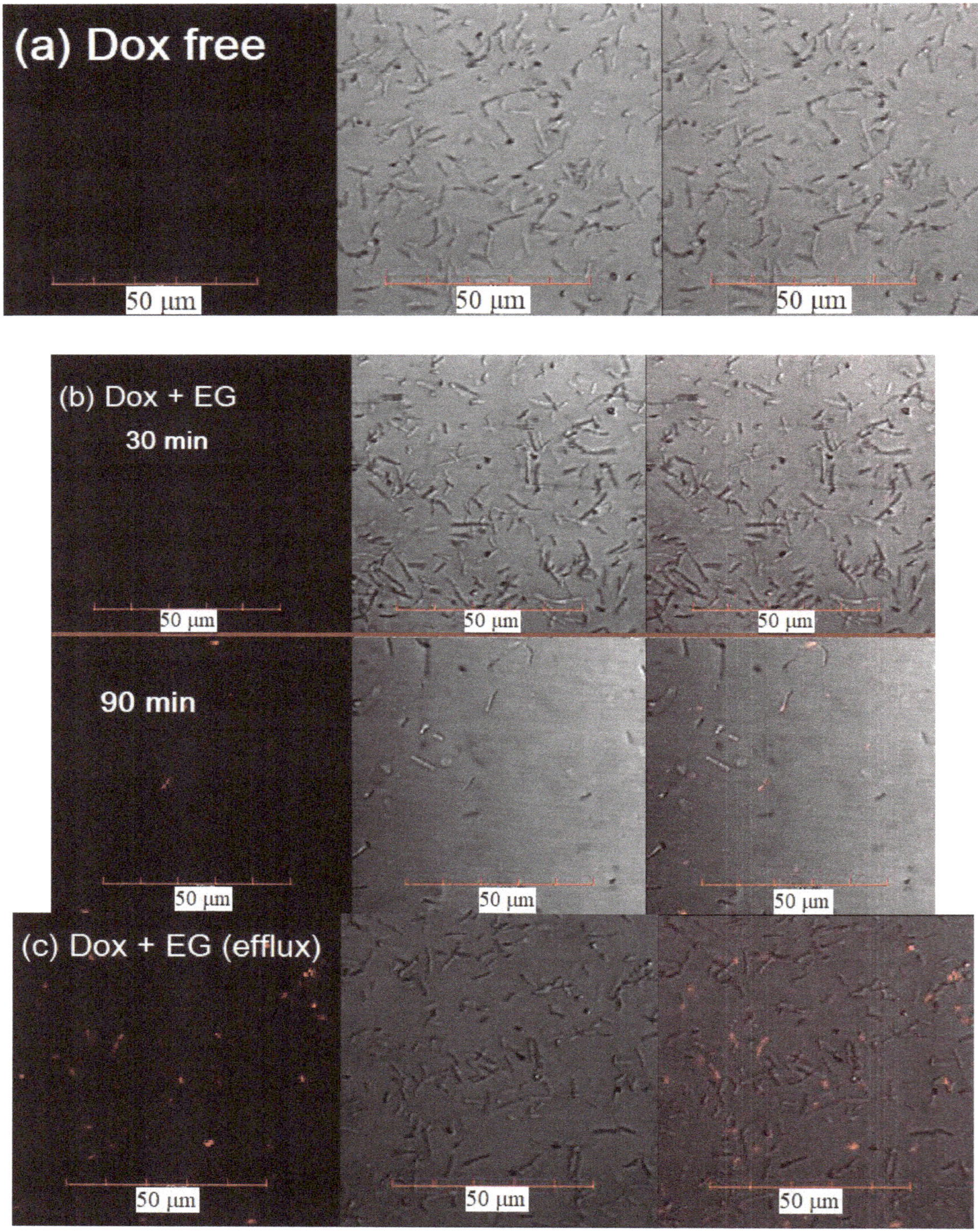

Figure 8. *Cont.*

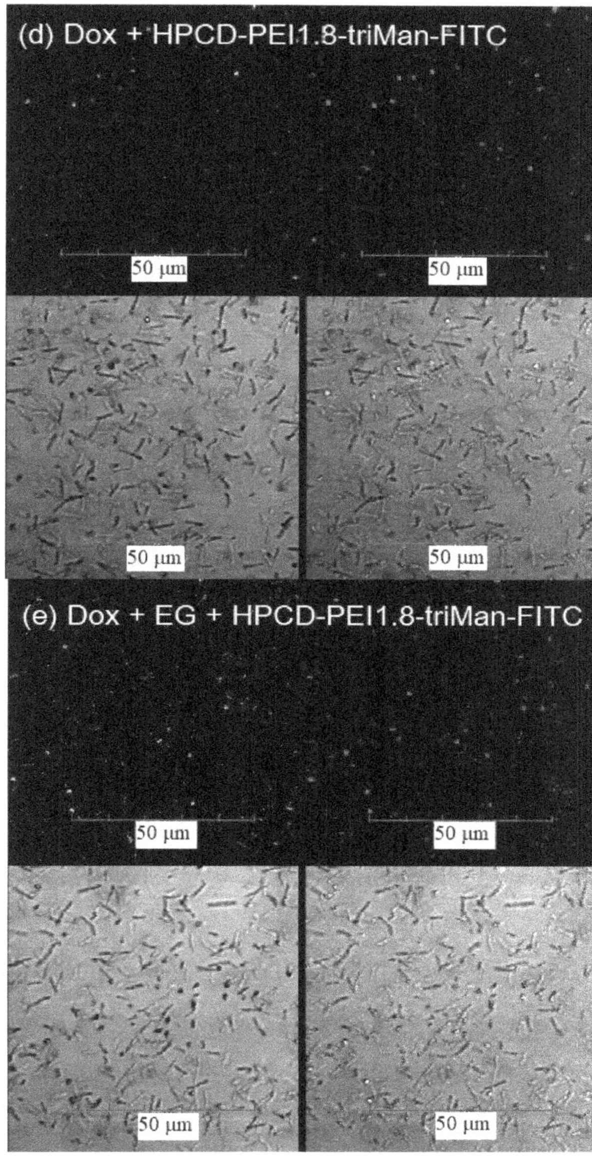

Figure 8. Confocal laser scanning images of *B. subtilis* with adsorbed and absorbed Dox and FITC-labeled ligands based on PEI, HPCD, and triMan. (**a**) Dox-free. (**b**) Simultaneously added Dox + EG (0.1 mg/mL) as a membrane destroyer and protein efflux inhibitor. (**c**) Dox + EG (0.1 mg/mL) only as a protein efflux inhibitor, where EG is added in advance and then washed using SDS. (**d**) Dox + FITC-HPCD-PEI1.8-triMan. (**e**) Dox + EG + FITC-HPCD-PEI1.8-triMan. The scale segment is 50 μm (division value is 10 μm); 30 min of incubation (except for point (**b**)). C_0(Dox) = 1.3 μg/mL. C_0(FITC) = 0.2 μg/mL; 3–4 channels are shown: red, Dox; green, FITC; gray, transmission light mode; and overlay. Emission fluorescence spectra of FITC and Dox obtained by CLSM are presented in Figure S10, λ_{em} = 488 nm (multiline Argon laser).

Efflux inhibition by EG without its effect on the membrane permeability. EG preincubated with cells for an hour and then removed from the solution and desorbed from the membrane (by cells washing with 0.01% SDS, a control turbidimetry and FTIR experiments showing that SDS in 0.01% has no effect on the membrane; results are presented in Figures S8b and S9 and in Methods section) can act primarily as an inhibitor of efflux proteins and, a to lesser extent, destroy the membrane. The maximum effect of EG inside the cells and washed outside is reached after 30 min (Figure 8c, Table S1). EG effectively "turned off" the pump proteins and a significant amount of Dox was able to penetrate into the cells Visually, the presence of Dox outside the cells is noticeably less than in the previous case, which means that the efflux efficiency is achieved in a few tens of minutes, while the EG starts to increase the membrane permeability immediately (Figure 8b), which is consistent with the change of A600 from incubation time (Figure S8a).

The effect of molecular containers (ligands) on the accumulation of Dox. As discussed above, the use of a molecular container is expected to increase the degree of adsorption and concentration of the drug in the vicinity of the cells surface, penetration, and stabilization and maintain the concentration level above the threshold for a long time (at least a few days) According to experimental data, due to the adsorption of polymer particles on the cell membrane, the accumulation of Dox increases with a time of 0–15–30–90 min (Figure 8d,e, Table S1). The ligand effect is especially noticeable when comparing micrographs of free Dox accumulation (Figure 8a,d).

The greatest effect is achieved for the polymer ligand 2b grafted with HPCD tori since Dox is more efficiently loaded into the hydrophobic cavities of the CD, as earlier reported for similar conjugates [25,86,106]. Furthermore, CD additionally enhances penetration through the membrane [52].

The *additive effect* is observed in the system (Dox + EG + FITC-HPCD-PEI1.8-triMan), in which the penetration of Dox increases due to the addition of the above effects, while the greatest contribution is made by efflux inhibitor EG. Moreover, the ligand loads both EG and Dox, and since the distribution of components is uniform for such double-inclusion complexes (earlier shown for levofloxacin and EG [25]), both EG and Dox molecules are likely to be next to each other at once. The first one will create a defect in the membrane where the target substance, Dox, will penetrate directly through the hole formed nearby Thus, the complex formulation antibiotic+adjuvant in a drug delivery system is an effective way to accumulate the drug in the target bacteria.

2.4.3. Fluorescent Determination of MF and EG Adsorbed and Absorbed by Cells

CLSM allows semi-quantitative observation of the studied processes of increased penetration due to defects in the membrane and inhibition of efflux. Fluorescence spectroscopy was used to quantify the effects of EG. The essence of the experiment was to incubate cells with MF or MF+EG during 14 h and then study the distribution of substances inside/on the cell surface or in solution depending on the time and correlate the results with the visualization obtained by CLSM.

Table 5 shows the concentrations of MF and EG depending on the incubation time of substances with *B. subtilis*. The concentration of MF in the solution decreases due to adsorption on the surface of the bacterial cell wall and penetration of fluoroquinolone into the cells. At the same time, the concentration of absorbed MF did not exceed 2.8 µM and decreased due to efflux, which was observed also in a confocal microscope.

However, EG accelerates the penetration of MF due to increased membrane permeability, and in 15 min, the maximum concentration of the antibiotic is reached inside the cells (4.4 µM). EG does not affect maximum time statistically significantly. Thus, EG allowed the accumulation of MF by 60% more.

Table 5. Concentrations of MF and EG in solution adsorbed on the cell wall surface and absorbed by *B. subtilis* cells (10^6 CFU) depending on the composition of the system and incubation time. Incubation in medium LB (pH 7.2), 37 °C. Fluorescent detection of MF and EG: λ_{exc} = 280 nm, λ_{emi}(EG) = 320 nm, λ_{emi}(MF) = 465 nm–22 °C.

Time	PEI10-triMan 0.5 mg/mL + MF 47 µM			PEI10-triMan 0.5 mg/mL + MF 47 µM + EG 6 mM					
	C in Solution	C Adsorbed on the Cell Wall Surface	C Absorbed Inside the Cell	C in Solution		C Adsorbed on the Cell Wall Surface		C Absorbed Inside the Cell	
	MF, µM			MF, µM	EG, mM	MF, µM	EG, mM	MF, µM	EG, mM
0 min	47	0	0	47	6	0	0	0	0
15 min	29 ± 3	15 ± 3	2.8 ± 0.3	15 ± 2	0.20 ± 0.05	27 ± 3	5.8 ± 0.3	4.4 ± 0.2	0.030 ± 0.005
45 min	15 ± 2	30 ± 4	1.6 ± 0.2	10 ± 1	0.10 ± 0.02	34 ± 4	5.9 ± 0.3	3.3 ± 0.2	0.029 ± 0.006
2 h	10 ± 2	35 ± 4	1.2 ± 0.2	12 ± 1	0.10 ± 0.02	32 ± 4	5.9 ± 0.1	3.1 ± 0.2	0.026 ± 0.004
14 h	12 ± 2	33 ± 4	1.7 ± 0.3	6 ± 1	0.010 ± 0.005	40 ± 5	6.0 ± 0.1	0.5 ± 0.1	0.0010 ± 0.0005

Pump proteins work in cells, throwing MF and EG out. Approximately 0.5% of the total concentration of EG is absorbed, and the remaining molecules are adsorbed on the membrane. EG absorbed inside the cells inhibits pump proteins; however, this does not happen instantly but after tens of minutes, as shown using CLSM. After inhibition of efflux, EG allows maintaining twice the concentration of MF inside the cells for several hours. After prolonged incubation due to significant damage to the cell membrane and partial lysis, substances leak out (Figure S8a). A significant part of the cells in the EG system was destroyed, while free MF could not completely inhibit the growth of the number of bacteria. Thus, a way was found to increase the effectiveness of antibiotics by several times through the use of polymer ligands and adjuvants. The obtained data are consistent with CLSM data, which link the mechanism of penetration and efflux of Dox and MF.

2.5. Drug Delivery System Interactions with Macrophages

The key experiment demonstrating the therapeutic potential of the developed systems is the study of the interaction of delivery systems with macrophages. Seven days after adding PMA, we derived macrophage-like cells from THP-1 cell line as described before [88]. Most of these cells were CD206-positive according to immunocytochemical evaluation (Figure 9). We also determined that CD206-positive cells effectively phagocytosed the FITC-labeled polymer as shown in Figure 9. Cytometry assay determined that 95.5% macrophage-like cells were FITC-positive after adding HPCD-PEI1.8-triMan, and 61.7% were FITC-positive after adding HPCD-PEI1.8-Man (Figure 10). Primary culture of human dermal fibroblasts (HDF) was used as a negative control for CD206 receptors. Figures S11 and S12 show immunocytochemical evaluation of CD206, phagocytosis assay, and flow cytometry data on FITC+ distribution of fibroblasts. Fibroblasts express CD206 in small amounts and are able to non-specifically phagocytize mannosylated particles. Cytometry assay determined that 17.4% of human dermal fibroblasts (CD206-negative) cells were FITC-positive after adding HPCD-PEI1.8-triMan, and 4.3% were FITC-positive after adding HPCD-PEI1.8-Man (Figure S12). The phagocytic activity of fibroblasts against polymer mannosylated particles is low, which confirms the specificity of CD206 macrophage delivery systems to mannose recipes. Thus, the developed drug carriers fulfilled three standards: high macrophage uptake, high drug-carrying capacity, and prolonged and enhanced antibacterial effect with adjuvants, which determines the potential applicability for the use of complex combined drugs for medical purposes.

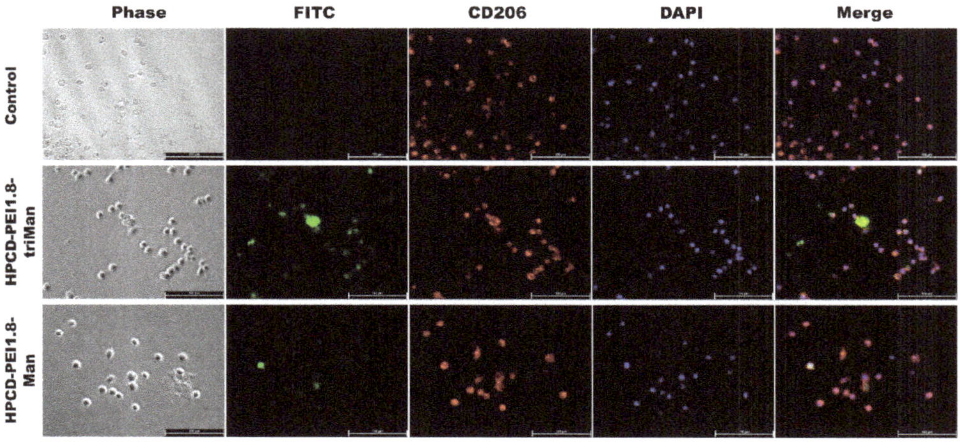

Figure 9. Immunocytochemical evaluation of CD206 in THP-1-derived macrophage-like cells (red channel). Phagocytosis assay with HPCD-PEI1.8-triMan and HPCD-PEI1.8-Man conjugated with a FITC fluorescent label after incubation for 40 min (green channel). Phase contrast microscopy, fluorescent microscopy, blue channel—nuclei stained with DAPI. Scale bar, 100 μm.

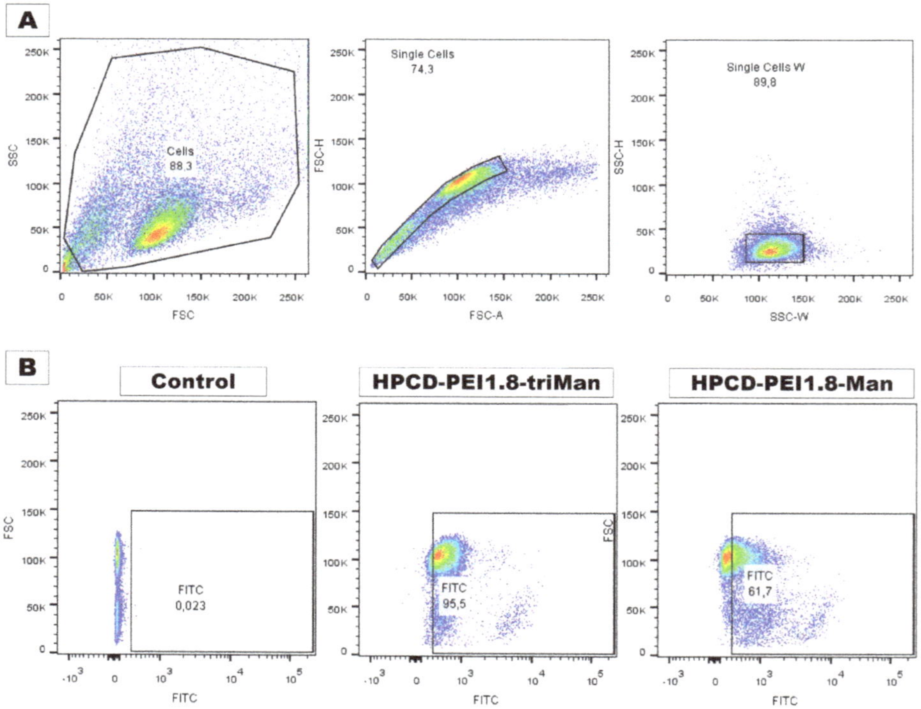

Figure 10. Flow cytometry of macrophage-like cells. (**A**) Gating strategy. Selection of single cells by forward-scattered (single cells) and side-scattered light (single cells W). (**B**) Flow cytometry data on FITC macrophage-fluorescence distribution for cells preincubated with HPCD-PEI1.8-triMan or HPCD-PEI1.8-Man and control cells.

3. Materials and Methods

3.1. Reagents

Carbonyldiimidazole–CDI was obtained from GL Biochem (Shanghai, China) Ltd. via an intermediary Himprocess (Moscow, Russia). ConA was purchased from Paneco-ltd (Moscow, Russia). D-mannose, PEI 1.8 kDa and PEI 10 kDa (branched); and fluorescein isothiocyanate (FITC), doxorubicin, DMF, DMSO, Et_3N, 2-hydroxypropyl-β-cyclodextrin (HPCD), 1 M 2,4,6-trinitrobenzenesulfonic acid, and moxifloxacin hydrochloride were obtained from Sigma Aldrich (St. Louis, MI, USA). Mono-(6-(1,6-hexamethylenediamine)-6-deoxy)-β-cyclodextrin (amCD) was supplied by Dayang Chem (Hangzhou, China) Co., Ltd. Mannotriose-di-(*N*-acetyl-D-glucosamine) was obtained from Dayang Chem (Hangzhou) Co., Ltd. Other chemicals, namely $NaBH_4$, salts, and acids, were from Reakhim Production (Moscow, Russia). All chemicals were of analytical grade or higher and were used without additional purification.

3.2. Synthesis and Purification of Mannosylated Conjugates

Activated HPCD was synthesized as described earlier [25]. Briefly, the slow drop-by-drop addition of HPCD solution in water 1 mL (0.4 mM) to the 10 mL CDI solution in DMSO (1.5 mM) led to the formation of the HPCD–CDI_7 composition at the output.

Ligands **1** (Figure 1a), **2a**, and **3a** (Figure 1b): Three specimens containing (1) 180 mg of PEI1.8 and 580 mg of triMan-GlcNAc_2, (2) 100 mg of PEI10 and 500 mg of triMan-GlcNAc_2, and (3) 150 mg of amCD and 250 mg of triMan-GlcNAc_2 were dissolved in 5 mL 0.01, 0.1, and 0.001 M HCl, respectively, followed by immediate alkalinization (1 M NaOH) to pH 9–10 and incubation at 50 °C for 4 h (here and further, the Biosan ES-20/60, 160 rpm shaker incubator was used) and subsequent reduction of Shiff bases. Samples of solutions from systems 2 and 3 were taken: conjugates **2a** and **3a**. Further crosslinking of PEI amino groups in systems 2 and 3 with activated HPCD was carried out by addition of 7- and 12-fold molar excess of HPCD–CDI_7 relative to PEI chains, respectively, followed by incubation at the same temperature for an hour. Samples **2b** and **3b** were obtained. Conjugates purification was carried out by dialysis at 22 °C against water in three stages of 12 h with water replacement: ligands **2a**, **2b**—cut-off 3 kDa; ligands **3a**, **3b**—cut-off 6–8 kDa; ligand **1**—cut-off 2 kDa. All samples were freeze-dried for two days at −60 °C (Edwards 5, BOC Edwards, UK). The purification of the samples was performed by HPLC gel filtration in a Knauer chromatography system (Knauer, Berlin, Germany) on Diasfer-110-C18 column (BioChemMack, Moscow, Russia): grains—6 μm, size 4 × 150 mm. The eluent was CH_3CN-H_2O (90:10, v:v); the elution rate was 0.8 mL/min, 25 °C. Due to high polydispersity of the conjugates, fractions corresponding to maxima ± 0.5 min by retention time were used in the experiments. The chromatogram of the resulting conjugates is shown in supplemental materials (Figure S4). The degree of mannosylation was calculated according to spectrophotometric titration of amino groups (before and after mannosylation) with 2,4,6-trinitrobenzenesulfonic acid [25,101]. Mainly, primary amino groups were detected.

3.3. FTIR Spectroscopy

ATR-FTIR spectra of samples' solutions were recorded using a Bruker Tensor 27 spectrometer equipped with a liquid-nitrogen-cooled MCT (mercury cadmium telluride) detector. Samples were placed in a thermostatic cell BioATR-II with ZnSe ATR element (Bruker, Germany). The IR spectrometer was purged with a constant flow of dry air (Jun-Air, USA). FTIR spectra were acquired from 900 to 3000 cm^{-1} with 1 cm^{-1} spectral resolution. For each spectrum, 50–70 scans were accumulated at 20 kHz scanning speed and averaged. Spectral data were processed using the Bruker software system Opus 8.2.28 (Bruker, Germany), which includes linear blank subtraction, baseline correction, differentiation (second order, 9 smoothing points), min–max normalization, and atmosphere compensation [25,118]. If necessary, 5-point Savitsky–Golay smoothing was used to remove noise. Peaks were identified by standard Bruker picking-peak procedure.

3.4. NMR Spectroscopy

Next, 12 mg of the sample was dissolved in 700 μL of D_2O. ^1H NMR spectra of the solutions were recorded on a Bruker Avance 400 spectrometer (Germany) with an operating frequency of 400 MHz. Chemical shifts are given in ppm on the δ scale relative to hexamethyldisiloxane as an internal standard. The analysis and processing of NMR spectra were performed in the program MestReNova v. 12.0.0–20080 (Mestrelab Research S.L.).

3.5. Dynamic Light Scattering (DLS)

The particle sizes and zeta potentials were measured using a Zetasizer Nano S «Malvern» (England) (4mW He–Ne laser, 633 nm, scattering angle 173°) in 0.01 M PBS (pH 7.4). The experiment was performed three times in a temperature-controlled cell at 25 °C Autocorrelation functions of intensity fluctuations of light scattering were obtained using the correlation of the Correlator system K7032-09 «Malvern» (Worcestershire, UK). Polydispersity indices were calculated based on particle size distributions (Table 1). Experimental data were processed using «Zetasizer Software» (v. 8.02).

3.6. Mathematical Calculations and Equations

(1) Calculation of the dissociation constants ConA ligand. Consider the equilibrium: ConA + n ligand ↔ ConA nligand, where $K_d = [\text{ligand}]^n \cdot [\text{ConA}]/[\text{ConA nligand}]$.

(2) First, fitting of the curves of change of the analytical signal ξ (peak intensity or peak position in FTIR native or second-derivative spectra) versus concentration of the ligand was carried out using the Hill equations: (1) $\xi = \xi_\infty \cdot [\text{ligand}]^n/([\text{ligand}]^n + K^n)$, where ξ_∞—horizontal asymptote; (2) $\xi = \xi_0 + (\xi_\infty - \xi_0) \cdot [\text{ligand}]^n/([\text{ligand}]^n + K^n)$ Second, calculation of n and K_d values was performed by Hill's linearization in n-binding site model: $\lg (\theta/(1 - \theta)) = n \cdot \lg [L] - \lg K_d$, where $\theta = | (\xi - \xi_0)/(\xi_\infty - \xi_0) |$ is a fraction of the bound substance. n in Hill equation—number of ligand molecules per one ConA subunit.

(3) Calculation of the dissociation constants drug ligand. Consider the equilibrium: drug + ligand ↔ drug ligand, where $K_d = [\text{ligand}]\,[\text{drug}]/[\text{drug ligand}]$. The calculations were performed similarly to the above point with the difference in calculating the number of drug molecules (MF) per conjugate (ligand) molecule. The number of MF molecules retained by one conjugate molecule in CD tori (N_{CD}) is assumed to be equal to 90% of the total number of CD tori. The number of MF molecules held by one conjugate molecule by a polymer matrix ($N_{polymer}$) is calculated based on the drug ligand ratio, when the degree of binding θ is 95%, minus the drug molecules in the CD tori.

(4) Calculations for the following items are performed when $C_{ligand\_total} = 0.0001$ M, $C_{MF\_total} = 0.001$ M.

(5) Entrapment efficiency of MF into conjugates [18,24,119]: EE (%) = 100 · (MF amount in form of complex with ligand)/(total MF amount). $K_d = N \cdot (C_{MF\_total} - C_{MF\_bounded}) \cdot (C_{ligand\_total} - C_{MF\_bounded}/N)/C_{MF\_bounded}$. From the last equation, find $C_{MF\_bounded}$ and then, $EE = C_{MF\_bounded}/C_{MF\_total}$.

(6) Loading capacity of drug MF into conjugates [18,24,119]: LC (%) = 100 · (mass of loaded MF)/(mass of sample) = 100 · (number of drug molecules per ligand molecule) · (molar mass of MF)/(molar mass of ligand). The number of drug molecules per ligand molecule is calculated in paragraph 2.

(7) Statistical analysis of obtained data was carried out using the Student's *t*-test Origin 2022 software (OriginLab Corporation). Values are presented as the mean SD of three experiments.

3.7. Confocal Laser Scanning Microscopy

B. subtilis cells (NCIB 8054, overnight culture, 10^7 CFU) were centrifuged twice (Eppendorf centrifuge 5415C, 10 min, 12,000× *g*) and washed with 0.01 M PBS from the culture medium. Next, the cells were incubated for 15–30–90 min at 37 °C with 500 μL of solutions indicated in the caption to Figure 8 and Table S1. A special case was a pre-incubation of EG with cells for 1 h and washing with 0.01% SDS and a two-time washing-centrifugation of PBS and subsequent incubation with MF 15-30-90 min to study the efflux due to EG penetrating into the cells. The cells were centrifuged twice with PBS washing. The cell centrifuge was suspended in 200 μL of PBS, followed by the addition 100 μL of a 16% agarose solution at 45 °C to solidify the cell suspension in the wells of a fluorescent plate.

Fluorescence images were obtained by confocal laser scanning microscope (CLSM) Olympus FluoView FV1000 equipped with both spectral version scan unit with emission detectors and transmitted light detector. CLSM is based on a motorized inverted microscope Olympus IX82. The excitation wavelength 488 nm (multiline Argon laser) and dry objective lens Olympus UPLSAPO 40X NA 0.90 were used for the measurements. Laser power, sampling speed, and averaging were the same for all image acquisitions. The scan area was 80×80 μm^2. FITC and Dox fluorescence were collected using the emission windows set at 510–540 nm and at 590–680 nm, respectively (emission fluorescence spectra are presented on Figure S10). The signals were adjusted to the linear range of the detectors. Olympus FV10 ASW 1.7 software was used for acquisition of the images. FITC fluorescence is shown in green, Dox is red, and the image on the light is gray.

3.8. Determination of MF and EG Concentrations Absorbed and Adsorbed by Cells

B. subtilis cells (NCIB 8054, overnight culture, 10^6 CFU) were centrifuged three times (Eppendorf centrifuge 5415C, 10 min, 12,000× *g*) and washed with 0.01 M PBS from the culture medium. Next, the cells were incubated for 15–45–120 min and 14 h at 37 °C with 200 μL of solutions indicated in Table 4. The cells were centrifuged (10 min, 12,000× *g*) and washed with PBS. The supernatant was used to determine MF and EG not bound to cells (C in solution). The cell centrifuge was suspended in 200 μL of PBS containing 0.02–0.05% SDS, followed by incubation at 37 °C 30 min. By centrifugation (10 min, 12,000× *g*), 2 fractions were obtained: a supernatant containing EG and MF corresponding to adsorbed substances (C adsorbed) and a cellular centrifugate washed from non-absorbed substances. The second fraction, resuspended in 1000 μL of PBS, was exposed to an ultrasonic probe for 30 min. Then, the amount of absorbed substances was determined in the cell lysate (C absorbed). Cells were also incubated with ligands without MF and EG, and these data were used as a control and background. Fluorescence of MF and EG was measured using a Varian Cary Eclipse spectrofluorometer (Agilent Technologies, Santa Clara, CA, USA) at 20 °C: λ_{exc} = 280 nm, λ_{emi}(EG) = 320 nm, λ_{emi}(MF) = 465 nm, slit 10/10 nm.

Control experiment on the effect of SDS on cell membrane permeability and drug absorption.

1. First, 2 mL of cell suspension (5×10^6 CFU/mL) was added to 2 cuvettes, and then, 20 mL of 10% SDS was added to the first one. Optical absorption at 600 nm as a correlate of CFU and membrane integrity was measured for two hours. Relative to the control solution, the CFU change did not exceed 10% of the initial value (Figure S8b).
2. FTIR spectroscopy data. The experiment consisted of incubating cells with Dox with and without the addition of SDS, then separating the supernatant and cells and determining the integrity of the membrane; then, Dox content was determined. The observed parameter is the intensity and positions of the characteristic peaks of the aromatic system of doxorubicin. In the case of a significant effect of SDS on membrane permeability, changes in peak intensity would be observed in the IR spectrum.

Technique. The cells were washed from the medium as described above, then suspended in 500 μL PBS (5×10^7 CFU/mL). Next, the cell suspension was incubated at 37 °C for two hours with 0.1 mg/mL of Dox-MCD (1:2) with and without addition of SDS to the final concentration 0.1%. By centrifugation (10 min, 12,000× *g*) 2 fractions were obtained with subsequent registration of FTIR spectra: a cellular centrifugate suspended in 50 μL PBS (Figure S9a,b) and the supernatant. The intensity of the characteristic peaks of doxorubicin (C–C_{arom} 1400–1550 cm^{-1}) practically do not depend on the addition of SDS both inside the cells (Figure S9b) as well as in the supernatant.

On the other hand, the FTIR spectra of bacteria (Figure S9a) contain information about the structural organization of cells [88,119]. Lipids and lipoproteins are the contribution of valence vibrations of CH_3–, CH_2– groups (2820–2980 cm^{-1}), valence oscillations C=O (amide 1), and deformation N-H (amide II)—for phospholipids 1240 cm^{-1} PO_2^-. According to the literature data [119], these peaks are the most sensitive to lipid membrane state, which was proven by cells incubation with surfactants. Since the characteristic peaks of the lipid membrane, surface proteins, and glycans practically do not change from the addition of 0.1% SDS, it can be assumed that there was no violation of the integrity of the membrane. Otherwise, amides I, II, and III (very sensitive to membrane state, 1450–1700 cm^{-1}) would have decreased [119] due to the greater availability of IR radiation of internal proteins and DNA. Therefore, the concentration of SDS studied with a 2–10-fold additional excess (0.1%) for 1–2 h does not have a significant effect on the amount of the drug absorbed by the cells.

3.9. Antibacterial Activity of MF in Free, Complex, and Enhanced with Adjuvants Forms

The strain used in this study was *Bacillus subtilis* (NCIB 8054) from the National Resource Center All-Russian collection of industrial microorganisms SIC "Kurchatov Institute"). The culture was cultivated for 18–20 h at 37 °C to CFU $\approx 1 \times 10^7$ (determined by A_{600}) in liquid nutrient medium Luria–Bertani (pH 7.2) with stirring at 160 rpm. The study of the antibacterial effect of MF in liquid media was carried out using an overnight bacterial culture. The cell suspension was diluted to CFU $\approx 3 \times 10^6 – 5 \times 10^6$ with LB liquid medium (pH 7.2), and 400 μL of sample (indicated in Figures 7 and S4) was added to 4 mL of culture. The resulting systems were incubated at a temperature of 32 °C and stirred at 160 rpm in an ES-20 BIOSAN incubator shaker (Latvia) for 2–4 days. Aliquots were selected at certain intervals to measure absorption at 600 nm to determine the number of viable cells.

3.10. Experiments with Macrophages

3.10.1. Cell Lines and Modelling

For the macrophages CD206+ phagocytose assay, a human monocyte cell line THP-1 (or human dermal fibroblasts (HDF) used as a negative control for CD206-) was used. Cells were obtained from the bank of cell lines of Lomonosov Moscow State University. THP-1 cells were cultured on RPMI-1640 (Gibco) supplemented with GlutaMAX™ Supplement (Gibco) buffered with 10 mM HEPES pH 7.4 containing 10% heat-inactivated FBS (Gibco) and 1% antimycotic antibiotic (HyClone) at 37°C and 5% CO_2. Macrophage-like cells were derived by adding to THP-1 (0.5 million cells/mL) and 100 nM phorbol 12-myristate 13-acetate (PMA, p8139, Sigma) for 72 h. After 72 h, the medium was replaced, and cells were cultured for another 96 h.

3.10.2. Phagocytosis Assay

Dry samples of HPCD-PEI1.8-triMan and HPCD-PEI1.8-Man conjugated with a FITC fluorescent label were diluted in PBS (PanEco, Moscow, Russia) with 10% DMSO. Macrophage-like cells were washed with Hanks' solution, and HPCD-PEI1.8-triMan and HPCD-PEI1.8-Man were added at final concentration of 0.10 mg/mL and 0.089 mg/mL, respectively, in serum-free media. As a control, we used macrophage-like cells without adding polymer. For cytometry assay, cells were incubated for 40 min, washed trice with Hank's solution, and scraped in Versen's solution. Cell suspension was centrifuged at 300× *g* for 5 min at +4 °C, and the cell pellet was resuspended in a PBS solution. For immunocytochemistry assay, cells were incubated for 4 min, washed with PBS, and fixed with 4% formaldehyde solutions.

3.10.3. Cytometry

Cytometry was performed on a BD FACSAria™ III Cell Sorter instrument, the fluorescence of the FITC-conjugated polymer captured by the cells was analyzed, the change in the percentage of cells fluorescing in the FITC channel was determined, and cells not incubated in the sample solution were evaluated (control).

3.10.4. Immunocytochemistry

To block nonspecific binding sites, cells were incubated with a 10% solution of normal goat serum in PBS with the 1% bovine serum albumin BSA for 1 h at RT. Then, the samples were incubated with a solution of anti-CD206 antibodies (ab64693, Abcam, 1:100) or rabbit polyclonal control IgG (910801, Biolegend) as a control for 2 h at RT and subsequently with goat-anti-rabbit antibody conjugated with Alexa594 (A11037, Invitrogen, 1:1000). The nuclei were labeled with DAPI (Sigma-Aldrich, St. Louis, MO, USA). Samples were analyzed with a Leica DM6000B fluorescent microscope equipped with a Leica DFC 360FX camera (Leica Microsystems GmbH, Wetzlar, Germany).

4. Conclusions

Currently, a promising direction in the treatment of complicated infectious diseases of the respiratory tract (including resistant tuberculosis) is the development of complex combined forms of drugs based on fluoroquinolones. In this paper, two focuses on solving the problem of bacterial resistance are considered: (i) the addressed delivery of fluoroquinolones to macrophages by targeting mannose receptors CD206 and (ii) the selective effect on bacteria (and in the future on cancer cells) due to adjuvant-increased penetration of drugs and inhibition of efflux and, as a consequence, accumulation of drugs in the area of pathogens localization. Firstly, it has been shown that oligo- and polymer conjugates, i.e., mannan grafted with cyclodextrins and mannosylated PEI with grafted CDs, are high-capacity molecular containers capable of loading 8–20% by weight of the therapeutic agent and delivering it to macrophage mannose receptors, which is shown on the model protein ConA and cells. Flow cytometry has shown that annotated polymers are actively absorbed by macrophages: with linear mannose, 62% of cells phagocytized a ligand, and a high-affinity trimannoside conjugate absorbed 95.5% of macrophage cells. Secondly, CLSM was used to study the mechanism of action of an adjuvant as an agent that increases the permeability of the membrane and an efflux inhibitor together and individually as well as a molecular container. Efflux management is important for overcoming multidrug resistance. Therefore, the conducted research opens up new opportunities to influence complex macrophage-associated infections. The data obtained in combination with previous studies form the basis for choosing the optimal delivery system and adjuvant for a specific therapy task and possibly a specific patient.

Supplementary Materials: The following supporting information can be downloaded at: https://www.mdpi.com/article/10.3390/ph15101172/s1, Figure S1. Fourier transform infrared spectra of PEI10-N=CH-triMan and PEI10-triMan. Aqueous solutions. T = 22 °C; Figure S2. Size distribution by volume determined by dynamic light scattering method: (a) amCD-triMan, (b) PEI1.8-triMan, (c) HPCD-PEI1.8-triMan, (d) PEI10-triMan, (e) HPCD-PEI10-triMan. T = 25 °C; Figure S3. Zeta-potential distribution determined by dynamic light scattering method: (a) amCD-triMan, (b) PEI1.8-triMan, (c) HPCD-PEI1.8-triMan, (d) PEI10-triMan, (e) HPCD-PEI10-triMan. T = 25 °C; Figure S4. High-performance liquid chromatogram of mannosylated drug delivery systems. Knauer chromatography system (Knauer, Berlin, Germany) on Diasfer-110-C18 column (BioChemMack, Moscow, Russia): grains—6 μm, size 4 × 150 mm. The eluent was CH_3CN-H_2O (90:10, v:v); the elution rate was 0.8 mL/min, 25 °C; Figure S5. Fourier transform infrared spectra of ConA and ConA complexes with PEI10-triMan. C_0(ConA subunit) = 0.1 mM. Natrium-phosphate buffer solution (0.02 M, pH 6.4). $C(Ca^{2+}) = C(Mn^{2+})$ = 1 mM. T = 22 °C; Figure S6. FTIR spectra of MF and MF complexes with PEI1.8-triMan. 0.01 M HCl. T = 22 °C; Figure S7. Cell viability of *B. subtilis* as a function of incubation time with drug samples. Major active component is MF; adjuvants—EG, menthol and safrole. CFU (0 h) = 8 × 10^6. pH 7.4 (0.01 M PBS), 32 °C; Table S1. Confocal laser scanning images of *B. subtilis* with adsorbed and absorbed Dox- and FITC-labeled ligands based on PEI, HPCD, and triMan. A time comparison (15–30–90 min) and dependence on the presence of the adjuvant EG are presented. The scale segment is 50 μm (division value is 10 μm). 3—4 channels are shown: red, Dox; green, FITC;

gray, transmission light mode; and overlay. Figure S8. The dependence of *B. subtilis* colony-forming units (CFU) on the incubation time with (a) EG (0.1 mg/mL), (b) SDS (0.1%). Medium LB (pH 7.2), 37 °C; Figure S9. Fourier transform infrared spectra of centrifuged *B. subtilis* followed by resuspension in PBS (0.01 M, pH 7.4): (a) cells alone; (b) cells hour-incubated with 0.1 mg/mL Dox-MCD and 0.1 mg/mL Dox-MCD with 0.1% SDS. CFU = 3×10^6. T = 22 °C; Figure S10. Emission fluorescence spectra of FITC and Dox obtained by CLSM. λ_{em} = 488 nm (multiline Argon laser); Figure S11 Immunocytochemical evaluation of CD206 in primary culture of human dermal fibroblasts (HDF) (red channel–low intensity). Phagocytosis assay with HPCD-PEI1.8-triMan and HPCD-PEI1.8-Man conjugated with a FITC fluorescent label after incubation for 40 min (green channel). Phase contrast microscopy, fluorescent microscopy, blue channel—nuclei stained with DAPI. Scale bar, 100 µm; Figure S12. Flow cytometry data on distribution of the number of human dermal fibroblast cells by the intensity of absorbed FITC-labeled conjugates HPCD-PEI1.8-triMan and HPCD-PEI1.8-Man.

Author Contributions: I.D.Z. performed the experiments, analyzed the data, and wrote the manuscript; A.A.E. conducted CLSM experiments; R.A.P. conducted NMR experiments; M.A.V. and O.A.G. conducted and conceptualized experiments with macrophages; N.G.B. conceptualized and conducted experiments on microbiology; E.V.K. analyzed the data, supervised the project, and edited the manuscript. All authors have read and agreed to the published version of the manuscript.

Funding: This research was funded by Russian Science Foundation, grant number 22-24-00604.

Institutional Review Board Statement: Not applicable.

Informed Consent Statement: Not applicable.

Data Availability Statement: The data presented in this study are available in the main text and Supplementary Materials.

Acknowledgments: The work was performed using equipment (FTIR spectrometer Bruker Tensor 27 and Jasco J-815 CD Spectrometer) of the program for the development of Moscow State University.

Conflicts of Interest: The authors declare no conflict of interest.

Abbreviations

CLSM, confocal laser scanning microscopy; CD, cyclodextrin; Dox, doxorubicin; EE, entrapment efficiency; EG, eugenol; HPCD, (2-hydroxypropyl)-β-cyclodextrin; LC, loading capacity; MF, moxifloxacin; PEI, polyethyleneimine; triMan, mannotriose residue.

References

1. Zamora-Mora, V.; Fernández-Gutiérrez, M.; González-Gómez, Á.; Sanz, B.; Román, J.S.; Goya, G.F.; Hernández, R.; Mijangos, C. Chitosan Nanoparticles for Combined Drug Delivery and Magnetic Hyperthermia: From Preparation to in Vitro Studies. *Carbohydr. Polym.* **2017**, *157*, 361–370. [CrossRef] [PubMed]
2. Atkinson, R.L.; Blank, R.C.; Loper, J.F.; Schumacher, D.; Lutes, R.A. Combined Drug Treatment of Obesity. *Obes. Res.* **1995**, *3* (Suppl. 4), 497S–500S. [CrossRef] [PubMed]
3. Jain, K.; Kesharwani, P.; Gupta, U.; Jain, N.K. A Review of Glycosylated Carriers for Drug Delivery. *Biomaterials* **2012**, *33*, 4166–4186. [CrossRef] [PubMed]
4. Jain, N.K.; Mishra, V.; Mehra, N.K. Targeted Drug Delivery to Macrophages. *Expert Opin. Drug Deliv.* **2013**, *10*, 353–367. [CrossRef] [PubMed]
5. Stebbins, N.D.; Ouimet, M.A.; Uhrich, K.E. Antibiotic-Containing Polymers for Localized, Sustained Drug Delivery. *Adv. Drug Deliv. Rev.* **2014**, *78*, 77–87. [CrossRef]
6. Tao, F.; Ma, S.; Tao, H.; Jin, L.; Luo, Y.; Zheng, J.; Xiang, W.; Deng, H. Chitosan-Based Drug Delivery Systems: From Synthesis Strategy to Osteomyelitis Treatment—A Review. *Carbohydr. Polym.* **2021**, *251*, 117063. [CrossRef]
7. Mavromoustakos, T.; Tzakos, A.G. *Supramolecules in Drug Discovery and Drug Delivery*; Humana Press: Totowa, NJ, USA, 2021. ISBN 9781071609194.
8. Le-Deygen, I.M.; Skuredina, A.A.; Kudryashova, E.V. Drug Delivery Systems for Fluoroquinolones: New Prospects in Tuberculosis Treatment. *Russ. J. Bioorg. Chem.* **2017**, *43*, 487–501. [CrossRef]
9. Neyfakh, A.A.; Bidnenko, V.E.; Chen, L.B. Efflux-Mediated Multidrug Resistance in Bacillus Subtilis: Similarities and Dissimilarities with the Mammalian System. *Proc. Natl. Acad. Sci. USA* **1991**, *88*, 4781–4785. [CrossRef]

10. Macêdo, N.S.; Silveira, Z.D.S.; Patrícia, P.; Cordeiro, M.; Douglas, H.; Coutinho, M.; Pinto, J.; Júnior, S.; José, L.; Júnior, Q.; et al. Inhibition of Staphylococcus Aureus Efflux Pump by O-Eugenol and Its Toxicity in Drosophila Melanogaster Animal Model. *BioMed Res. Int.* **2022**, *2022*, 1440996. [CrossRef]
11. Muniz, D.F.; dos Santos Barbosa, C.R.; de Menezes, I.R.A.; de Sousa, E.O.; Pereira, R.L.S.; Júnior, J.T.C.; Pereira, P.S.; de Matos, Y.M.L.S.; da Costa, R.H.S.; de Morais Oliveira-Tintino, C.D.; et al. In Vitro and in Silico Inhibitory Effects of Synthetic and Natural Eugenol Derivatives against the NorA Efflux Pump in Staphylococcus Aureus. *Food Chem.* **2021**, *337*, 127776. [CrossRef]
12. Van Bambeke, F.; Balzi, E.; Tulkens, P.M. Antibiotic Efflux Pumps. *Biochem. Pharmacol.* **2000**, *60*, 457–470. [CrossRef]
13. Li, X.Z.; Nikaido, H. Efflux-Mediated Drug Resistance in Bacteria. *Drugs* **2004**, *64*, 159–204. [CrossRef] [PubMed]
14. Chen, X.Y.; Qian, F.; Wang, Y.Y.; Liu, Y.; Sun, Y.; Bin Zha, W.; Hao, K.; Zhou, F.; Wang, G.J.; Zhang, J.W. Ginsenoside 20(S)-Rh2 Promotes Cellular Pharmacokinetics and Intracellular Antibacterial Activity of Levofloxacin against Staphylococcus Aureus through Drug Efflux Inhibition and Subcellular Stabilization. *Acta Pharmacol. Sin.* **2021**, *42*, 1930–1941. [CrossRef] [PubMed]
15. Keith, J.W.; Pamer, E.G. Enlisting Commensal Microbes to Resist Antibiotic-Resistant Pathogens. *J. Exp. Med.* **2019**, *216*, 10–19. [CrossRef]
16. Hortle, E.; Oehlers, S.H. Host-Directed Therapies Targeting the Tuberculosis Granuloma Stroma. *Pathog. Dis.* **2020**, *78*, ftaa015. [CrossRef]
17. Hemaiswarya, S.; Doble, M. Synergistic Interaction of Eugenol with Antibiotics against Gram Negative Bacteria. *Phytomedicine* **2009**, *16*, 997–1005. [CrossRef]
18. Kayaci, F.; Ertas, Y.; Uyar, T. Enhanced Thermal Stability of Eugenol by Cyclodextrin Inclusion Complex Encapsulated in Electrospun Polymeric Nanofibers. *J. Agric. Food Chem.* **2013**, *61*, 8156–8165. [CrossRef]
19. Ulanowska, M.; Olas, B. Biological Properties and Prospects for the Application of Eugenol—A Review. *Int. J. Mol. Sci.* **2021**, *22*, 3671. [CrossRef]
20. Tadtong, S.; Watthanachaiyingcharoen, R.; Kamkaen, N. Antimicrobial Constituents and Synergism Effect of the Essential Oils from Cymbopogon Citratus and Alpinia Galanga. *Nat. Prod. Commun.* **2014**, *9*, 277–280. [CrossRef]
21. Cardoso, N.N.R.; Alviano, C.S.; Blank, A.F.; Romanos, M.T.V.; Fonseca, B.B.; Rozental, S.; Rodrigues, I.A.; Alviano, D.S. Synergism Effect of the Essential Oil from Ocimum Basilicum Var. Maria Bonita and Its Major Components with Fluconazole and Its Influence on Ergosterol Biosynthesis. *Evid.-Based Complement. Altern. Med.* **2016**, *2016*, 5647182. [CrossRef]
22. Jafri, H.; Banerjee, G.; Khan, M.S.A.; Ahmad, I.; Abulreesh, H.H.; Althubiani, A.S. Synergistic Interaction of Eugenol and Antimicrobial Drugs in Eradication of Single and Mixed Biofilms of Candida Albicans and Streptococcus Mutans. *AMB Express* **2020**, *10*, 185. [CrossRef] [PubMed]
23. Zlotnikov, I.D.; Belogurova, N.G.; Krylov, S.S.; Semenova, M.N.; Semenov, V.V.; Kudryashova, E.V. Plant Alkylbenzenes and Terpenoids in the Form of Cyclodextrin Inclusion Complexes as Antibacterial Agents and Levofloxacin Synergists. *Pharmaceuticals* **2022**, *15*, 861. [CrossRef] [PubMed]
24. Boire, N.A.; Riedel, S.; Parrish, N.M. Essential Oils and Future Antibiotics: New Weapons against Emerging 'Superbugs'? *J. Anc. Dis. Prev. Rem.* **2013**, *1*, 105. [CrossRef]
25. Zlotnikov, I.D.; Kudryashova, E.V. Spectroscopy Approach for Highly—Efficient Screening of Lectin—Ligand Interactions in Application for Mannose Receptor and Molecular Containers for Antibacterial Drugs. *Pharmaceuticals* **2022**, *15*, 625. [CrossRef]
26. Herman, A.; Tambor, K.; Herman, A. Linalool Affects the Antimicrobial Efficacy of Essential Oils. *Curr. Microbiol.* **2016**, *72*, 165–172. [CrossRef]
27. Lichtenstein, E.P.; Liang, T.T.; Schulz, K.R.; Schnoes, H.K.; Carter, G.T. Insecticidal and Synergistic Components Isolated from Dill Plants. *J. Agric. Food Chem.* **1974**, *22*, 658–664. [CrossRef]
28. Samet, A.V.; Shevchenko, O.G.; Rusak, V.V.; Chartov, E.M.; Myshlyavtsev, A.B.; Rusanov, D.A.; Semenova, M.N.; Semenov, V.V. Antioxidant Activity of Natural Allylpolyalkoxybenzene Plant Essential Oil Constituents. *J. Nat. Prod.* **2019**, *82*, 1451–1458. [CrossRef]
29. Jiso, A.; Khemawoot, P.; Techapichetvanich, P.; Soopairin, S.; Phoemsap, K.; Damrongsakul, P.; Wongwiwatthananukit, S.; Vivithanaporn, P. Drug-Herb Interactions among Thai Herbs and Anticancer Drugs: A Scoping Review. *Pharmaceuticals* **2022**, *15*, 146. [CrossRef]
30. Peyron, P.; Vaubourgeix, J.; Poquet, Y.; Levillain, F.; Botanch, C.; Bardou, F.; Daffé, M.; Emile, J.F.; Marchou, B.; Cardona, P.J.; et al. Foamy Macrophages from Tuberculous Patients' Granulomas Constitute a Nutrient-Rich Reservoir for M. Tuberculosis Persistence. *PLoS Pathog.* **2008**, *4*, e1000204. [CrossRef]
31. Sharma, R.; O'Sullivan, K.M.; Holdsworth, S.R.; Bardin, P.G.; King, P.T. Visualizing Macrophage Extracellular Traps Using Confocal Microscopy. *J. Vis. Exp.* **2017**, *2017*, e56459. [CrossRef]
32. Bowdish, D.M.E.; Sakamoto, K.; Kim, M.J.; Kroos, M.; Mukhopadhyay, S.; Leifer, C.A.; Tryggvason, K.; Gordon, S.; Russell, D.G. MARCO, TLR2, and CD14 Are Required for Macrophage Cytokine Responses to Mycobacterial Trehalose Dimycolate and Mycobacterium Tuberculosis. *PLoS Pathog.* **2009**, *5*, e1000474. [CrossRef] [PubMed]
33. Ercoli, G.; Fernandes, V.E.; Chung, W.Y.; Wanford, J.J.; Thomson, S.; Bayliss, C.D.; Straatman, K.; Crocker, P.R.; Dennison, A.; Martinez-Pomares, L.; et al. Intracellular Replication of Streptococcus Pneumoniae inside Splenic Macrophages Serves as a Reservoir for Septicaemia. *Nat. Microbiol.* **2018**, *3*, 600–610. [CrossRef] [PubMed]
34. Gordon, S. The Role of the Macrophage in Immune Regulation. *Res. Immunol.* **1998**, *149*, 685–688. [CrossRef]

35. Valledor, A.F.; Comalada, M.; Santamaría-Babi, L.F.; Lloberas, J.; Celada, A. Macrophage Proinflammatory Activation and Deactivation. A Question of Balance. *Adv. Immunol.* **2010**, *108*, 1–20. [CrossRef] [PubMed]
36. Brundu, S.F.A. Polarization and Repolarization of Macrophages. *J. Clin. Cell. Immunol.* **2015**, *6*, 2. [CrossRef]
37. Cassetta, L.; Pollard, J.W. Repolarizing Macrophages Improves Breast Cancer Therapy. *Cell Res.* **2017**, *27*, 963–964. [CrossRef]
38. Kaku, Y.; Imaoka, H.; Morimatsu, Y.; Komohara, Y.; Ohnishi, K.; Oda, H.; Takenaka, S.; Matsuoka, M.; Kawayama, T.; Takeya, M.; et al. Overexpression of CD163, CD204 and CD206 on Alveolar Macrophages in the Lungs of Patients with Severe Chronic Obstructive Pulmonary Disease. *PLoS ONE* **2014**, *9*, e87400. [CrossRef]
39. East, L.; Isacke, C.M. The Mannose Receptor Family. *Biochim. Biophys. Acta Gen. Subj.* **2002**, *1572*, 364–386. [CrossRef]
40. Feinberg, H.; Jégouzo, S.A.F.; Lasanajak, Y.; Smith, D.F.; Drickamer, K.; Weis, W.I.; Taylor, M.E. Structural Analysis of Carbohydrate Binding by the Macrophage Mannose Receptor CD206. *J. Biol. Chem.* **2021**, *296*, 100368. [CrossRef]
41. Filatova, L.Y.; Klyachko, N.L.; Kudryashova, E.V. Targeted Delivery of Anti-Tuberculosis Drugs to Macrophages: Targeting Mannose Receptors. *Russ. Chem. Rev.* **2018**, *87*, 374–391. [CrossRef]
42. Zlotnikov, I.D.; Kudryashova, E.V. Mannose Receptors of Alveolar Macrophages as a Target for the Addressed Delivery of Medicines to the Lungs. *Russ. J. Bioorg. Chem.* **2022**, *48*, 46–75. [CrossRef]
43. Zlotnikov, I.D.; Kudryashova, E.V. Computer Simulation of the Receptor–Ligand Interactions of Mannose Receptor CD206 in Comparison with the Lectin Concanavalin A Model. *Biochemistry* **2022**, *87*, 54–69. [CrossRef] [PubMed]
44. Zlotnikov, I.D.; Vanichkin, D.A.; Kudryashova, E.V. Methods for Determining the Parameters of Receptor-Ligand Interactions on the Model of Concanavalin A and Mannosylated Chitosans Promising Carriers for Drug Delivery to Alveolar Macrophages. *Biotekhnologiya* **2021**, *37*, 28–40. [CrossRef]
45. Van Landschoot, A.; Loontiens, F.G.; De Bruyne, C.K. Binding of Manno-oligosaccharides to Concanavalin A: Substitution Titration with a Fluorescent-Indicator Ligand. *Eur. J. Biochem.* **1980**, *103*, 307–312. [CrossRef] [PubMed]
46. Naismith, J.H.; Field, R.A. Structural Basis of Trimannoside Recognition by Concanavalin A. *J. Biol. Chem.* **1996**, *271*, 972–976. [CrossRef]
47. Mislovičová, D.; Masárová, J.; Švitel, J.; Mendichi, R.; Šoltés, L.; Gemeiner, P.; Danielsson, B. Neoglycoconjugates of Mannan with Bovine Serum Albumin and Their Interaction with Lectin Concanavalin A. *Bioconjug. Chem.* **2002**, *13*, 136–142. [CrossRef]
48. Mandal, D.K.; Kishore, N.; Brewer, C.F. Thermodynamics of Lectin-Carbohydrate Interactions. Titration Microcalorimetry Measurements of the Binding of N-Linked Carbohydrates and Ovalbumin to Concanavalin A. *Biochemistry* **1994**, *33*, 1149–1156. [CrossRef]
49. Chen, P.; Zhang, X.; Jia, L.; Prud'Homme, R.K.; Szekely, Z.; Sinko, P.J. Optimal Structural Design of Mannosylated Nanocarriers for Macrophage Targeting. *J. Control. Release* **2014**, *194*, 341–349. [CrossRef]
50. Nimje, N.; Agarwal, A.; Saraogi, G.K.; Lariya, N.; Rai, G.; Agrawal, H.; Agrawal, G.P. Mannosylated Nanoparticulate Carriers of Rifabutin for Alveolar Targeting. *J. Drug Target.* **2009**, *17*, 777–787. [CrossRef]
51. Nahar, M.; Dubey, V.; Mishra, D.; Mishra, P.K.; Dube, A.; Jain, N.K. In Vitro Evaluation of Surface Functionalized Gelatin Nanoparticles for Macrophage Targeting in the Therapy of Visceral Leishmaniasis. *J. Drug Target.* **2010**, *18*, 93–105. [CrossRef]
52. Skuredina, A.A.; Tychinina, A.S.; Le-Deygen, I.M.; Belogurova, N.G.; Kudryashova, E.V. Regulation of Properties of Lipid Membranes by Interaction with 2-Hydroxypropyl β-Cyclodextrin: Molecular Details. *Russ. J. Bioorg. Chem.* **2020**, *46*, 692–701. [CrossRef]
53. Vyas, S.P.; Khatri, K. Liposome-Based Drug Delivery to Alveolar Macrophages. *Expert Opin. Drug Deliv.* **2007**, *4*, 95–99. [CrossRef] [PubMed]
54. Kelly, C.; Jefferies, C.; Cryan, S.-A. Targeted Liposomal Drug Delivery to Monocytes and Macrophages. *J. Drug Deliv.* **2011**, *2011*, 727241. [CrossRef]
55. Alizadeh, D.; Zhang, L.; Hwang, J.; Schluep, T.; Badie, B. Tumor-Associated Macrophages Are Predominant Carriers of Cyclodextrin-Based Nanoparticles into Gliomas. *Nanomed. Nanotechnol. Biol. Med.* **2010**, *6*, 382–390. [CrossRef] [PubMed]
56. Rodríguez-Lavado, J.; De La Mata, M.; Jiménez-Blanco, J.L.; García-Moreno, M.I.; Benito, J.M.; Díaz-Quintana, A.; Sánchez-Alcázar, J.A.; Higaki, K.; Nanba, E.; Ohno, K.; et al. Targeted Delivery of Pharmacological Chaperones for Gaucher Disease to Macrophages by a Mannosylated Cyclodextrin Carrier. *Org. Biomol. Chem.* **2014**, *12*, 2289–2301. [CrossRef] [PubMed]
57. Mahor, S.; Dash, B.C.; O'Connor, S.; Pandit, A. Mannosylated Polyethyleneimine-Hyaluronan Nanohybrids for Targeted Gene Delivery to Macrophage-like Cell Lines. *Bioconjug. Chem.* **2012**, *23*, 1138–1148. [CrossRef] [PubMed]
58. Bendas, G.; Krause, A.; Schmidt, R.; Vogel, J.; Ulrich Rothe Selectins as New Targets for Immunoliposome-Mediated Drug Delivery. A Potential Way of Anti-Inflammatory Therapy. *Pharm. Acta Helv.* **1998**, *73*, 19–26. [CrossRef]
59. Raviv, L.; Jaron-Mendelson, M.; David, A. Mannosylated Polyion Complexes for in Vivo Gene Delivery into CD11c+ Dendritic Cells. *Mol. Pharm.* **2015**, *12*, 453–462. [CrossRef]
60. Lopukhov, A.V.; Yang, Z.; Haney, M.J.; Bronich, T.K.; Sokolsky-Papkov, M.; Batrakova, E.V.; Klyachko, N.L.; Kabanov, A.V. Mannosylated Cationic Copolymers for Gene Delivery to Macrophages. *Macromol. Biosci.* **2021**, *21*, 2000371. [CrossRef]
61. Kunjachan, S.; Gupta, S.; Dwivedi, A.K.; Dube, A.; Chourasia, M.K. Chitosan-Based Macrophage-Mediated Drug Targeting for the Treatment of Experimental Visceral Leishmaniasis. *J. Microencapsul.* **2011**, *28*, 301–310. [CrossRef]
62. Jiang, H.L.; Kim, Y.K.; Arote, R.; Jere, D.; Quan, J.S.; Yu, J.H.; Choi, Y.J.; Nah, J.W.; Cho, M.H.; Cho, C.S. Mannosylated Chitosan-Graft-Polyethylenimine as a Gene Carrier for Raw 264.7 Cell Targeting. *Int. J. Pharm.* **2009**, *375*, 133–139. [CrossRef]

63. Yu, W.; Liu, C.; Liu, Y.; Zhang, N.; Xu, W. Mannan-Modified Solid Lipid Nanoparticles for Targeted Gene Delivery to Alveolar Macrophages. *Pharm. Res.* **2010**, *27*, 1584–1596. [CrossRef] [PubMed]
64. Titov, D.V.; Gening, M.L.; Tsvetkov, Y.E.; Nifantiev, N.E. Conjugates of Cyclooligosaccharide Scaffolds and Carbohydrate Ligands: Methods for Synthesis and the Interaction with Lectins. *Russ. J. Bioorg. Chem.* **2013**, *39*, 451–487. [CrossRef] [PubMed]
65. Yang, C.; Li, H.; Goh, S.H.; Li, J. Cationic Star Polymers Consisting of α-Cyclodextrin Core and Oligoethylenimine Arms as Nonviral Gene Delivery Vectors. *Biomaterials* **2007**, *28*, 3245–3254. [CrossRef] [PubMed]
66. Samal, S.K.; Dash, M.; Van Vlierberghe, S.; Kaplan, D.L.; Chiellini, E.; van Blitterswijk, C.; Moroni, L.; Dubruel, P. Cationic Polymers and Their Therapeutic Potential. *Chem. Soc. Rev.* **2012**, *41*, 7147–7194. [CrossRef]
67. Chaubey, P.; Mishra, B.; Mudavath, S.L.; Patel, R.R.; Chaurasia, S.; Sundar, S.; Suvarna, V.; Monteiro, M. Mannose-Conjugated Curcumin-Chitosan Nanoparticles: Efficacy and Toxicity Assessments against Leishmania Donovani. *Int. J. Biol. Macromol.* **2018**, *111*, 109–120. [CrossRef]
68. Freichels, H.; Wagner, M.; Okwieka, P.; Meyer, R.G.; Mailänder, V.; Landfester, K.; Musyanovych, A. (Oligo)Mannose Functionalized Hydroxyethyl Starch Nanocapsules: En Route to Drug Delivery Systems with Targeting Properties. *J. Mater. Chem. B* **2013**, *1*, 4338–4348. [CrossRef]
69. Yao, W.; Jiao, Y.; Luo, J.; Du, M.; Zong, L. Practical Synthesis and Characterization of Mannose-Modified Chitosan. *Int. J. Biol. Macromol.* **2012**, *50*, 821–825. [CrossRef]
70. Wang, S.K.; Liang, P.H.; Astronomo, R.D.; Hsu, T.L.; Hsieh, S.L.; Burton, D.R.; Wong, C.H. Targeting the Carbohydrates on HIV-1: Interaction of Oligomannose Dendrons with Human Monoclonal Antibody 2G12 and DC-SIGN. *Proc. Natl. Acad. Sci. USA* **2008**, *105*, 3690–3695. [CrossRef]
71. Ikehara, Y.; Shiuchi, N.; Kabata-Ikehara, S.; Nakanishi, H.; Yokoyama, N.; Takagi, H.; Nagata, T.; Koide, Y.; Kuzushima, K.; Takahashi, T.; et al. Effective Induction of Anti-Tumor Immune Responses with Oligomannose-Coated Liposome Targeting to Intraperitoneal Phagocytic Cells. *Cancer Lett.* **2008**, *260*, 137–145. [CrossRef]
72. Ito, M.; Murakami, K.; Yoshino, M. Antioxidant Action of Eugenol Compounds: Role of Metal Ion in the Inhibition of Lipid Peroxidation. *Food Chem. Toxicol.* **2005**, *43*, 461–466. [CrossRef] [PubMed]
73. Ali, S.M.; Khan, A.A.; Ahmed, I.; Musaddiq, M.; Ahmed, K.S.; Polasa, H.; Rao, L.V.; Habibullah, C.M.; Sechi, L.A.; Ahmed, N. Antimicrobial Activities of Eugenol and Cinnamaldehyde against the Human Gastric Pathogen Helicobacter Pylori. *Ann. Clin. Microbiol. Antimicrob.* **2005**, *4*, 20. [CrossRef] [PubMed]
74. Joardar, A.; Meher, G.; Bag, B.P.; Chakraborty, H. Host-Guest Complexation of Eugenol in Cyclodextrins for Enhancing Bioavailability. *J. Mol. Liq.* **2020**, *319*, 114336. [CrossRef]
75. Leite, A.M.; Lima, E.D.O.; De Souza, E.L.; Diniz, M.D.F.F.M.; Trajano, V.N.; De Medeiros, I.A. Inhibitory Effect of β-Pinene, α-Pinene and Eugenol on the Growth of Potential Infectious Endocarditis Causing Gram-Positive Bacteria. *Rev. Bras. Ciencias Farm. J. Pharm. Sci.* **2007**, *43*, 121–126. [CrossRef]
76. Cotmore, J.M.; Burke, A.; Lee, N.H.; Shapiro, I.M. Respiratory Inhibition of Isolated Rat Liver Mitochondria by Eugenol. *Arch. Oral Biol.* **1979**, *24*, 565–568. [CrossRef]
77. Bucak, C.D.; Kürekci, C.; Dinç, C.Ö. Carrying System Formula for Eugenol Encapsulation: Glycodendritic Polyamine Dextran-G2.5, Synthesis and in Vitro Antibacterial Activity. *Polym. Bull.* **2021**, *78*, 601–620. [CrossRef]
78. Pramod, K.; Ansari, S.H.; Ali, J. Eugenol: A Natural Compound with Versatile Pharmacological Actions. *Nat. Prod. Commun.* **2010**, *5*, 1999–2006. [CrossRef]
79. Wang, T.; Li, B.; Si, H.; Lin, L.; Chen, L. Release Characteristics and Antibacterial Activity of Solid State Eugenol/β-Cyclodextrin Inclusion Complex. *J. Incl. Phenom. Macrocycl. Chem.* **2011**, *71*, 207–213. [CrossRef]
80. Garg, A.; Gupta, B.; Prakash, R.; Singh, S. Preparation and Characterization of Hydroxypropyl-β-Cyclodextrin Inclusion Complex of Eugenol: Differential Pulse Voltammetry and 1H-NMR. *Chem. Pharm. Bull.* **2010**, *58*, 1313–1319. [CrossRef]
81. Valdivieso-Ugarte, M.; Gomez-Llorente, C.; Plaza-Díaz, J.; Gil, Á. Antimicrobial, Antioxidant, and Immunomodulatory Properties of Essential Oils: A Systematic Review. *Nutrients* **2019**, *11*, 2786. [CrossRef]
82. Gao, J.Q.; Zhao, Q.Q.; Lv, T.F.; Shuai, W.P.; Zhou, J.; Tang, G.P.; Liang, W.Q.; Tabata, Y.; Hu, Y.L. Gene-Carried Chitosan-Linked-PEI Induced High Gene Transfection Efficiency with Low Toxicity and Significant Tumor-Suppressive Activity. *Int. J. Pharm.* **2010**, *387*, 286–294. [CrossRef] [PubMed]
83. Tang, G.P.; Guo, H.Y.; Alexis, F.; Wang, X.; Zeng, S.; Lim, T.M.; Ding, J.; Yang, Y.Y.; Wang, S. Low Molecular Weight Polyethylenimines Linked by β-Cyclodextrin for Gene Transfer into the Nervous System. *J. Gene Med.* **2006**, *8*, 736–744. [CrossRef] [PubMed]
84. Ruan, G.X.; Chen, Y.Z.; Yao, X.L.; Du, A.; Tang, G.P.; Shen, Y.Q.; Tabata, Y.; Gao, J.Q. Macrophage Mannose Receptor-Specific Gene Delivery Vehicle for Macrophage Engineering. *Acta Biomater.* **2014**, *10*, 1847–1855. [CrossRef] [PubMed]
85. Taguchi, T.; Kitajima, K.; Niimi, T.; Muto, Y.; Yokoyama, S.; Inoue, S.; Inoue, Y. Complete assignments of 13C NMR resonances to all the carbon atoms of the trimannosido-di-N-acetylchitobiosyl structure in a pentaantennary decasaccharide glycopeptide. *Carbohydr. Res.* **1995**, *275*, 185–191. [CrossRef]
86. Skuredina, A.A.; Tychinina, A.S.; Le-Deygen, I.M.; Golyshev, S.A.; Belogurova, N.G.; Kudryashova, E.V. The Formation of Quasi-Regular Polymeric Network of Cross-Linked Sulfobutyl Ether Derivative of β-Cyclodextrin Synthesized with Moxifloxacin as a Template. *React. Funct. Polym.* **2021**, *159*, 104811. [CrossRef]
87. Mi, R.P.; Ki, O.H.; In, K.H.; Myung, H.C.; Jae, W.N.; Yun, J.C.; Chong, S.C. Degradable Polyethylenimine-Alt-Poly(Ethylene Glycol) Copolymers as Novel Gene Carriers. *J. Control. Release* **2005**, *105*, 367–380. [CrossRef]

88. Legal, J.M.; Manfait, M.; Theophanides, T. Applications of FTIR Spectroscopy in Structural Studies of Cells and Bacteria. *J. Mol Struct.* **1991**, *242*, 397–407. [CrossRef]
89. Szabó, Z.I.; Deme, R.; Mucsi, Z.; Rusu, A.; Mare, A.D.; Fiser, B.; Toma, F.; Sipos, E.; Tóth, G. Equilibrium, Structural and Antibacterial Characterization of Moxifloxacin-β-Cyclodextrin Complex. *J. Mol. Struct.* **2018**, *1166*, 228–236. [CrossRef]
90. Cameron, K.S.; Fletcher, D.; Fielding, L. An NMR Study of Cyclodextrin Complexes of the Steroidal Neuromuscular Blocker Drug Rocuronium Bromide. *Magn. Reson. Chem.* **2002**, *40*, 251–260. [CrossRef]
91. Honary, S.; Zahir, F. Effect of Zeta Potential on the Properties of Nano-Drug Delivery Systems—A Review (Part 1). *Trop. J. Pharm. Res.* **2013**, *12*, 255–264. [CrossRef]
92. Griesser, J.; Burtscher, S.; Köllner, S.; Nardin, I.; Prüfert, F.; Bernkop-Schnürch, A. Zeta Potential Changing Self-Emulsifying Drug Delivery Systems Containing Phosphorylated Polysaccharides. *Eur. J. Pharm. Biopharm.* **2017**, *119*, 264–270. [CrossRef] [PubMed]
93. Weis, W.I.; Drickamer, K. Structural Basis of Recognition Lectin-Carb Ohydrate. *Anna Rev. Biochen.* **1996**, *65*, 441–473. [CrossRef] [PubMed]
94. Ghotbi, Z.; Haddadi, A.; Hamdy, S.; Hung, R.W.; Samuel, J.; Lavasanifar, A. Active Targeting of Dendritic Cells with Mannan-Decorated PLGA Nanoparticles. *J. Drug Target.* **2011**, *19*, 281–292. [CrossRef] [PubMed]
95. Dam, T.K.; Roy, R.; Das, S.K.; Oscarson, S.; Brewer, C.F. Binding of Multivalent Carbohydrates to Concanavalin A and Dioclea Grandiflora Lectin. Thermodynamic Analysis of the 'Multivalency Effect'. *J. Biol. Chem.* **2000**, *275*, 14223–14230. [CrossRef] [PubMed]
96. Kaushik, S.; Mohanty, D.; Surolia, A. The Role of Metal Ions in Substrate Recognition and Stability of Concanavalin A: A Molecular Dynamics Study. *Biophys. J.* **2009**, *96*, 21–34. [CrossRef]
97. Pflumm, M.N.; Wang, J.L.; Edelman, G.M. Conformational Changes in Concanavalin A. *J. Biol. Chem.* **1971**, *246*, 4369–4370. [CrossRef]
98. Loontiens, F.G.; Clegg, R.M.; Van Landschoot, A. Some Physicochemical Aspects of Oligosaccharide Binding to Concanavalin A and Wheat Germ Agglutinin. *J. Biosci.* **1983**, *5*, 105–120. [CrossRef]
99. Bryce, R.A.; Hillier, I.H.; Naismith, J.H. Carbohydrate-Protein Recognition: Molecular Dynamics Simulations and Free Energy Analysis of Oligosaccharide Binding to Concanavalin A. *Biophys. J.* **2001**, *81*, 1373–1388. [CrossRef]
100. Satoh, T.; Suzuki, K.; Yamaguchi, T.; Kato, K. Structural Basis for Disparate Sugar-Binding Specificities in the Homologous Cargo Receptors ERGIC-53 and VIP36. *PLoS ONE* **2014**, *9*, e87963. [CrossRef]
101. Le-Deygen, I.M.; Mamaeva, P.V.; Skuredina, A.A.; Kudryashova, E.V. A Spectral Approach to Study Interaction between Chitosan Modified with Mannose and Concavalin A for the Creation of Address Delivery Systems of Antituberculosis Drugs. *Moscow Univ. Chem. Bull.* **2020**, *75*, 213–217. [CrossRef]
102. Díaz-Moscoso, A.; Guilloteau, N.; Bienvenu, C.; Méndez-Ardoy, A.; Jiménez Blanco, J.L.; Benito, J.M.; Le Gourriérec, L.; Di Giorgio, C.; Vierling, P.; Defaye, J.; et al. Mannosyl-Coated Nanocomplexes from Amphiphilic Cyclodextrins and PDNA for Site-Specific Gene Delivery. *Biomaterials* **2011**, *32*, 7263–7273. [CrossRef] [PubMed]
103. Weatherman, R.V.; Kiessling, L.L. Fluorescence Anisotropy Assays Reveal Affinities of C- and O-Glycosides for Concanavalin A. *J. Org. Chem.* **1996**, *61*, 534–538. [CrossRef] [PubMed]
104. Novak, U.; Grdadolnik, J. The Hydration of Concanavalin A Studied by Infrared Spectroscopy. *J. Mol. Struct.* **2017**, *1135*, 138–143. [CrossRef]
105. Ogura, K.; Kobayashi, M.; Nakayama, M.; Miho, Y. In-Situ FTIR Studies on the Electrochemical Oxidation of Histidine and Tyrosine. *J. Electroanal. Chem.* **1999**, *463*, 218–223. [CrossRef]
106. Skuredina, A.A.; Le-Deygen, I.M.; Belogurova, N.G.; Kudryashova, E.V. Effect of Cross-Linking on the Inclusion Complex Formation of Derivatized β-Cyclodextrins with Small-Molecule Drug Moxifloxacin. *Carbohydr. Res.* **2020**, *498*, 108183. [CrossRef]
107. Skuredina, A.A.; Le-Deygen, I.M.; Uporov, I.V.; Kudryashova, E.V. A Study of the Physicochemical Properties and Structure of Moxifloxacin Complex with Methyl-β-Cyclodextrin. *Colloid J.* **2017**, *79*, 668–676. [CrossRef]
108. Sukhoverkov, K.V.; Le-Deygen, I.M.; Egorov, A.M.; Kudryashova, E.V. Physicochemical Properties of the Inclusion Complex of Moxifloxacin with Hydroxypropyl-β-Cyclodextrin Synthesized by RESS. *Russ. J. Phys. Chem. B* **2018**, *12*, 1193–1204. [CrossRef]
109. Elshafie, H.S.; Sakr, S.H.; Sadeek, S.A.; Camele, I. Biological Investigations and Spectroscopic Studies of New Moxifloxacin/Glycine-Metal Complexes. *Chem. Biodivers.* **2019**, *16*, e1800633. [CrossRef]
110. Al Omari, M.M.H.; Jaafari, D.S.; Al-Sou'od, K.A.; Badwan, A.A. Moxifloxacin hydrochloride. *Profiles Drug Subst. Excip. Relat. Methodol.* **2014**, *39*, 299–431.
111. Jelić, R.; Tomović, M.; Stojanović, S.; Joksović, L.; Jakovljević, I.; Djurdjević, P. Study of Inclusion Complex of β-Cyclodextrin and Levofloxacin and Its Effect on the Solution Equilibria between Gadolinium(III) Ion and Levofloxacin. *Monatshefte Chem.* **2015**, *146*, 1621–1630. [CrossRef]
112. Chen, M.; Diao, G.; Zhang, E. Study of Inclusion Complex of β-Cyclodextrin and Nitrobenzene. *Chemosphere* **2006**, *63*, 522–529. [CrossRef] [PubMed]
113. Banjare, M.K.; Behera, K.; Banjare, R.K.; Pandey, S.; Ghosh, K.K. Inclusion Complexation of Imidazolium-Based Ionic Liquid and β-Cyclodextrin: A Detailed Spectroscopic Investigation. *J. Mol. Liq.* **2020**, *302*, 112530. [CrossRef]
114. Fernandes, C.M.; Carvalho, R.A.; Pereira da Costa, S.; Veiga, F.J.B. Multimodal Molecular Encapsulation of Nicardipine Hydrochloride by β-Cyclodextrin, Hydroxypropyl-β-Cyclodextrin and Triacetyl-β-Cyclodextrin in Solution. Structural Studies by 1H NMR and ROESY Experiments. *Eur. J. Pharm. Sci.* **2003**, *18*, 285–296. [CrossRef]

115. Kost, B.; Brzeziński, M.; Zimnicka, M.; Socka, M.; Wielgus, E.; Słowianek, M.; Biela, T. PLA Stereocomplexed Microspheres Modified with Methyl-β-Cyclodextrin as an Atropine Delivery System. Synthesis and Characterization. *Mater. Today Commun.* **2020**, *25*, 101605. [CrossRef]
116. Le-Deygen, I.M.; Skuredina, A.A.; Uporov, I.V.; Kudryashova, E.V. Thermodynamics and Molecular Insight in Guest–Host Complexes of Fluoroquinolones with β-Cyclodextrin Derivatives, as Revealed by ATR-FTIR Spectroscopy and Molecular Modeling Experiments. *Anal. Bioanal. Chem.* **2017**, *409*, 6451–6462. [CrossRef]
117. Wang, B.; Ye, Z.; Tang, Y.; Liu, H.; Lin, Q.; Chen, H.; Nan, K. Loading of Antibiotics into Polyelectrolyte Multilayers after Self-Assembly and Tunable Release by Catechol Reaction. *J. Phys. Chem. C* **2016**, *120*, 6145–6155. [CrossRef]
118. Deygen, I.M.; Kudryashova, E.V. New Versatile Approach for Analysis of PEG Content in Conjugates and Complexes with Biomacromolecules Based on FTIR Spectroscopy. *Colloids Surf. B Biointerfaces* **2016**, *141*, 36–43. [CrossRef] [PubMed]
119. Corte, L.; Tiecco, M.; Roscini, L.; Germani, R.; Cardinali, G. FTIR Analysis of the Metabolomic Stress Response Induced by N-Alkyltropinium Bromide Surfactants in the Yeasts Saccharomyces Cerevisiae and Candida Albicans. *Colloids Surf. B Biointerfaces* **2014**, *116*, 761–771. [CrossRef] [PubMed]

pharmaceuticals

MDPI

Review

Antibiotics and Carbohydrate-Containing Drugs Targeting Bacterial Cell Envelopes: An Overview

Federico Riu [1,†], Alessandro Ruda [2,†], Roberta Ibba [1,*], Simona Sestito [3], Ilenia Lupinu [1], Sandra Piras [1], Göran Widmalm [2] and Antonio Carta [1]

[1] Department of Medicine, Surgery and Pharmacy, University of Sassari, Via Muroni 23/A, 07100 Sassari, Italy; federicoriu1@gmail.com (F.R.); ilenialup@gmail.com (I.L.); piras@uniss.it (S.P.); acarta@uniss.it (A.C.)

[2] Department of Organic Chemistry, Arrhenius Laboratory, Stockholm University, S-106 91 Stockholm, Sweden; alessandro.ruda@su.se (A.R.); goran.widmalm@su.se (G.W.)

[3] Department of Chemical, Physical, Mathematical and Natural Sciences, University of Sassari, Via Vienna 2, 07100 Sassari, Italy; ssestito@uniss.it

* Correspondence: ribba@uniss.it

† These authors contributed equally to this work.

Abstract: Certain bacteria constitute a threat to humans due to their ability to escape host defenses as they easily develop drug resistance. Bacteria are classified into gram-positive and gram-negative according to the composition of the cell membrane structure. Gram-negative bacteria have an additional outer membrane (OM) that is not present in their gram-positive counterpart; the latter instead hold a thicker peptidoglycan (PG) layer. This review covers the main structural and functional properties of cell wall polysaccharides (CWPs) and PG. Drugs targeting CWPs are discussed, both noncarbohydrate-related (β-lactams, fosfomycin, and lipopeptides) and carbohydrate-related (glycopeptides and lipoglycopeptides). Bacterial resistance to these drugs continues to evolve, which calls for novel antibacterial approaches to be developed. The use of carbohydrate-based vaccines as a valid strategy to prevent bacterial infections is also addressed.

Keywords: carbohydrate-based drugs; cell wall polysaccharides; bacterial membrane; antibiotic resistance; carbohydrate-based vaccines

Citation: Riu, F.; Ruda, A.; Ibba, R.; Sestito, S.; Lupinu, I.; Piras, S.; Widmalm, G.; Carta, A. Antibiotics and Carbohydrate-Containing Drugs Targeting Bacterial Cell Envelopes: An Overview. *Pharmaceuticals* **2022**, *15*, 942. https://doi.org/10.3390/ph15080942

Academic Editors: Federica Sodano and Maria Grazia Rimoli

Received: 27 June 2022
Accepted: 20 July 2022
Published: 29 July 2022

1. Introduction

1.1. Bacteria

1.1.1. Gram-Positive and Gram-Negative Bacteria: General Remarks

Bacteria are microscopic single-celled organisms that are listed among the oldest known life forms on Earth. They can be found in the soil, ocean water, ice, and underneath the earth's crust; that is because these microorganisms are often essential to other organisms' life [1–3]. The worldwide bacterial population is estimated to reach 2×10^{30} cells, with a major role in the ecosystem [4]. Moreover, nitrogen-fixing bacteria are essential for the natural growth of several plant species [5,6]. The bacterial fermentation process is also fundamental for the digestive systems of sheep and cattle. In the food industry area, bacteria are involved in the manufacturing of many products, including dairy products, baked goods, and alcoholic beverages [7]. In industrial microbiology, microorganisms are grown on a large scale to produce compounds such as antibiotics, enzymes, and various chemicals, while in biotechnology, genetically-modified organisms are used to synthesize, for example, human proteins for medical use. Bacteria also live in the human and animal bodies, particularly on the skin, airways, oral cavity, and the digestive, reproductive, and urinary systems, normally without causing any harm [8]. They populate the resident flora, or "microbiota", are essential to facilitate food digestion or prevent the proliferation of other more dangerous bacteria, and acting as saprophytes [9–11]. Figure 1 depicts the representative genera of the constitutive bacterial flora in humans [12–15]. The intestinal microbiota plays a central role in the development of the immune system and the modulation of its function. This has led to investigations of the mechanisms that link autoimmune diseases or allergies to changes in the microbiome [16–19]. Currently, the connections between the human body and the microbiome are not limited to the gastrointestinal or immune system but concern almost all other systems. Alzheimer's disease, multiple sclerosis, and autism have been studied for a putative role of microbiota in their development [20–23]. Based on the bacterium-host relationship, these microorganisms can be divided into four categories: symbionts, that live and multiply in contact with the host without causing damage and establishing a beneficial reciprocal relationship; diners that live and multiply in contact with the host without causing damage; pathogens that take advantage of the host causing diseases, from mild to serious ones; and opportunists, which normally are harmless but can cause disease, even serious ones if a weakening of the defense system of the organism occurs. Among them, pathogens represent a small percentage of the entire known bacterial population.

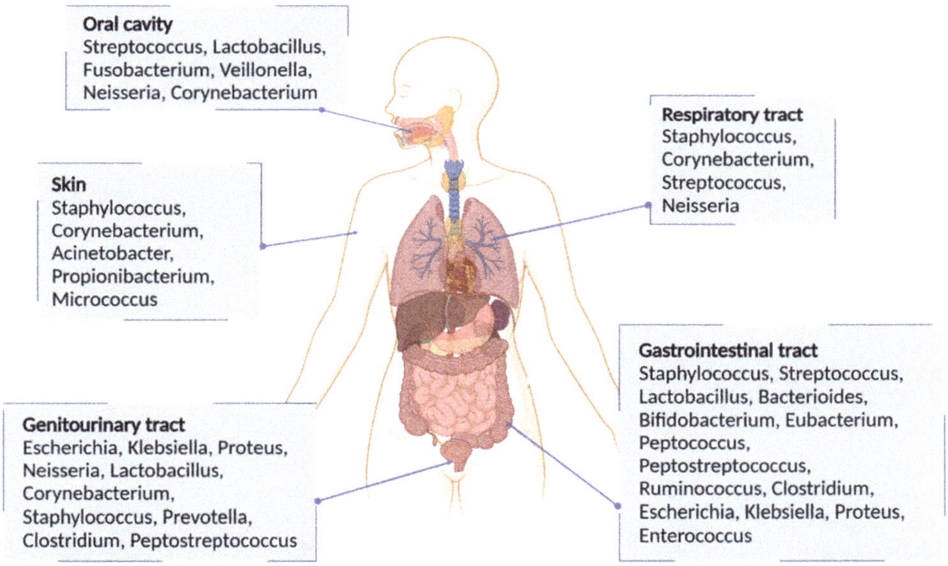

Figure 1. Schematic representation of target human districts of the main gram-positive and gram-negative bacteria. Created with BioRender.com (Accessed on 29 April 2022).

At the beginning of the 20th century, infectious diseases originating from bacterial and viral pathogens were the main cause of death. A decrease in incidence occurred through the years thanks to a greater understanding of the infectious process, the improvement of public health services, vaccination campaigns, and last but not least, the discovery of antibiotics, which made the managing of infectious diseases possible.

1.1.2. Morphology of Gram-Positive and Gram-Negative Bacteria

Bacteria are unicellular organisms that are formed by a prokaryotic cell, with dimensions ranging from 0.2 to 2 μm. The cell is surrounded by the plasma membrane, which regulates the transport of nutrients and waste substances. The plasma membrane is covered by a wall with different characteristics, leading to the well-known division into gram-positive and gram-negative bacteria. In the gram-negative cell wall, a lipid bilayer membrane, the outer membrane (OM) is present, while the above-mentioned plasma membrane is the inner membrane (IM). In the inter-membrane space, there is a thin layer of peptidoglycan (or murein), referred to as the periplasm, a complex polymer that is made up of amino sugars and amino acids that completely envelop the cell. The outer membrane acts as a barrier against the diffusion of many compounds, making the cell more resistant to toxic substances. Gram-positive bacteria do not have an outer membrane and their wall is mainly composed of a thick layer of peptidoglycan [24].

The different composition of the bacterial wall determines the different coloration that is taken up by the bacteria with the dye that was created by the Danish microbiologist Hans Christian Gram in 1884 and thus allows one to discern the two groups of bacteria. Gram stain is still used in normal clinical practice, as the first step in the screening of a biological sample to identify the pathogenic microorganisms present [25].

In addition to the above-described cell wall, some bacteria possess additional layers of coating (S layer, capsule, or glycocalyx) and surface appendages (pili, fimbriae, flagella) The capsule or mucous layer consists of polysaccharide or protein substances that are secreted on the cell surface by bacteria for several functions, including adhesion (e.g., on solid surfaces generating biofilms) and protection.

1.1.3. The Cytoplasmic Membrane

The cytoplasmic membrane is a thin barrier that surrounds the cell and separates the cytoplasm from the extracellular environment. The membrane is made of a phospholipid bilayer: phospholipids contain both hydrophobic (fatty acid) and hydrophilic (glycerol-phosphate) components. In the phospholipid membrane, the fatty acids point inward toward each other to form the hydrophobic environment, while the hydrophilic portion remains exposed to the external environment or the cytoplasm [26]. The membrane has a thickness of 6–8 nm and can be seen using an electron microscope: it appears as two dark-colored lines (glycerophosphate) that are separated by a lighter area (fatty acids) The structure is fluid and the proteins that are immersed in the phospholipid bilayer can undergo both rotational and lateral movement. The cytoplasmic membranes of some bacteria are strengthened by molecules called hopanoids, which are rigid planar molecules and structural analogs of sterols [27]. Many membrane proteins are firmly embedded in the membrane and are named integral membrane proteins [28]. Other proteins have one portion that is anchored in the membrane and extra-membrane regions that point into or out of the cell. The so-called peripheral membrane proteins remain firmly associated with the membrane surfaces [29]. The peripheral proteins interact with integral membrane proteins in important cellular processes such as energy metabolism and transport [29].

The main function of the cytoplasmic membrane is to prevent the passive leakage of solutes into or out of the cell; it acts as a gateway for the transport of nutrients into and wastes out of the cell (permeability barrier). The membrane is also an anchor for many proteins, glucides, lipids, and complexes of the same. Some are enzymes that catalyze bioenergetic reactions whereas other transport solutes into and out of the cell [26].

Salts, sugars, amino acids, nucleotides, and many other substances need to be transported by proteins that, besides ferrying substances across the membrane, accumulate solutes against the concentration gradient. One substance that can cross the membrane in both directions is water, but its movement is accelerated by dedicated transport proteins, called "aquaporins" [30].

1.1.4. Pathogenic Bacteria

As previously mentioned, disease-causing bacteria are referred to as pathogens. However, in some cases, it is not possible to draw a clear distinction between pathogenic and non-pathogenic bacteria since the encounter of the human organism with the pathogen does not always generate disease. The severity and outcome depend on the complex relationship that is established between the microorganisms and the human organism. Under normal conditions, the thousands of bacteria that make up the microbiota are harmless to our health; indeed, they assist and help maintain it. The immune defenses of our body, therefore, help us to avoid the invasion of microbiota microorganisms, but when the body is compromised for various reasons, invasion can occur, and serious illness appears. Hence, the same beneficial bacteria can become pathogenic. Table 1 shows the different human pathogenic bacteria based on morphology and gram stain [24].

Table 1. Pathogenic bacteria and their associated infections. Examples of gram-negative and gram-positive pathogenic bacteria and their associated infections.

GRAM NEGATIVE	ILLNESSES
Cocci	
Moraxella catarrhalis	Respiratory tract infections
Neisseria gonorrhoeae	Gonorrhea
Neisseria meningitidis	Meningococcal Meningitis
Chlamydia trachomatis	Chlamydia
Cocci-bacilli	
Rickettsia prowazekii	Epidemic typhus
Rickettsia typhi	Murine typhus
Cocci-bacilli and bacilli	
Brucella	Brucellosis
Bordetella	Whooping cough
Legionella	Pneumonia
Bacilli	
Escherichia coli	Gastroenteric and urinary tract infections
Salmonella typhi	Typhoid fever
Salmonella spp.	Diarrhea
Haemophilus influenzae	Otitis, bronchitis, pneumonia, meningitis
Flagellated bacilli	
Helicobacter pylori	Stomach and duodenal ulcers
Vibrions	
Vibrio cholerae	Cholera
Spirochetes	
Borrelia burgdorferi	Lyme disease
Leptospira	Leptospirosis
Non-fermenting Bacilli	
Pseudomonas aeruginosa	Lungs, blood or heart valves infections
GRAM POSITIVE	ILLNESSES
Cocci	
Staphylococcus aureus, S. epidermis	Skin, heart valve and bone infections, pneumonia
Enterococcus faecalis, E. faecium	Hospital Intestinal infections
Streptococcus B: S. agalactiae	Infections of various systems
Streptococcus A: S. pyogenes	Necrotizing fasciitis, Acute rheumatic fever, Acute glomerulus nephritis, Scarlet fever
Streptococcus pneumoniae	Pneumococcal infections
Bacilli	
Clostridium perfrigens	Food poisoning
Clostridium tetani	Tetanus
Clostridium botulinum	Botulism
Corynebacterium diphtheriae	Diphtheria
Corynebacterium jeikeium	Severe infections in the hospitalized patient
Actinomyces	Actinomycosis
Bacillus anthracis	Anthrax
Listeria monocytogenes	Listeriosis

Table 1. *Cont.*

GRAM NEGATIVE	ILLNESSES
Acid-fast bacilli	
Mycobacterium avium complex	Lung infections
Mycobacterium tuberculosis	Tuberculosis
Mycobacterium Leprae	Leprosy
Nocardia spp.	Nocardiosis

1.2. *Bacterial Cell Wall Polysaccharides (CWPs) and Peptidoglycans (PGs): Structure and Functions*

From a structural point of view, the bacterial cell wall can be classified into two different types, depending on gram-positive or gram-negative bacteria, as depicted in Figure 2. The main difference between the two cell walls is the absence of a second bilayer membrane in the gram-positive bacteria, which is compensated by a thicker peptidoglycan (PG) layer. The bilayer membrane in gram-negative bacteria is highly asymmetric; the inner leaflet is composed of phospholipids while the outer leaflet mainly contains lipopolysaccharides (LPS) whose polysaccharide chains occupy the extracellular space (Figure 2). [31,32] The PG layer in gram-positive bacteria is a fundamental component of the cell wall placed outside the cytoplasmic membrane of almost every bacterium [33–35], with very rare exceptions (e.g., *Mycoplasmas, Planctomyces, Orientia* (*Rickettsia*) *tsutsugamushi*) [36,37]. Its primary role is to maintain a sustained turgor of the cell wall, preserving the integrity and counteracting the osmotic pressure of the cell [38,39], but it also represents the anchor point for proteins and cell wall polysaccharides (e.g., teichoic acids) [40,41]. Together with the assembly of bacterial peptidoglycan hydrolases, the PG is also deeply involved in the process of cell division and cell growth [42]. Given its almost ubiquitous nature, higher eukaryotes evolved several PG recognition molecules (e.g., CD13, Nod1-2, lysozymes, and amidases) representing the first defense line against bacterial infections [38,43,44]. Moreover, PG represents a robust target for almost all effective antibiotics inhibiting bacterial cell wall synthesis [39]. The presence of such a layer is of fundamental importance for the integrity of the cell. Indeed, any degradation (e.g., by lysozymes) during cell growth or inhibition of its biosynthetic path leads irremediably to cell lysis. In addition to morphological functions, the cell wall undergoes recycling where it is constantly broken down, turned over, and remodeled [45–47].

PG is of major importance in gram-positive bacteria given its higher abundance which accounts for almost half of the cell wall mass, while in gram-negative bacteria, the PG layer is rather thin and not directly exposed to the extracellular space due to the presence of the outer membrane. Owing to its close to the omnipresent occurrence, PG has been one of the major antibiotic targets in recent years (e.g., penicillin, cephalosporins, carbapenems, vancomycin, teicoplanin) [48].

1.2.1. Chemical Structure and Variability of PG and Correlated Functions

The basic scaffold of the PG layer, as depicted in Figure 3, consists of polymeric glycan strands of two alternating β-(1→4)-linked sugars: *N*-acetylmuramic acid (MurNAc) and *N*-acetylglucosamine (GlcNAc). The different strands are arranged parallel to each other and covalently connected via peptide linkers that are substituted at the D-lactyl group of each MurNAc residue. The most common peptide linker is composed of L-Ala-γ-D-Glu-*meso*-A2pm (or L-Lys)-D-Ala-D-Ala. The last D-alanine residue is lost in the mature macromolecule and the previous D-Ala condenses with the corresponding D-Ala residue of another strand's linker, thus creating the distinctive net of the PG layer.

Variations from the general structure of PG have largely been divided into three categories, *viz.*, of: (i) the glycan strand structure and substitution pattern, (ii) the lipid linker structure and condensation point, and (iii) the glycosylation point by extracellular glycans. The most commonly reported modification in the first case is *N*-deacetylation and *O*-acetylation patterns at both GlcNAc and MurNAc, *N*-glycolylation of MurNAc residues, the modification of MurNAc to muramic δ-lactam, or formation of 1,6-anhydro MurNAc.

The extent of these substitutions is species-dependent and has been extensively reported in previous literature works [49]. Table 2 shows the most common modifications of the PG repeating unit that have been discovered so far.

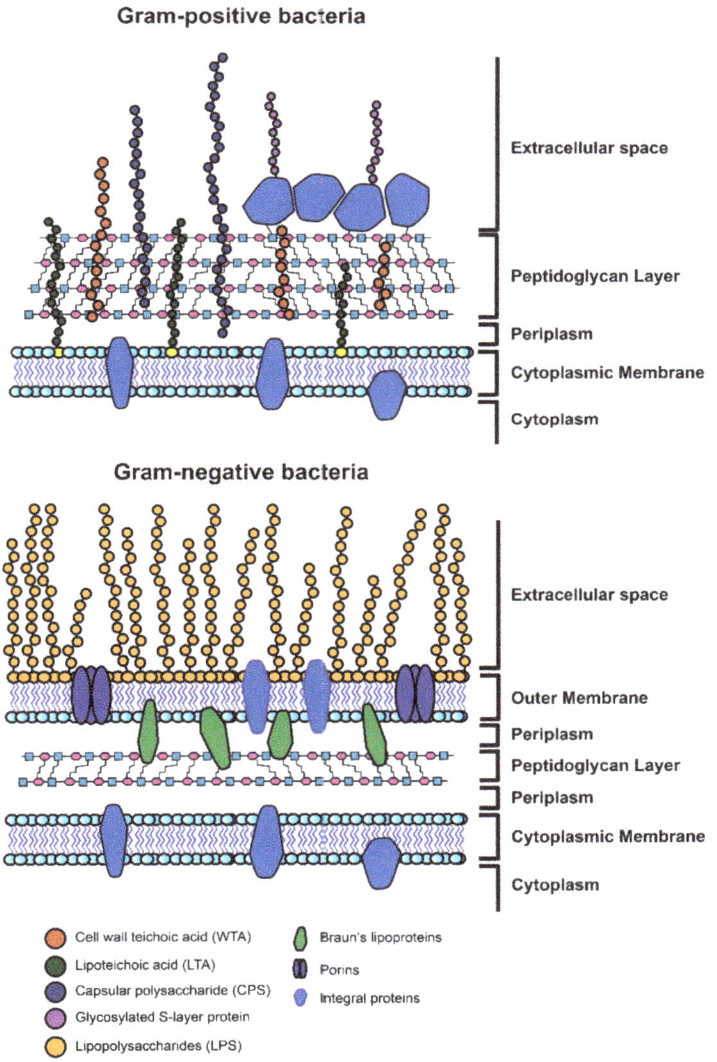

Figure 2. Schematic representation of the cell wall structure of gram-positive and gram-negative bacteria.

N-deacetylated sugars in the PG layer of *Bacillus* strains have been reported extensively for both residues [50–52], and more recently also for Streptococcal bacteria (e.g., *S. pneumoniae, S. pyogenes, S. iniae*) [53–55]. Deacetylation most likely occurs on the mature peptidoglycan since deacetylated precursors have not yet been reported in species containing PG deacetylases, and the latter enzymes have precise extra-cytoplasmic localization [54].

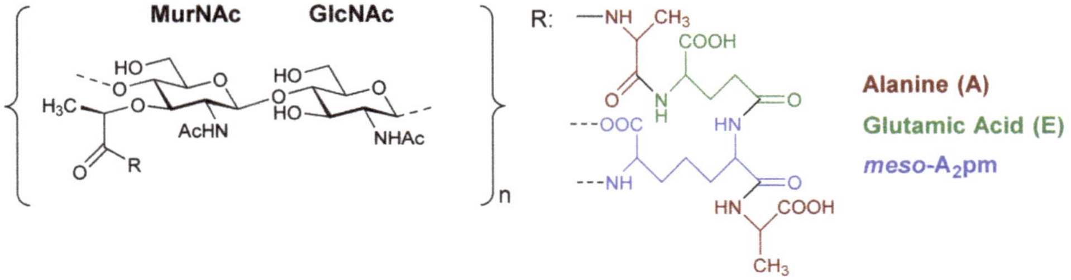

Figure 3. Key structural elements that are representative of the peptidoglycan region. The strands of PG are linked via peptide chains that condense with each other forming amide bonds via either the meso-A2pm carboxyl group or amine group.

Table 2. Known modification of the PG repeating unit.

Repeating Unit	Modifications
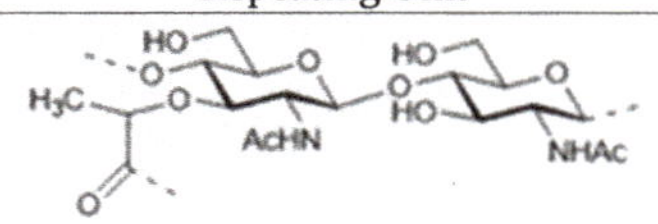	Unmodified unit
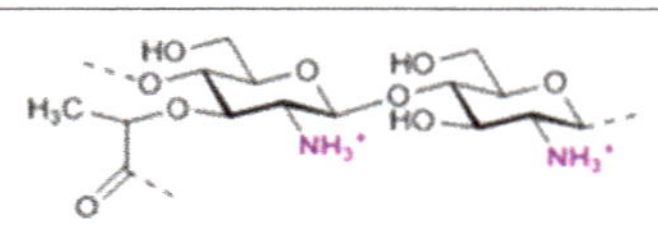	*N*-Deacetylation
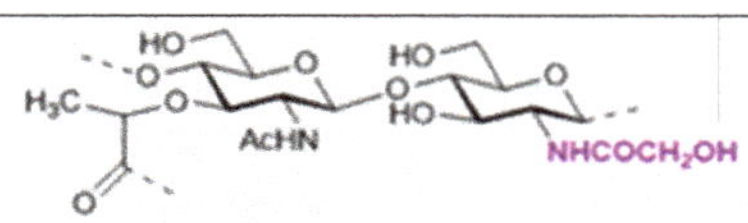	Glycolylation
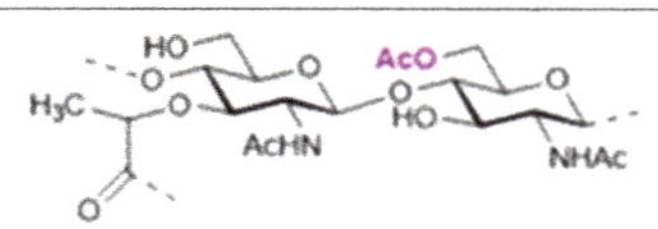	O-Acetylation
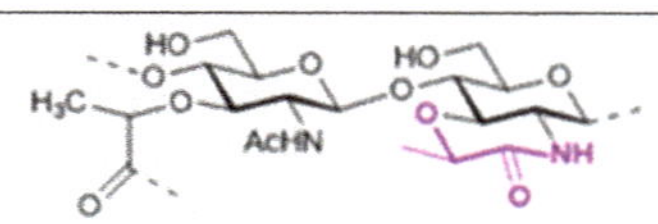	MurN-δ-lactam
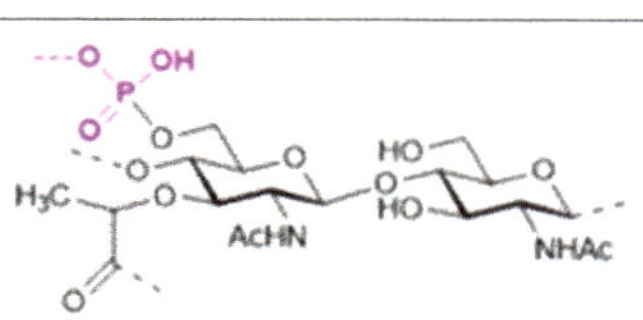	MurNAc-6-phosphate
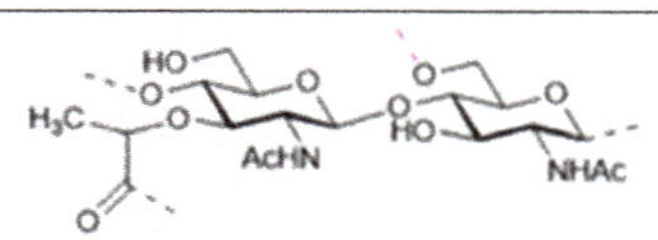	Linkage at GlcNAc

It is well known that *N*-deacetylation plays a key role in the protection of certain bacteria from host antimicrobial proteins such as lysozymes, which catalyzes the breakage of the β-(1→4)-linkage between GlcNAc and MurNAc leading to hydrolyzed PG layers and bacterial lysis [56]. This was shown to be related to the increased positive charge of the PG layer which decreased the efficiency of the secondary (non-enzymatic) mechanism of action of lysozymes [57], where these proteins cross the cell wall and bind to negatively charged bacterial membranes leading to its permeabilization. The lack of *N*-acetyl groups has been shown to provide a poor substrate for lysozyme binding, while upon chemical *N*-acetylation of the substrate lysozyme activity seems to be restored [54,58–60].

N-glycolylated muramic acids in PG, first described in *Mycobacterium smegmatis* [61], are used in bacterial taxonomy to classify *Actinomycetales*, a class of bacteria from which important antibiotics are isolated (e.g., streptomycin, actinomycin, streptothricin) [62]. Such a modification is widely present in the Mycolata taxon, a group of bacteria containing mycolic acids in their cell walls (e.g., *Mycobacterium, Rhodococcus, Tsukamurella, Gordonia, Nocardia, Skermania, Dietzia*) [63–65]. The modification is introduced during the synthesis of the UDP-precursors and generally is present only on a fraction of the muramic acid residues. Therefore, bacteria containing *N*-glycolylated muramic acids also contain MurNAc residues in the PG layer. The role of the *N*-glycoloyl group as a substituent has been the object of speculation but is still to be clarified. It has been proposed that the extra hydroxyl would engage in hydrogen bonding within the cell envelope with stabilizing effects [66].

Intramolecular amide bond formation between the amino group and the carbonyl of the lactyl group in MurNAc generates the muramic acid δ-lactam modification which appears to be quite abundant in *Bacillus* species and *Clostridium sporogenes* spores. For this modification to occur, the lactyl group must not have any peptide attached and the MurNAc residue must be *N*-deacetylated. This process is generally handled by just two enzymes. In *B. subtilis* for instance, an amidase CwlD removes the peptide that is linked to MurNAc and the peptidoglycan MurNAc *N*-deacetylase PdaA cleaves the acetyl group of the sugar residue [67]. Their role seems to be related to the spores' ability to complete the germination process to produce viable cells. The muramic acid δ-lactam modification is employed by germination-specific hydrolases as a marker to recognize the spore's PG [68–70].

The *O*-acetylation in PG appears to be uniquely present at the MurNAc residues [71,72]. First reported in *Micrococcus luteus* and *S. faecalis*, it is present in both gram-positive and gram-negative bacteria (e.g., *B. cereus, Staphylococcus aureus, Enterococcus hirae,* and *S. pneumoniae* for the gram-positive; *Neisseria gonorrhoeae, Neisseria meningitidis, H. pylori,* and *Proteus mirabilis* for the gram-negative ones). As a structural variation, *O*-acetylation is more prevalent than *N*-deacetylation and the extent of it varies between <20% and 70% depending on the species and strain that is being considered [73–75]. The source of acetyl groups for the *O*-acetylation has been a matter of discussion over the years. It has been proposed to be extracted from the N-2 position of GlcNAc or MurNAc [76]. But there are cases where acetylation in MurNAc6Ac is present in fully *N*-acetylated PG (e.g., *P. mirabilis*); alternatively, it was suggested to be obtained from peptidoglycan turnover products [73]. More recent theories have proposed acetyl-coenzyme A (CoA) or acetyl phosphate to be the source [77].

O-acetylation of PG has an important role in the ability of bacteria to produce disease. The first experimental evidence suggested this substitution to be responsible for the increased resistance towards hen egg-white lysozyme hydrolytic activity [72]. Further shreds of evidence showed the same resistance to egg white lysozyme on 14 strains of *P. mirabilis* containing acetylated PG, and further species lacking *O*-acetylation patterns were shown to be more sensitive to exogenous lysozymes [42,77,78]. Therefore, *O*-acetylation is a contributing factor offering resistance against antimicrobial proteins. Interestingly, the degree of *O*-acetylation seems to be negatively affected by β-lactam antibiotics that target proteins that are related to transpeptidation reactions in the synthesis of peptidoglycans. *O*-acetylation of peptidoglycan appears to play a not-yet resolved role in the cross-linking reaction [79–81].

1.2.2. Cell Wall Polysaccharides: Structure and Functions

Gram-positive and gram-negative bacteria produce extracellular polysaccharides which can either be covalently bound to the cell envelope or surface proteins or be a slime that is weakly attached to the bacterial cell wall. Being present at the interface between microbes and mammalian hosts–other than representing recognition elements for pathogens–these glycoconjugates often contain antigenic moieties that trigger an immunogenic response in hosts. For this reason, these glycans and their biosynthetic pathways have experienced increased attention from the scientific community during the last decades.

As a result of Red Queen evolutionary dynamics [82], the structures of bacterial glycans present the highest variability compared to other non-bacterial classes of glycans and cannot, therefore, be easily generalized. As previously reported, the term "Red Queen" effect takes inspiration from Lewis Carroll's quote "it takes all the running you can do, to keep in the same place" borrowed to explain the glycan structural variation that contribute to the bacterial diversity in nature. As above mentioned, bacteria have historically been divided based on the outcome of the gram staining procedure. In gram-negative bacteria, glycans are mostly present in the form of lipopolysaccharides (LPSs). A typical structure of an LPS polymer is represented in Figure 4.

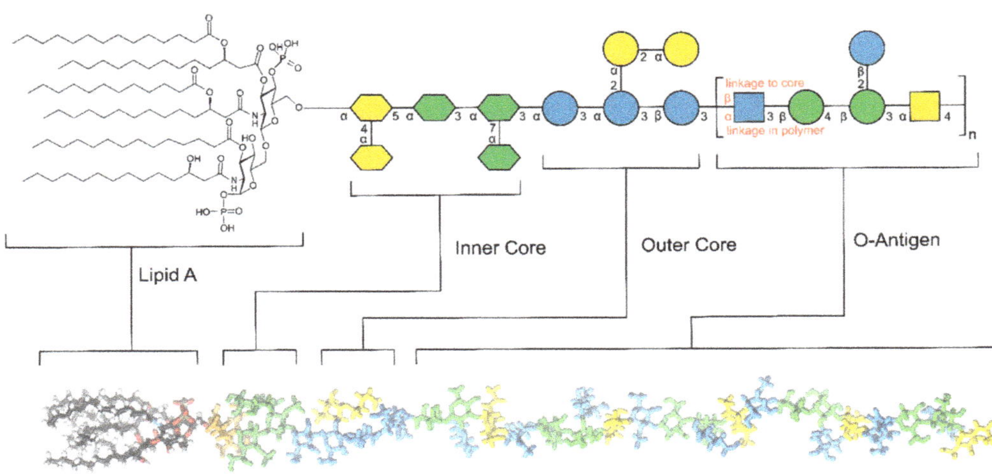

Figure 4. Lipopolysaccharide with its O-antigen from *E. coli* O6 built with *LPS modeler* [83] and rendered in VMD. Figure adapted from Furevi [84].

LPS is generally divided into three sections: the lipid A, the core oligosaccharide (inner and outer core), and the O-antigen polysaccharide. The lipid A (or endotoxin) is the innermost and hydrophobic region of the lipopolysaccharide and possesses a conserved structure that is composed of two β-(1→6)-linked glucosamine residues often phosphorylated at the reducing end and non-reducing end residues in positions 1 and 4′, respectively. Acyl groups are condensed at the hydroxyl and amine groups in a variable number [85,86]. Due to this common architecture, most of the lipid A moieties are detected at picomolar levels by receptors that are present on immune system macrophages and endothelial animal cells [86], and a multitude of enzymes that are linked to the lipid A biosynthesis (e.g., LpxC) have over the years been validated as a target for newly developed antibiotics.

Discovered in the late 1800s and isolated from *E. coli*, it was confirmed that LPS was responsible for the pathogenic (endotoxic) properties, which correlated to pathological outcomes (e.g., fever and septic shock). Lipid A is part of pathogen-associated molecular patterns (PAMPs), i.e., molecules that can activate innate immune systems when they are recognized by toll-like receptors (TLRs) or other pattern recognition receptors (PRRs).

Roughly 10^6 lipid A residues are present in a single cell of *Escherichia coli* [87]. The fatty acids are somehow variable in structure and chain length but only one type of fatty acid is amide-linked, i.e., (*R*)-3-hydroxy fatty acids. The length of these fatty acids is 10–20 carbon atoms per chain and if the third carbon atom carries a hydroxyl group it is most often substituted by an additional fatty acid in at least one of the sugar residues [88–91].

Some species have been shown to contain a keto group in place of the hydroxyl in the *N*-acyl fatty acids [92,93]. Besides that, (*R*)-3-hydroxy fatty acids are predominant also for the ester-bound fatty acids, but these are still more variable and species-dependent. These are fatty acids from myristic and lauric acids and their hydroxylated derivatives, to even more unique structures such as *cis*-11-octadecenoic acid, 3-hydroxy-5-dodecenoic acid, and iso-2,3-dihydroxytetradecanoic acid [94].

The lipid A is covalently linked to heteropolysaccharides via a core region consisting of two parts, viz., first an inner core containing one to four units of Kdo (3-deoxy-D-manno-oct-2-ulosonic acid) [95], as well as L-*glycero*-D-*manno*-heptose residues, which are found in many inner cores. Additional charged elements such as phosphate and uronic acids may be present in the inner core structure providing a binding site for divalent cations (Mg^{2+} and Ca^{2+}) with the effect of stabilizing the outer membrane [96].

The second portion of the core, referred to as the outer core is slightly more variable but structural modifications are still limited if compared to the O-antigen. This last complex repeating unit determines the serological and antigenic properties of the LPS and possesses a high structural variability. This diversity is reflected in the wide range of residues composing the repeating unit, which includes uronic acids, amino sugars, methylated and acetylated derivatives, deoxy sugars, but also non-sugar moieties such as amino acids and phosphate groups. The variability is further extended by the linkage diversity between the different residues in the branched and non-branched repeating units. The number of sugars in each repeating unit is typically four or five, although up to seven sugar residues have been reported for *E. coli* [97]. The total number of residues in a single chain can add up to several hundred, as for some bacterial species up to one hundred repeating units have been shown to be the outcome of the O-antigen biosynthesis. Hence, even a single bacterial species can account for hundreds of different O-antigen structures. *E. coli* is one of the most characterized species in terms of serotypes having–to date–197 reported O-antigens [97].

Over the years databases have been created to collect the different O-antigen structures of various bacterial species from a structural and genetic point of view [98–100]. Gram-positive bacteria contain a wide range of glycans on their cell wall. Similarly to glycans from gram-negative bacteria, these are important structural elements with significant patho/physiological relevance, other than representing key surface units for cellular recognition and signaling. Some of the most common and well-studied glycans from gram-positive bacteria are the cell wall teichoic acids (WTAs), which consist of copolymers of glycerol phosphate or ribitol phosphate and carbohydrates that are linked together by phosphodiester bridges.

These polyanionic glycopolymers play critical roles in the cell with functions that include cell morphology and division, autolytic activity, antibiotic resistance, metal ion homeostasis, phage-mediated horizontal gene transfer, and protection of bacteria from host defense peptides and antimicrobial peptides [101,102]. For their importance, these structures and their biosynthetic pathways represent attractive targets for the design of antibiotics and vaccines.

The WTA polymers can account for up to 60% of the cell wall mass containing between 40 to 60 polyol repeats. Structurally, WTA can be divided into two parts: the main chain polymer and a linkage disaccharide that acts as a bridge between the main chain and the peptidoglycan (Figure 5) [101,103,104].

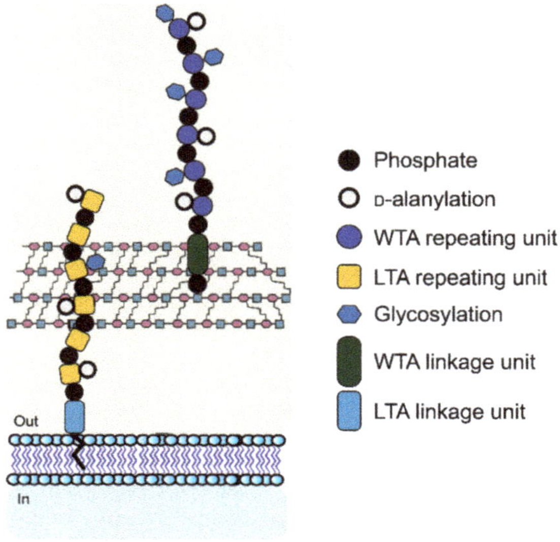

Figure 5. Schematic representation of LTA and WTA in the gram-positive cell wall. The WTA extends beyond the PG layer while LTA may not be able to become elongated past it.

The disaccharide unit is highly conserved across species and its structure is β-D-ManNAc-(1→4)-α-D-GlcNAc-1-*P*, where *P* represents a phosphodiester group that is linked to the hydroxyl group at position 6 of *N*-acetyl muramic acid of the PG layer. The *N*-acetyl-β-D-mannosamine residue is substituted in position O4 with one to two glycerol 3-phosphate units [105,106].

The most common WTA repeating units contain ribitol 5-phosphate (RboP) or glycerol 3-phosphate (GroP), but there is great variability in the WTA monomer structure with more unusual moieties [107–109]. Additional structural diversity may arise from the presence of substituents that are attached to the hydroxyl groups of the polyol repeating units, e.g., D-alanine, monosaccharides such as Glc or GlcNAc, or oligosaccharides [110].

The presence of these substituents is important in bacterial defense against a host and antibiotics. The D-alanylation, for instance, has been proven to protect against host-defense mechanisms [111,112]. A reduction or complete absence of D-alanine residues was shown to be correlated with an increased susceptibility to phagocytes, neutrophils, lysostaphin, and lysozymes, together with glycopeptide antibiotics and cationic antimicrobial peptides [113–117]. These structural changes may thus be correlated to the overall change in the charge of the membrane [116]. While the specific removal of β-D-GlcNAc modification has been reported to increase sensitivity to β-lactam antibiotics [118].

The second most characterized cell-wall glycopolymer in gram-positive species are the lipoteichoic acids (LTAs), which, similar to WTA, are zwitterionic polymers, but with a simpler structure compared to WTA. LTA typically consists of a polyglycerolphosphate (PGP) chain that is linked to the bacterial membrane via a glycolipid anchor (see Figure 5) [119]. As in WTA, the backbone of LTA is modified with alanine or glycosyl residues. Their structures also contain an initial linkage unit which is mostly represented by a disaccharide that is linked to diacylglycerol (DAG). In *S. aureus* and *B. subtilis*, the disaccharide unit is *β*-D-Glc(1→6)-*β*-D-Glc.

LTA can coexist with WTA in the same bacteria and their functions are generally similar. In bacteria containing only LTA moieties, the lack of these may lead–in addition to the same roles that are observed for WTA–to morphological defects such as the increased size of the bacteria or temperature-sensitive growth phenotypes [120,121]. Besides their location in the membrane, where LTA are tethered to the membrane via van der Waals forces and WTA are covalently attached, the LTA generally do not exceed in length past the peptidoglycan membrane while WTA extends further out into the extracellular space.

While WTA and LTA structures are possibly the most common and well-known membrane glycans in gram-positive bacteria, the cell walls of many gram-positive bacteria of the genera *Streptococcus*, *Enterococcus* and *Lactococcus* lack these polyanionic structures and have evolved glycopolymers that are characterized by the presence of rhamnose. These rhamnose-containing CWPs (RhaCWPs) are of particular interest due to the total absence of L-rhamnose in humans, which makes their biosynthetic pathway an attractive therapeutic target. Indeed, L-rhamnose is very often essential for bacterial virulence or viability [122].

First reported by Rebecca Lancefield in the 1930s, these structures have historically been used in the serological classification of streptococci, which initially were used to define Group A–E streptococci and later expanded to more than 20 serogroups [123,124], Figure 6 [125].

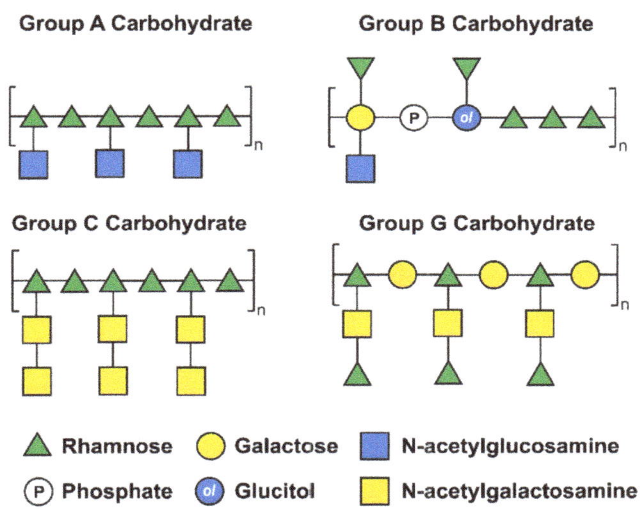

Figure 6. Different types of streptococcal glycan repeating units according to the Lancefield classification. The representation follows the shapes and colors of the SNFG representation for carbohydrates [125].

Over the years, the classification became superfluous since it was found to be unable to differentiate between different species. It is still commonly accepted that a single streptococcal species can express different Group antigens (e.g., *Streptococcus dysgalactiae* subsp. *equisimilis* strains) [126]. Similarly, to WTA, the RhaCWP comprises up to 60% of the dry cell wall mass. They are mainly localized on the outermost surface of the cell wall, but they can also be found intercalated in the peptidoglycan net [127–129].

Streptococcal species seem to lack orthologues of the genes that are associated with the expression of WTA (e.g., TagB, TagD, TagF) [130], while Lactococcal and Enterococcal species were reported to contain orthologues for WTA and both WTA and LTA biosynthesis, respectively, thus having a more heterogeneous expression of glycans on their surface [131–135]. The main constituents of the RhaCWP are rhamnose with variable combinations of Glc, GlcNAc, Gal, GalNAc, and phosphate groups, that are differently linked [136]. Even though the structure of the polymers may vary significantly across species, some exceptions show structural similarities, such as GAC and GCC. These different motifs reflect the discriminatory ability of the Lancefield scheme that is based on the diverse structural elements that are present in the different groups. Some of the known structures for the Lancefield Groups A, B, C, and G are shown in Figure 6. Variations of these structures have been reported in different works and serve as homologs of WTA in terms of biological functions [137].

The first studies on the biological functions of RhaCWP indicated a strong relationship between these and the structural identity of the cell wall [138]. The lack of RhaCWP expression was shown to lead to overgrowth and cell division abnormalities [138–141]. Inhibition of the UDP-GlcNAc:lipid phosphate transferases in *S. galactiae* [140] and *S. pyogenes* [142] was shown to cause important morphological alterations such as a reduced level of cross-linked PG, increased chain length, and mislocalization of PG hydrolases.

RhaCWP also represents an important phage receptor in many species, as demonstrated in several studies, forming epitopes in conjunction with side-chains of the polymer (e.g., GlcNAc, Glc) [143–145]. Additional roles are associated with the impact on virulence of the bacteria, modulated by structural modification on the RhaCWP, still maintaining unaltered bacterial physiology [133,142,146] and cell wall morphology [142]. Furthermore, knowledge of the biosynthetic pathways of RhaCWP facilitates the identification of attractive targets for developing new antimicrobial agents with limited risk of side effects due to off-target actions, since humans completely lack L-rhamnose in their system [147–149].

2. Drugs Targeting CWPs and PGs

2.1. Antibiotics

Common therapeutic strategies focus on the direct inhibition of the biosynthesis of the cell wall polysaccharides, blocking bacterial growth and host infection. Some widespread and well-known drugs that are recognized as inhibitors of cell wall biosynthesis are *β-lactam drugs* (e.g., penicillin) and *glycopeptides* (e.g., vancomycin and teicoplanin). Other antibacterial compounds are known for a different mechanism of action, viz., by disrupting the membrane structures of bacteria, as carried out by, e.g., polymyxins and daptomycin [150–152].

2.2. Carbohydrate-Based Antibiotics Targeting CWPs and PGs: Glycopeptides and Lipoglycopeptides

Glycopeptides and lipoglycopeptides are the representative antibiotics that are known to target the synthetic pathway of the bacterial cell wall [153]. Glycopeptide antibiotics are very important in the antibacterial treatment of gram-positive and gram-negative bacteria. In particular, lipoglycopeptides can dimerize, enhancing the binding affinity (hence the overall potency) for the peptidoglycan cell wall. The membrane anchoring is mainly achieved by the hydrophobic moieties [154].

Vancomycin, depicted in Figure 7, is a glycopeptide drug, whose name comes from the word "vanquish". In the 1950s, it was found in *Streptomyces orientalis* (now *Amycolaptosis*) resident in the Borneo jungle [155]. It was demonstrated in 1958 to be clinically important in the treatment of gram-positive penicillin-resistant staphylococci. Since then, it has been the best treatment for hospitalized patients with severe infections [153]. Due to its potential toxicity and the rise of semisynthetic penicillin that is potent on staphylococci, the interest in vancomycin decreased. However, some drug-resistant gram-positive bacteria became a threat, such as methicillin-resistant *Staphylococcus aureus* (MRSA); hence, vancomycin returned as an antibiotic. In routine treatments, vancomycin has been indicated for important infections such as meningitis, endocarditis, osteomyelitis, bacteremia, and skin infections that are caused by gram-positive strains. In particular, it is crucial in cases of intolerance-allergy to β-lactam antibiotics, or in the case of MRSA [153].

In the 1980s, vancomycin-resistant strains of gram-positive bacteria spread, for example, vancomycin-resistant enterococci (VRE), even if vancomycin is recognized to have a broad spectrum of activity [156]. Vancomycin showed its potency against almost every strain of *Staphylococcus aureus*, including the methicillin-resistant ones. *Leuconostoc* spp., *Listeria monocytogenes*, and *Lactobacillus* spp. are the ones that are not sensitive to vancomycin efficacy [153]. The broad spectrum of vancomycin includes most *S. epidermidis*, all *Streptococcus pneumoniae* [157], and all *Streptococcus pyogenes* that have been tested, most of

Streptococcus bovis and *Streptococcus viridans*, and *Enterococcus* (only a few resistant) strains. Diphtheroids, the few strains of *Listeria monocytogenes*, most strains of *Clostridium species* (but some strains of *C. ramosum* are slightly resistant), half of the strains of *Actinomyces* species are also vancomycin-sensitive. All strains of the gram-negative bacillus *Flabacterium meningosepticum* resulted in resistance [158].

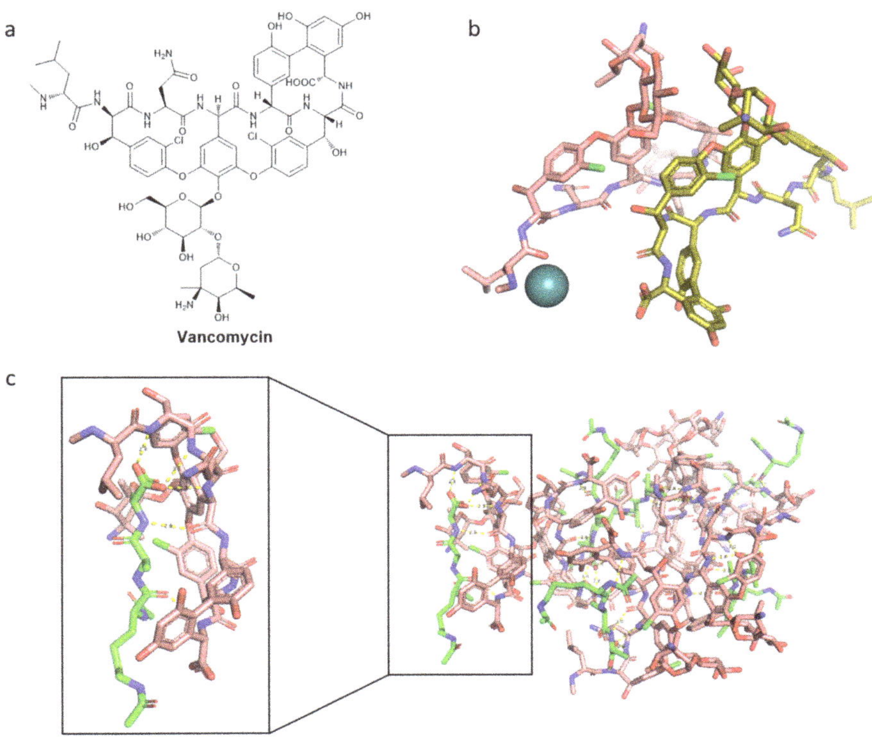

Figure 7. (**a**) 2D structure of vancomycin. (**b**) PyMOL 3D structure of two vancomycin molecules with a Zn^{2+} ion (PDB code 5M2K, ethanediol was removed); (**c**) 3D representation of the complex of vancomycin (colored in salmon) and D-Ala-D-Ala (in green, PDB ID: IFVM).

Considering the mechanism of action of vancomycin, it has a different mode of inhibition of cell wall biosynthesis in comparison to β-lactams. It prevents the polymerization of the phosphodisaccharide-pentapeptide-lipid complex, in the second biosynthetic step consisting of cross-linking between the peptide chains [159]. Vancomycin strongly binds to peptides presenting a D-alanyl-D-alanine repeating entity at the free carboxyl end [160]. Vancomycin is a bulky drug; hence, it creates an extensive steric hindrance when it is bound to the peptide, not allowing for binding to the substrate site on the enzyme peptidoglycan synthetase. Thus, it blocks the peptidoglycan biosynthesis. As its secondary mode of action, vancomycin was also more recently shown to interfere with the cell membrane permeability [161,162].

Vancomycin (Figure 7) is still essential in the treatment of methicillin-resistant staphylococci infections, with the help of other important antibiotics: first- and second-generation cephalosporins, co-trimoxazole, clindamycin, or rifampicin. However, as already mentioned, it is relatively toxic, with a high incidence of side effects (e.g., red man syndrome). It is also not intramuscularly administrable, it is expensive, and not so active against enterococci [163].

Another important glycopeptide is teicoplanin. It was isolated from the actinomycete *Actinoplanes teichomyceticus* [164], with the original name of teichomycin A2. Teicoplanins refer to five structurally similar glycopeptide antibiotics, possessing a tetracyclic backbone, which is reminiscent of the tricyclic motif of vancomycin. It is a reliable alternative to vancomycin, with different advantages in comparison to the latter due to its clinical efficacy at lower dosages in severe staphylococcal infections [163]. Other advantages are the lower incidence of side effects, such as the already-described red man syndrome and the lower toxicity (in particular in co-administration with an aminoglycoside) [165]. However, teicoplanin has been commercialized in Europe and Asia, but not in the USA [153].

Many types of gram-positive bacteria are sensitive to teicoplanin, which has bacteriostatic and bactericidal activity). Some families are involved in in vitro sensitivity, such as staphylococci (even the β-lactamase-producing and methicillin-resistant strains; *S. aureus*, *S. epidermidis*, *S. hominis*, *S. haemolyticus*, *S. saprophyticus*), streptococci (*S. pyogenes*, *S. pneumoniae*, *S. agalactiae*, *S. bovis*, *S. milleri*, *S. mitis*, *S. sanguis*, etc.), enterococci, and several anaerobic ones (*Clostridium difficile*, *Clostridium perfringens*, *Propionibacterium acnes*, *Listeria monocytogenes*, and *Corynebacterium jeikeium*).

Teicoplanin, depicted in Figure 8, shares the same mechanism of action with vancomycin: saturation by aspecific binding to the outer layers of PG, then binding to the terminal amino acyl-D-alanyl-D-alanine in PG. It follows a blockade of PG elongation, inhibiting the biosynthetic step of transglycosylation and interfering with the function of transpeptidases [166–168]. This glycopeptide has an "open pocket" conformation; the "seam" is represented by the heptapeptide backbone [169]. The unit *N*-acyl-D-alanyl-D-alanine fits into the pocket [170]. In particular, teicoplanin inhibits the biosynthesis of the cell wall and cellular growth. Most likely, cell death is induced by cellular hydrolytic enzymes. The PG of gram-negative bacteria is sensitive to teicoplanin, but the outer lipid membrane repels this bulky and polar drug [166,170].

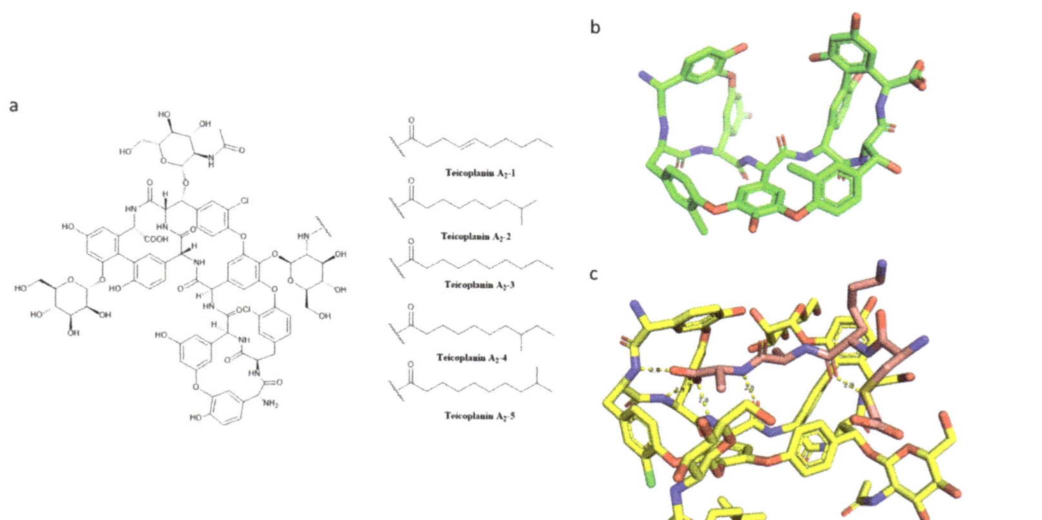

Figure 8. (**a**) 2D structure of teicoplanin, in its different forms. (**b**) 3D structure of teicoplanin aglycone (TAG: PDB ID 6TOV), DMS (dimethyl sulfoxide was removed from the original pdb file of the crystal structure) [171], and (**c**) teicoplanin in complex with D-Ala-D-Ala residue with polar contacts (PDB ID: 3VFJ).

In a comparison of teicoplanin to vancomycin, the first is several times more potent against enterococci compared to the latter. The relationship with other antibiotics could be different; the co-administration with rifampicin is additive or sub-additive, and the mix with ampicillin antagonizes the β-lactam-based antibiotic. However, teicoplanin has a certified, but also combination-dependent, synergic effect on gram-positive cocci with other antibiotics, such as aminoglycosides, imipenem, and fosfomycin.

As above mentioned, vancomycin is crucial for antibacterial treatment, mainly for the worldwide spreading of MRSA [172,173]. Clinical failure is step by step increasing for MRSA treatment [174]; hence, it brought about the more effective semi-synthetic (lipo)glycopeptide antibiotics (with prolonged efficacy, thus less frequent dosage), such as telavancin, dalbavancin, and oritavancin [153,175]. A disadvantage is their accessibility as intravenous administration and their limited use because of their high cost. They mainly have a therapeutic indication for skin infections. The side effects are also a significant issue for these semi-synthetic lipopolysaccharide antibiotics: QT (the interval between the start of Q wave to the end of T wave measured on an electrocardiogram) prolongation and nephrotoxicity for telavancin and an increase of coagulation episodes for telavancin and oritavancin [153]. Telavancin (TD-6424, or Vibativ) is a vancomycin-derived lipoglycopeptide (for the decylaminoethyl substituent) which represents a valuable drug against the increased spreading of MRSA infections, and in general, infections that are caused by resistant gram-positive bacteria [175,176]. 2D and 3D structures of telavancin are presented in Figure 9. To potentiate the vancomycin antimicrobial activity, telavancin was developed in the context to improve the hydrophobicity at the disaccharide amine moiety. In this case, the hydrophobic portion is represented by a decylaminoethyl side chain on the vancosamine part and a phosphonomethyl aminomethyl functional group in position 4 on aminoacid 7 [177]. Telavancin offers different advantages, more related to specificity and pharmacodynamics properties. An increased affinity profile is offered by the glycopeptide moiety toward the D-Ala-D-Ala-containing peptidoglycan intermediates, and the negatively charged phosphonic acid moiety provides a more efficient urinary excretion [177,178]. The drug shows fewer problems after infusion and could be administered just once per day [176].

FDA-approved telavancin is used for complicated skin and soft tissue infections (cSSTIs), hospital-acquired bacterial pneumonia (HABP), and ventilator-associated bacterial pneumonia (VABP) that is caused by susceptible isolates of *S. aureus*, without current specific treatment. Different trials were conducted on telavancin for MRSA bacteremia. A randomized Phase II clinical trial (NCT00062647) confirmed that all the patients were cured [179]. It was also investigated for the treatment of *Staphylococcus aureus* bacteremia (SAB) [180] in a Phase III clinical trial, randomized, open-label, non-inferiority trial (NCT02208063), which failed due to insufficient statistical data [181,182]. More specific trials are needed to assess further the efficacy of telavancin, investigating the patients with complicated bacteremia more extensively [182]. Its mechanism of action, recognized to have dual activity, is mainly due to the interaction between the carboxylate binding pocket with terminal D-Ala-D-Ala residues and the decylaminoethyl side chain with the cell membrane [178,182]. Focusing on the D-Ala-D-Ala terminus, telavancin inhibits the peptidoglycan biosynthesis binding to *N*-acetyl-glucosamine-*N*-muramylpentapeptide of the peptidoglycan. The consequences of telavancin are the inhibition of different biosynthetic steps, such as the transglycosylation (PG polymerization) and the subsequent transpeptidation (generation of cross-links). In MRSA cells, telavancin is 10-fold more active than vancomycin in the inhibition of peptidoglycan synthesis [178]. The second mechanism of action is achieved by the decylaminoethyl side chain, which targets the cell membrane. This affects the cell membrane potential, through rapid concentration-dependent dissipation. The consequence is the generation of membrane pores [178], which leads to the extracellular leakage of adenosine triphosphate (ATP) and K^+ ions. This second activity is specific for bacterial membranes, not for mammalian biomembranes.

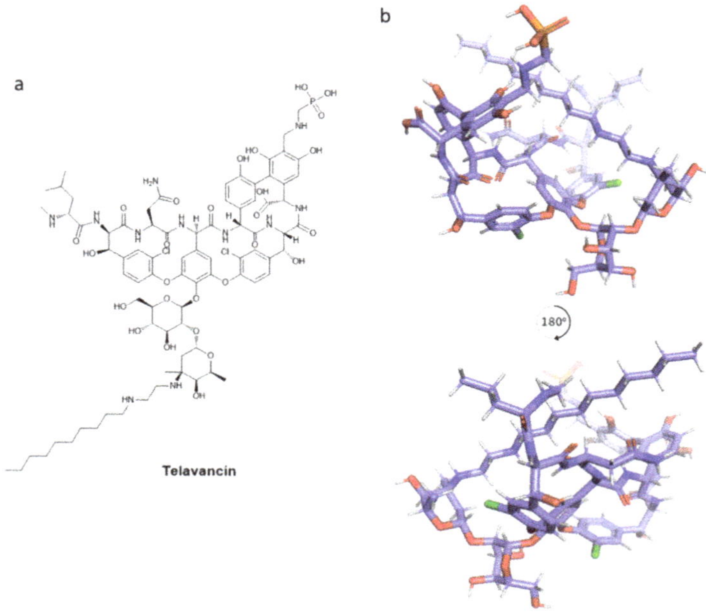

Figure 9. (**a**) Telavancin 2D structure and (**b**) 3D conformation of the minimized structure of telavancin (purple) that was obtained with PyMOL.

Clinical and surveillance studies involving the antibiotic spectrum and activity of telavancin have been described [183–185]. In general, telavancin showed remarkable in vitro activity against skin and skin structure infections and pneumonia that is provoked by gram-positive bacteria (pneumonia could not be treated with daptomycin because it does not target lung tissue) [176,183]. Telavancin is particularly active against bacteria that are resistant to methicillin, daptomycin, and linezolid. It is also indicated for the treatment of vancomycin-intermediate *S. aureus* (VISA) and heterogeneous VISA, *Clostridium* species, including *Clostridium perfringens* and *Clostridium difficile*, *Peptococcus anaerobius*, and *Bacillus anthracis*. A synergic activity was demonstrated for the co-administration of telavancin with gentamycin and rifampicin [185–187]. To visualize telavancin, a structure representation was generated through energy minimization via Chem3D and then Avogadro (algorithm Steepest descent, force field MMFF94s), resulting in potential energy minimized 3D structure (Figure 9).

Oritavancin (previously LY333328, then KIMYRSA™, or ORBACTIV®) is a "long-acting" intravenously administered lipopolyglycopeptide that is approved by FDA for adult patients with acute bacterial skin and skin structure infections (ABSSSIs) that are caused by gram-positive bacteria. It is known for its antibiotic activity, efficacy, and safety. Other minor, but important advantages are a simpler preparation, and shorter infusion time (1 h) in a lower volume (250 mL) [188]. The initial novelty of oritavancin was its activity against vancomycin-resistant *Enterococcus faecalis* [189]. With its different hydrophobic and hydrophilic groups, it differs from vancomycin having an additional unsubstituted saccharide moiety and an aromatic lipophilic side chain [189]. It has a higher potency compared to vancomycin [190].

As for the mechanism of action, it acts by killing rapidly growing bacteria in a concentration-dependent way (unlike vancomycin) [154]. The inhibition of the peptidoglycan synthesis is achieved via different pathways: transglycosylation blockade, transpeptidation (cross-linking) inhibition, and membrane potential perturbation (depolarization and increased permeability) [189]. As for the spectrum of activity of oritavancin, it acts against gram-positive bacteria, remarkably against nosocomial infections that are caused by vancomycin-resistant enterococci (VRE) [191–194].

Other notable incidences were found against Micrococcus species, *Listeria monocytogenes*, and Corynebacterium species, but also anaerobic gram-positive bacteria such as *Clostridium difficile*, *Clostridium perfringens*, *Peptostreptococcus* species, *Peptococcus* species, and *Propionibacterium acnes* [191]. Oritavancin has a well-known synergy with gentamicin against VRE, and with gentamicin, linezolid, moxifloxacin, and rifampin [176,195]. There is no crystal structure involving oritavancin in the Protein Data Bank (PDB) [196]; hence, energy minimization via Chem3D and then Avogadro was performed, and the resulting conformation is displayed in Figure 10.

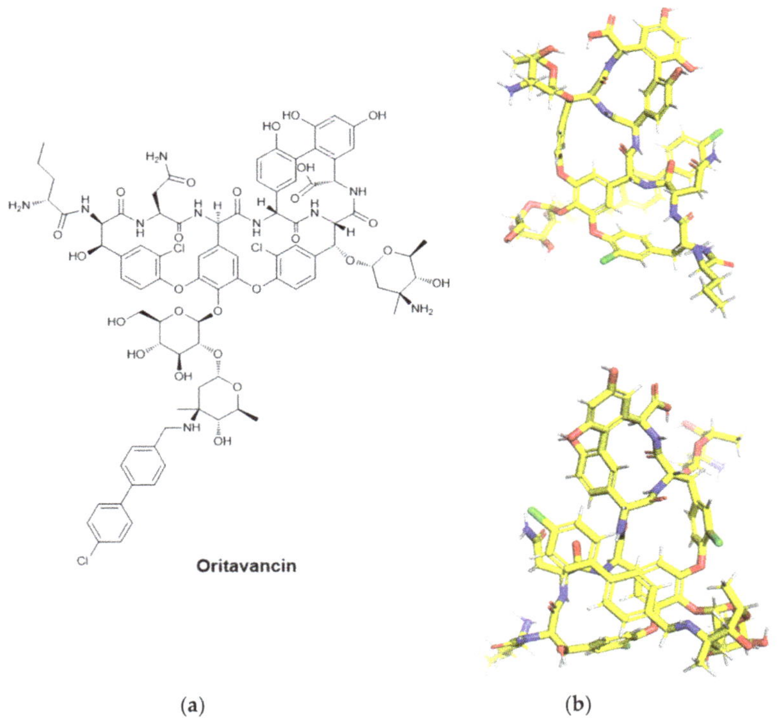

Figure 10. (**a**) 2D representation of oritavancin; (**b**) 3D energy minimized structure of oritavancin (yellow), made with PyMOL.

The last semi-synthetic lipoglycopeptide that is presented is dalbavancin (Figure 11), which has a therapeutic indication for ABSSSI that is caused by gram-positive bacteria. In 2005, a Phase II clinical trial study demonstrated that dalbavancin is beneficial as an alternative to vancomycin for catheter-related bloodstream infections, even if more data are needed in support of this [197]. Dalbavancin has a good safety profile [192], but data are needed to evaluate its importance in MRSA bacteremia [182]. A Phase II clinical trial focused on the comparison between dalbavancin and the standard of care (SOC) therapy for the treatment of complicated bacteremia or right-sided endocarditis that is caused by *S. aureus* will be completed in August 2023 [182]. In Figure 11b,c, several structures of dalbavancin are presented as blue sticks, in complex with human serum albumin (HAS); their association with HAS is essential for clinical use [198].

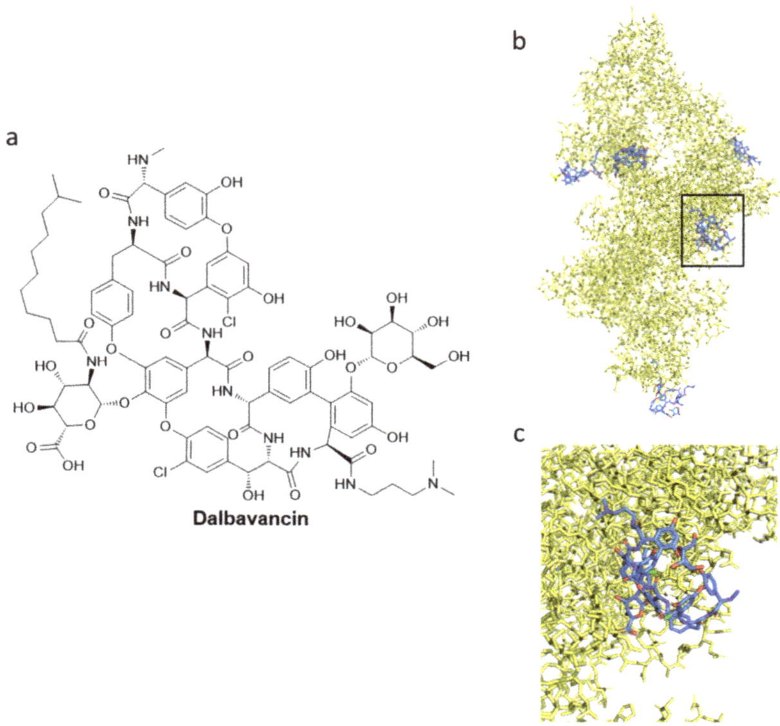

Figure 11. (**a**) Chemical structure of dalbavancin; (**b**,**c**) crystal structure (entire structure and close-up, respectively) of dalbavancin (blue) in complex with human serum albumin (HAS, yellow), PDB ID: 6M5E [198]. Figure made by ChemDraw and PyMOL.

2.3. Antibiotics Targeting CWPs and PGs: β-lactams, Polymyxins, and Daptomycin

Due to the complexity of bacterial diseases (especially MDR ones), the treatment is carried out by combination therapies to gain activity through several mechanisms of action [199]. Penicillin, the "wonder drug", is the oldest pure antibiotic worldwide that is available [200]. Penicillins (depicted in Figure 12) possess a β-lactam ring as the main chemical scaffold of the molecule.

The peptidoglycan (PG) layer generates cross-links at its C-terminus thanks to specific enzymes, transpeptidases (TPDs) [201]. The cross-linked PG polymer is essential for maintaining the membrane shape and prevents cellular osmotic rupture [150,151,201]. TPDs are PBP (penicillin-binding protein) enzymes because they can be inhibited by β-lactams such as penicillins [201]. β-lactams bind and inhibit PBP enzymes ending in bacteriolysis, hence cell death [150,151,201]. Penicillin-derived compounds are divided into natural (e.g., penicillin G), penicillinase-resistant (e.g., oxacillin), amino-, (e.g., amoxicillin), carboxy- (e.g., carbenicillin), and ureido- (e.g., piperacillin) penicillins. Natural penicillins are mostly active against gram-positive bacteria (streptococci, enterococci, and some non-β-lactamases-producing staphylococci), while synthetic ones (e.g., aminopenicillins) can be active also against some gram-negative bacteria such as non-resistant *H. influenza*, *N. gonorrhoeae*, and *E. coli*. Even if penicillin resistance is an increasing problem, this old class of antibiotics continues to lead to antibacterial therapy [202].

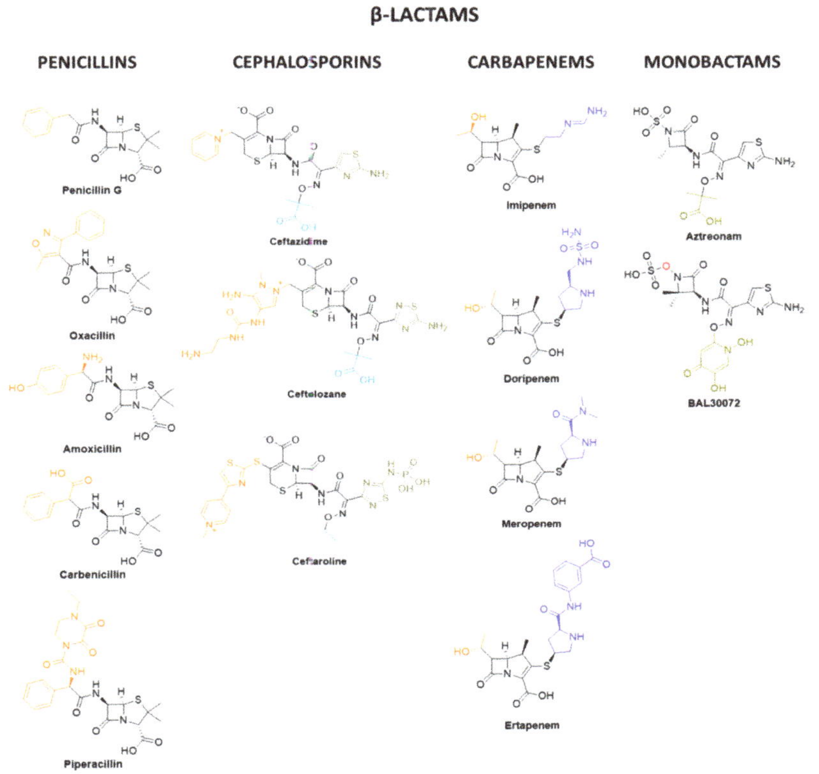

Figure 12. Structure of representative penicillins (penicillin G, oxacillin, amoxicillin, carbenicillin, and piperacillin); cephalosporins (ceftazidime, ceftolozane, and ceftaroline); carbapenems (imipenem, doripenem, meropenem, and ertapenem); and monobactams (aztreonam and BAL30072).

Moving forward into the family of β-lactam compounds, cephalosporins are important due to their activity against gram-negative bacilli. There are five generations of cephalosporins. Examples of cephalosporin-based therapies are available against most *Enterobacteriaceae* (third-generation ceftazidime + avibactam, their combined effect resulting in β-lactamase inhibition), resistant gram-negative bacilli such as *P. aeruginosa* (ceftolozane + tazobactam, another combination used for β-lactamase inhibition), and methicillin-resistant *S. aureus* (MRSA, with fifth generation ceftaroline as the only available treatment) [203].

Imipenem, doripenem, meropenem, and ertapenem (reported in Figure 12) are some of the drugs composing the well-known class of the carbapenems [203]. They derive from the naturally occurring antibiotic thienamycin, extracted by the soil microorganism *Streptomyces cattleya* [204,205]. They have the broadest spectrum activity compared to other β-lactam classes and are particularly important in case of severe sepsis that is caused by antibiotic-resistant bacteria. In general, they are active against both gram-positive (except methicillin-resistant *S. aureus*, *Enterococcus faecium*, and *Enterococcus faecalis*) and gram-negative bacteria as well as anaerobic species. [206]. The beneficial effect of carbapenems is reflected also in the significant post-antibiotic effect against gram-negative bacteria, such as *P. aeruginosa* [207,208]. Severe problems in using carbapenems for the treatment of infections are represented by the spread of carbapenemase-producing gram-negative bacilli [203].

Monobactams were developed in concomitance to the decrease of the efficacy of β-lactam antibiotics, mainly caused by extended-spectrum β-lactamases (ESBLs). These β-lactam-based drugs are structurally monocyclic and which makes them hydrolysis-stable compared to other β-lactam and are, therefore, interesting for new therapies [209,210]. Aztreonam (AZN, Figure 12) is currently the only available representative of this antibiotic family. It is only active against aerobic gram-negative bacilli [203]. Unfortunately, it is nephrotoxic and weakly immunogenic and is also very expensive. As with other β-lactams, aztreonam binds some transpeptidases and PBPs (only gram-negative ones), leading to cell lysis [211]. The potent activity against many gram-negative strains (also MDR) was demonstrated by BAL30072 (Figure 12), a derivative that entered Phase I clinical trials. However, BAL30072 has not progressed after Phase I in clinical trials in 2014 because of serious side effects, such as high alanine transaminase (ALT) levels in blood [212,213].

The class of lipopeptides is a broad-spectrum antibacterial-based family that is crucial to facing menaces of the highly-spreading MRSA [214] and the expansion of vancomycin-resistance MRSA infections [215–221].

Polymyxins are drugs of last resort but are still effective on dangerous bacteria such as carbapenem-resistant *Enterobacteriaceae*, MDR *Pseudomonas aeruginosa*, and *Acinetobacter baumannii* [216], also fosfomycin, ceftazidime/avibactam and meropenem–vaborbactam are an efficacious treatment for the above-mentioned bacteria [222,223]. They are known to disrupt the outer membrane of gram-negative bacteria [224], causing some antibiotic resistance problems [217]. Polymyxins [218,225] are non-ribosomal cyclic lipopeptides being secondary metabolites of a bacterium, *Paenibacillus polymyxa*; in particular, colistin (polymyxin E) [226,227], derives from *Bacillus polymyxa* var. *colistinus*. Structurally, they are decapeptides, consisting of a cyclic heptapeptide that is linked to a linear tripeptide side-chain acylated at the N-terminus by a fatty acid tail [228]. As for their mechanism of action, the primary electrostatic interaction with lipid A disrupts the outer membrane, following hydrophobic insertion of the fatty acyl chain of polymyxin into the membrane, making the bacterium succumb [229]. Attempts to extend the ring structure, generated by the insertion of additional Dab residues, produced compounds that are significantly less effective than polymyxins [27].

Daptomycin (Cubicin®), derived from *Streptomyces roseosporus* [229,230] has an in vitro selectivity against many MDR and susceptible gram-positive bacteria [231], including methicillin-sensitive and methicillin-resistant *S. aureus* (MRSA), streptococci, and *Enterococcus* (both vancomycin-susceptible strains and VRE) [153]. Daptomycin is not effective against gram-negative bacteria. In Europe, it is administered to adults intravenously once per day for cSSTI and complicated skin and skin structure infections whereas in the US it is used for the treatment of *S. aureus* bacteremia with or without right-sided infective endocarditis [153]. Its unique mechanism of action is extracellular and engages an oligomerization that is calcium-dependent [224]. Then, it binds to and alters the outer membrane, with a subsequent cellular potassium efflux. The final consequence is the depolarization of the membrane [153]. Figure 13 reports the 2D structure of polymyxins B, E (colistin), and daptomycin.

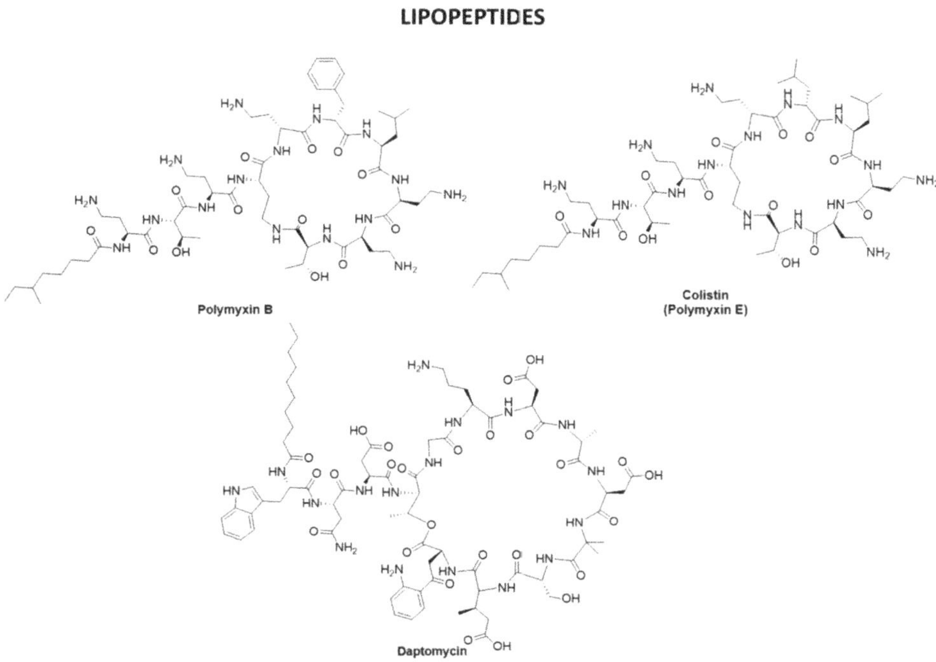

Figure 13. Structure of polymyxin B, colistin (polymyxin E), and daptomycin.

3. Antibiotic Drug Resistance

3.1. Consumption and Therapy

Since the discovery of penicillin in 1928, different classes of antibacterial drugs have been developed; unfortunately, bacteria mutate into resistant strains overcoming the mechanism of action of a specific drug. The increasing use of antibacterial agents made them gradually become less potent, also decreasing the therapeutic options that are available for patients that are affected by resistant strains. In the beginning, the problem was solved through the identification of new classes and the modification of existing antibiotics with limited cross-resistance to already in-use drugs. From the 1980s, the major scientific challenges in antibiotic discovery programs pushed large pharmaceutical companies to abandon this research area. Later, antibacterial resistance slowly became one of the main concerns in global clinics, pushing the scientific community, including some pharmaceutical companies, to re-invest money and efforts in the field. The main strategy that was pursued was to improve the potency of known antibiotic treatments, but new approaches to antibacterial drug discovery were also applied [232].

Recently, the World Health Organization (WHO) initiated a drug antibacterial drug development mapping activity and published a global clinical antibacterial pipeline report. The study highlighted that the clinical pipeline is dominated by derivatives of established classes and most development candidates display limited innovation, suggesting that the finding of new antibacterial agents without pre-existing cross-resistance should be prioritized in public funding strategies [233].

Regarding the preclinical phase, a review that was published in 2020 focusing on discovery and preclinical development projects has found, as of 1 May 2019, 407 antibacterial projects that were carried out from 314 different institutions, most of which are small and medium-sized enterprises (SMEs). The study evidenced that the preclinical pipeline contains 135 projects focusing on direct-acting small molecules belonging to innovative classes, directed towards new targets, or possessing new mechanisms of action, with high levels of diversity, and remarkable scientific concepts [234].

Together with the scarce availability of innovative drugs, the antibiotic resistance crisis has its roots in overuse; epidemiological evidence indicates a direct relationship between antibiotic consumption and the emergence and dissemination of resistant bacteria strains. Despite the warnings, antibiotics are overprescribed worldwide, and, in some countries, antibiotics are unregulated, cheap, and available without a prescription, promoting their overuse. Inappropriate prescribing play a major role in antibiotic resistance and it has been estimated that antibiotic therapy (treatment indication, choice of agent, or duration of the treatment) is incorrect in 30% to 50% of the cases, strongly contributing to the development of resistant bacteria. For instance, sub-inhibitory and sub-therapeutic antibiotic concentrations support genetic alterations, such as changes in gene expression, horizontal gene transfer (HGT), and mutagenesis. Besides promoting the development of antibiotic resistance, the incorrect prescription may also expose patients to potential complications of antibiotic therapy. The extensive agricultural use of antibiotic agents as supplements in livestock to promote growth and to prevent infection were discovered to cause an ecosystem alteration that was related to the rising of resistant bacteria. Moreover, via the handling or consuming food, antibiotics reach consumers as well as resistant bacteria that developed in farms. A schematic representation of the main causes of the antibiotic resistance crisis is depicted in Figure 14. Most of these antibiotics are taken by humans from food; moreover, resistant bacteria in farm animals reach consumers through meat products. In addition, antibiotics that are used in livestock are excreted through urine and stool, and, together with sprayed antibiotics as pesticides, pollute the environment and alter the ecology by increasing the rate of resistant versus susceptible microorganisms [235].

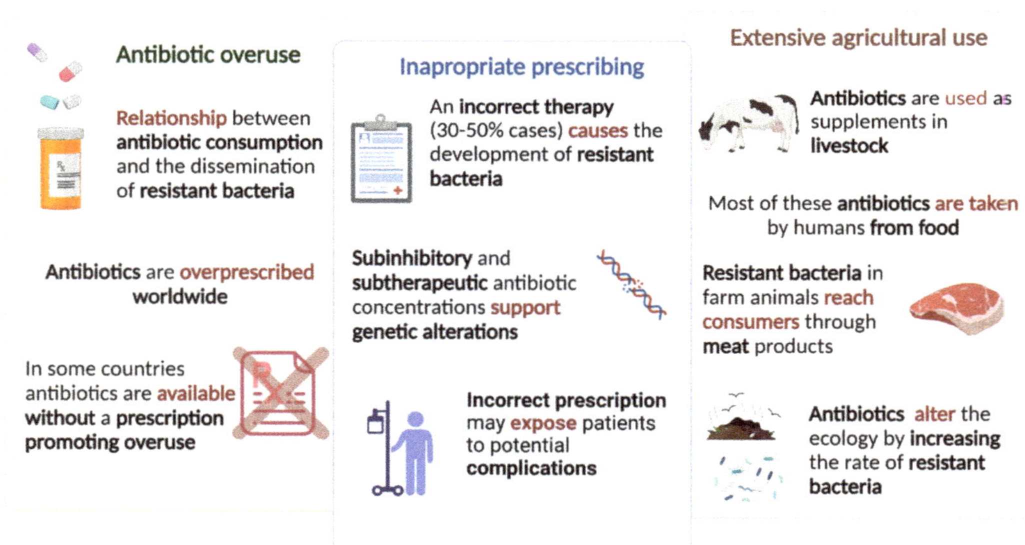

Figure 14. Summary of the roots of the antibiotic resistance crisis. Created with BioRender.com (Accessed on 29 April 2022).

An emerging tool, too long underestimated, that is useful to reduce the development of drug resistance is the prevention of bacterial infections and bacterial spread. Prevention can be pursued by developing new vaccines (vide infra Section 3.3) [236,237]. The prophylactic use of vaccines by reducing the diffusion of pathogens lowers antibiotic use and reduces multidrug resistance. This new role of vaccines in fighting drug resistance has been validate by several studies [238,239].

3.2. Drug Resistance Involving the Outer Membrane (OM)

Antibiotic resistance consists of a decreased sensitivity for a specific antibiotic, which allows the survival of the bacterium. Resistance to antibiotics is a natural phenomenon that has been noted since the introduction of penicillin in 1940 [240]. Very soon thereafter, the key question was raised, i.e., whether bacteria are inherently resistant to drugs or become resistant as a result of encountering the 'antibiotic' compounds. In search of an answer, it has been shown that bacteria can mutually transfer drug resistance to each other by passing genetic material between species. This process mainly occurs through the horizontal transfer of plasmids or bacteriophages, also known as horizontal gene transfer (HGT). Bacteriophages, i.e., viruses that are able to infect bacteria only, are also able to directly affect the bacterial genome. Moreover, the adaptation and evolution of microorganisms should also be taken into account in describing an integral view of the evolution of drug resistance [241].

There are four broad mechanisms that are indicated to lead to phenotypic resistance evolution in bacteria. A schematic representation of the bacterial mechanisms that lead to the evolution of phenotypic resistance is shown in Figure 15. The first one consists of the mutation of the target, resulting in the drug-binding inhibition. Multiple bacterial processes have been exploited as antimicrobial targets such as cell wall synthesis, nucleic acid synthesis, and metabolic pathways. All these targets could be modified by the bacteria to become resistant to the corresponding drug. A second option is related to the alteration of existing genes or the acquisition of new genes encoding drug efflux pumps. These efflux pumps extrude different classes of toxic agents, including a wide range of antibiotics, from the intracellular to the extracellular environment and their upregulation represents a crucial mechanism for acquired bacterial resistance. Also, the acquisition of new genes encoding enzymes that are able to modify or hydrolyze the specific drug could lead to a resistant phenotype. This process sequentially led to the acquisition of genes encoding for penicillinases, cephalosporinases, extended-spectrum β-lactamases, and carbapenemases. These enzymes are responsible for the decreased susceptibility of the related antibiotic class, the last of which has currently no solution. Lastly, the remodeling of the membrane could prevent drug influx at the cell surface, thus impeding access to the target that is located in the internal compartments [242–244].

Most antibiotics that are used in the 20th century bind specific targets that are exposed in the internal compartments of the bacterial cell, thus altering bacterial survival processes.

In gram-negative species, the impermeable LPS-bacterial outer membrane (OM) strongly limits the access into the bacterium; entry of nutrients and other water-soluble molecules, drugs included, is controlled through the "major" (i.e., the most present) porins, aqueous pore that is abundantly expressed and largely integrated into the OM. Porins present a central luminal space that serves as an aqueous pore, which selectively allows the passage of hydrophilic molecules/drugs and whose expression is finely regulated [245]. For instance, in *E. coli*, the two genes OmpC and OmpF encode for the major porins; they are structurally almost identical, but their expression is modulated based on environmental conditions, such as hypo/hyperosmotic conditions (Figure 16).

In general, the OM proteome is frequently subjected to remodeling through two main pathways: the degradation of the present membrane components and the insertion of new proteins. Of the two biochemical processes, the latter has been quite unraveled; newly synthesized porin precursors move from the cytoplasm, across the inner membrane and the periplasmic space before arriving at the OM, where the β-barrel assembly machinery (BAM) complex integrates the porin precursors. Moreover, porin production is regulated in response to environmental changes by a complex system of regulators at both transcriptional and post-transcriptional levels. These regulators include many factors, such as small noncoding RNAs (e.g., micF, micC, micA), or two-component signaling systems, such as OmpR and CpxR [242].

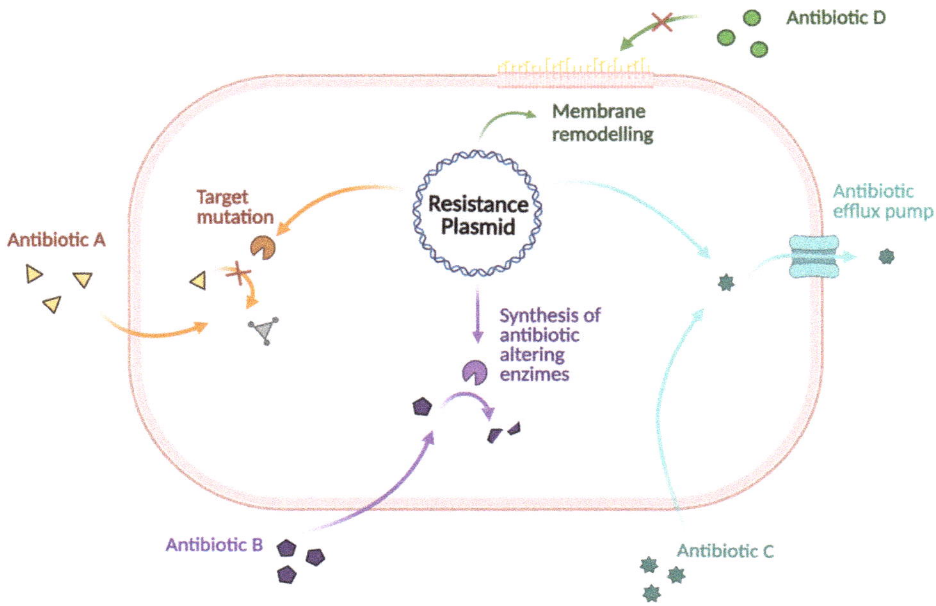

Figure 15. Scheme of mechanisms that lead phenotypic resistance evolution in bacteria. The highlighted paths include the mutation of the target (orange), the synthesis of antibiotic-altering enzymes (purple), the synthesis of genes encoding drug efflux pumps (light blue), and the membrane remodeling (green). Created with BioRender.com (Accessed on 29 April 2022).

Therefore, OM makes gram-negative bacteria, which are considered urgent pathogens by WHO, intrinsically less sensitive to a wide range of antibiotics compared to their gram-positive counterparts. For instance, gram-negative bacteria show natural resistance to vancomycin and teicoplanin. These two big tricyclic glycopeptides ($M_W \approx 1.45$ and 1 7 kDa, respectively) interact with the terminal D-alanyl-D-alanine sequence on the pentapeptide side-chains of the peptidoglycan precursor, hence inhibiting peptidoglycan synthesis. Porins in the OM of gram-negative bacteria selectively allow hydrophilic molecules with $M_W < 600$ Da, thus impeding vancomycin and teicoplanin to reach their target. Therefore, they are considered gram-positive selective antibiotics [246].

Reduction in the number of porins and/or mutations modulating the selectivity of the porin channel could also contribute to limiting drug uptake. For instance, carbapenem resistance in some strains of *Enterobacteriaceae* was aimed at the reduction or abrogation of certain porins production. Conversely, imipenem and cephalosporins resistance in *E. aerogenes* and resistance to β-lactams and tetracycline in *Neisseria gonorrhoeae* were related to a modification in porins selectivity [243].

Since antibiotic resistance frequently results in a complex phenotype, the specific targeting of membrane remodeling, i.e., the alteration of membrane components (proteins or lipids) to allow bacterial adaptation to a new environmental condition, could offer a valid strategy to restore bacterial sensitivity to the available antibiotics. For instance, both the BAM and the transcriptional/post-transcriptional regulators could serve as a drug target, offering an innovative strategy to target OM remodeling.

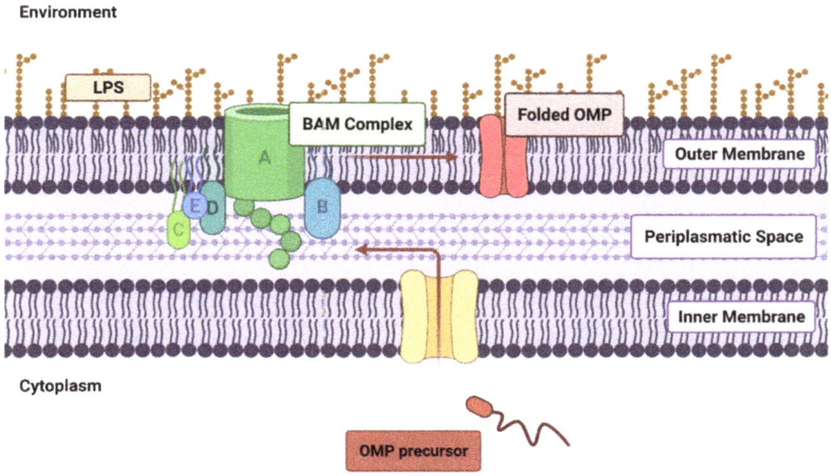

Figure 16. Biogenesis of bacterial OMP. Newly synthesized OMP precursors (red) move from the cytoplasm across the inner membrane and the periplasmic space. The β-barrel assembly machinery (BAM) complex (green), located in the OM, integrates the OMP precursors, creating the folded OMP (orange). Created with BioRender.com (Accessed on 29 April 2022).

Notably, OM has been identified as a potential target for nanoparticles that are designed for this purpose. For instance, it was recently reported that carbohydrate-coated silica nanoparticles were used to target *E. coli*. The study showed that gluconamide moieties on a nanoparticle coating interact with LPS molecules in the OM, allowing for penetration into the membrane without causing cytotoxic effects, thereby enabling drug delivery in a controlled way [247].

3.3. Carbohydrate-Based Antibacterial Vaccines

Vaccination is one of the key achievements in the fight against infectious diseases that has improved the length and the quality of human lives. It is the most efficient and least expensive strategy to prevent infectious diseases. It is based on the principle that stimulating an antigen-specific immune response provides protection from the respective eventual infection. The induced immune response includes the production of specific antibodies with long life and the formation of long-lasting memory B- and T-cells and it will be activated upon encounter with the pathogen [248].

Although vaccines were discovered and evolved from 1796 and onward it was only in the 1920s that the carbohydrate-based vaccines were developed. Avery and Heidelberger discovered that capsular polysaccharides (CPSs) from *Streptococcus pneumoniae* are immunoreactive components of the pathogen. This initiated the use of CPSs as antigens in the development of carbohydrate-based vaccines. Microbial pathogens carry a variously composed pattern of polysaccharides on their surface, based on oligosaccharide repeating units, in part functioning to evade the host immune system.

CPS antigens were discovered to give an age-dependent immune response and to be thymus-independent antigens and thus do not activate T-cells. The sole B-cell activation results in the production of low-affinity antibodies [249]. Protein-conjugated polysaccharides, zwitterionic polysaccharides and glycolipids, being thymus-dependent antigens, are capable of stimulating both B- and T-cells thereby generating high-affinity antibodies and immunological memory. The first CPS-based vaccine was a six-valent vaccine that was developed to prevent *S. pneumoniae* and was approved in 1947 but it was only in the 1970–80s that the more efficient CPS-conjugate vaccines were developed and approved [250,251]. The immunogenicity and efficacy of carbohydrate-based vaccines were effectively enhanced by coupling CPSs to carrier proteins. The proteins that were generally selected as carriers are denatured bacterial toxoids.

At present, several glycoconjugate vaccines are licensed and are commonly used to prevent *Haemophilus influenzae* type b, *Neisseria meningitidis*, *Streptococcus pneumoniae*, and *Salmonella typhi* [252]; a selection of FDA-approved vaccines are listed in Table 3, https://www.fda.gov/vaccines-blood-biologics/vaccines (Accessed on 29 April 2022).

Table 3. Approved carbohydrate-based vaccines; a selection from FDA-approved vaccines. https://www.fda.gov/vaccines-blood-biologics/vaccines (Accessed on 29 April 2022).

Pathogen	Vaccine	Manufacturer and Trade Name
Haemophilus influenzae type b (Hib)	Glycoconjugate, polysaccharide with tetanus toxoid (TT)	Sanofi Pasteur (ActHIB®); GlaxoSmithKline Biologicals (Hiberix®)
	Diphtheria toxoid (DT), TT and acellular pertussis adsorbed, inactivated poliovirus and Hib–TT conjugate vaccine	Sanofi Pasteur (Pentacel®)
	Hib conjugate (meningococcal protein conjugate)	Merck & Co (PedvaxHIB®)
	Hib conjugate (meningococcal protein conjugate) and hepatitis B (recombinant) vaccine	Merck & Co (Comvax®)
Neisseria meningitidis A, C, Y and W-135	Glycoconjugate, meningococcal polysaccharide with DT	Sanofi Pasteur (Menactra®)
	Meningococcal polysaccharide	Sanofi Pasteur (Menomune-A/C/Y/W-135®)
	De-O-acetylated polysaccharide (C11 strain of MenC)	Pfizer (NeisVac-C®)
	Capsular oligosaccharide (C11 strain of MenC)	GSK (Menjugate®)
Salmonella typhi	Vi capsular polysaccharide	Sanofi Pasteur (TYPHIm Vi®)
Streptococcus pneumoniae 4, 6B, 9V, 14, 18C, 19F and 23F	Pneumococcal polysaccharide 7-valent–CRm197 conjugate	Wyeth Pharmaceuticals (Prevnar®)
Streptococcus pneumoniae 1, 2, 3, 4, 5, 6B, 7F, 8, 9N, 9V, 10A, 11A, 12F, 14, 15B, 17F, 18C,19F, 19A, 20, 22F, 23F and 33F	Pneumococcal polysaccharide, 23-valent	Merck & Co (Pneumovax 23®)

The zwitterionic polysaccharides (ZPSs) contain both positive and negative charges in their repeating units. ZPSs activate the T-cells via electrostatic interactions, but only a small number of bacteria express zwitterionic polysaccharides on their surface. Therefore, chemical modification of a polysaccharide to incorporate also positively charged functional groups was performed to obtain a ZPS [253,254].

Glycolipids were discovered to have an immunomodulatory activity that is capable of activating a singular subset of T-lymphocytes, also known as invariant natural killer T (iNKT)-cells [255,256]. The immune potentiating properties of iNKT cells are currently being explored for the development of potential vaccines against metastasis, autoimmune, and infectious diseases [257,258]. The successful development of CPS-based vaccines first and glycoconjugate vaccines later stimulated the research on preventing further diseases such as cancer or infections that are caused by viruses, fungi, protozoan parasites, helminths, or other bacteria using carbohydrate-based vaccines [259–261].

4. Conclusions

Bacterial drug resistance is a worldwide threat to human health and both gram-positive and gram-negative bacteria can bypass host defenses to avoid the chemotherapeutic treatments of infection. The bacterial cell membrane functions as a barrier to the surroundings and whereas gram-positive bacteria have only a single cell membrane that is surrounded by a thick layer of peptidoglycan. Those that are defined as gram-negative contain two membranes, with only a thin peptidoglycan layer in between, and lipopolysaccharides in the outer leaflet of the outer membrane. Cell wall polysaccharides such as lipoteichoic acids and wall teichoic acids further complement the architecture of gram-positive bacteria. Therapeutic strategies to combat bacterial infection rely on the inhibition of cell wall biosynthesis employing β-lactam drugs such as penicillins or glycopeptides such as vancomycin and teicoplanin, whereas other approaches function by disrupting the membrane of bacteria, a mode of action that is used by polymyxins and daptomycin. Carbohydrate-based drugs that are used to treat infection that are caused by gram-positive bacteria include vancomycin, which blocks peptidoglycan biosynthesis. However, as vancomycin-resistant enterococci have emerged, treatment of these using existing antimicrobial agents poses significant difficulties. Furthermore, although drugs that are utilized in targeting gram-positive bacteria have been used in strategies for combating infections from gram-negative bacteria, by employing a combined use of rifampicin and colistin in combination with β-lactam drugs, the challenge of restraining and battling antibiotic resistance in gram-negative bacteria is paramount in future research.

Author Contributions: Conceptualization, F.R. and R.I.; writing—original draft preparation, F.R., A.R., R.I., I.L., S.S. and S.P.; writing—review and editing, F.R., R.I., A.R. and G.W.; supervision, G.W. and A.C.; project administration, R.I., G.W. and A.C.; funding acquisition, A.C. All authors have read and agreed to the published version of the manuscript.

Funding: This research was funded by Regione Autonoma della Sardegna (Sardinia, Italy), through "Legge Regionale 7 Agosto 2007:CRP1_574, 22/41 del 2017" grant number RASSR01499; and by University of Sassari, grant number FARPIRAS2020.

Institutional Review Board Statement: Not applicable.

Informed Consent Statement: Not applicable.

Data Availability Statement: Data sharing not applicable.

Acknowledgments: The authors acknowledge the funders for the financial support.

Conflicts of Interest: The authors declare no conflict of interest.

Abbreviations

Outer membrane (OM); inner membrane (IM); peptidoglycan (PG); cell wall polysaccharides (CWPs); *N*-acetylmuramic acid (MurNAc); *N*-acetylglucosamine (GlcNAc); lipopolysaccharide (LPS); pathogen-associated molecular patterns (PAMPs); toll-like receptors (TLRs); pattern recognition receptors (PRRs); cell wall teichoic acid (WTA); ribitol 5-phosphate (RboP); glycerol 3-phosphate (GroP); lipoteichoic acids (LTAs); polyglycerolphosphate (PGP); diacylglycerol (DAG); rhamnose-containing CWPs (RhaCWPs); multidrug-resistant (MDR); methicillin-resistant *Staphylococcus aureus* (MRSA); vancomycin-resistant enterococci (VRE); complicated skin and soft tissue infections (cSSTIs); hospital-acquired bacterial pneumonia (HABP); ventilator-associated bacterial pneumonia (VABP); *Staphylococcus aureus* bacteremia (SAB); vancomycin intermediate *S. aureus* (VISA); acute bacterial skin and skin structure infections (ABSSSIs); horizontal gene transfer (HGT); Protein Data Bank (PDB); standard of care (SOC); human serum albumin (HAS); transpeptidases (TPDs); penicillin-binding protein (PBP); extended-spectrum β-lactamases (ESBLs); alanine transaminase (ALT); capsular polysaccharide (CPS); zwitterionic polysaccharide (ZPS); invariant natural killer T (iNKT); small and medium-sized enterprises (SMEs); β-barrel assembly machinery (BAM); outer membrane protein (OMP).

References

1. Shi, Y.; Jiang, Y.; Wang, S.; Wang, X.; Zhu, G. Biogeographic Distribution of Comammox Bacteria in Diverse Terrestrial Habitats. *Sci. Total Environ.* **2020**, *717*, 137257. [CrossRef] [PubMed]
2. Mckay, R.M.L.; Prášil, O.; Pechar, L.; Lawrenz, E.; Rozmarynowycz, M.J.; Bullerjahn, G.S. Freshwater Ice as Habitat: Partitioning of Phytoplankton and Bacteria between Ice and Water in Central European Reservoirs. *Environ. Microbiol. Rep.* **2015**, *7*, 887–898. [CrossRef] [PubMed]
3. Sul, W.J.; Oliver, T.A.; Ducklow, H.W.; Amaral-Zettlera, L.A.; Sogin, M.L. Marine Bacteria Exhibit a Bipolar Distribution. *Proc. Natl. Acad. Sci. USA* **2013**, *110*, 2342–2347. [CrossRef] [PubMed]
4. Louca, S.; Mazel, F.; Doebeli, M.; Parfrey, L.W. A Census-Based Estimate of Earth's Bacterial and Archaeal Diversity. *PLoS Biol.* **2019**, *17*, e3000106. [CrossRef]
5. Barney, B.M. Aerobic Nitrogen-Fixing Bacteria for Hydrogen and Ammonium Production: Current State and Perspectives. *Appl. Microbiol. Biotechnol.* **2020**, *104*, 1383–1399. [CrossRef]
6. Reis, V.M.; Teixeira, K.R.D.S. Nitrogen Fixing Bacteria in the Family Acetobacteraceae and Their Role in Agriculture. *J. Basic Microbiol.* **2015**, *55*, 931–949. [CrossRef]
7. Kurth, J.M.; den Camp, H.J.M.O.; Welte, C.U. Several Ways One Goal—Methanogenesis from Unconventional Substrates. *Appl. Microbiol. Biotechnol.* **2020**, *104*, 6839–6854. [CrossRef]
8. Tropini, C.; Earle, K.A.; Huang, K.C.; Sonnenburg, J.L. The Gut Microbiome: Connecting Spatial Organization to Function. *Cell Host Microbe* **2017**, *21*, 433–442. [CrossRef]
9. Lombardo, R.; Tema, G.; Cornu, J.N.; Fusco, F.; McVary, K.; Tubaro, A.; de Nunzio, C. The Urothelium, the Urinary Microbioma and Men LUTS: A Systematic Review. *Minerva Urol. E Nefrol.* **2020**, *72*, 712–722. [CrossRef]
10. Annalisa, N.; Alessio, T.; Claudette, T.D.; Erald, V.; Antonino, D.L.; Nicola, D.D. Gut Microbioma Population: An Indicator Really Sensible to Any Change in Age, Diet, Metabolic Syndrome, and Life-Style. *Mediat. Inflamm.* **2014**, *2014*, 901308. [CrossRef]
11. Requena, T.; Velasco, M. The Human Microbiome in Sickness and in Health. *Rev. Clínica Española* **2021**, *221*, 233–240. [CrossRef] [PubMed]
12. Swidsinski, A.; Ladhoff, A.; Pernthaler, A.; Swidsinski, S.; Loening–Baucke, V.; Ortner, M.; Weber, J.; Hoffmann, U.; Schreiber, S.; Dietel, M.; et al. Mucosal Flora in Inflammatory Bowel Disease. *Gastroenterology* **2002**, *122*, 44–54. [CrossRef] [PubMed]
13. Hart, A.L.; Stagg, A.J.; Frame, M.; Graffner, H.; Glise, H.; Falk, P.; Kamm, M.A. The Role of the Gut Flora in Health and Disease, and Its Modification as Therapy. *Aliment. Pharmacol. Ther.* **2002**, *16*, 1383–1393. [CrossRef] [PubMed]
14. Larsen, B.; Monif, G.R.G. Understanding the Bacterial Flora of the Female Genital Tract. *Clin. Infect. Dis.* **2001**, *32*, e69–e77. [CrossRef] [PubMed]
15. Byrd, A.L.; Belkaid, Y.; Segre, J.A. The Human Skin Microbiome. *Nat. Rev. Microbiol.* **2018**, *16*, 143–155. [CrossRef]
16. Kataoka, K. The Intestinal Microbiota and Its Role in Human Health and Disease. *J. Med. Investig.* **2016**, *63*, 27–37. [CrossRef]
17. Kåhrström, C.T.; Pariente, N.; Weiss, U. Intestinal Microbiota in Health and Disease. *Nature* **2016**, *535*, 47. [CrossRef]
18. Patterson, E.E.; Ryan, P.M.; Cryan, J.F.; Dinan, T.G.; Paul Ross, R.; Fitzgerald, G.F.; Stanton, C.E. Gut Microbiota, Obesity and Diabetes. *Postgrad. Med. J.* **2016**, *92*, 286–300. [CrossRef]
19. Dabke, K.; Hendrick, G.; Devkota, S. The Gut Microbiome and Metabolic Syndrome. *J. Clin. Investig.* **2019**, *129*, 4050–4057. [CrossRef]
20. Spielman, L.J.; Gibson, D.L.; Klegeris, A. Unhealthy Gut, Unhealthy Brain: The Role of the Intestinal Microbiota in Neurodegenerative Diseases. *Neurochem. Int.* **2018**, *120*, 149–163. [CrossRef]
21. Jiang, C.; Li, G.; Huang, P.; Liu, Z.; Zhao, B. The Gut Microbiota and Alzheimer's Disease. *J. Alzheimer's Dis.* **2017**, *58*, 1–15. [CrossRef]
22. Kesika, P.; Suganthy, N.; Sivamaruthi, B.S.; Chaiyasut, C. Role of Gut-Brain Axis, Gut Microbial Composition, and Probiotic Intervention in Alzheimer's Disease. *Life Sci.* **2021**, *264*, 118627. [CrossRef]
23. Mirza, A.; Forbes, J.D.; Zhu, F.; Bernstein, C.N.; Van Domselaar, G.; Graham, M.; Waubant, E.; Tremlett, H. The Multiple Sclerosis Gut Microbiota: A Systematic Review. *Mult. Scler. Relat. Disord.* **2020**, *37*, 101427. [CrossRef]
24. Riedel, S.; Morse, S.A.; Mietzner, T.; Miller, S. *Jawetz Melnick & Adelbergs Medical Microbiology*, 28th ed.; McGraw Hill: Columbus, OH, USA, 2019.
25. Madigan, M.T.; Martinko, J.M.; Bender, K.S.; Buckley, D.H.; Stahl, D.A. *Brock Biology of Microorganisms*, 14th ed.; Pearson Education, Inc.: London, UK, 2015.
26. Singer, S.J.; Nicolson, G.L. The Fluid Mosaic Model of the Structure of Cell Membranes. *Science* **1972**, *175*, 720–731. [CrossRef]
27. Sáenz, J.P.; Grosser, D.; Bradley, A.S.; Lagny, T.J.; Lavrynenko, O.; Broda, M.; Simons, K. Hopanoids as Functional Analogues of Cholesterol in Bacterial Membranes. *Proc. Natl. Acad. Sci. USA* **2015**, *112*, 11971–11976. [CrossRef]
28. Mbamala, E.C.; Ben-Shaul, A.; May, S. Domain Formation Induced by the Adsorption of Charged Proteins on Mixed Lipid Membranes. *Biophys. J.* **2005**, *88*, 1702–1714. [CrossRef]
29. Whited, A.M.; Johs, A. The Interactions of Peripheral Membrane Proteins with Biological Membranes. *Chem. Phys. Lipids* **2015**, *192*, 51–59. [CrossRef]
30. Tanghe, A.; van Dijck, P.; Thevelein, J.M. Why Do Microorganisms Have Aquaporins? *Trends Microbiol.* **2006**, *14*, 78–85. [CrossRef]
31. Horne, J.E.; Brockwell, D.J.; Radford, S.E. Role of the Lipid Bilayer in Outer Membrane Protein Folding in Gram-Negative Bacteria. *J. Biol. Chem.* **2020**, *295*, 10340–10367. [CrossRef]

32. Henderson, J.C.; Zimmerman, S.M.; Crofts, A.A.; Boll, J.M.; Kuhns, L.G.; Herrera, C.M.; Trent, M.S. The Power of Asymmetry: Architecture and Assembly of the Gram-Negative Outer Membrane Lipid Bilayer. *Annu. Rev. Microbiol.* **2016**, *70*, 255–278. [CrossRef]
33. Chattaway, F. Microbial Cell Walls and Membranes. *Biochem. Educ.* **1982**, *10*, 33. [CrossRef]
34. Mengin-Lecreulx, D.; Lemaitre, B. Structure and Metabolism of Peptidoglycan and Molecular Requirements Allowing Its Detection by the Drosophila Innate Immune System. *J. Endotoxin Res.* **2005**, *11*, 105–111. [CrossRef] [PubMed]
35. Vollmer, W.; Höltje, J.V. Morphogenesis of *Escherichia coli*. *Curr. Opin. Microbiol.* **2001**, *4*, 625–633. [CrossRef]
36. Moulder, J.W. Why Is Chlamydia Sensitive to Penicillin in the Absence of Peptidoglycan? *Infect. Agents. Dis.* **1993**, *2*, 87–99. [PubMed]
37. Tamura, A.; Ohashi, N.; Urakami, H.; Miyamura, S. Classification of *Rickettsia tsutsugamushi* in a New Genus, *Orientia* Gen. Nov., as *Orientia tsutsugamushi* Comb. Nov. *Int. J. Syst. Bacteriol.* **1995**, *45*, 589–591. [CrossRef]
38. Schleifer, K.H.; Kandler, O. Peptidoglycan Types of Bacterial Cell Walls and Their Taxonomic Implications. *Bacteriol. Rev.* **1972**, *36*, 407–477. [CrossRef] [PubMed]
39. Doyle, R.J.; Dziarski, R. The Bacterial Cell: Peptidoglycan. *J. Clin. Pathol.* **2002**, *1*, 137–153. [CrossRef]
40. Dramsi, S.; Magnet, S.; Davison, S.; Arthur, M. Covalent Attachment of Proteins to Peptidoglycan. *FEMS Microbiol. Rev.* **2008**, *32*, 307–320. [CrossRef]
41. Neuhaus, F.C.; Baddiley, J. A Continuum of Anionic Charge: Structures and Functions of d -Alanyl-Teichoic Acids in Gram-Positive Bacteria. *Microbiol. Mol. Biol. Rev.* **2003**, *67*, 686–723. [CrossRef]
42. Vollmer, W.; Joris, B.; Charlier, P.; Foster, S. Bacterial Peptidoglycan (Murein) Hydrolases. *FEMS Microbiol. Rev.* **2008**, *32*, 259–286. [CrossRef]
43. Dziarski, R. Recognition of Bacterial Peptidoglycan by the Innate Immune System. *Cell. Mol. Life Sci.* **2003**, *60*, 1793–1804. [CrossRef]
44. Rosenthal, R.S.; Dziarski, R. Isolation of Peptidoglycan and Soluble Peptidoglycan Fragments. *Methods* **1994**, *235*, 253–285.
45. Park, J.T.; Uehara, T. How Bacteria Consume Their Own Exoskeletons (Turnover and Recycling of Cell Wall Peptidoglycan). *Microbiol. Mol. Biol. Rev.* **2008**, *72*, 211–227. [CrossRef]
46. Reith, J.; Mayer, C. Peptidoglycan Turnover and Recycling in Gram-Positive Bacteria. *Appl. Microbiol. Biotechnol.* **2011**, *92*, 1–11. [CrossRef]
47. Mayer, C.; Kluj, R.M.; Mühleck, M.; Walter, A.; Unsleber, S.; Hottmann, I.; Borisova, M. Bacteria's Different Ways to Recycle Their Own Cell Wall. *Int. J. Med. Microbiol.* **2019**, *309*, 151326. [CrossRef]
48. Kohanski, M.A.; Dwyer, D.J.; Collins, J.J. How Antibiotics Kill Bacteria: From Targets to Networks. *Nat. Rev. Microbiol.* **2010**, *8*, 423–435. [CrossRef]
49. Vollmer, W. Structural Variation in the Glycan Strands of Bacterial Peptidoglycan. *FEMS Microbiol. Rev.* **2008**, *32*, 287–306. [CrossRef]
50. Zipperle, G.F.; Ezzell, J.W.; Doyle, R.J. Glucosamine Substitution and Muramidase Susceptibility in Bacillus Anthracis. *Can. J. Microbiol.* **1984**, *30*, 553–559. [CrossRef]
51. Hayashi, H.; Araki, Y.; Ito, E. Occurrence of Glucosamine Residues with Free Amino Groups in Cell Wall Peptidoglycan from Bacilli as a Factor Responsible for Resistance to Lysozyme. *J. Bacteriol.* **1973**, *113*, 592–598. [CrossRef]
52. Atrih, A.; Bacher, G.; Allmaier, G.; Williamson, M.P.; Foster, S.J. Analysis of Peptidoglycan Structure from Vegetative Cells of *Bacillus subtilis* 168 and Role of PBP 5 in Peptidoglycan Maturation. *J. Bacteriol.* **1999**, *181*, 3956–3966. [CrossRef]
53. Ohno, N.; Yadomae, T.; Miyazaki, T. Identification of 2-Amino-2-Deoxyglucose Residues in the Peptidoglucan of *Streptococcus pneumoniae*. *Carbohydr. Res.* **1982**, *107*, 152–155. [CrossRef]
54. Vollmer, W.; Tomasz, A. The PgdA Gene Encodes for a Peptidoglycan N-Acetylglucosamine Deacetylase in *Streptococcus pneumoniae*. *J. Biol. Chem.* **2000**, *275*, 20496–20501. [CrossRef] [PubMed]
55. Rush, J.S.; Parajuli, P.; Ruda, A.; Li, J.; Pohane, A.A.; Zamakhaeva, S.; Rahman, M.M.; Chang, J.C.; Gogos, A.; Kenner, C.W.; et al. PplD Is a De-N-Acetylase of the Cell Wall Linkage Unit of Streptococcal Rhamnopolysaccharides. *Nat. Commun.* **2022**, *13*, 590. [CrossRef] [PubMed]
56. Schindler, M.; Sharon, N.; Assaf, Y.; Chipman, D.M. Mechanism of Lysozyme Catalysis: Role of Ground-State Strain in Subsite D in Hen Egg-White and Human Lysozymes. *Biochemistry* **1977**, *16*, 423–431. [CrossRef]
57. Ragland, S.A.; Criss, A.K. From Bacterial Killing to Immune Modulation: Recent Insights into the Functions of Lysozyme. *PLoS Pathog.* **2017**, *13*, e1006512. [CrossRef]
58. Amano, K.; Hayashi, H.; Araki, Y.; Ito, E. The Action of Lysozyme on Peptidoglycan with N-Unsubstituted Glucosamine Residues. *Eur. J. Biochem.* **1977**, *76*, 299–307. [CrossRef]
59. Amano, K.-I.; Araki, Y.; Ito, E. Effect of N-Acyl Substitution at Glucosamine Residues on Lysozyme-Catalyzed Hydrolysis of Cell-Wall Peptidoglycan and Its Oligosaccharides. *Eur. J. Biochem.* **1980**, *107*, 547–553. [CrossRef]
60. Westmacott, D.; Perkins, H.R. Effects of Lysozyme on *Bacillus cereus* 569: Rupture of Chains of Bacteria and Enhancement of Sensitivity to Autolysins. *J. Gen. Microbiol.* **1979**, *115*, 1–11. [CrossRef]
61. Adam, A.; Petit, J.F.; Wietzerbin-Falszpan, J.; Sinay, P.; Thomas, D.W.; Lederer, E. L'acide N-Glycolyl-Muramique, Constituant Des Parois de *Mycobacterium smegmatis*: Identification Par Spectrometrie de Masse. *FEBS Lett.* **1969**, *4*, 87–92. [CrossRef]

62. Uchida, K.; Kudo, T.; Suzuki, K.I.; Nakase, T. A New Rapid Method of Glycolate Test by Diethyl Ether Extraction, Which Is Applicable to a Small Amount of Bacterial Cells of Less than One Milligram. *J. Gen. Appl. Microbiol.* **1999**, *45*, 49–56. [CrossRef]
63. Sutcliffe, I.C. Cell Envelope Composition and Organisation in the Genus *Rhodococcus*. *Antonie Van Leeuwenhoek Int. J. Gen. Mol. Microbiol.* **1998**, *74*, 49–58. [CrossRef]
64. Azuma, I.; Thomas, D.; Adam, A.; Ghuysen, J.; Bonaly, R.; Petit, J.; Lederer, E. Occurence of N-Glucosylmuramic Acid in Bacterial Cell Walls. *Biochim. Biophys. Acta* **1970**, *20*, 444–451. [CrossRef]
65. Waksman, S.A.; Schatz, A.; Reynolds, D.M. Production of Antibiotic Substances by Actinomycetes. *Ann. N. Y. Acad. Sci.* **2010**, *1213*, 112–124. [CrossRef]
66. Brennan, P.J.; Nikaido, H. The Envelope of Mycobacteria. *Annu. Rev. Biochem.* **1995**, *64*, 29–63. [CrossRef]
67. Sekiguchi, J.; Akeo, K.; Yamamoto, H.; Khasanov, F.K.; Alonso, J.C.; Kuroda, A. Nucleotide Sequence and Regulation of a New Putative Cell Wall Hydrolase Gene, CwlD, Which Affects Germination in *Bacillus subtilis*. *J. Bacteriol.* **1995**, *177*, 5582–5589. [CrossRef]
68. Popham, D.L.; Helin, J.; Costello, C.E.; Setlow, P. Muramic Lactam in Peptidoglycan of *Bacillus subtilis* Spores Is Required for Spore Outgrowth but Not for Spore Dehydration or Heat Resistance. *Proc. Natl. Acad. Sci. USA* **1996**, *93*, 15405–15410. [CrossRef]
69. Woodward, R.; Yi, W.; Li, L.; Zhao, G.; Eguchi, H.; Sridhar, P.R.; Guo, H.; Song, J.K.; Motari, E.; Cai, L.; et al. In Vitro Bacterial Polysaccharide Biosynthesis: Defining the Functions of Wzy and Wzz. *Nat. Chem. Biol.* **2010**, *6*, 418–423. [CrossRef]
70. Chirakkal, H.; O'Rourke, M.; Atrih, A.; Foster, S.J.; Moir, A. Analysis of Spore Cortex Lytic Enzymes and Related Proteins in *Bacillus subtilis* Endoscope Germination. *Microbiology* **2002**, *148*, 2383–2392. [CrossRef]
71. Abrams, A. O-Acetyl Groups in the Cell Wall of *Streptococcus faecalis*. *J. Biol. Chem.* **1958**, *230*, 949–959. [CrossRef]
72. Brumfitt, W.; Wardlaw, A.C.; Park, J.T. Development of Lysozyme-Resistance in *Micrococcus lysodiekticus* and Its Association with an Increased O-Acetyl Content of the Cell Wall. *Nature* **1958**, *181*, 1783–1784. [CrossRef]
73. Clarke, A.J.; Dupont, C. O-Acetylated Peptidoglycan: Its Occurrence, Pathobiological Significance, and Biosynthesis. *Can. J. Microbiol.* **1992**, *38*, 85–91. [CrossRef] [PubMed]
74. Clarke, A.J. Extent of Peptidoglycan O Acetylation in the Tribe Proteeae. *J. Bacteriol.* **1993**, *175*, 4550–4553. [CrossRef] [PubMed]
75. Weadge, J.T.; Pfeffer, J.M.; Clarke, A.J. Identification of a New Family of Enzymes with Potential O-Acetylpeptidoglycan Esterase Activity in Both Gram-Positive and Gram-Negative Bacteria. *BMC Microbiol.* **2005**, *5*, 1–15. [CrossRef] [PubMed]
76. Dupont, C.; Clarke, A.J. Evidence for N → O Acetyl Migration as the Mechanism for O Acetylation of Peptidoglycan in *Proteus mirabilis*. *J. Bacteriol.* **1991**, *173*, 4318–4324. [CrossRef] [PubMed]
77. Bera, A.; Herbert, S.; Jakob, A.; Vollmer, W.; Götz, F. Why Are Pathogenic Staphylococci so Lysozyme Resistant? The Peptidoglycan O-Acetyltransferase OatA Is the Major Determinant for Lysozyme Resistance of *Staphylococcus aureus*. *Mol. Microbiol.* **2005**, *55*, 778–787. [CrossRef]
78. Dupont, C.; Clarke, A.J. Dependence of Lysozyme-Catalyzed Solubilization of *Proteus mirabilis* Peptidoglycan on the Extent of O-Acetylation. *Eur. J. Biochem.* **1991**, *195*, 763–769. [CrossRef]
79. Sauvage, E.; Kerff, F.; Terrak, M.; Ayala, J.A.; Charlier, P. The Penicillin-Binding Proteins: Structure and Role in Peptidoglycan Biosynthesis. *FEMS Microbiol. Rev.* **2008**, *32*, 234–258. [CrossRef]
80. Burghaus, P.; Johannsen, L.; Naumann, D.; Labischinski, H.; Bradaczek, H.; Giesbrecht, P. The Influence of Different Antibiotics on the Degree of O-Acetylation of Staphylococcal Cell Walls. In *The Target of Penicillin*; De Gruyter: Berlin, Germany, 1983; pp. 317–322. [CrossRef]
81. Sidow, T.; Johannsen, L.; Labischinski, H. Penicillin-Induced Changes in the Cell Wall Composition of *Staphylococcus aureus* before the Onset of Bacteriolysis. *Arch. Microbiol.* **1990**, *154*, 73–81. [CrossRef]
82. Varki, A. Nothing in Glycobiology Makes Sense, except in the Light of Evolution. *Cell* **2006**, *126*, 841–845. [CrossRef]
83. Lee, J.; Patel, D.S.; Ståhle, J.; Park, S.-J.; Kern, N.R.; Kim, S.; Lee, J.; Cheng, X.; Valvano, M.A.; Holst, O.; et al. CHARMM-GUI Membrane Builder for Complex Biological Membrane Simulations with Glycolipids and Lipoglycans. *J. Chem. Theory Comput.* **2019**, *15*, 775–786. [CrossRef]
84. Furevi, A. Structural and Conformational Analysis of Bacterial Polysaccharides Using NMR Spectroscopy. Ph.D. Thesis, Stockholm University, Stockholm, Sweden, 2022.
85. Brade, H.; Opal, S.M.; Vogel, S.N.; Morrison, D.C. *Endotoxin in Health and Disease*; Marcel Dekker: New York, NY, USA, 1999.
86. Raetz, C.R.H.; Whitfield, C. Lipopolysaccharide Endotoxins. *Annu. Rev. Biochem.* **2002**, *71*, 635–700. [CrossRef]
87. Wyckoff, T.J.O.; Raetz, C.R.H. The Active Site of *Escherichia coli* UDP-N-Acetylglucosamine Acyltransferase. *J. Biol. Chem.* **1999**, *274*, 27047–27055. [CrossRef]
88. Rietschel, E.T. *Handbook of Endotoxin, Vol.1*; Elsevier: Amsterdam, The Netherlands, 1985.
89. Lugtenberg, B.; Van Alphen, L. Molecular Architecture and Functioning of the Outer Membrane of Escherichia Coli and Other Gram-Negative Bacteria. *BBA Rev. Biomembr.* **1983**, *737*, 51–115. [CrossRef]
90. Weintraub, A.; Zähringer, U.; Wollenweber, H.W.; Seydel, U.; Rietschel, E. Structural Characterization of the Lipid A Component of *Bacteroides fragilis* Strain NCTC 9343 Lipopolysaccharide. *Eur. J. Biochem.* **1989**, *183*, 425–431. [CrossRef]
91. Wollenweber, H.W.; Rietschel, E.T. Analysis of Lipopolysaccharide (Lipid A) Fatty Acids. *J. Microbiol. Methods* **1990**, *11*, 195–211. [CrossRef]
92. Salimath, P.V.; Weckesser, J.; Strittmatter, W.; Mayer, H. Structural Studies on the Non-Toxic Lipid A from *Rhodopseudomonas sphaeroides* ATCC 17023. *Eur. J. Biochem.* **1983**, *136*, 195–200. [CrossRef]

93. Weckesser, J.; Mayer, H. Different Lipid A Types in Lipopolysaccharides of Phototrophic and Related Non-Phototrophic Bacteria. *FEMS Microbiol. Lett.* **1988**, *54*, 143–153. [CrossRef]
94. Moran, A.P.; Zähringer, U.; Seydel, U.; Scholz, D.; Stutz, P.; Rietschel, E.T. Structural Analysis of the Lipid A Component of *Campylobacter jejuni* CCUG 10936 (Serotype O:2) Lipopolysaccharide. *Eur. J. Biochem.* **1991**, *198*, 459–469. [CrossRef]
95. Zhou, X.-Y.; Li, L.-X.; Zhang, Z.; Duan, S.-C.; Huang, Y.-W.; Luo, Y.-Y.; Mu, X.-D.; Chen, Z.-W.; Qin, Y.; Hu, J.; et al. Chemical Synthesis and Antigenic Evaluation of Inner Core Oligosaccharides from *Acinetobacter baumannii* Lipopolysaccharide. *Angew. Chem. Int. Ed.* **2022**, *61*, e202204420. [CrossRef]
96. Vaara, M. Agents That Increase the Permeability of the Outer Membrane. *Microbiol. Rev.* **1992**, *56*, 395–411. [CrossRef]
97. Liu, B.; Furevi, A.; Perepelov, A.V.; Guo, X.; Cao, H.; Wang, Q.; Reeves, P.R.; Knirel, Y.A.; Wang, L.; Widmalm, G. Structure and Genetics of *Escherichia coli* O Antigens. *FEMS Microbiol. Rev.* **2020**, *44*, 655–683. [CrossRef]
98. Lundborg, M.; Modhukur, V.; Widmalm, G. Glycosyltransferase Functions of *E. coli* O-Antigens. *Glycobiology* **2009**, *20*, 366–368. [CrossRef]
99. Li, Y.; Huang, J.; Wang, X.; Xu, C.; Han, T.; Guo, X. Genetic Characterization of the O-Antigen and Development of a Molecular Serotyping Scheme for *Enterobacter cloacae*. *Front. Microbiol.* **2020**, *11*, 1–12. [CrossRef]
100. Toukach, P.V.; Egorova, K.S. Carbohydrate Structure Database Merged from Bacterial, Archaeal, Plant and Fungal Parts. *Nucleic Acids Res.* **2016**, *44*, D1229–D1236. [CrossRef]
101. Brown, S.; Santa Maria, J.P.; Walker, S. Wall Teichoic Acids of Gram-Positive Bacteria. *Annu. Rev. Microbiol.* **2013**, *67*, 313–336. [CrossRef]
102. Weidenmaier, C.; Peschel, A. Teichoic Acids and Related Cell-Wall Glycopolymers in Gram-Positive Physiology and Host Interactions. *Nat. Rev. Microbiol.* **2008**, *6*, 276–287. [CrossRef]
103. Swoboda, J.G.; Campbell, J.; Meredith, T.C.; Walker, S. Wall Teichoic Acid Function, Biosynthesis, and Inhibition. *ChemBioChem* **2009**, *11*, 35–45. [CrossRef] [PubMed]
104. Schaefer, K.; Owens, T.W.; Kahne, D.; Walker, S. Substrate Preferences Establish the Order of Cell Wall Assembly in *Staphylococcus aureus*. *J. Am. Chem. Soc.* **2018**, *140*, 2442–2445. [CrossRef]
105. Araki, Y.; Ito, E. Linkage Units in Cell Walls of Gram-Positive Bacteria. *Crit. Rev. Microbiol.* **1989**, *17*, 121–135. [CrossRef] [PubMed]
106. Kojima, N.; Araki, Y.; Ito, E. Structure of the Linkage Units between Ribitol Teichoic Acids and Peptidoglycan. *J. Bacteriol.* **1985**, *161*, 299–306. [CrossRef]
107. Endl, J.; Seidl, H.P.; Fiedler, F.; Schleider, K.H. Chemical Composition and Structure of Cell Wall Teichoic Acids of Staphylococci. *Arch. Microbiol.* **1983**, *135*, 215–223. [CrossRef] [PubMed]
108. Endl, J.; Seidl, P.H.; Fiedler, F.; Schleifer, K.H. Determination of Cell Wall Teichoic Acid Structure of Staphylococci by Rapid Chemical and Serological Screening Methods. *Arch. Microbiol.* **1984**, *137*, 272–280. [CrossRef]
109. Naumova, I.B.; Shashkov, A.S.; Tul'Skaya, E.M.; Streshinskaya, G.M.; Kozlova, Y.I.; Potekhina, N.V.; Evtushenko, L.I.; Stackebrandt, E. Cell Wall Teichoic Acids: Structural Diversity, Species Specificity in the Genus Nocardiopsis, and Chemotaxonomic Perspective. *FEMS Microbiol. Rev.* **2001**, *25*, 269–283. [CrossRef] [PubMed]
110. Shashkov, A.S.; Potekhina, N.V.; Naumova, I.B.; Evtushenko, L.I.; Widmalm, G. Cell Wall Teichoic Acids of Actinomadura Viridis VKM Ac-1315T. *Eur. J. Biochem.* **1999**, *262*, 688–695. [CrossRef]
111. Kristian, S.A.; Datta, V.; Weidenmaier, C.; Kansal, R.; Fedtke, I.; Peschel, A.; Gallo, R.L.; Nizet, V. D-Alanylation of Teichoic Acids Promotes Group A *Streptococcus* Antimicrobial Peptide Resistance, Neutrophil Survival, and Epithelial Cell Invasion. *J. Bacteriol.* **2005**, *187*, 6719–6725. [CrossRef]
112. Kristian, S.A.; Lauth, X.; Nizet, V.; Goetz, F.; Neumeister, B.; Peschel, A.; Landmann, R. Alanylation of Teichoic Acids Protects *Staphylococcus aureus* against Toll-like Receptor 2-Dependent Host Defense in a Mouse Tissue Cage Infection Model. *J. Infect. Dis.* **2003**, *188*, 414–423. [CrossRef]
113. Collins, L.V.; Kristian, S.A.; Weidenmaier, C.; Faigle, M.; Van Kessel, K.P.M.; Van Strijp, J.A.G.; Goütz, F.; Neumeister, B.; Peschel, A. *Staphylococcus aureus* Strains Lacking D-Alanine Modifications of Teichoic Acids Are Highly Susceptible to Human Neutrophil Killing and Are Virulence Attenuated in Mice. *J. Infect. Dis.* **2002**, *186*, 214–219. [CrossRef]
114. Peschel, A.; Otto, M.; Jack, R.W.; Kalbacher, H.; Jung, G.; Götz, F. Inactivation of the Dlt Operon in *Staphylococcus aureus* Confers Sensitivity to Defensins, Protegrins, and Other Antimicrobial Peptides. *J. Biol. Chem.* **1999**, *274*, 8405–8410. [CrossRef]
115. Peschel, A.; Vuong, C.; Otto, M.; Gotz, F. The D-Alanine Residues of *Staphylococcus aureus* Teichoic Acids Alter the Susceptibility to Vancomycin and the Activity of Autolytic Enzymes. *Antimicrob. Agents Chemother.* **2000**, *44*, 2845–2847. [CrossRef]
116. Vadyvaloo, V.; Arous, S.; Gravesen, A.; Héchard, Y.; Chauhan-Haubrock, R.; Hastings, J.W.; Rautenbach, M. Cell-Surface Alterations in Class IIa Bacteriocin-Resistant Listeria Monocytogenes Strains. *Microbiology* **2004**, *150*, 3025–3033. [CrossRef]
117. Walter, J.; Loach, D.M.; Alqumber, M.; Rockel, C.; Hermann, C.; Pfitzenmaier, M.; Tannock, G.W. D-Alanyl Ester Depletion of Teichoic Acids in *Lactobacillus reuteri* 100-23 Results in Impaired Colonization of the Mouse Gastrointestinal Tract. *Environ. Microbiol.* **2007**, *9*, 1750–1760. [CrossRef] [PubMed]
118. Brown, S.; Xia, G.; Luhachack, L.G.; Campbell, J.; Meredith, T.C.; Chen, C.; Winstel, V.; Gekeler, C.; Irazoqui, J.E.; Peschel, A.; et al. Methicillin Resistance in *Staphylococcus aureus* Requires Glycosylated Wall Teichoic Acids. *Proc. Natl. Acad. Sci. USA* **2012**, *109*, 18909–18914. [CrossRef] [PubMed]
119. Fisher, W. Glycolipids, Phosphoglycolipids, and Sulfoglycolipids. In *Handbook of Lipid Research, Vol.6*; Morris, K., Ed.; Prenum Press: New York, NY, USA, 1990; pp. 123–234.

120. Gründling, A.; Schneewind, O. Synthesis of Glycerol Phosphate Lipoteichoic Acid in *Staphylococcus aureus*. *Proc. Natl. Acad. Sci. USA* **2007**, *104*, 8478–8483. [CrossRef] [PubMed]
121. Webb, A.J.; Karatsa-Dodgson, M.; Gründling, A. Two-Enzyme Systems for Glycolipid and Polyglycerolphosphate Lipoteichoic Acid Synthesis in Listeria Monocytogenes. *Mol. Microbiol.* **2009**, *74*, 299–314. [CrossRef]
122. Mäki, M.; Renkonen, R. Biosynthesis of 6-Deoxyhexose Glycans in Bacteria. *Glycobiology* **2004**, *14*, 1R–15R. [CrossRef]
123. Facklam, R. What Happened to the Streptococci: Overview of Taxonomic and Nomenclature Changes. *Clin. Microbiol. Rev.* **2002**, *15*, 613–630. [CrossRef]
124. Köhler, W. The Present State of Species within the Genera *Streptococcus* and *Enterococcus*. *Int. J. Med. Microbiol.* **2007**, *297*, 133–150. [CrossRef]
125. Neelamegham, S.; Aoki-Kinoshita, K.; Bolton, E.; Frank, M.; Lisacek, F.; Lütteke, T.; O'Boyle, N.; Packer, N.H.; Stanley, P.; Toukach, P.; et al. Updates to the Symbol Nomenclature for Glycans Guidelines. *Glycobiology* **2019**, *29*, 620–624. [CrossRef]
126. McMillan, D.J.; Vu, T.; Bramhachari, P.V.; Kaul, S.Y.; Bouvet, A.; Shaila, M.S.; Karmarkar, M.G.; Sriprakash, K.S. Molecular Markers for Discriminating *Streptococcus pyogenes* and *S. dysgalactiae* Subspecies *equisimilis*. *Eur. J. Clin. Microbiol. Infect. Dis.* **2010**, *29*, 585–589. [CrossRef]
127. Swanson, J.; Gotschlich, E.C. Electron Microscopic Studies on Streptococci II. Group A Carbohydrate. *J. Exp. Med.* **1973**, *138*, 245–258. [CrossRef]
128. Wagner, M.; Wagner, B.; Ryc, M. An Electron Microscopic Study of the Location of Peptidoglycan in Group A and C Streptococcal Cell Walls. *J. Gen. Microbiol.* **1978**, *108*, 283–294. [CrossRef] [PubMed]
129. Wagner, B.; Wagner, M.; Kubin, V.; Ryc, M. Immunoelectron Microscopic Study of the Location of Group-Specific and Protein Type-Specific Antigens of Group B Streptococci. *J. Gen. Microbiol.* **1980**, *118*, 95–105. [CrossRef] [PubMed]
130. Sutcliffe, I.C.; Black, G.W.; Harrington, D.J. Bioinformatic Insights into the Biosynthesis of the Group B Carbohydrate in *Streptococcus agalactiae*. *Microbiology* **2008**, *154*, 1354–1363. [CrossRef]
131. Chapot-Chartier, M.P. Interactions of the Cell-Wall Glycopolymers of Lactic Acid Bacteria with Their Bacteriophages *Front. Microbiol.* **2014**, *5*, 1–10. [CrossRef] [PubMed]
132. Hancock, L.E.; Gilmore, M.S. The Capsular Polysaccharide of *Enterococcus faecalis* and Its Relationship to Other Polysaccharides in the Cell Wall. *Proc. Natl. Acad. Sci. USA* **2002**, *99*, 1574–1579. [CrossRef] [PubMed]
133. Teng, F.; Singh, K.V.; Bourgogne, A.; Zeng, J.; Murray, B.E. Further Characterization of the Epa Gene Cluster and Epa Polysaccharides of *Enterococcus faecalis*. *Infect. Immun.* **2009**, *77*, 3759–3767. [CrossRef] [PubMed]
134. Thurlow, L.R.; Thomas, V.C.; Hancock, L.E. Capsular Polysaccharide Production in *Enterococcus faecalis* and Contribution of CpsF to Capsule Serospecificity. *J. Bacteriol.* **2009**, *191*, 6203–6210. [CrossRef]
135. Theilacker, C.; Holst, O.; Lindner, B.; Huebner, J.; Kaczyński, Z. The Structure of the Wall Teichoic Acid Isolated from *Enterococcus faecalis* Strain 12030. *Carbohydr. Res.* **2012**, *354*, 106–109. [CrossRef]
136. Pritchard, D.G.; Coligan, J.E.; Speed, S.E.; Gray, B.M. Carbohydrate Fingerprints of Streptococcal Cells. *J. Clin. Microbiol.* **1981**, *13*, 89–92. [CrossRef]
137. Mistou, M.Y.; Sutcliffe, I.C.; Van Sorge, N.M. Bacterial Glycobiology: Rhamnose-Containing Cell Wall Polysaccharides in Gram-Positive Bacteria. *FEMS Microbiol. Rev.* **2016**, *40*, 464–479. [CrossRef]
138. McCarty, M. The Lysis of Group A Hemolytic Streptococci by Extracellular Enzymes of *Streptomyces albus*. *J. Exp. Med.* **1952**, *96*, 569–581. [CrossRef] [PubMed]
139. Tsukioka, Y.; Yamashita, Y.; Nakano, Y.; Oho, T.; Koga, T. Identification of a Fourth Gene Involved in DTDP-Rhamnose Synthesis in *Streptococcus mutans*. *J. Bacteriol.* **1997**, *179*, 4411–4414. [CrossRef] [PubMed]
140. Caliot, É.; Dramsi, S.; Chapot-Chartier, M.P.; Courtin, P.; Kulakauskas, S.; Péchoux, C.; Trieu-Cuot, P.; Mistou, M.Y. Role of the Group B Antigen of *Streptococcus agalactiae*: A Peptidoglycan-Anchored Polysaccharide Involved in Cell Wall Biogenesis. *PLoS Pathog.* **2012**, *8*, e1002756. [CrossRef] [PubMed]
141. Van der Beek, S.L.; Le Breton, Y.; Ferenbach, A.T.; Chapman, R.N.; van Aalten, D.M.F.; Navratilova, I.; Boons, G.J.; Mciver, K.S.; van Sorge, N.M.; Dorfmueller, H.C. GacA Is Essential for Group A *Streptococcus* and Defines a New Class of Monomeric DTDP-4-Dehydrorhamnose Reductases (RmlD). *Mol. Microbiol.* **2015**, *98*, 946–962. [CrossRef]
142. Van Sorge, N.M.; Cole, J.N.; Kuipers, K.; Henningham, A.; Aziz, R.K.; Kasirer-Friede, A.; Lin, L.; Berends, E.T.M.; Davies, M.R.; Dougan, G.; et al. The Classical Lancefield Antigen of Group A *Streptococcus* Is a Virulence Determinant with Implications for Vaccine Design. *Cell Host Microbe* **2014**, *15*, 729–740. [CrossRef] [PubMed]
143. Krause, R.M. Studies on Bacteriophages of Hemolytic Streptococci. I. Factors Influencing the Interaction of Phage and Susceptible Host Cell. *J. Exp. Med.* **1957**, *106*, 365–384. [CrossRef]
144. Fischetti, V.A.; Zabriskie, J.B. Studies on Streptococcal Bacteriophages II. Adsorpion Studies on Group A and Group C Streptococcal Bacteriophages. *J. Exp. Med.* **1967**, *127*, 489–505. [CrossRef]
145. Shibata, Y.; Yamashita, Y.; Van Der Ploeg, J.R. The Serotype-Specific Glucose Side Chain of Rhamnose-Glucose Polysaccharides Is Essential for Adsorption of Bacteriophage M102 to *Streptococcus mutans*. *FEMS Microbiol. Lett.* **2009**, *294*, 68–73. [CrossRef]
146. Xu, Y.; Singh, K.V.; Qin, X.; Murray, B.E.; Weinstock, G.M. Analysis of a Gene Cluster of *Enterococcus faecalis* Involved in Polysaccharide Biosynthesis. *Infect. Immun.* **2000**, *68*, 815–823. [CrossRef]

147. Alphey, M.S.; Pirrie, L.; Torrie, L.S.; Boulkeroua, W.A.; Gardiner, M.; Sarkar, A.; Maringer, M.; Oehlmann, W.; Brenk, R.; Scherman, M.S.; et al. Allosteric Competitive Inhibitors of the Glucose-1-Phosphate Thymidylyltransferase (RmlA) from *Pseudomonas aeruginosa*. *ACS Chem. Biol.* **2013**, *8*, 387–396. [CrossRef]
148. Sewell, E.W.C.; Brown, E.D. Taking Aim at Wall Teichoic Acid Synthesis: New Biology and New Leads for Antibiotics. *J. Antibiot.* **2014**, *67*, 43–51. [CrossRef]
149. Campbell, J.; Singh, A.K.; Santa Maria, J.P.; Kim, Y.; Brown, S.; Swoboda, J.G.; Mylonakis, E.; Wilkinson, B.J.; Walker, S. Synthetic Lethal Compound Combinations Reveal a Fundamental Connection between Wall Teichoic Acid and Peptidoglycan Biosyntheses in *Staphylococcus aureus*. *ACS Chem. Biol.* **2011**, *6*, 106–116. [CrossRef]
150. Nicolau, D.P. Carbapenems: A Potent Class of Antibiotics. *Expert Opin. Pharmacother.* **2008**, *9*, 23–37. [CrossRef]
151. Zhanel, G.G.; Wiebe, R.; Dilay, L.; Thomson, K.; Rubinstein, E.; Hoban, D.J.; Noreddin, A.M.; Karlowsky, J.A. Comparative Review of the Carbapenems. *Drugs* **2007**, *67*, 1027–1052. [CrossRef]
152. Tenover, F.C. Mechanisms of Antimicrobial Resistance in Bacteria. *Am. J. Med.* **2006**, *119*, S3–S10. [CrossRef]
153. Watkins, R.R. *A Primer on Antimicrobials*; Elsevier Inc.: Amsterdam, The Netherlands, 2022.
154. Belley, A.; McKay, G.A.; Arhin, F.F.; Sarmiento, I.; Beaulieu, S.; Fadhil, I.; Parr, T.R.; Moeck, G. Oritavancin Disrupts Membrane Integrity of *Staphylococcus aureus* and Vancomycin-Resistant Enterococci to Effect Rapid Bacterial Killing. *Antimicrob. Agents Chemother.* **2010**, *54*, 5369–5371. [CrossRef]
155. Griffith, R.S. Vancomycin Use—An Historical Review. *J. Antimicrob. Chemother.* **1984**, *14*, 1–5. [CrossRef]
156. Wilhelm, M.P.; Estes, L. Symposium on Antimicrobial Agents—Part XII. Vancomycin. *Mayo Clin. Proc.* **1999**, *74*, 928–935. [CrossRef]
157. Jacobs, M.R. Emergence of Multiple Resistant Pneumococci. *J. Occup. Environ. Med.* **1980**, *22*, 559. [CrossRef]
158. Aber, R.C.; Wennersten, C.; Moellering, R.C. Antimicrobial Susceptibility of Flavobacteria. *Antimicrob. Agents Chemother.* **1978**, *14*, 483–487. [CrossRef]
159. Anderson, J.S.; Matsuhashi, M.; Haskin, M.A.; Strominger, J.L. Lipid-Phosphoacetylmuramyl-Pentapeptide and Lipid-Phosphodisaccharide-Pentapeptide: Presumed Membrane Transport Intermediates in Cell Wall Synthesis. *Proc. Natl. Acad. Sci. USA* **1965**, *53*, 881–889. [CrossRef]
160. Perkins, H.R. Specificity of Combination between Mucopeptide Precursors and Vancomycin or Ristocetin. *Biochem. J.* **1969**, *111*, 195–205. [CrossRef]
161. Hancock, R.; Fitz-James, P.C. Some Differences in the Action of Penicillin, Bacitracin, and Vancomycin on *Bacillus megaterium*. *J. Bacteriol.* **1964**, *87*, 1044–1050. [CrossRef]
162. Jordan, D.C.; Mallory, H.D. Site of Action of Action of Vancomycin on *Staphylococcus aureus*. *Antimicrob. Agents Chemother.* **1964**, *10*, 489–494.
163. Campoli-Richards, D.M.; Brogden, R.N.; Faulds, D. Teicoplanin. A Review of Its Antibacterial Activity, Pharmacokinetic Properties and Therapeutic Potential. *Drugs* **1990**, *40*, 449–486. [CrossRef]
164. Parenti, F.; Beretta, G.; Berti, M.; Arioli, V. Teichomycins, New Antibiotics from Actinoplanes Teichomyceticus Nov. Sp. I. Description of the Producer Strain, Fermentation Studies and Biological Properties. *J. Antibiot.* **1978**, *31*, 276–283. [CrossRef] [PubMed]
165. Janknegt, R. Teicoplanin in Perspective A Critical Comparison with Vancomycin. *Pharm. Weekbl. Sci.* **1991**, *13*, 153–160. [CrossRef] [PubMed]
166. Parenti, F. Structure and Mechanism of Action of Teicoplanin. *J. Hosp. Infect.* **1986**, *7*, 79–83. [CrossRef]
167. Somma, S.; Gastaldo, L.; Corti, A. Teicoplanin, a New Antibiotic from *Actinoplanes teichomyceticus* Nov. Sp. *Antimicrob. Agents Chemother.* **1984**, *26*, 917–923. [CrossRef]
168. Reynolds, P.E.; Somner, E.A. Comparison of the Target Sites and Mechanisms of Action of Glycopeptide and Lipoglycodepsipeptide Antibiotics. *Drugs Exp. Clin. Res.* **1990**, *16*, 385–389.
169. Barna, J.C.J.; Williams, D.H.; Stone, D.J.M.; Leung, T.W.C.; Doddrell, D.M. Structure Elucidation of the Teicoplanin Antibiotics. *J. Am. Chem Soc.* **1984**, *106*, 4895–4902. [CrossRef]
170. Parenti, F. Glycopeptide Antibiotics. *J. Clin. Pharmacol.* **1988**, *28*, 136–140. [CrossRef] [PubMed]
171. Bolognino, I.; Carrieri, A.; Purgatorio, R.; Catto, M.; Caliandro, R.; Carrozzini, B.; Belviso, B.D.; Majellaro, M.; Sotelo, E.; Cellamare, S.; et al. Enantiomeric Separation and Molecular Modelling of Bioactive 4-Aryl-3,4-Dihydropyrimidin-2(1H)-One Ester Derivatives on Teicoplanin-Based Chiral Stationary Phase. *Separations* **2022**, *9*, 7. [CrossRef]
172. Kirst, H.A.; Thompson, D.G.; Nicas, T.I. Historical Yearly Usage of Vancomycin. *Antimicrob. Agents Chemother* **1998**, *42*, 1303–1304. [CrossRef]
173. Pakyz, A.L.; Macdougall, C.; Oinonen, M.; Polk, R.E. Trends in Antibacterial Use in US Academic Health Centers. *Arch. Intern. Med.* **2008**, *168*, 2254–2260. [CrossRef]
174. Steinkraus, G.; White, R.; Friedrich, L. Vancomycin MIC Creep in Non-Vancomycin-Intermediate *Staphylococcus aureus* (VISA), Vancomycin-Susceptible Clinical Methicillin-Resistant *S. Aureus* (MRSA) Blood Isolates from 2001 05. *J. Antimicrob. Chemother.* **2007**, *60*, 788–794. [CrossRef]
175. King, A.; Phillips, I.; Kaniga, K. Comparative in Vitro Activity of Telavancin (TD-6424), a Rapidly Bactericidal, Concentration-Dependent Anti-Infective with Multiple Mechanisms of Action against Gram-Positive Bacteria. *J. Antimicrob. Chemother.* **2004**, *53*, 797–803. [CrossRef]

176. Saravolatz, L.D.; Stein, G.E.; Johnson, L.B. Telavancin: A Novel Lipoglycopeptide. *Clin. Infect. Dis.* **2009**, *49*, 1908–1914. [CrossRef]
177. Leadbetter, M.R.; Adams, S.M.; Bazzini Kevin Krause, B.M.; T Lam, B.M.; Linsell, M.S.; Kelly Quast, M.; Shaw, J.; Soriano, E.; Terry Wu, S.X.; Christensen, B.G.; et al. Hydrophobic Vancomycin Derivatives with Improved ADME Properties: Discovery of Telavancin (TD-6424). *J. Antibiot.* **2004**, *57*, 326–336. [CrossRef]
178. Higgins, D.L.; Chang, R.; Debabov, D.V.; Leung, J.; Wu, T.; Krause, K.M.; Sandvik, E.; Hubbard, J.M.; Kaniga, K.; Schmidt, D.E.; et al. Telavancin, a Multifunctional Lipoglycopeptide, Disrupts Both Cell Wall Synthesis and Cell Membrane Integrity in Methicillin-Resistant *Staphylococcus aureus*. *Antimicrob. Agents Chemother.* **2005**, *49*, 1127–1134. [CrossRef]
179. Stryjewski, M.E.; Lentnek, A.; O'Riordan, W.; Pullman, J.; Tambyah, P.A.; Miró, J.M.; Fowler, V.G.; Barriere, S.L.; Kitt, M.M.; Corey, G.R. A Randomized Phase 2 Trial of Telavancin versus Standard Therapy in Patients with Uncomplicated *Staphylococcus aureus* Bacteremia: The ASSURE Study. *BMC Infect. Dis.* **2014**, *14*, 289. [CrossRef]
180. Holland, T.L.; Chambers, H.F.; Boucher, H.W.; Corey, G.R.; Coleman, R.; Castaneda-Ruiz, B.; Fowler, V.G. Considerations for Clinical Trials of *Staphylococcus aureus* Bloodstream Infection in Adults. *Clin. Infect. Dis.* **2019**, *68*, 865–872. [CrossRef] [PubMed]
181. A Phase 3 Telavancin *Staphylococcus aureus* (*S. aureus*) Bacteremia Trial. 2020. Available online: https://Clinicaltrials.Gov/Ct2/Show/Results/NCT02208063 (accessed on 15 February 2022).
182. Burgin, D.J.; Liu, R.; Hsieh, R.C.; Heinzinger, L.R.; Otto, M. Investigational Agents for the Treatment of Methicillin-Resistant *Staphylococcus aureus* (MRSA) Bacteremia: Progress in Clinical Trials. *Expert Opin. Investig. Drugs* **2022**, *31*, 263–279. [CrossRef] [PubMed]
183. Draghi, D.C.; Benton, B.M.; Krause, K.M.; Thornsberry, C.; Pillar, C.; Sahm, D.F. Comparative Surveillance Study of Telavancin Activity against Recently Collected Gram-Positive Clinical Isolates from across the United States. *Antimicrob. Agents Chemother.* **2008**, *52*, 2383–2388. [CrossRef] [PubMed]
184. Saravolatz, L.D.; Pawlak, J.; Johnson, L.B. Comparative Activity of Telavancin against Isolates of Community-Associated Methicillin-Resistant *Staphylococcus aureus*. *J. Antimicrob. Chemother.* **2007**, *60*, 406–409. [CrossRef]
185. Goldstein, E.J.C.; Citron, D.M.; Merriam, C.V.; Warren, Y.A.; Tyrrell, K.L.; Fernandez, H.T. In Vitro Activities of the New Semisynthetic Glycopeptide Telavancin (TD-6424), Vancomycin, Daptomycin, Linezolid, and Four Comparator Agents against Anaerobic Gram-Positive Species and *Corynebacterium* Spp. *Antimicrob. Agents Chemother.* **2004**, *48*, 2149–2152. [CrossRef]
186. Leuthner, K.D.; Cheung, C.M.; Rybak, M.J. Comparative Activity of the New Lipoglycopeptide Telavancin in the Presence and Absence of Serum against 50 Glycopeptide Non-Susceptible Staphylococci and Three Vancomycin-Resistant *Staphylococcus aureus*. *J. Antimicrob. Chemother.* **2006**, *58*, 338–343. [CrossRef]
187. Leonard, S.N.; Vidaillac, C.; Rybak, M.J. Activity of Telavancin against *Staphylococcus aureus* Strains with Various Vancomycin Susceptibilities in an In Vitro Pharmacokinetic/Pharmacodynamic Model with Simulated Endocardial Vegetations. *Antimicrob. Agents Chemother.* **2009**, *53*, 2928–2933. [CrossRef]
188. Heo, Y.A. Oritavancin (KIMYRSA™) in Acute Bacterial Skin and Skin Structure Infections: A Profile of Its Use in the USA. *Drugs Ther. Perspect.* **2022**, *38*, 57–63. [CrossRef]
189. Zhanel, G.G.; Schweizer, F.; Karlowsky, J.A. Oritavancin: Mechanism of Action. *Clin. Infect. Dis.* **2012**, *54*, S214–S219. [CrossRef]
190. Guskey, M.T.; Tsuji, B.T. A Comparative Review of the Lipoglycopeptides: Oritavancin, Dalbavancin, and Telavancin. *Pharmacotherapy* **2010**, *30*, 80–94. [CrossRef]
191. Mendes, R.E.; Farrell, D.J.; Sader, H.S.; Jones, R.N. Oritavancin Microbiologic Features and Activity Results from the Surveillance Program in the United States. *Clin. Infect. Dis.* **2012**, *54*, 203–216. [CrossRef]
192. Billeter, M.; Zervos, M.J.; Chen, A.Y.; Dalovisio, J.R.; Kurukularatne, C. Dalbavancin: A Novel Once-Weekly Lipoglycopeptide Antibiotic. *Clin. Infect. Dis.* **2008**, *46*, 577–583. [CrossRef]
193. Mendes, R.E.; Farrell, D.J.; Sader, H.S.; Flamm, R.K.; Jones, R.N. Baseline Activity of Telavancin against Gram-Positive Clinical Isolates Responsible for Documented Infections in U.S. Hospitals (2011–2012) as Determined by the Revised Susceptibility Testing Method. *Antimicrob. Agents Chemother.* **2015**, *59*, 702–706. [CrossRef]
194. Saravolatz, L.D.; Pawlak, J.; Johnson, L.B. In Vitro Susceptibilities and Molecular Analysis of Vancomycin-Intermediate and Vancomycin-Resistant *Staphylococcus aureus* Isolates. *Clin. Infect. Dis.* **2012**, *55*, 582–586. [CrossRef]
195. Belley, A.; Neesham-Grenon, E.; Arhin, F.F.; McKay, G.A.; Parr, T.R.; Moeck, G. Assessment by Time-Kill Methodology of the Synergistic Effects of Oritavancin in Combination with Other Antimicrobial Agents against *Staphylococcus aureus*. *Antimicrob. Agents Chemother.* **2008**, *52*, 3820–3822. [CrossRef]
196. Berman, H.M.; Battistuz, T.; Bhat, T.N.; Bluhm, W.F.; Bourne, P.E.; Burkhardt, K.; Feng, Z.; Gilliland, G.L.; Iype, L.; Jain, S.; et al. The Protein Data Bank. *Acta Crystallogr. Sect. D Biol. Crystallogr.* **2002**, *58*, 899–907. [CrossRef]
197. Raad, I.; Darouiche, R.; Vazquez, J.; Lentnek, A.; Hachem, R.; Hanna, H.; Goldstein, B.; Henkel, T.; Seltzer, E. Efficacy and Safety of Weekly Dalbavancin Therapy for Catheter-Related Bloodstream Infection Caused by Gram-Positive Pathogens. *Clin. Infect. Dis.* **2005**, *40*, 374–380. [CrossRef]
198. Ito, S.; Senoo, A.; Nagatoishi, S.; Ohue, M.; Yamamoto, M.; Tsumoto, K.; Wakui, N. Structural Basis for the Binding Mechanism of Human Serum Albumin Complexed with Cyclic Peptide Dalbavancin. *J. Med. Chem.* **2020**, *63*, 14045–14053. [CrossRef]
199. Xu, X.; Xu, L.; Yuan, G.; Wang, Y.; Qu, Y.; Zhou, M. Synergistic Combination of Two Antimicrobial Agents Closing Each Other's Mutant Selection Windows to Prevent Antimicrobial Resistance OPEN. *Sci. Rep.* **2018**, *8*, 7237. [CrossRef]
200. Henderson, J.W. The Yellow Brick Road to Penicillin: A Story of Serendipity. *Mayo Clin. Proc.* **1997**, *72*, 683–687. [CrossRef]

201. Terrak, M. Peptidoglycan Glycosyltransferase Inhibition: New Perspectives for an Old Target. *Anti-Infect. Agents Med. Chem.* **2008**, *7*, 180–192. [CrossRef]
202. Miller, E.L. The Penicillins: A Review and Update. *J. Midwifery Women's Health* **2002**, *47*, 426–434. [CrossRef]
203. Watkins, R.R. *Chapter 5—A Primer on Antimicrobials*; Elsevier Inc.: Amsterdam, The Netherlands, 1941.
204. Kahan, F.M.; Kropp, H.; Sundelof, J.G.; Birnbaum, J. Thienamycin: Development of Imipenem-Cilastatin. *J. Antimicrob. Chemother.* **1983**, *12*, 1–35. [CrossRef]
205. Geddes, A.M.; Stille, W. Imipenem: The First Thienamycin Antibiotic. *Rev. Infect. Dis.* **1985**, *7*, S353–S356. [CrossRef]
206. Papp-Wallace, K.M.; Endimiani, A.; Taracila, M.A.; Bonomo, R.A. Carbapenems: Past, Present, and Future. *Antimicrob. Agents Chemother.* **2011**, *55*, 4943–4960. [CrossRef]
207. Chambers, H.F. Other Beta-Lactam Antibiotics. In *Mandell, Douglas, and Bennett's Principles and Practice of Infectious Diseases*, 5th ed.; Mandell, G.L., Bennett, J.E., Dolin, R., Eds.; Churchill Livingstone: Philadelphia, PA, USA, 2000; Volume 1, pp. 291–293.
208. Neu, H.C.; Novelli, A.; Chin, N.X. In Vitro Activity and Beta-Lactamase Stability of a New Carbapenem, SM-7338. *Antimicrob. Agents Chemother.* **1989**, *33*, 1009–1018. [CrossRef]
209. Sakurai, Y.; Yoshida, Y.; Saitoh, K.; Nemoto, M.; Yamaguchi, A.; Sawai, T. Characteristics of Aztreonam as a Substrate, Inhibitor and Inducer for Beta-Lactamases. *J. Antibiot.* **1990**, *43*, 403–410. [CrossRef] [PubMed]
210. Sykes, R.B.; Bonner, D.P. Discovery and Development of the Monobactams. *Rev. Infect. Dis.* **1985**, *7*, S579–S593. [CrossRef]
211. Gutmann, L.; Vincent, S.; Billot-Klein, D.; Acar, J.F.; Mrèna, E.; Williamson, R. Involvement of Penicillin-Binding Protein 2 with Other Penicillin-Binding Proteins in Lysis of *Escherichia coli* by Some β-Lactam Antibiotics Alone and in Synergistic Lytic Effect of Amdinocillin (Mecillinam). *Antimicrob. Agents Chemother.* **1986**, *30*, 906–912. [CrossRef]
212. Paech, F.; Messner, S.; Spickermann, J.; Wind, M.; Schmitt-Hoffmann, A.H.; Witschi, A.T.; Howell, B.A.; Church, R.J.; Woodhead, J.; Engelhardt, M.; et al. Mechanisms of Hepatotoxicity Associated with the Monocyclic β-Lactam Antibiotic BAL30072. *Arch. Toxicol.* **2017**, *91*, 3647–3662. [CrossRef] [PubMed]
213. Prasad, N.K.; Seiple, I.B.; Cirz, R.T.; Rosenberg, O.S. Leaks in the Pipeline: A Failure Analysis of Gram-Negative Antibiotic Development from 2010 to 2020. *Antimicrob. Agents Chemother.* **2022**, *66*, e00054-22. [CrossRef]
214. Li, J.; Milne, R.W.; Nation, R.L.; Turnidge, J.D.; Coulthard, K. Stability of Colistin and Colistin Methanesulfonate in Aqueous Media and Plasma as Determined by High-Performance Liquid Chromatography. *Antimicrob. Agents Chemother.* **2003**, *47*, 1364–1370. [CrossRef] [PubMed]
215. Aeschlimann, J.R.; Hershberger, E.; Rybak, M.J. Analysis of Vancomycin Population Susceptibility Profiles, Killing Activity, and Postantibiotic Effect against Vancomycin-Intermediate *Staphylococcus aureus*. *Antimicrob. Agents Chemother.* **1999**, *43*, 1914–1918. [CrossRef]
216. Baddour, L.M.; Wilson, W.R.; Bayer, A.S.; Fowler, V.G.; Tleyjeh, I.M.; Rybak, M.J.; Barsic, B.; Lockhart, P.B.; Gewitz, M.H.; Levison, M.E.; et al. Infective Endocarditis in Adults: Diagnosis, Antimicrobial Therapy, and Management of Complications: A Scientific Statement for Healthcare Professionals from the American Heart Association. *Circulation* **2015**, *132*, 1435–1486. [CrossRef]
217. Stevens, D.L.; Bisno, A.L.; Chambers, H.F.; Everett, E.D.; Dellinger, P.; Goldstein, E.J.C.; Gorbach, S.L.; Hirschmann, J.V.; Kaplan, E.L.; Montoya, J.G.; et al. Practice Guidelines for the Diagnosis and Management of Skin and Soft-Tissue Infections. *Clin. Infect. Dis.* **2005**, *41*, 1373–1406. [CrossRef]
218. Gemmell, C.G.; Edwards, D.I.; Fraise, A.P.; Gould, F.K.; Ridgway, G.L.; Warren, R.E. Guidelines for the Prophylaxis and Treatment of Methicillin-Resistant *Staphylococcus aureus* (MRSA) Infections in the UK. *J. Antimicrob. Chemother.* **2006**, *57*, 589–608. [CrossRef]
219. Mermel, L.A.; Farr, B.M.; Sherertz, R.J.; Raad, I.I.; O'Grady, N.; Harris, J.S.; Craven, D.E. Guidelines for the Management of Intravascular Catheter-Related Infections. *Clin. Infect. Dis.* **2001**, *32*, 1249–1272. [CrossRef]
220. Horstkotte, D.; Follath, F.; Gutschik, E.; Lengyel, M.; Oto, A.; Pavie, A.; Soler-Soler, J.; Thiene, G.; Von Graevenitz, A. Guidelines on Prevention, Diagnosis and Treatment of Infective Endocarditis Executive Summary: The Task Force on Infective Endocarditis of the European Society of Cardiology. *Eur. Heart J.* **2004**, *25*, 267–276. [CrossRef]
221. Elliott, T.S.J.; Foweraker, J.; Gould, F.K.; Perry, J.D.; Sandoe, J.A.T. Guidelines for the Antibiotic Treatment of Endocarditis in Adults: Report of the Working Party of the British Society for Antimicrobial Chemotherapy. *J. Antimicrob. Chemother.* **2004**, *54*, 971–981. [CrossRef]
222. Eliopoulos, G.M. Antimicrobial Agents for Treatment of Serious Infections Caused by Resistant *Staphylococcus aureus* and Enterococci. *Eur. J. Clin. Microbiol. Infect. Dis.* **2005**, *24*, 826–831. [CrossRef]
223. FDA. FDA SSSI Guidance for Industry Acute Bacterial Skin and Skin Structure Infections: Developing Drugs for Treatment. In *US Food and Drug Administration. Guidance for Industry. Acute Bacterial Skin and Skin Structure Infections: Developing Drugs for Treatment*; FDA: Washington, DC, USA, 2013.
224. Tally, F.P.; DeBruin, M.F. Development of Daptomycin for Gram-Positive Infections. *J. Antimicrob. Chemother.* **2000**, *46*, 523–526. [CrossRef] [PubMed]
225. Weems, J.J. The Many Faces of *Staphylococcus aureus* Infection: Recognizing and Managing Its Life-Threatening Manifestations. *Postgraduate Med.* **2001**, *110*, 24–36. [CrossRef] [PubMed]
226. King, M.D.; Humphrey, B.J.; Wang, Y.F.; Kourbatova, E.V.; Ray, S.M. Annals of Internal Medicine Article *Staphylococcus aureus* USA 300 Clone as the Predominant Cause of Skin and Soft-Tissue Infections. *Ann. Intern. Med.* **2006**, *144*, 309–318. [CrossRef]

227. Li, J.; Rayner, C.R.; Nation, R.L.; Owen, R.J.; Spelman, D.; Tan, K.E.; Liolios, L. Heteroresistance to Colistin in Multidrug-Resistant *Acinetobacter baumannii*. *Antimicrob. Agents Chemother.* **2006**, *50*, 2946–2950. [CrossRef]
228. Cathy, A.; Petti, M.D.A.; Vance, G.; Fowler, M.D., Jr. Optimal Treatment of Complicated Skin and Skin Structure Infections. *Infect. Dis. Clin. N. Am.* **2002**, *16*, 413–435.
229. Eisenstein, B.I.; Oleson, F.B., Jr.; Baltz, R.H. Daptomycin: From the Mountain to the Clinic, with Essential Help from Francis Tally, MD. *Clin. Infect. Dis.* **2010**, *50*, S10–S15. [CrossRef]
230. Fenton, C.; Keating, G.M.; Curran, M.P. Daptomycin. *Drugs* **2004**, *64*, 445–455. [CrossRef]
231. Streit, J.M.; Jones, R.N.; Sader, H.S. Daptomycin Activity and Spectrum: A Worldwide Sample of 6737 Clinical Gram-Positive Organisms. *J. Antimicrob. Chemother.* **2004**, *53*, 669–674. [CrossRef]
232. Niccolai, D.; Tarsi, L.; Thomas, R.J. The Renewed Challenge of Antibacterial Chemotherapy. *Chem. Commun.* **1997**, *24*, 2333–2342. [CrossRef]
233. Theuretzbacher, U.; Gottwalt, S.; Beyer, P.; Butler, M.; Czaplewski, L.; Lienhardt, C.; Moja, L.; Paul, M.; Paulin, S.; Rex, J.H.; et al. Analysis of the Clinical Antibacterial and Antituberculosis Pipeline. *Lancet Infect. Dis.* **2019**, *19*, e40–e50. [CrossRef]
234. Theuretzbacher, U.; Outterson, K.; Engel, A.; Karlén, A. The Global Preclinical Antibacterial Pipeline. *Nat. Rev. Microbiol.* **2020**, *18*, 275–285. [CrossRef] [PubMed]
235. Ventola, C.L. The Antibiotic Resistance Crisis: Part 1: Causes and Threats. *Pharm. Ther.* **2015**, *40*, 277–283.
236. Buchy, P.; Ascioglu, S.; Buisson, Y.; Datta, S.; Nissen, M.; Tambyah, P.A.; Vong, S. Impact of Vaccines on Antimicrobial Resistance. *Int. J. Infect. Dis.* **2020**, *90*, 188–196. [CrossRef]
237. Micoli, F.; Bagnoli, F.; Rappuoli, R.; Serruto, D. The Role of Vaccines in Combatting Antimicrobial Resistance. *Nat. Rev. Microbiol.* **2021**, *19*, 287–302. [CrossRef]
238. Marchetti, F.; Prato, R.; Viale, P. Survey among Italian Experts on Existing Vaccines' Role in Limiting Antibiotic Resistance. *Hum. Vaccines Immunother.* **2021**, *17*, 4283–4290. [CrossRef]
239. Jansen, K.U.; Anderson, A.S. The Role of Vaccines in Fighting Antimicrobial Resistance (AMR). *Hum. Vaccines Immunother.* **2018**, *14*, 2142–2149. [CrossRef]
240. Lobanovska, M.; Pilla, G. Penicillin's Discovery and Antibiotic Resistance: Lessons for the Future? *Yale J. Biol. Med.* **2017**, *90*, 135–145.
241. Guevara Salazar, J.A.; Morán Díaz, J.R.; Ramírez Segura, E.; Trujillo Ferrara, J.G. What Are the Origins of Growing Microbial Resistance? Both Lamarck and Darwin Were Right. *Expert Rev. Anti-Infect. Ther.* **2021**, *19*, 563–569. [CrossRef]
242. Rosas, N.C.; Lithgow, T. Targeting Bacterial Outer-Membrane Remodelling to Impact Antimicrobial Drug Resistance. *Trends Microbiol.* **2021**, *30*, 544–552. [CrossRef]
243. Reygaert, W.C. An Overview of the Antimicrobial Resistance Mechanisms of Bacteria. *AIMS Microbiol.* **2018**, *4*, 482–501. [CrossRef]
244. Li, X.-Z.; Plésiat, P.; Nikaido, H. The Challenge of Efflux-Mediated Antibiotic Resistance in Gram-Negative Bacteria. *Clin. Microbiol. Rev.* **2015**, *28*, 337–418. [CrossRef]
245. Ghai, I.; Ghai, S. Understanding Antibiotic Resistance via Outer Membrane Permeability. *Infect. Drug Resist.* **2018**, *11*, 523–530. [CrossRef]
246. Johnson, A.P.; Uttley, A.H.C.; Woodford, N.; George, R.C. Resistance to Vancomycin and Teicoplanin: An Emerging Clinical Problem. *Clin. Microbiol. Rev.* **1990**, *3*, 280–291. [CrossRef]
247. Capeletti, L.B.; de Oliveira, J.F.A.; Loiola, L.M.D.; Galdino, F.E.; da Silva Santos, D.E.; Soares, T.A.; de Oliveira Freitas, R.; Cardoso, M.B. Gram-Negative Bacteria Targeting Mediated by Carbohydrate–Carbohydrate Interactions Induced by Surface-Modified Nanoparticles. *Adv. Funct. Mater.* **2019**, *29*, 1–11. [CrossRef]
248. Plotkin, S. History of Vaccination. *Proc. Natl. Acad. Sci. USA* **2014**, *111*, 12283–12287. [CrossRef]
249. Weintraub, A. Immunology of Bacterial Polysaccharide Antigens. *Carbohydr. Res.* **2003**, *338*, 2539–2547. [CrossRef]
250. Sun, L.; Middleton, D.R.; Wantuch, P.L.; Ozdilek, A.; Avci, F.Y. Carbohydrates as T-Cell Antigens with Implications in Health and Disease. *Glycobiology* **2016**, *26*, 1029–1040. [CrossRef]
251. Werdelin, O.; Meldal, M.; Jensen, T. Processing of Glycans on Glycoprotein and Glycopeptide Antigens in Antigen-Presenting Cells. *Proc. Natl. Acad. Sci. USA* **2002**, *99*, 9611–9613. [CrossRef]
252. Khatun, F.; Toth, I.; Stephenson, R.J. Immunology of Carbohydrate-Based Vaccines. *Adv. Drug Deliv. Rev.* **2020**, *165–166*, 117–126. [CrossRef]
253. Gallorini, S.; Berti, F.; Mancuso, G.; Cozzi, R.; Tortoli, M.; Volpini, G.; Telford, J.L.; Beninati, C.; Maione, D.; Wack, A. Toll-like Receptor 2 Dependent Immunogenicity of Glycoconjugate Vaccines Containing Chemically Derived Zwitterionic Polysaccharides. *Proc. Natl. Acad. Sci. USA* **2009**, *106*, 17481–17486. [CrossRef]
254. Gallorini, S.; Berti, F.; Parente, P.; Baronio, R.; Aprea, S.; D'Oro, U.; Pizza, M.; Telford, J.L.; Wack, A. Introduction of Zwitterionic Motifs into Bacterial Polysaccharides Generates TLR2 Agonists Able to Activate APCs. *J. Immunol.* **2007**, *179*, 8208–8215. [CrossRef]
255. Carreño, L.J.; Saavedra-Ávila, N.A.; Porcelli, S.A. Synthetic Glycolipid Activators of Natural Killer T Cells as Immunotherapeutic Agents. *Clin. Transl. Immunol.* **2016**, *5*, e69. [CrossRef]
256. Kim, S.; Lalani, S.; Parekh, V.V.; Wu, L.; Van Kaer, L. Glycolipid Ligands of Invariant Natural Killer T Cells as Vaccine Adjuvants. *Expert Rev. Vaccines* **2008**, *7*, 1519–1532. [CrossRef]

257. Avci, F.Y.; Li, X.; Tsuji, M.; Kasper, D.L. Carbohydrates and T Cells: A Sweet Twosome. *Semin. Immunol.* **2013**, *25*, 146–151. [CrossRef]
258. Cavallari, M.; Stallforth, P.; Kalinichenko, A.; Rathwell, D.C.K.; Gronewold, T.M.A.; Adibekian, A.; Mori, L.; Landmann, R.; Seeberger, P.H.; De Libero, G. Asemisynthetic Carbohydrate-Lipid Vaccine That Protects against S. Pneumoniae in Mice. *Nat. Chem. Biol.* **2014**, *10*, 950–956. [CrossRef]
259. Oppenheimer, S.B.; Alvarez, M.; Nnoli, J. Carbohydrate-Based Experimental Therapeutics for Cancer, HIV/AIDS and Other Diseases. *Acta Histochem.* **2008**, *110*, 6–13. [CrossRef]
260. Yin, Z.; Huang, X. Recent Development in Carbohydrate Based Anticancer Vaccines. *J. Carbohydr. Chem.* **2012**, *31*, 143–186. [CrossRef]
261. Hossain, F.; Andreana, P.R. Developments in Carbohydrate-Based Cancer Therapeutics. *Pharmaceuticals* **2019**, *12*, 84. [CrossRef]

pharmaceuticals

MDPI

Review

GCase Enhancers: A Potential Therapeutic Option for Gaucher Disease and Other Neurological Disorders

Macarena Martínez-Bailén [1,2,*], Francesca Clemente [1], Camilla Matassini [1] and Francesca Cardona [1,*]

1 Dipartimento di Chimica "Ugo Schiff" (DICUS), Università degli Studi di Firenze, Via della Lastruccia 3-13, 50019 Sesto Fiorentino, Italy; francesca.clemente@unifi.it (F.C.); camilla.matassini@unifi.it (C.M.)

2 Glycosystems Laboratory, Instituto de Investigaciones Químicas (IIQ), CSIC—Universidad de Sevilla, Av. Américo Vespucio 49, 41092 Sevilla, Spain

* Correspondence: macarena.martinez@iiq.csic.es (M.M.-B.); francesca.cardona@unifi.it (F.C.)

Abstract: Pharmaceutical chaperones (PCs) are small compounds able to bind and stabilize misfolded proteins, allowing them to recover their native folding and thus their biological activity. In particular, lysosomal storage disorders (LSDs), a class of metabolic disorders due to genetic mutations that result in misfolded lysosomal enzymes, can strongly benefit from the use of PCs able to facilitate their translocation to the lysosomes. This results in a recovery of their catalytic activity. No PC for the GCase enzyme (lysosomal acid-β-glucosidase, or glucocerebrosidase) has reached the market yet, despite the importance of this enzyme not only for Gaucher disease, the most common LSD, but also for neurological disorders, such as Parkinson's disease. This review aims to describe the efforts made by the scientific community in the last 7 years (since 2015) in order to identify new PCs for the GCase enzyme, which have been mainly identified among glycomimetic-based compounds.

Keywords: Gaucher disease; lysosomal storage disorders; pharmacological chaperones; allosteric chaperones; iminosugars; Parkinson disease

Citation: Martínez-Bailén, M.; Clemente, F.; Matassini, C.; Cardona, F. GCase Enhancers: A Potential Therapeutic Option for Gaucher Disease and Other Neurological Disorders. *Pharmaceuticals* **2022**, *15*, 823. https://doi.org/10.3390/ph15070823

Academic Editors: Federica Sodano and Maria Grazia Rimoli

Received: 10 June 2022
Accepted: 29 June 2022
Published: 2 July 2022

1. Introduction

Lysosomal storage disorders (LSDs) are a group of more than 70 genetic metabolic diseases characterized by lysosomal dysfunction due to gene mutations that encode for lysosomal enzymes. The gene mutations affect the function of the encoded protein, which is not correctly folded and thus undergoes endoplasmic reticulum-associated protein degradation (ERAD). As a consequence, the protein-misfolding phenomenon causes a lack of catalytic activity, leading to accumulation of unmetabolized substrates in the lysosomes [1,2].

Pharmacological chaperones (PCs) are small compounds able to bind and stabilize misfolded lysosomal enzymes, facilitating their translocation to the lysosome, which results in the recovery of catalytic activity. The great majority of PCs identified to date bind to the catalytic site acting as inhibitors, but once the enzyme arrives at the lysosome, the PC dissociation permits substrate turnover. These inhibitors act as a template, stabilizing the native folding state in the endoplasmic reticulum (ER) by occupying the active site of the mutant enzyme, with a consequent final counterintuitive effect of enhancing lysosomal enzyme catalytic activity [3,4] (Figure 1). However, more recently, non-competitive inhibitors acting as PCs have also been discovered.

Gaucher disease (GD) is the most common LSD of glycosphingolipids and originates from mutations in the *GBA* gene (chromosome 1q21−22), which encodes for the lysosomal enzyme acid-β-glucosidase (glucocerebrosidase or GCase, EC 3.2.1.45, MIM*606463). GCase is the enzyme responsible for the hydrolysis of glucosylceramide (GlcCer) to glucose and ceramide in the lysosomes. Due to its heterogeneity, GD is classified into three types: type 1, primarily not associated with neurological symptoms, and neuronopathic types 2 and 3 [5].

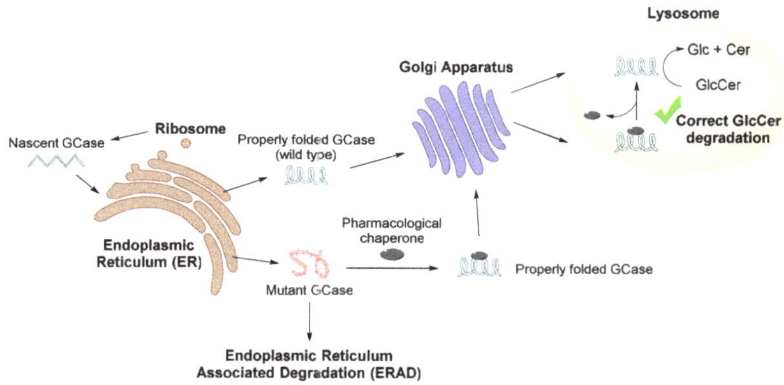

Figure 1. Schematic representation of the action of a pharmacological chaperone (PC) on mutant GCase, which recovers a correct GlcCer degradation.

More recently, a connection between GCase activity and onset of Parkinson's disease (PD), the second most common neurological disorder, was highlighted. GCase mutations were shown to be a major risk factor for the onset of PD, since people carrying heterozygous mutations in the *GBA* gene do not develop GD, but have an increased risk of developing PD [6–8]. More interestingly, the loss of GCase function contributes to the pathogenesis of PD, even independently of the genetic mutation [9], suggesting that modulation of GCase activity may be the key therapeutic target for both pathologies.

Over the past 20 years, pharmacological chaperone therapy (PCT) has been investigated as a potential treatment for LSDs, including GD, and has been the subject of numerous reviews [3,10–20]. However, few of these have been limited to GD, i.e., also include other LSDs, and do not cover recent years. Moreover, since the GD–PD connection emerged only recently, the potential use of PCT for treating PD has not been extensively described before.

Iminosugars are among the most studied classes of compounds as potential PCs for GD, and include structural classes belonging to polyhydroxylated pyrrolidines, piperidines, pyrrolizidines, indolizidines, and nortropanes [21]. In particular, isofagomine (IFG, **1**, Figure 2) is the compound that reached the most advanced clinical trial, but this stopped at phase II in 2018. IFG is a competitive inhibitor of human lysosomal GCase ($K_i = 0.016 \pm 0.009$ μM; $IC_{50} = 0.06$ μM) [22], and was found to increase mutant GCase activity 1.6-fold at 10 μM after incubation in fibroblasts bearing homozygous N370S missense mutations (Figure 2) [23]. However, IFG was not effective in reducing the accumulation of lipid substrates when administered to adult patients with type 1 GD, probably due to its high hydrophilicity, which hampers efficient transport to the cells. To overcome this issue, several alkylated piperidine iminosugars have been proposed.

6-Nonyl IFG (**4**) [24,25], α-1-*C*-nonyl-DIX (**5**) [26], and *N*-nonyl-deoxynojirimycin (NN-DNJ, **6**) [27,28] were among the most effective chaperones identified for GD until 2015, being able to strongly enhance the mutated GCase activity in N370S GD fibroblasts, as shown in Figure 2. It is worth noting that the position of the alkyl chain is crucial for the pharmacological chaperoning activity dependent on the peculiar hydroxypiperidine skeleton. Indeed, while for deoxynojirimycin (DNJ, **3**) analogues, *N*-alkylation renders good PCs for GCase (e.g., compound **6**, Figure 2), in the case of IFG (**1**) and 1,5-dideoxy-1,5-iminoxylitol (DIX, **2**) analogues, the alkylation of the carbon adjacent to nitrogen furnishes better PCs for GD (e.g., compounds **4** and **5**, Figure 2). Despite these encouraging results, no PC for the GCase enzyme has reached the drug market yet.

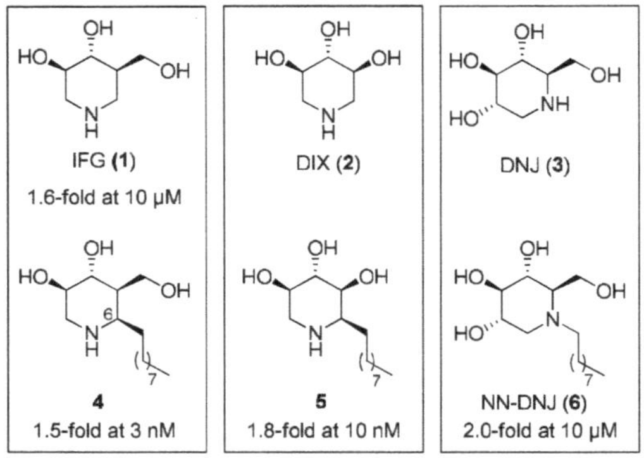

Figure 2. GCase activity increase observed in N370S GD fibroblasts in the presence of IFG (**1**), DIX (**2**), and DNJ (**3**) and their alkylated derivatives (compounds **4**–**6**).

This review focuses on the efforts made by the scientific community in the last 7 years (since 2015) in order to identify new PCs for GCase. Both inhibitory compounds, historically the first pharmacological chaperones identified, and allosteric enhancers will be discussed in this account. While the former, covered in the first section, are lysosomal enzyme inhibitors with a counterintuitive effect at sub-inhibitory concentrations corresponding to chaperone behavior, the latter are commonly discovered through a high-throughput screening approach or drug repurposing, and will be the subject of the second section. The final effect of both types of PC is an increase in enzymatic activity. We have chosen to organize the inhibitory chaperone section according to compound structure: iminosugars (six-membered, five-membered, bicyclic, and multivalent iminosugars), sugar analogues, and miscellaneous examples. Although a huge number of strong GCase inhibitors have been reported, the discussion will be limited to those inhibitors that have been tested in vitro on mutated human cells.

The second section, which is more innovative in its conception, will cover GCase enhancements and mechanistic studies relative to the very few and cutting-edge examples of allosteric chaperones.

The last section of this review is devoted to relevant examples of GCase chaperones assayed in PD animal models.

2. Inhibitory Chaperones

2.1. Iminosugars

2.1.1. Six-Membered Iminosugars

NN-DNJ (**6**, Figure 2) was reported as the first iminosugar with potential for the treatment of GD [27]. Since then, further investigation has been performed in the search for piperidine-derived compounds with potential applications as chaperones for GD. Parmeggiani et al. [29] developed an iminosugar-based library composed of polyhydroxylated piperidines, pyrrolidines, and pyrrolizidines (see Sections 2.1.2 and 2.1.3 for pyrrolidine and pyrrolizidine analogues, respectively). Regarding the piperidine derivatives, they were prepared from a carbohydrate-derived aldehyde and incorporated different moieties, such as a lipophilic chain or a benzyl group, either on the endocyclic or the exocyclic nitrogen atom. More polar groups (aminomethyl, carboxylic acid, or polar-ending lipophilic chain) were also investigated. When evaluated towards GCase, the best inhibitory activities were obtained for derivatives **7** and **8** bearing an eight-carbon alkyl chain (IC_{50} = 30 and 40 μM, respectively). Both compounds showed a measurable chaperoning activity towards human fibroblasts derived from Gaucher patients bearing the N370S/RecNcil mutations (1.25- and 1.5-fold enzyme-activity enhancement at 100 μM, respectively) (Figure 3).

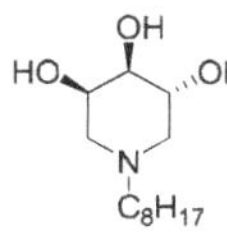

7
IC_{50} = 30 μM
(1.25-fold at 100 μM)

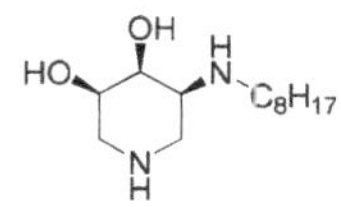

8
IC_{50} = 40 μM
(1.5-fold at 100 μM)

Figure 3. Inhibitory activity and GCase enhancement of alkylated glycomimetics **7** and **8**.

A previous study pointed out that sp^2-iminosugars behave as strong competitive inhibitors of GCase and have potential as PCs for GD [30]. Mena-Barragán et al. [31] reported the synthesis of several DNJ-based sp^2-iminosugars incorporating an orthoester fragment to switch the nature of the aglycone moiety from hydrophobic to hydrophilic in the pH 7 to pH 5 window, which had a dramatic effect on enzyme-binding affinity, thus maximizing the chaperone over the inhibitory behavior. The authors chose the orthoester group due to its chemical acid lability and stability in neutral conditions, properties that allow for optimal performance in a useful pH window. Compounds **9–11** (Figure 4) behaved as selective GCase inhibitors and were 1.3- to 200-fold better ligands of this enzyme than ambroxol (IC_{50} = 41.5 μM, see Section 2.3), a non-glycomimetic PC under clinical trial for Gaucher disease [32], at neutral pH in the endoplasmic reticulum (ER). At an acidic pH (lysosome), the product **12** (Figure 4) obtained from the hydrolysis of **9–11**, was 3-fold weaker ligand than ambroxol.

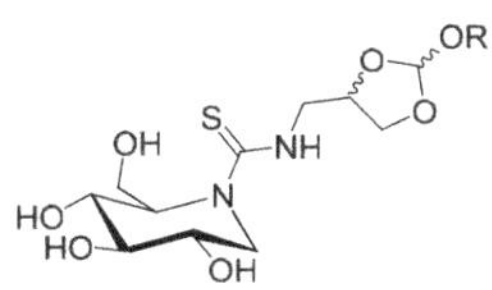

9, R = *n*-butyl, IC_{50} = 0.20 μM
10, R = *n*-octyl, IC_{50} = 0.15 μM
11, R = *n*-dodecyl, IC_{50} = 32.6 μM

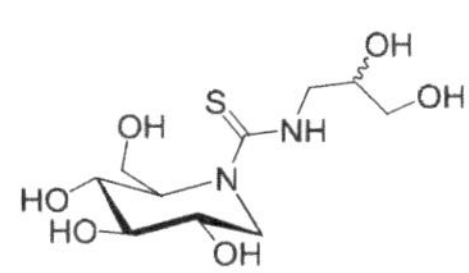

12, IC_{50} > 1000 μM

Figure 4. GCase inhibitory activity of DNJ-based sp^2-iminosugars **9–12**.

When tested in cell patients with the N188S/G193W mutation associated with the type 3 phenotype of GD, compound **9**, bearing a butoxy chain, showed better enhancing activity (6-fold activity increase at 50 μM) than the octyl and dodecyl analogues **10** and **11**. Compound **9** additionally proved to be efficient in enhancing GCase activity in V230G/R296X or F213I/F213I type 2 GD fibroblasts. However, in type 1 GD fibroblasts (N370S/N370S), the enhancement was more modest (1.5-fold enhancement). In the case of L444P/L444P fibroblasts, no activity rescue was observed.

Apart from DNJ (**3**, Figure 2), other six-membered glycomimetics, such as DIX (**2**, Figure 2), were studied by Sevsěk et al. [33] for the preparation of guanidine (**13**) and urea (**14**) DIX-derived compounds (Figure 5) bearing different lipophilic chains at the exocyclic nitrogen atom. The prepared compounds were evaluated against a panel of commercial glycosidases and human recombinant enzymes: GCase and β-galactocerebrosidase (GALC). Regarding the results for GCase, only the guanidine derivatives were potent and selective inhibitors in the low nanomolar range (IC_{50} = 17–245 nM), probably due to the positive charge of the guanidinium group, as suggested by the authors. The guanidine derivatives were better inhibitors than the simple *N*-nonyl DIX (IC_{50} = 752 nM), and this result agrees with the docking studies performed by the authors, which highlighted additional cation–π interactions and extra hydrogen bonding in the guanidine function. The length of the lipophilic chain in the molecule also affected the inhibition: compounds with longer chains were better inhibitors. The authors carried out experiments using fibroblasts homozygous for the N370S mutation, and compound **13a** (Figure 5) displayed measurable chaperone activity (1.2-fold at 1 μM).

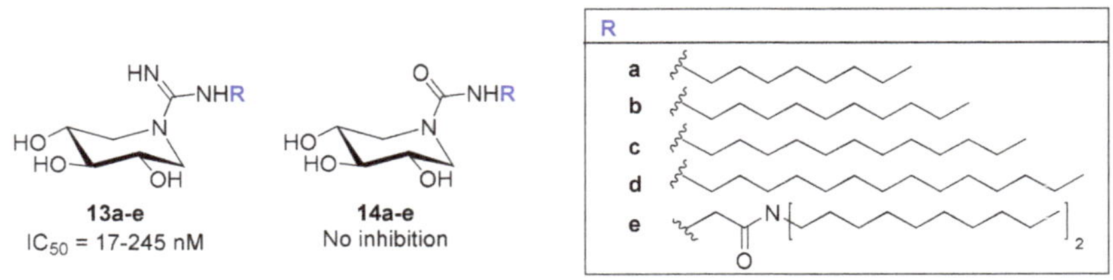

Figure 5. GCase inhibitory activity of DIX–derived guanidine and urea compounds **13** and **14**.

Génisson and coworkers reported how ring expansion of pyrrolidine derivatives (**15** and **17**, Figure 6) to the piperidine ring system (**16** and **18**, Figure 6) led to a selective targeting of GlcCer formation or hydrolysis. The authors identified in *C*-octyl piperidine **16** and 2-*O*-heptyl piperidine **18** good GCase inhibitors (IC_{50} = 11 nM and 4 nM, respectively) and PCs with low cytotoxicity. A remarkable rescue of 3-fold GCase activity was observed at 100 nM for both compounds **16** and **18** after incubation (3 days) with homozygous N370S GD fibroblasts. On the contrary, the five-membered octyl derivatives **15** and **17** showed inhibitory activity towards the glucosylceramide synthase enzyme, involved in the synthesis of GlcCer, but behaved as weaker GCase inhibitors than their corresponding six-membered counterparts (Figure 6). Molecular docking studies showed that piperidines **16** and **18** bound the active site similarly to isofagomine (IFG (**1**), Figure 2) [34].

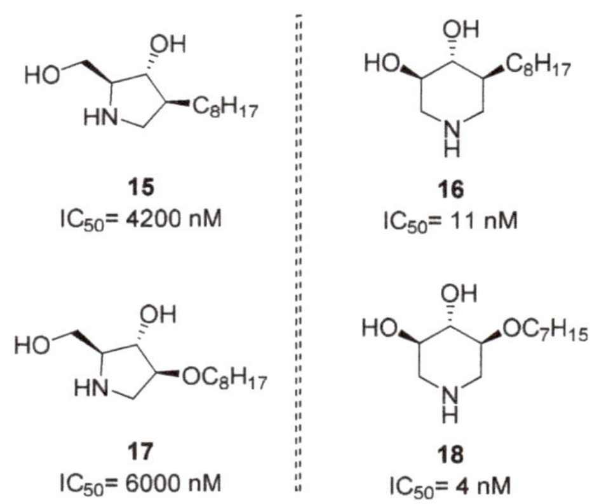

Figure 6. GCase inhibitory activity of alkylated pyrrolidines **15** and **17** and their piperidine analogues **16** and **18** obtained through ring expansion.

The role of alkyl chain length in six-membered glycomimetics on PC activity was also investigated by Cardona and coworkers, who reported the synthesis of trihydroxypiperidines **19** and **20** (Figure 7) alkylated at C-2 with both configurations and chains of different lengths [35,36]. The new compounds, assayed towards GCase and other lysosomal enzymes, proved to be good and selective GCase inhibitors with IC_{50} in the micromolar range. Selectivity towards GCase over other lysosomal enzymes is important to avoid undesired side effects. An interesting trend in the inhibitory activity was highlighted, which was dependent on the configuration at C-2 and chain length. In particular, piperidine **20d** (Figure 7) was the best inhibitor among the whole series, with an IC_{50} = 1.5 μM.

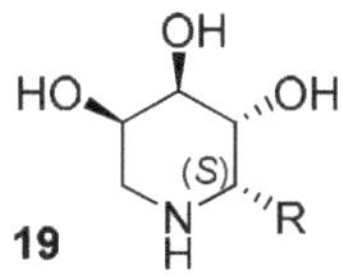

R= octyl (a)
R= nonyl (b)
R= undecyl (c)
R= dodecyl (d)
R= tridecyl (e)

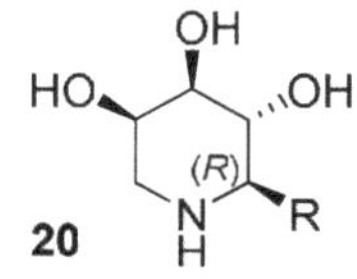

19a IC_{50}= 93.5 μM
19b IC_{50}= 143 μM
19c IC_{50}= 100 μM
19d IC_{50}= 23.3 μM
19e IC_{50}= 160 μM

20a IC_{50}= 29.3 μM
20b IC_{50}= 8.0 μM
20c IC_{50}= 7.0 μM
20d IC_{50}= 1.5 μM
20e IC_{50}= 9.0 μM

Figure 7. GCase inhibitory activity of the trihydroxypiperidine glycomimetics **19** and **20**.

All compounds were able to rescue the activity of mutated GCase from 1.2- to 1.9-fold in fibroblasts derived from Gaucher patients bearing the N370S/RecNcil mutations. The best GCase rescue was obtained with the octyl compound with the *R* configuration at C-2 **20a** (Figure 7), which was able to enhance GCase activity up to 1.9-fold at 50 μM in fibroblasts bearing the N370S mutation. Remarkably, in the presence of compound **20a**, an 80% activity rescue (1.8-fold GCase enhancement) was measured in fibroblasts bearing the homozygous L444P mutation, which is often resistant to most PCs. With a longer chain, compound **20d** was the best inhibitor of the series, but not the best pharmacological chaperone. This effect can be explained in light of the fact that during in vitro assays on cell lines, other factors play a crucial role, such as cell permeability and cell viability. The improved potency provided by a longer alkyl moiety was also concurrent with low cell viability. These data demonstrated, analogously to what happens with IFG and DIX derivatives, that the presence of a lipophilic chain of proper length plays a pivotal role on the chaperoning properties and that the alkylation of the carbon adjacent to nitrogen furnishes better PCs for GD also in the case of these trihydroxypiperidine iminosugars.

2.1.2. Five-Membered Iminosugars

Five-membered iminosugars (polyhydroxylated pyrrolidines) are a group of glycomimetics that—in comparison with their six-membered analogues—have been less studied in the field of glycosidase inhibition. This family of compounds sometimes shows problems of selectivity due to their higher conformational flexibility with respect to piperidines, but the attachment of different functional groups to the pyrrolidine core can lead to more potent and selective glycosidase inhibitors [37].

Using a D-arabinose-derived nitrone as key intermediate, Parmeggiani et al. [29] developed a library of polyhydroxylated pyrrolidines, analogues of the naturally occurring iminosugar 1,4-dideoxy-1,4-imino-D-arabinitol (DAB, **21**, Figure 8), in their *N*- or *C*-alkylated version (chain length of 4, 8, or 12 carbon atoms). The best results in terms of GCase inhibition were obtained with pyrrolidines bearing a longer chain (8 or 12 atom carbons, compounds **22–27**, Figure 8), thus probably suggesting that these compounds better mimic the natural substrate of GCase. Unfortunately, despite the good inhibitory activity displayed for compounds **22–27** (IC_{50} = 0.72–150 μM, Figure 8), they did not show any chaperone activity when tested on cell lines.

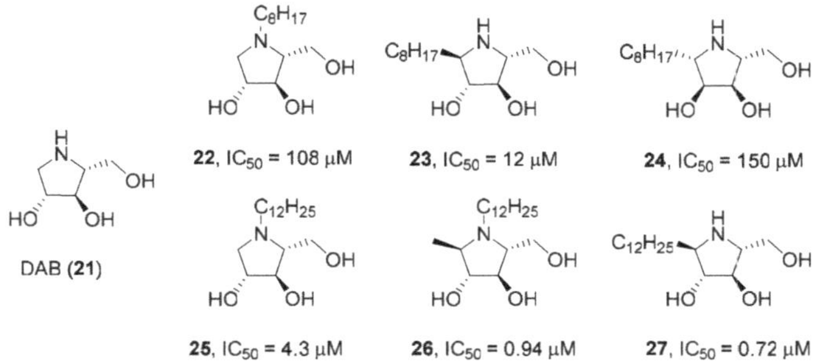

Figure 8. *N*- and *C*-alkylated pyrrolidine derivatives analogues of DAB (**1**) as GCase inhibitors.

Conversely, Kato et al. [38] prepared DAB (**21**, Figure 8) and LAB (1,4-dideoxy-1,4-imino-L-arabinitol, ***ent*-21**, Figure 9) *C*-alkylated pyrrolidines bearing a butyl or octyl chain (compounds **23**, **28**, ***ent*-23** and ***ent*-28**, Figure 9). The inhibition results towards GCase showed that the best candidates were the DAB-configured analogues, and among them, the insertion of an octyl group (**23**) gave the best inhibitory activity. So, in a next step, the length of the alkyl chain was changed, ranging from 4 to 13 carbon atoms (derivatives **27** and **29–35**, Figure 9). The results showed that the longer the alkyl chain is, the stronger is the inhibition, with optimal results obtained for the tridecyl analogue **35** with an IC_{50} value of 0.77 µM. The derivatives **29**, **32**, and **35** were evaluated as chaperones in fibroblasts with the homozygous N370S mutation and compared with isofagomine (IFG (**1**), Figure 2), used as a positive control. Compound **35** was the best chaperone candidate, with a GCase rescue similar to that observed with IFG (around 1.5-fold at 5 µM) but at a 10-fold lower concentration (0.5 µM).

DAB (**21**), R = H, No inhibition
28, R = C_4H_9, No inhibition
29, R = C_5H_{11}, IC_{50} = 34 µM
30, R = C_6H_{13}, IC_{50} = 30 µM
31, R = C_7H_{15}, IC_{50} = 38 µM
23, R = C_8H_{17}, IC_{50} = 6.2 µM
32, R = C_9H_{19}, IC_{50} = 3.3 µM
33, R = $C_{10}H_{21}$, IC_{50} = 1.9 µM
34, R = $C_{11}H_{23}$, IC_{50} = 2.3 µM
27, R = $C_{12}H_{25}$, IC_{50} = 1.5 µM
35, R = $C_{13}H_{27}$, IC_{50} = 0.77 µM

LAB (***ent*-21**), R = H, No inhibition
***ent*-28**, R = C_4H_9, No inhibition
***ent*-23**, R = C_8H_{17}, No inhibition

Figure 9. DAB and LAB alkylated analogues and their activity towards GCase.

The presence of the hydroxymethyl moiety in the pyrrolidine skeleton makes an important contribution to glycosidase inhibition [39], but its removal can lead to more hydrophobic compounds that can better cross the cell membranes. Indeed, Castellan et al. [40] reported the synthesis of a couple of enantiomeric pyrrolidines, **36** and ***ent*-36** (Figure 10), and their evaluation as GCase inhibitors and activity enhancers in fibroblasts bearing the homozygous N370S mutation. These compounds can be also considered DAB (**21**, Figure 8) and LAB (***ent*-21**, Figure 9) analogues, respectively, but lacking the hydroxymethyl moiety. The removal of this group was advantageous for GCase inhibition, as both compounds were more potent inhibitors than the corresponding hydroxymethylated counterparts reported by Kato et al. (compounds **36** vs. **23** and ***ent*-36** vs. ***ent*-23**) [38].

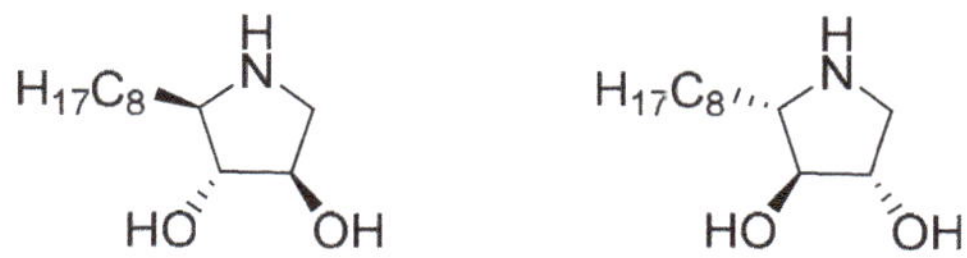

36, IC_{50} = 0.78 μM　　　***ent*-36**, IC_{50} = 59.6 μM

Figure 10. Compounds **36** and ***ent*-36** lacking the hydroxymethyl moiety of DAB and LAB.

Compound **36**, with a non-competitive inhibition mode, was two orders of magnitude more potent than competitive ***ent*-36**. The inhibitory mode of both compounds was further confirmed by the application of Theorell's graphical method [41] and through docking simulations, which showed a tight alignment of ***ent*-36** with NN-DNJ (**6**, Figure 2) and IFG (**1**, Figure 2) and the retention of the main hydrogen-bond interactions within the GCase active site.

The chaperone activity of both compounds was measured in fibroblasts with the homozygous N370S mutation. Compound **36** gave the best results, showing an activity enhancement comparable to that observed for NN-DNJ (**6**) and IFG (**1**), used as reference compounds, i.e., 1.5-fold or 2-fold at 3 μM independently of the chosen cell lines. Indeed, it is worth noting that intracellular enzymatic enhancement may highly depend on the chosen cell lines, given the same N370S mutation. Nevertheless, the enhancement of ***ent*-36** (1.6-fold at 3 μM) is also significant, despite being a less potent competitive inhibitor (IC_{50} = 59.6 μM), demonstrating that not always the more potent inhibitors are also the best chaperone candidates.

Not only the insertion of alkyl chains on the pyrrolidine core can improve the biological properties towards the GCase enzyme. The introduction of (hetero)aromatic moieties exploiting copper(I)-catalyzed alkyne-azide cycloaddition (CuAAC) can also afford more potent and selective GCase inhibitors [42], probably due to the establishment of additional interactions with the enzyme, allowing at the same time improved cell permeability and cell trafficking. In this context, Martínez-Bailén et al. [43] reported the synthesis of a set of epimeric dihydroxypyrrolidines containing an aromatic moiety attached to the triazole generated in the CuAAC reaction. This new family of inhibitors exhibited a tunable biological activity towards GCase in the function of the substitution pattern of the aromatic ring, the 3,5-disubstituted derivatives being the most potent inhibitors (compounds **37–56**, Figure 11). The best inhibitors of the library (compounds **44**, **45**, and **53**) were able to increase the in vitro activity of GCase in Gaucher patient fibroblasts with the homozygous N370S mutation. Specifically, compound **45** increased the GCase activity 2-fold at 10 μM concentration, an enhancement that is comparable to that observed for isofagomine used as positive control at the same concentration. The crystal structure of complex **53**:GCase and docking experiments revealed key interactions between the inhibitor and the enzyme that supported the importance of the aryltriazole moiety and was in agreement with the improved trafficking of GCase observed in immunolabeling experiments performed by the authors.

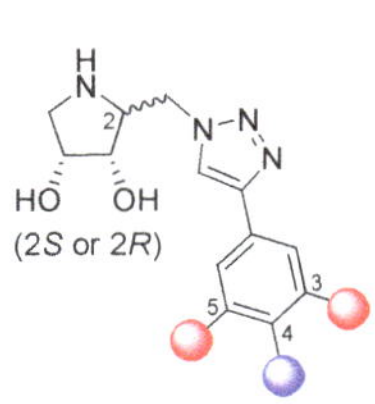

(2*S* or 2*R*)

Compound				IC_{50} (μM)	
2*S*	2*R*			2*S*	2*R*
37	**47**	H	H	11	28
38	**48**	H	OH	27	34
39	**49**	H	CH_2OH	102	187
40	**50**	H	OMe	39	31
41	**51**	H	F	23	24
42	**52**	F	H	3.8	5.7
43	**53**	Cl	H	1.5	1.3
44	**54**	Br	H	0.83	1.4
45	**55**	CF_3	H	0.95	1.6
46	**56**	OMe	H	8.3	7.3

Figure 11. Dihydroxypyrrolidines **37–56** decorated with an aryltriazole moiety.

2.1.3. Bicyclic Iminosugars

Calystegines are polyhydroxy nortropane alkaloids that display GCase selectivity over other lysosomal enzymes [44]. García-Moreno et al. [45] reported the synthesis and biological evaluation of several amphiphilic glycomimetics presenting a rigid nortropane skeleton based on 1,6-anhydro-L-idonojirimycin, formulated as the corresponding inclusion complexes with β-cyclodextrin (βCD). The insertion of a terminal polyfluorinated fragment allowed the improvement of the complex stability and GCase-rescue capabilities due to additional interactions with the protein surface according to medicinal chemistry principles [46,47], improving membrane crossing, also including the blood–brain barrier (BBB). The authors encapsulated their fluorinated derivatives in form of β-cyclodextrin complexes, taking advantage of the affinity of this kind of compound for the cavity of β-cyclodextrin and avoiding the intrinsic aggregation previously observed for polyfluorinated derivatives [48], thus gaining in drug availability and cell permeability.

The new compounds presented an alkyl or fluorinated chain of different lengths, and all of them showed strong competitive inhibition towards the mammalian β-glucosidase enzymes selectively. The encapsulated βCD complexes (K_i = 0.09–8.2 μM) behaved as stronger inhibitors than compounds **57–60** (K_i = 0.33–90 μM) (Figure 12) alone (without encapsulation). In the case of GCase, the complexes were better inhibitors at pH 7 than at pH 5, which is a positive feature for a chaperone candidate.

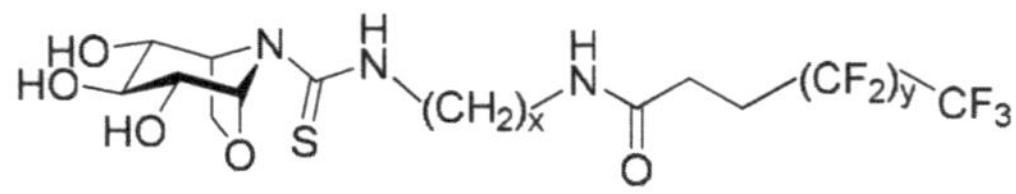

57, x = 4; y = 3 **59**, x = 9; y = 3
58, x = 4; y = 5 **60**, x = 9; y = 5

Figure 12. Polyfluorinated nortropanes **57–60**.

The complexes of **57–60** with βCD were tested as chaperones in fibroblasts of Gaucher patients bearing the following mutations: N370S/N370S or N370S/84GG (non-neuronopathic, type 1 GD), V230G/R296X or L444P/P415R (acute neuronopathic, type 2 GD), and N188S/G183W or L444P/L444P (neuronopathic, type 3 GD). The best enhancements observed for each mutation are summarized in Table 1. From the results, it can be concluded that the complex **58**:βCD was the one with the widest activity spectrum (Table 1, entries 1, 2, and 4–6). Further immunolabeling experiments demonstrated that the chaperone-action mechanism involved rescuing and trafficking of GCase, as the recovered enzyme was colocalized in the lysosome and no longer in the ER after treatment of the cells with **58**:βCD. The results on the L444P/L444P lines are remarkable (Table 1, entry 6), since these cell lines are resistant to most PCs.

Table 1. Enhancements observed for each Gaucher disease variant.

Entry	GCase Variant	Severity	Best Enhancement	Concentration	Complex
1	N370S/N370S	Type I	1.5-fold	2 μM	**58**:βCD
2	N370S/84GG		2.8-fold	20 μM	**58**:βCD
3	V230G/R296X	Type II	2.7-fold	20 μM	**57**:βCD
4	L444P/P415R		2.9-fold	20 μM	**58**:βCD
5	N188S/G183W	Type III	1.8-fold	20 μM	**57**:βCD and **58**:βCD
6	L444P/L444P		2.8-fold	20 μM	**58**:βCD

Pyrrolizidines are another class of bicyclic iminosugars that has been widely studied in the field of glycosidase inhibition [49]. Parmeggiani et al. [29] completed their screening on pyrrolidines and piperidines with several pyrrolizidine iminosugars obtained from nitrones. Unfortunately, the great majority of pyrrolizidines did not impart any relevant inhibition towards GCase, except for compounds **61** and **62**, the latter bearing a phenyl-substituted triazole moiety, thus showing the importance of the incorporation of an aromatic group in the molecule (Figure 13). However, when tested on cell lines, these compounds did not give any enzymatic rescue.

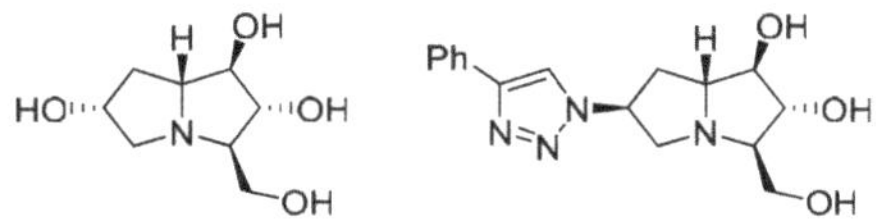

61, IC_{50} = 58 μM **62**, IC_{50} = 100 μM

Figure 13. Pyrrolizidine iminosugars **61** and **62**.

Mena-Barragán et al. [50] prepared a family of bicyclic iminothiazolidine-sp^2-iminosugars derived from D-fagomine, DAB (**21**, Figure 8) and LAB (***ent*-21**, Figure 9) bearing alkyl chains of different lengths (4, 8, or 16 carbon atoms). The bicyclic compounds **63–66**, ***ent*-65**, and ***ent*-66** (Figure 14) were better inhibitors than the corresponding monocyclic counterparts, especially those with a longer alkyl chain (octyl and hexadecyl), and showed IC_{50} values towards GCase in the low micromolar range (IC_{50} = 4.1–12.4 μM). Interestingly, the compounds were one order or magnitude more potent inhibitors at pH 7 than at pH 5.

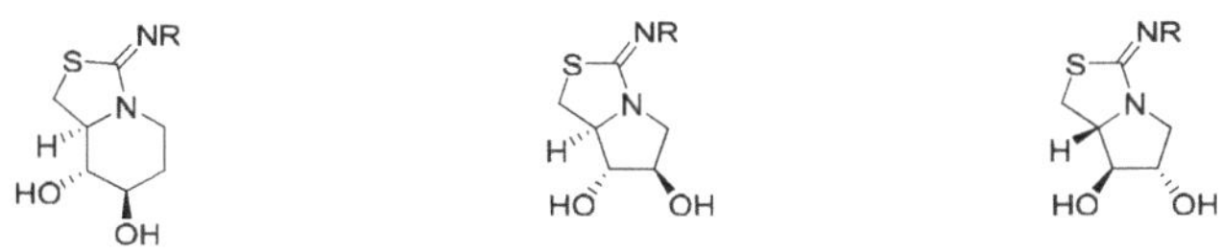

63, R = C_8H_{17}, IC_{50} = 12.4 μM **65**, R = C_8H_{17}, IC_{50} = 7.3 μM ***ent*-65**, R = C_8H_{17}, IC_{50} = 8.6 μM
64, R = -$(CH_2)_{16}OH$, IC_{50} = 6.8 μM **66**, R = -$(CH_2)_{16}OH$, IC_{50} = 4.9 μM ***ent*-66**, R = -$(CH_2)_{16}OH$, IC_{50} = 4.1 μM

Figure 14. Bicyclic iminothiazolidine-sp^2-iminosugars.

These compounds were tested as chaperones in fibroblasts of GD patients bearing the N188S/G183W, V230G/R296X or L444P/L444P mutations, which are associated with type 2 or type 3 neuronopathic variants. Unfortunately, only compound **63** was able to increase the GCase activity by 1.3-fold in the N188S/G183W and V230G/R296X fibroblasts at 2 μM. Regarding the homozygous L444P variant, none of the compounds was effective.

Ortiz Mellet and coworkers reported the synthesis of a series of sp^2-iminosugars with different structural requirements, in order to evaluate the effect of these modifications on the inhibitory activity towards GCase and their chaperoning capabilities in GD cell lines [51]. When tested on a panel of commercial glycosidases and human GCase, the best chaperone candidates were compounds **67–70** (Figure 15). The GCase-chaperoning capabilities of these compounds were evaluated in type 3 GD fibroblasts with the N188S/G183W mutation, and the results showed that the non-reducing derivatives behaved as better chaperones than the corresponding reducing analogues (**69** and **70** vs. **67** and **68**, Table 2, entries 3 and 4 vs. 1 and 2). The presence of a sulfur atom in the iminosugars also led to an improvement in chaperone behavior with respect to the oxygenated counterparts (**68** and **70** vs. **67** and **69**, Table 2, entries 2 and 4 vs. 1 and 3). It is worth highlighting that all derivatives showed higher GCase inhibitory activity at neutral pH (Table 2), which is in favor of a good chaperone candidate, as a higher binding affinity in the ER is required to ensure the correct folding of the enzyme. Compounds **67**, **68**, and **69** had previously been tested in other GD cell lines, displaying moderate-to-good chaperoning effects [31,52,53], but the new derivative **70**, which presented the most suitable structure (non-reducing and sulfured-sp^2-iminosugar), emerged as the most active chaperone candidate reported to date for a neuronopathic Gaucher mutation, as it enhanced GCase activity even in the picomolar range (1.7-fold at 20 pM).

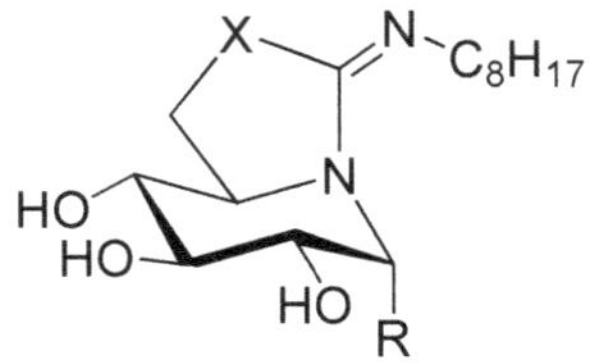

67, X=O, R = OH
68, X=S, R = OH
69, X=O, R = H
70, X=S, R= H

Figure 15. sp^2-Iminosugars **67–70**.

Table 2. K_i (in μM) and GCase enhancement for compounds **67–70**.

Entry	Compound	K_i (μM) [a]		Enhancement (N188S/G183W)
		pH 7	pH 5	
1	**67**	15.1	54.4	3-fold at 20 μM
2	**68**	0.26	1.05	3.2-fold at 2 μM
3	**69**	1.7	6.3	3.2-fold at 200 nM
4	**70**	0.013	0.059	3-fold at 2 nM

[a] Inhibition was competitive in all cases.

2.1.4. Multivalent Iminosugars

The multivalent display of carbohydrates on adequate platforms can afford more potent glycosidase inhibitors than the corresponding monovalent counterparts. Despite many efforts to develop multivalent glycosidase inhibitors in recent years [54–59], this strategy has been proved successful mainly in the field of α-mannosidases [54–61], with few examples of multivalent derivatives targeting the glycosidases implicated in the LSDs [62–64]. More recently, some of us have described multivalent derivatives with potential as pharmacological chaperones in fibroblasts of Fabry patients [65]. Regarding GCase and Gaucher disease, the first examples of multivalent PCs were reported by Compain and coworkers 10 years ago. The authors identified the iminosugar–cyclodextrin conjugate **71** [66] (Figure 16) and the tetravalent DNJ derivative **72** [67] (Figure 16) as potent PCs able to maximally increase GCase activity in N370S Gaucher fibroblasts up to 2.4-fold and 3.3-fold at 10 μM, respectively (Table 3, entries 1 and 2) [67]. Furthermore, the trivalent acetyl-DNJ-derivative **74** (Figure 16), although being a worse inhibitor, showed higher enhancement in GCase activity than its corresponding unprotected analogue **73** (Figure 16) (3.0-fold vs. 2.4-fold, Table 3, entries 4 vs. 3) in N370S GD fibroblasts at one order of magnitude lower concentration (1 μM vs. 10 μM, Table 3, entries 4 vs. 3) [67]. The authors argued that compound **74**, with improved permeability and cellular uptake with respect to **73**, may undergo acetyl hydrolysis in vivo, providing the first evidence of the potential of prodrugs in PCT.

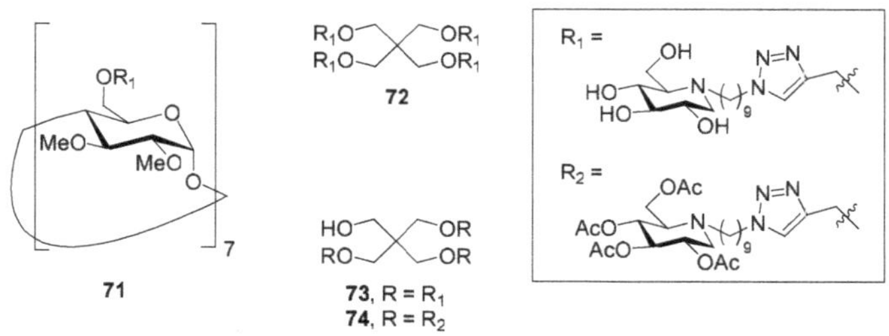

Figure 16. Multivalent iminosugars **71–74**.

Table 3. GCase enhancement shown by compounds **71–74** in N370S Gaucher fibroblasts.

Entry	Compound	Enhancement
1	**71**	2.4-fold at 10 μM
2	**72**	3.3-fold at 10 μM
3	**73**	2.4-fold at 10 μM
4	**74**	3.0-fold at 1 μM

Later on, Laigre et al. [68] prepared a set of DNJ-based multivalent clusters with free OH (compounds **72**, Figure 16, and **76**, Figure 17) or in their polyacetylated version (compounds **75** and **77**, Figure 17) in order to better investigate the prodrug effect. They also incorporated a morpholine moiety in some derivatives to study the chaperone effect in Gaucher disease by synthesizing lysosome-targeting probes [69]. The GCase inhibition and enhancement results are summarized in Table 4.

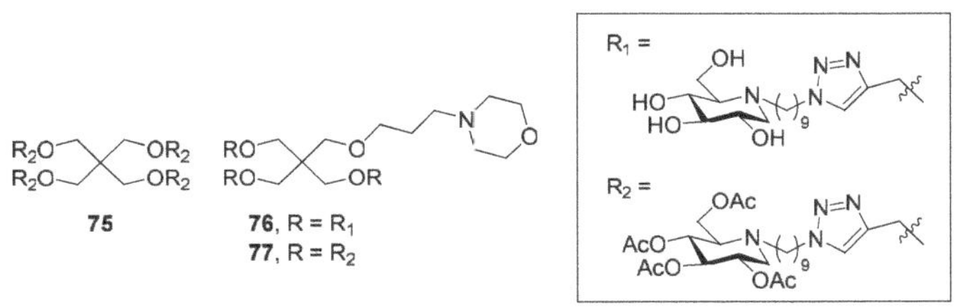

Figure 17. Multivalent iminosugars **75–77**.

Table 4. IC_{50} values (in nM) and enhancements observed in N370S Gaucher fibroblasts (at 1 μM).

Entry	Compound	IC_{50} (nM)	Enhancement
1	**72**	500	2.5-fold
2	**75**	n.i.	1.8-fold
3	**76**	109	2.0-fold
4	**77**	n.i.	2.1-fold

n.i. = No inhibition detected at 0.1 mM.

The best enhancement was obtained for **72** (2.5-fold at 1 μM, Table 4, entry 1) although this compound was not the best inhibitor. Unfortunately, no prodrug effect was observed, as all the peracetylated derivatives showed decreased or similar chaperoning activity with respect to the deprotected analogues (**75** vs. **72** and **77** vs. **76**, Table 4, entries 2 vs. 1 and 4 vs. 3). In addition, the insertion of the morpholine moiety did not further improve GCase-activity enhancement (**76** vs. **72** and **77** vs. **75**, Table 4, entries 3 vs. 1, and 4 vs. 2). Since the replacement of an iminosugar unit (both deprotected and peracetylated) with a morpholine moiety resulted in comparable chaperoning activity, it can be argued that the morpholine moiety acts as an iminosugar surrogate, rather than a lysosome-targeting probe.

Stauffert et al. [70] developed a library of dimers derived from DIX (**2**, Figure 2) and its enantiomer with the aim of simultaneously targeting the active site and a secondary binding site of GCase. The library was composed of five heterodimers (compounds **78a–e**) and two homodimers (compounds **79** and ***ent*-79**) of both enantiomers (Figure 18). The corresponding monovalent counterparts (**80–83** and ***ent*-80** and ***ent*-83**) were also prepared and evaluated towards GCase (Figure 18 and Table 5).

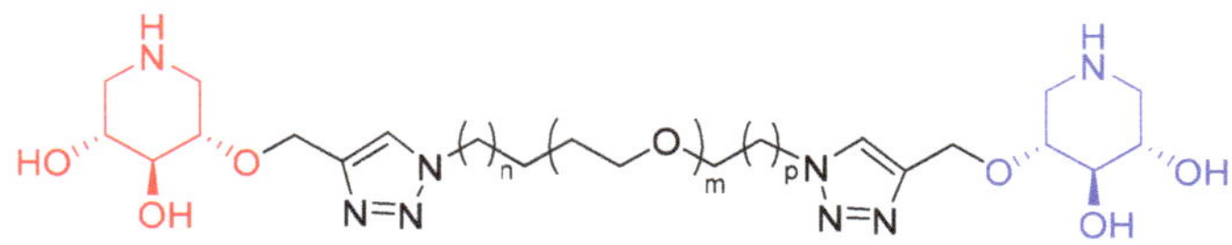

Heterodimers **78**

78a, n = 4, m = 0, p = 0
78b, n = 8, m = 0, p = 0
78c, n = 3, m = 1, p = 5
78d, n = 3, m = 3, p = 1
78e, n = 7, m = 4, p = 1

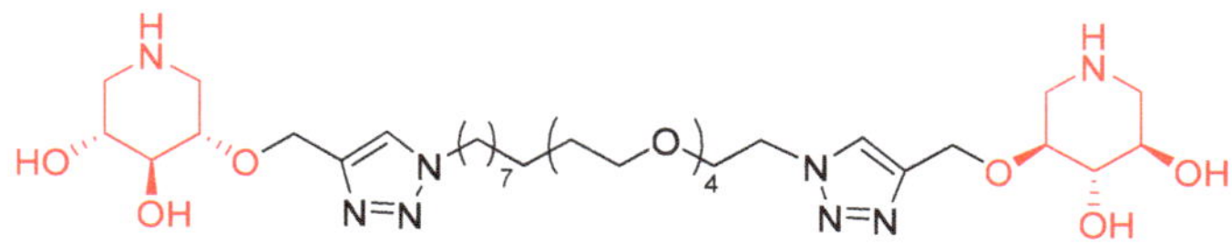

Homodimer **79**

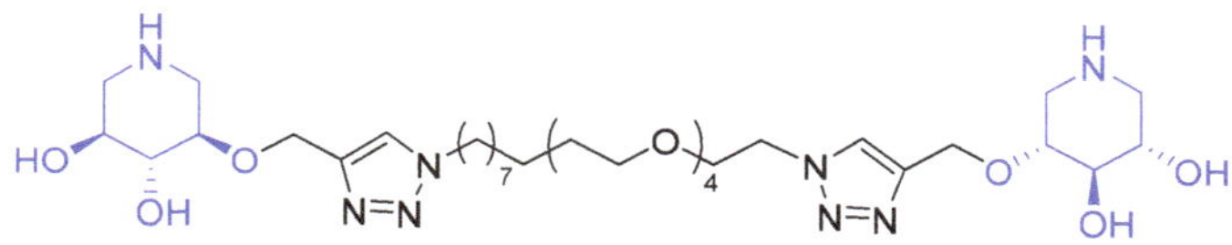

Homodimer ***ent*-79**

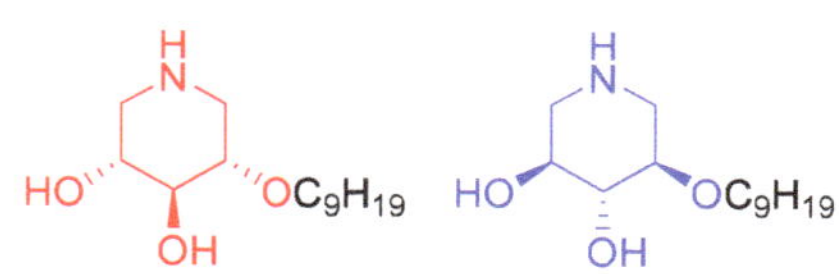

80 ***ent*-80**

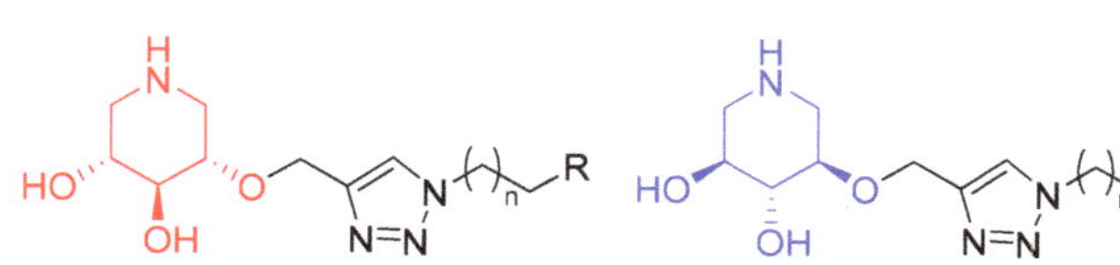

81, n = 5, R = OH
82, n = 9, R = OH
83, n = 7, R = CH_3

***ent*-83**, n = 7, R = CH_3

Figure 18. Homo- and heterodimers derived from DIX (**2**) and its enantiomer.

Table 5. IC_{50} values (in nM) and enhancements observed in G202R/G202R fibroblasts.

Entry	Compound	Valence	IC_{50} (nM)		Enhancement	Concentration (nM)
			pH 5.2	pH 7.0		
1	**78a**	2	1800	223	1.5-fold	2000
2	**78b**	2	23.6	32.5	1.7-fold	2000
3	**78c**	2	57.1	28.5	2.2-fold	2000
4	**78d**	2	1100	362	1.6-fold	2000
5	**78e**	2	46.6	17.0	2.2-fold	100
6	**79**	2	3.6	3.3	2.5-fold	100
7	***ent*-79**	2	162,000	94,000	-	-
8	**80** [a]	1	4.3	5.7	1.8-fold	1
9	**81**	1	292	182	1.6-fold	2000
10	**82** [a]	1	24.3	19.9	3.6-fold	100
11	**83**	1	4.8	4.3	2.2-fold	10
12	***ent*-80**	1	75,000	42,000	-	300
13	***ent*-83**	1	58,000	19,000	-	2000

All compounds behave as non-competitive inhibitors, unless otherwise noted. [a] Competitive inhibitors. "-": No enhancement observed.

All compounds resulted to be GCase inhibitors in the micromolar or nanomolar range, and some of them exhibited a better inhibition at pH 7, which means a higher affinity towards GCase at the neutral pH of the ER. With the exception of compounds **80** (Table 5, entry 8) and **82** (Table 5, entry 10), all the new derivatives were non-competitive inhibitors. The non-competitive inhibition mode observed for these compounds indicated that they were able to bind GCase in a secondary binding site. The best inhibition results at pH 7 were obtained for monomers **80**, **82**, and **83** (Table 5, entries 8, 10 and 11) and dimers **78e** and **79** (Table 5, entries 5 and 6). All compounds were tested as GCase enhancers in a panel of fibroblasts from GD patients with three different genotypes (N370S/N370S, L444P/L444P or G202R/G202R), and good results were observed only with G202R/G202R fibroblasts. In the monomeric series, the best performances were obtained for compounds **80**, **82**, and **83** (Table 5, entries 8, 10 and 11), **82** (3.6-fold at 100 nm, Table 5, entry 10) being the best enhancer, even if this compound is not the most potent inhibitor of the series. Among the dimeric analogues, heterodimer **78e** and homodimer **79** gave the most outstanding results (2.2-fold and 2.5-fold at 100 nM, respectively, Table 5, entries 5 and 6), but they did not surpass the enhancement observed for the corresponding monomeric counterpart **82** (Table 5, entry 10). Decreased glucosylceramide levels and colocalization studies further confirmed the observed results. It is worth highlighting that the absolute configuration of the iminosugar is crucial for GCase enhancement, as dimers **78e** and **79** showed similar enhancements, while dimer ***ent*-79** was inactive (Table 5, entries 5–7).

Very recently, Cardona, Matassini and coworkers reported the synthesis of five new multivalent iminosugars, obtained through CuAAC reaction of an azido ending trihydroxypiperidine with several propargylated scaffolds (Figure 19), designed to investigate how different valences and topologies affect the GCase response [71].

The whole set of inhibitors showed GCase inhibition in the low micromolar range, with lowest IC_{50} values obtained for the tri- and hexavalent compounds **85** and **87** (IC_{50} = 7 μM and 6 μM, respectively). While negligible chaperoning activity was found for compounds **85**, **86**, and **88** in fibroblasts from GD patients bearing the N370S/RecNcil mutations (1.2- to 1.3-fold at 10–50 μM), a remarkable 2-fold activity enhancement at 10 μM was obtained for the trivalent **85** in GD patients' fibroblasts bearing the homozygous L444P mutation. This enhancement is one of the highest ever observed for a PC towards these PC refractory cell lines.

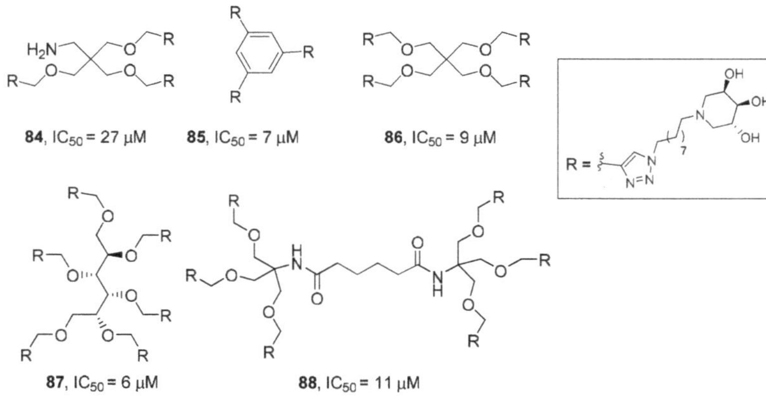

Figure 19. Multivalent iminosugars 84–88.

2.2. Sugar Analogues

Apart from iminosugars, other carbohydrate-derived analogues have been studied as possible chaperones for the treatment of GD. Castilla et al. [72] reported the synthesis of pyranoid-type glycomimetics with a *cis*-1,2-fused glucopyranose-2-alkylsulfanyl-1,3-oxazoline (Glc-PSO) structure. Based on previous results [73], which suggested that the nature of the moiety attached to the sulfur atom highly influenced the inhibitory potency against human GCase, the authors investigated the inhibitory properties of compounds incorporating functional groups of different nature (fluorine, iodine, methyl ester, and carboxylic acid, compounds **89–92**, Figure 20). These compounds can be considered conformationally locked *N*-glycosides, due to their bicyclic fused skeleton formed by a six-membered and a five-membered cycle, which imposes a skew-boat conformation on the pyranose ring. This conformation was found to be responsible for the bioactivity, since it imparts to the compounds a transition state mimic feature [74].

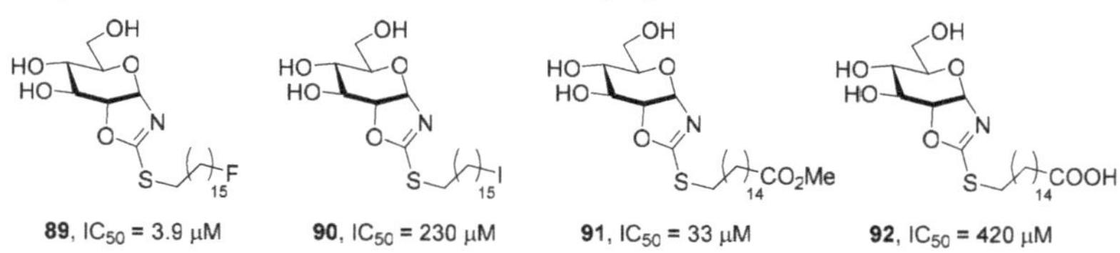

Figure 20. Pyranoid-type glycomimetics **89–92**.

The results showed a clear influence of the terminal group of the chain on inhibitory activity, with compound **89** incorporating a fluorine atom being the most active derivative with an IC_{50} value of 3.9 μM. They also observed an increase in IC_{50} value when the pH was changed from 7 to 5, which is in favor of a good chaperone candidate. The observed variability agrees with the initial hypothesis of the additional stabilization of the enzyme–inhibitor complex, due to interactions of this terminal group with the enzyme. The most active compounds, **89** and **91**, were evaluated as chaperones in fibroblasts of patients with the N370S/N370S or L444P/L444P mutations. The best results were obtained for compound **89** towards N370S/N370S-mutated fibroblasts, detecting an improvement of 62% at 30 μM, which was higher than that observed for ambroxol (see Section 2.3) at the same concentration, probably due to the overcoming of the inhibitory effect. Unfortunately, no improvement was observed in fibroblasts bearing the L444P/L444P mutation.

Navo et al. [75] developed a family of 3-amino-6,7-dihydroxy-3,5-bis(hydroxymethyl) hexahydropyrano[3,2-*b*]pyrrol-2(1*H*)-one derivatives (APP) obtained from tri-*O*-benzyl-2-nitro-D-galactal following a methodology previously reported by the authors [76]. The new derivatives incorporated a palmitoyl fragment to evaluate the effect of this hydrophobic moiety on the inhibitory profile. After a screening towards a panel of commercial glycosidases, derivatives **93–96** (Figure 21) were the most promising candidates among all the compounds prepared and were subsequently tested towards human GCase.

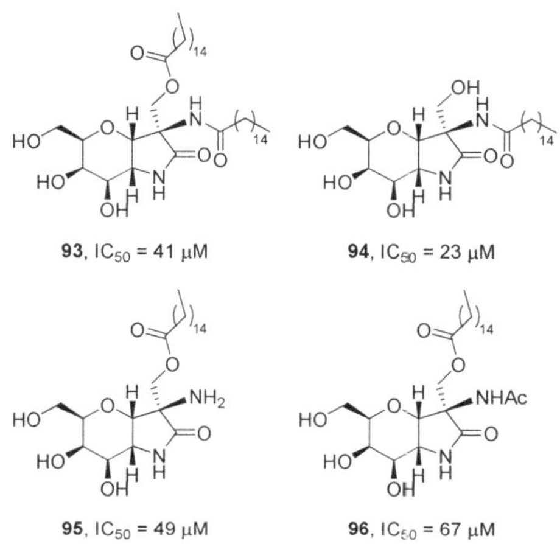

Figure 21. Pyranoid-type glycomimetics **93–96**.

From the results obtained, it was deduced that the presence of an *O*-acyl substituent in the aglycone was detrimental to inhibitory activity, as compound **94** (K_i = 20 μM at pH 7) was a 1.5 to 2.5-fold more potent inhibitor than derivatives **93**, **95**, and **96** (K_i = 33–52 μM at pH 7). The authors also observed that compounds **93–96** behaved as stronger inhibitors of GCase at pH 7 than at pH 5. These compounds were assayed as chaperones in fibroblasts of GD patients with the N370S/N370S, F213I/F213I, or L444P/L444P mutations, associated with type 1, 2, and 3 GD, respectively. Only compound **94** displayed a significant enhancement in GCase activity at 20 μM for the N370S/N370S and F213I/F213I mutations (1.3- and 1.5-fold increase, respectively).

To justify the better biological properties of compound **94** over derivatives **93**, **95**, and **96**, molecular dynamics (MD) simulations were performed and the calculations showed that the pyranoid glycone moiety established hydrogen bonding in a similar way to other competitive inhibitors of the glycomimetic family. The amide moiety also participated in the formation of hydrogen bonds with the enzyme, thus influencing the orientation of the aliphatic palmitoyl chain towards the hydrophobic region that gives access to the active site of the enzyme. This shift allowed the gain of additional interactions that reinforced interaction with the GCase enzyme.

2.3. Miscellaneous Examples

The use of activity-based probes (ABPs) can be helpful to understand the stabilization mechanism of GCase through inhibitor binding. Ben Bdira et al. [77] described the use of cyclophellitol-derived ABPs that bind the GCase active site irreversibly to the catalytic nucleophile E340. The amphiphilic ABPs studied in vitro and in situ displayed a fluorescent hydrophobic green or red BODIPY tail (compounds **97** and **98**, Figure 22). These compounds were able to enhance thermal stability of GCase and sensitivity to tryptic digestion. The findings pointed out that ABP labeling stabilizes GCase against proteolytic degradation in the lysosomes and did not avoid *N*-glycan modifications by lysosomal glycosidases. These types of compounds can be valuable marker tools to study the contribution of hydrophobic interactions to GCase stabilization by selective labeling of the active form of GCase in living cells and laboratory animals.

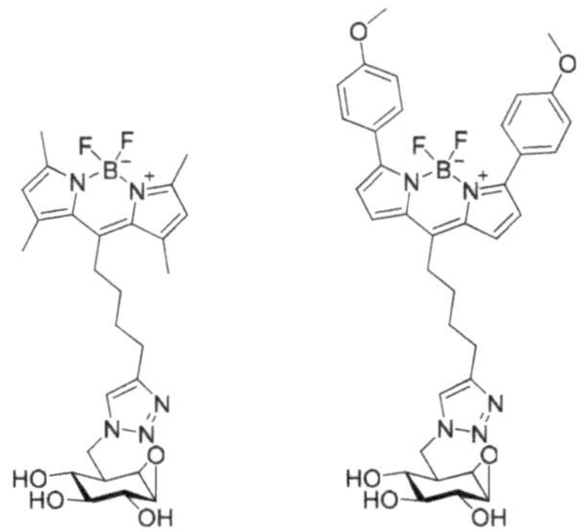

97, IC_{50} = 1.24 nM **98**, IC_{50} = 1.94 nM

Figure 22. Fluorescent cyclophellitol derivatives **97** and **98**.

In 2009, Maegawa et al. screened a library of 1040 FDA-approved drugs through a thermal denaturation assay using wild-type GCase, and identified ambroxol (ABX, **99**, Figure 23), a cough medicine, as a pH-dependent, mixed-type inhibitor of GCase [78]. In particular, ABX (**99**) showed its maximal inhibitory activity at neutral pH (8.1 μM), but turned out to be a negligible inhibitor of GCase at acidic pH (31 μM) that can be found in the lysosomes, a crucial tandem of properties for a good chaperone candidate [79].

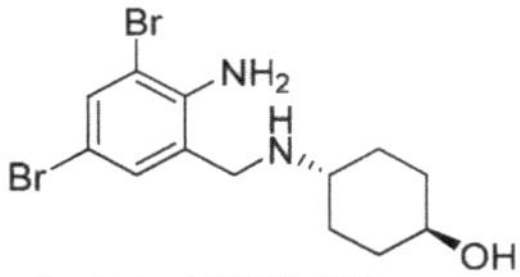

Ambroxol (ABX, **99**)

Figure 23. Ambroxol (ABX, **99**), a mixed-type GCase inhibitor.

In vitro screening of ABX demonstrated low cytotoxicity (60 μM in GD patients' fibroblasts) and good chaperoning activity with F213I/F213I, R120W/L444P, N188S/G193W, N370S/N370S, and F213I/L444P mutant cells, while no effects were observed with the L444P/L444P mutant cells or the control cell line. The chaperoning activity and toxicity of ABX (**99**) were also tested in a wild-type mouse model, showing increased GCase activity in some tissues (namely, cerebellum, heart, and spleen) and demonstrating that ABX (**99**) penetrates the BBB, exerting its activity as PC in the brain [80,81]. However, after a pilot study was carried out in 2013 to assess the tolerability and efficacy of ABX (**99**) as a PC in patients with symptomatic type 1 GD [82], this PC still appears stuck in phase II of the clinical trial [32]. In addition, ABX (**99**) is currently being evaluated in a phase II trial involving 75 patients with moderate PD [83], and preliminary data indicate good brain distribution and safety profile [84].

3. Allosteric Enhancers

The use of inhibitory chaperones can be quite challenging, considering that GCase needs to overcome the inhibitory effect in the lysosomes once it is properly folded in order to carry out its biological action. As such, the search for new PCs that do not bind the active site of the enzyme is an emerging alternative that has gained interest in the last few years. In particular, since ambroxol was first identified as a chaperone for GCase through high-throughput screening (HTS), this technique has turned out to be useful for the search of other non-carbohydrate-derived molecules with potential as allosteric binders of GCase [85].

Compounds NCGC607 (**100**) and NCGC758 (**101**) were discovered in a second HTS of a library of small molecules composed of 250,000 compounds (Figure 24) [86]. Aflaki et al. [87] studied the treatment of the cells of five patients with GD type 1 or 2 with or without PD with the noninhibitory salicylic acid derivative NCGC607 (**100**). The cells showed restored GCase activity and protein levels, reduced glycolipid storage, and reduced α-synuclein accumulation in patients with PD. These results also suggested that compound NCGC607 (**100**) chaperoned mutations other than N370S, as enhanced activity was observed in patients with type 2 GD with the L444P mutation. Compound NCGC758 (**101**), with a pyrazolopyrimidine moiety in its structure, also demonstrated beneficial properties for the treatment of GD.

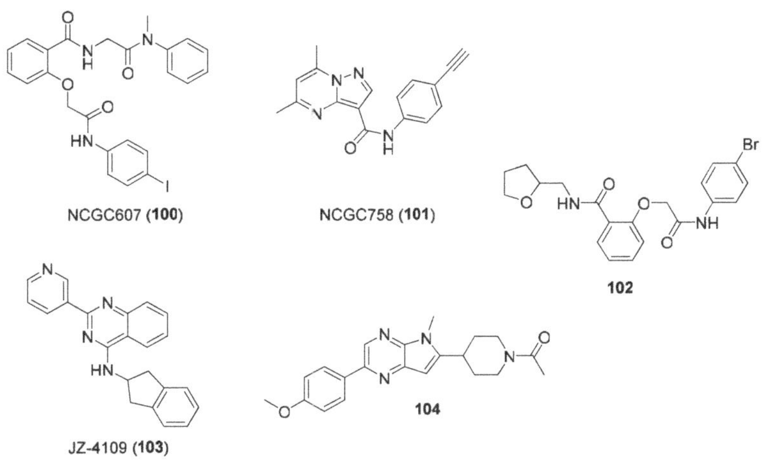

Figure 24. Non-carbohydrate-derived allosteric enhancers **100–104**.

Sun and Wang reported a screening of noninhibitory compounds and found that the racemic mixture **102** (Figure 24) behaved as a good enhancer (ca 8-fold at 10 μM) in mouse fibroblasts bearing the V394L/V394L mutation [88].

In this context, Zheng et al. identified a series of extremely potent non-iminosugar GCase modulators through a systematic structure–activity relationship (SAR) study on quinazoline derivatives. In particular, compound JZ-4109 (**103**, Figure 24) was able to significantly increase GCase protein abundance and GCase activity in homozygous N370S-mutant fibroblast cells from a type 1 GD patient (about 2-fold at 2 μM) [89]. Even more interestingly, the authors characterized (by native mass spectrometry and X-ray crystallography) an allosteric binding site for these modulators by conducting a covalent modification of a lysine residue close to a properly modified ligand. Through transmission electron microscopy (TEM) experiments, they also provided insight into the mechanism of action of **103**, suggesting that GCase stabilization is due to the ability of the ligand to induce GCase dimerization [90].

A further confirmation of the role played by dimerization in GCase activation was recently provided by Rodríguez Sarmiento and coworkers. Indeed, the authors observed through analytical ultracentrifugation (AUC) that GCase dimerization was induced by the pyrrolo[2,3-*b*]pyrazine **104** (Figure 24), identified by testing on a cellular assay 400 diverse heterocyclic compounds, selected according to the literature and filtered by molecular properties. The GCase activator that emerged from this study did not bind to the GCase catalytic site, but to a new pocket at a dimer interface, as supported by X-ray and surface plasmon resonance (SPR) competition experiments. These results gave important mechanistic insight on the field of GCase activation, paving the way to the development of a new generation of GCase activators with CNS drug-like properties, for the treatment of Gaucher and Parkinson's disease [91].

Other examples of allosteric GCase binders were provided by García Collazo et al. The authors reported an extensive biological screening of non-carbohydrate-derived compounds that were allosteric binders of mutated GCase. Chaperone assays on mutated homozygous L444P human fibroblasts showed moderate enhancements in the 6–50 μM concentration range. Among all compounds tested, the better enhancements (>2-fold) were observed for the synthetic compounds **105–108** and the commercially available derivative **109** (Figure 25) [92].

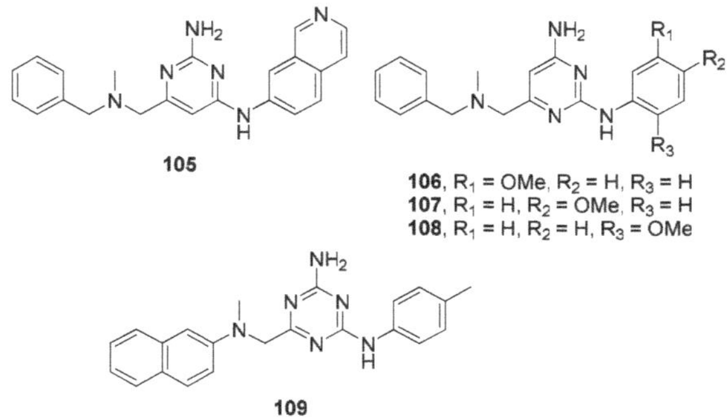

Figure 25. Non-carbohydrate derived GCase modulators **105–109**.

4. Studies in Parkinson's Animal Models

Parkinson disease (PD) is the second most common neurodegenerative disorder affecting approximately 1% of the population over 60 and 5% over 80–85 [93]. The pathological hallmarks of PD are formation of α-synuclein (α-syn) aggregates and marked loss of dopaminergic neurons in the *substantia nigra pars compacta* [94]. GCase mutations were shown to be a major risk factor of developing PD: approximately 7–12% of PD patients carry *GBA* mutations [95], and 25% of GD patients have a first- or second-degree relative with PD [96].

In addition, it emerged that enhancing GCase, also in the absence of GD mutations, may represent a valid therapeutic strategy for sporadic forms of PD [97]. The exact mechanism through which GCase deficiency contributes to PD pathogenesis is still unclear, but it includes the accumulation of α-syn, impaired lysosomal function, and ER-associated stress. Impairment of the autophagy–lysosomal pathway plays a principal role in PD pathogenesis and in particular its effect on abnormal accumulation of α-syn with formation of oligomers and fibrils [98]. The relevant examples of GCase chaperones in PD animal models are summarized in Table 6.

Several drug screenings have identified a number of candidates, including already known drugs (drug-repositioning or drug-repurposing strategy), such as ambroxol (ABX, **99**, Figure 23), a drug used in the treatment of respiratory diseases, and IFG (**1**, Figure 2), under clinical phase trials for the treatment of Gaucher disease. The fruit fly *Drosophila melanogaster* shows that development of PD in carriers of GD mutations results from the presence of mutant *GBA* alleles. Expression of mutated *GBA* in *Drosophila* resulted in dopaminergic neuronal loss, a progressive locomotor defect, and increased levels of ER stress. In a *Drosophila* model, both ABX (**99**) and IFG (**1**) have been shown to effectively reduce ER stress and the locomotor deficit, as reported by Schapira and Horowitz and their coworkers [99,100].

Moreover, ABX (**99**) treatment resulted in increased brain GCase activity in a transgenic mouse model expressing the heterozygous L444P mutation in the murine *GBA* gene, a PD model of healthy carrier of GD (*GBA*-PD model) [101].

Conversely, several studies have been performed on transgenic mice overexpressing human wild-type α-synuclein (Thy1-αSyn), which reproduced key features of Parkinson's disease in the absence of GCase mutations, including nonmotor deficits, pathological, biochemical, and molecular alterations, and at older ages, alterations in nigrostriatal dopaminergic neurons, including loss of striatal dopamine (PD models) [101,102]. Oral treatment with ABX (**99**) and IFG (**1**) increased wild-type GCase activity in the brain of Thy1-αSyn mice, inducing a decrease of α-syn accumulation in dopaminergic neurons of the substantia nigra and improving motor and nonmotor functions. The pharmacodynamic properties of these compounds in the brain of mouse models demonstrated their ability to cross the blood–brain barrier (BBB) [101,102]. Moreover, Migdalska-Richards et al. also reported the ability of ABX (**99**) to increase brain GCase activity of 20% in the midbrain of wild-type nonhuman primates after oral chronic treatment, demonstrating, once again, that ABX (**99**) crosses the blood–brain barrier (BBB) [103].

Table 6. Relevant examples of GCase chaperones in animal models bearing heterozygous *GBA* mutations, developing PD (*GBA*-PD) and in PD animal models.

Compound	Disease Models	Animal Model	Effect of Chaperone	Refs.
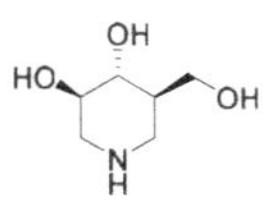 IFG (**1**)	*GBA*-PD	*Drosophila* with mutant *GBA* (N370S and L444P)	- reduce ER stress - reverse locomotor deficits	[99]
	PD	Transgenic mice overexpressing human WT α-syn	- improved motor function - reduced α-syn immunoreactivity - reduced α-syn aggregates	[102]
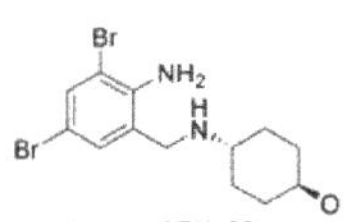 Ambroxol (ABX, **99**)	*GBA*-PD	*Drosophila* with mutant *GBA* (N370S and L444P)	- reduce ER stress - reverse locomotor deficits	[99,100]
	GBA-PD	Gba1$^{L444P/+}$ mice	- increased GCase activity in brains	[101]
	PD	Transgenic mice overexpressing human WT α-syn	- reduced α-syn levels	[101]
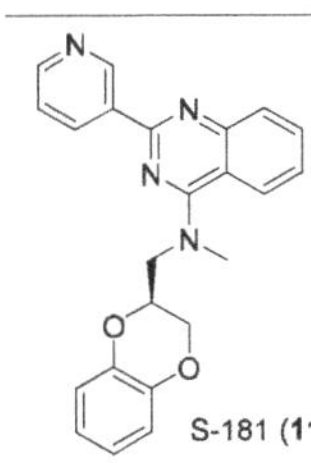 S-181 (**110**)	*GBA*-PD	Gba1$^{D409V/+}$ mice	- reduced GlcCer levels in brain - reduced α-syn levels in brain	[104]

Finally, Burbulla et al. [104] developed compound S-181 (**110**, Table 6), a quinazoline-derived modulator of GCase with good pharmacokinetic properties. Intraperitoneal single administration of S-181 into wild-type mice showed a maximal concentration of 2.9 μM in brain and a half-life of 20.7 h, with higher concentrations in brain than in plasma after 2 h. Compound S-181 (**110**) enhanced wild-type GCase activity in heterozygous D409V mice, a *GBA*-PD model, resulting in reduction of the lipid substrates and α-syn in brain.

In consideration of the preliminary results concerning the ability to cross the BBB, the compounds described in Table 6 represent promising therapeutics for future development.

5. Conclusions

The human lysosomal enzyme acid-β-glucosidase (also known as glucocerebrosidase or GCase) has recently become an increasingly interesting biological target to address not only the most common form of lysosomal storage disorders (LSDs), Gaucher disease (GD), but also much more common neurological pathologies, such as Parkinson's disease (PD). In this regard, pharmacological chaperone therapy (PCT), which employs small compounds able to bind and stabilize the misfolded lysosomal enzymes, facilitating their translocations to the lysosomes, and resulting in a net recovery of the compromised catalytic activity, has recently emerged and been placed on the market of the first PC for an LSD (e.g., Galafold, for the treatment of Fabry disease). However, there is no PC-based drug for the treatment of GD or PD, yet. Iminosugar-based glycomimetics are among the most investigated class of PC for GD. They are lysosomal GCase inhibitors that act as a template, stabilizing the native folding state of the enzyme in the endoplasmic reticulum (ER), with a consequent final counterintuitive effect of enhancing the lysosomal enzyme catalytic activity in the lysosomes.

Alkylated iminosugars have great potential as PCs for the GCase enzyme. It emerged that the position and length of the alkyl chain is crucial to impart the desired activity, which is not possible to be determined a priori with docking studies, since during in vitro tests with mutated cell lines, other important factors take place, such as membrane permeability and cell viability. Overall, an octyl alkyl chain seems to represent the optimal compromise between good affinity with the enzyme active site and good cell permeability, with negligible toxicity. Iminosugars of different types were identified recently as good PCs, also towards mutated cell lines (e.g., L444P, present in neurological forms of the disease), which were quite resistant to the majority of the PCs found in the past. The in vitro enzymatic enhancement observed with iminosugar-based PCs is usually around 2-fold, which is reached with different concentration of compounds ranging from micromolar to nanomolar. However, this aspect strongly depends on compound toxicity and cell permeability, so it is difficult to predict a general trend. Nevertheless, chaperoning activity observed at nanomolar concentration is in principle better.

In most of the reports, the selectivity towards GCase over other lysosomal glycosidases was also verified, which is important to avoid undesired side effects.

The main limitation of the competitive inhibitor-based PCs is that for their application in vivo, careful dosage tuning is needed to overcome the inhibitory effect. For this reason, more recently, allosteric enhancers have been identified through HTS screening or drug repurposing. This review aimed to describe the efforts made by the scientific community in the last 7 years (since 2015) in order to identify new PCs for the GCase enzyme. We deem that the identification of an effective PC for GCase is of fundamental importance not only for Gaucher disease but also for other diseases related to the misfolding of this enzyme, such as Parkinson's disease, and we are confident that the scientific community will find solutions based on this therapy to be placed on the market.

Author Contributions: Conceptualization, writing, and editing, M.M.-B., F.C. (Francesca Clemente), C.M. and F.C. (Francesca Cardona). All authors have read and agreed to the published version of the manuscript.

Funding: This research was funded by Università di Firenze and Fondazione CR Firenze ("Bando congiunto per il finanziamento di progetti competitivi sulle malattie neurodegenerative 2018"—project: A multidisciplinary approach to target Parkinson's disease in Gaucher related population, acronym MuTaParGa) and Regione Toscana (Bando Salute 2018—project: Late onset lysosomal storage disorders (LSDs) in the differential diagnosis of neurodegenerative diseases: development of new diagnostic procedures and focus on potential pharmacological chaperones (PCs), acronym LysoLate) (Fellowships for Macarena Martínez-Bailén and Francesca Clemente).

Institutional Review Board Statement: Not applicable.

Informed Consent Statement: Not applicable.

Data Availability Statement: Data sharing not applicable.

Conflicts of Interest: The authors declare no conflict of interest.

References

1. Platt, F.M.; d'Azzo, A.; Davidson, B.L.; Neufeld, E.F.; Tifft, C.J. Lysosomal storage diseases. *Nat. Rev. Dis. Primers* **2018**, *4*, 27. [CrossRef] [PubMed]
2. Marques, A.R.A.; Saftig, P. Lysosomal storage disorders—Challenges, concepts and avenues for therapy: Beyond rare diseases. *J. Cell Sci.* **2019**, *132*, jcs221739. [CrossRef] [PubMed]
3. Boyd, R.E.; Lee, G.; Rybczynski, P.; Benjamin, E.R.; Khanna, R.; Wustman, B.A.; Valenzano, K.J. Pharmacological Chaperones as Therapeutics for Lysosomal Storage Diseases. *J. Med. Chem.* **2013**, *56*, 2705–2725. [CrossRef] [PubMed]
4. Pereira, D.M.; Valentão, P.; Andrade, P.B. Tuning protein folding in lysosomal storage diseases: The chemistry behind pharmacological chaperones. *Chem. Sci.* **2018**, *9*, 1740–1752. [CrossRef]
5. Grabowski, G.A. *Advances in Gaucher Disease: Basic and Clinical Perspectives*; Future Medicine Ltd.: London, UK, 2013.
6. Sidransky, E.; Nalls, M.A.; Aasly, J.O.; Aharon-Peretz, J.; Annesi, G.; Barbosa, E.R.; Bar-Shira, A.; Berg, D.; Bras, J.; Brice, A.; et al. Multicenter Analysis of Glucocerebrosidase Mutations in Parkinson's Disease. *N. Engl. J. Med.* **2009**, *361*, 1651–1661. [CrossRef] [PubMed]
7. Alcalay, R.N.; Levy, O.A.; Waters, C.C.; Fahn, S.; Ford, B.; Kuo, S.-H.; Mazzoni, P.; Pauciulo, M.W.; Nichols, W.C.; Gan-Or, Z.; et al. Glucocerebrosidase activity in Parkinson's disease with and without *GBA* mutations. *Brain* **2015**, *138*, 2648–2658. [CrossRef] [PubMed]
8. Blanz, J.; Saftig, P. Parkinson's disease: Acid-glucocerebrosidase activity and alpha-synuclein clearance. *J. Neurochem.* **2016**, *139* (Suppl. 1), 198–215. [CrossRef] [PubMed]
9. Li, H.; Ham, A.; Ma, T.C.; Kuo, S.-H.; Kanter, E.; Kim, D.; Ko, H.S.; Quan, Y.; Sardi, S.P.; Li, A.; et al. Mitochondrial dysfunction and mitophagy defect triggered by heterozygous *GBA* mutations. *Autophagy* **2019**, *15*, 113–130. [CrossRef] [PubMed]
10. Yu, Z.; Sawkar, A.R.; Kelly, J.W. Pharmacologic chaperoning as a strategy to treat Gaucher disease. *FEBS J.* **2007**, *274*, 4944–4950. [CrossRef] [PubMed]
11. Parenti, G. Treating lysosomal storage diseases with pharmacological chaperones: From concept to clinics. *EMBO Mol. Med.* **2009**, *1*, 268–279. [CrossRef] [PubMed]
12. Benito, J.M.; García Fernández, J.M.; Ortiz Mellet, C. Pharmacological chaperone therapy for Gaucher disease: A patent review. *Expert Opin. Ther. Pat.* **2011**, *21*, 885–903. [CrossRef]
13. Trapero, A.; Llebaria, A. Glucocerebrosidase inhibitors for the treatment of Gaucher disease. *Future Med. Chem.* **2013**, *5*, 573–590. [CrossRef] [PubMed]
14. Parenti, G.; Pignata, C.; Vajro, P.; Salerno, M. New strategies for the treatment of lysosomal storage diseases (Review). *Int. J. Mol. Med.* **2013**, *31*, 11–20. [CrossRef] [PubMed]
15. Aymami, J.; Barril, X.; Rodríguez-Pascau, L.; Martinell, M. Pharmacological chaperones for enzyme enhancement therapy in genetic diseases. *Pharm. Pat. Anal.* **2013**, *2*, 109–124. [CrossRef] [PubMed]
16. Parenti, G.; Moracci, M.; Fecarotta, S.; Andria, G. Pharmacological chaperone therapy for lysosomal storage diseases. *Future Med. Chem.* **2014**, *6*, 1031–1045. [CrossRef]
17. Sánchez-Fernández, E.M.; García Fernández, J.M.; Ortiz Mellet, C. Glycomimetic-based pharmacological chaperones for lysosomal storage disorders: Lessons from Gaucher, G_{M1}-gangliosidosis and Fabry diseases. *Chem. Commun.* **2016**, *52*, 5497–5515. [CrossRef]
18. Tran, M.L.; Génisson, Y.; Ballereau, S.; Dehoux, C. Second-Generation Pharmacological Chaperones: Beyond Inhibitors. *Molecules* **2020**, *25*, 3145. [CrossRef] [PubMed]
19. Liguori, L.; Monticelli, M.; Allocca, M.; Hay Mele, B.; Lukas, J.; Cubellis, M.V.; Andreotti, G. Pharmacological Chaperones: A Therapeutic Approach for Diseases Caused by Destabilizing Missense Mutations. *Int. J. Mol. Sci.* **2020**, *21*, 489. [CrossRef] [PubMed]
20. Chiti, F.; Kelly, J.W. Small molecule protein binding to correct cellular folding or stabilize the native state against misfolding and aggregation. *Curr. Opin. Struct. Biol.* **2022**, *72*, 267–278. [CrossRef]
21. Compain, P.; Martin, O.R. *Iminosugars: From Synthesis to Therapeutic Applications*; Wiley VCH: New York, NY, USA, 2007.
22. Kuriyama, C.; Kamiyama, O.; Ikeda, K.; Sanae, F.; Kato, A.; Adachi, I.; Imahori, T.; Takahata, H.; Okamoto, T.; Asano, N. In vitro inhibition of glycogen-degrading enzymes and glycosidases by six-membered sugar mimics and their evaluation in cell cultures. *Bioorg. Med. Chem.* **2008**, *16*, 7330–7336. [CrossRef]
23. Chang, H.-H.; Asano, N.; Ishii, S.; Ichikawa, Y.; Fan, J.-Q. Hydrophilic iminosugar active-site-specific chaperones increase residual glucocerebrosidase activity in fibroblasts from Gaucher patients *FEBS J.* **2006**, *273*, 4082–4092. [CrossRef] [PubMed]
24. Zhu, X.; Sheth, K.A.; Li, S.; Chang, H.-H.; Fan, J.-Q. Rational Design and Synthesis of Highly Potent β-Glucocerebrosidase Inhibitors. *Angew. Chem. Int. Ed.* **2005**, *44*, 7450–7453. [CrossRef] [PubMed]
25. Fan, J.-Q.; Zhu, X.; Sheth, K. Hydroxy Piperidine Derivatives to Treat Gaucher Disease. U.S. Patent 7,741,340, 22 June 2010.
26. Compain, P.; Martin, O.R.; Boucheron, C.; Godin, G.; Yu, L.; Ikeda, K.; Asano, N. Design and Synthesis of Highly Potent and Selective Pharmacological Chaperones for the Treatment of Gaucher's disease. *ChemBiochem* **2006**, *7*, 1356–1359. [CrossRef]

27. Sawkar, A.R.; Cheng, W.-C.; Beutler, E.; Wong, C.-H.; Balch, W.E.; Kelly, J.W. Chemical chaperones increase the cellular activity of N370S β-glucosidase: A therapeutic strategy for Gaucher disease. *Proc. Natl. Acad. Sci. USA* **2002**, *99*, 15428–15433. [CrossRef]
28. Brumshtein, B.; Greenblatt, H.M.; Butters, T.D.; Shaaltiel, Y.; Aviezer, D.; Silman, I.; Futerman, A.H.; Sussman, J.L. Crystal Structures of Complexes of *N*-Butyl- and *N*-Nonyl-Deoxynojirimycin Bound to Acid β-Glucosidase: Insights into the mechanism of chemical chaperone action in Gaucher disease. *J. Biol. Chem.* **2007**, *282*, 29052–29058. [CrossRef] [PubMed]
29. Parmeggiani, C.; Catarzi, S.; Matassini, C.; D'Adamio, G.; Morrone, A.; Goti, A.; Paoli, P.; Cardona, F. Human Acid β-Glucosidase Inhibition by Carbohydrate Derived Iminosugars: Towards New Pharmacological Chaperones for Gaucher Disease. *ChemBiochem* **2015**, *16*, 2054–2064. [CrossRef] [PubMed]
30. Luan, Z.; Higaki, K.; Aguilar-Moncayo, M.; Ninomiya, H.; Ohno, K.; García-Moreno, M.I.; Ortiz Mellet, C.; García Fernández, J.M.; Suzuki, Y. Chaperone Activity of Bicyclic Nojirimycin Analogues for Gaucher Mutations in Comparison with *N*-(*n*-nonyl)-Deoxynojirimycin. *ChemBiochem* **2009**, *10*, 2780–2792. [CrossRef] [PubMed]
31. Mena-Barragán, T.; Narita, A.; Matias, D.; Tiscornia, G.; Nanba, E.; Ohno, K.; Suzuki, Y.; Higaki, K.; Garcia Fernández, J.M.; Ortiz Mellet, C. pH-Responsive Pharmacological Chaperones for Rescuing Mutant Glycosidases. *Angew. Chem. Int. Ed.* **2015**, *54*, 11696–11700. [CrossRef]
32. ClinicalTrials.gov. Available online: https://clinicaltrials.gov/ct2/show/NCT03950050 (accessed on 13 May 2022).
33. Sevšek, A.; Šrot, L.; Rihter, J.; Čelan, M.; van Ufford, L.Q.; Moret, E.E.; Martin, N.I.; Pieters, R.J. *N*-Guanidino Derivatives of 1,5-Dideoxy-1,5-imino-D-xylitol are Potent, Selective, and Stable Inhibitors of β-Glucocerebrosidase. *ChemMedChem* **2017**, *12*, 483–486. [CrossRef]
34. Baudoin-Dehoux, C.; Castellan, T.; Rodriguez, F.; Rives, A.; Stauffert, F.; Garcia, V.; Levade, T.; Compain, P.; Génisson, Y. Selective Targeting of the Interconversion between Glucosylceramide and Ceramide by Scaffold Tailoring of Iminosugar Inhibitors. *Molecules* **2019**, *24*, 354. [CrossRef] [PubMed]
35. Clemente, F.; Matassini, C.; Goti, A.; Morrone, A.; Paoli, P.; Cardona, F. Stereoselective Synthesis of C-2 Alkylated Trihydroxypiperidines: Novel Pharmacological Chaperones for Gaucher Disease. *ACS Med. Chem. Lett.* **2019**, *10*, 621–626. [CrossRef] [PubMed]
36. Clemente, F.; Matassini, C.; Faggi, C.; Giachetti, S.; Cresti, C.; Morrone, A.; Paoli, P.; Goti, A.; Martínez-Bailén, M.; Cardona, F. Glucocerebrosidase (GCase) activity modulation by 2-alkyl trihydroxypiperidines: Inhibition and pharmacological chaperoning. *Bioorg. Chem.* **2020**, *98*, 103740. [CrossRef]
37. Suzuki, K.; Nakahara, T.; Kanie, O. 3,4-Dihydroxypyrrolidine as Glycosidase Inhibitor. *Curr. Top. Med. Chem.* **2009**, *9* 34–57. [CrossRef]
38. Kato, A.; Nakagome, I.; Sato, K.; Yamamoto, A.; Adachi, I.; Nash, R.J.; Fleet, G.W.; Natori, Y.; Watanabe, Y.; Imahori, T.; et al. Docking study and biological evaluation of pyrrolidine-based iminosugars as pharmacological chaperones for Gaucher disease. *Org. Biomol. Chem.* **2016**, *14*, 1039–1048. [CrossRef]
39. Stocker, B.L.; Dangerfield, E.M.; Win-Mason, A.L.; Haslett, G.W.; Timmer, M.S.M. Recent Developments in the Synthesis of Pyrrolidine-Containing Iminosugars. *Eur. J. Org. Chem.* **2010**, *2010*, 1615–1637. [CrossRef]
40. Castellan, T.; Garcia, V.; Rodriguez, F.; Fabing, I.; Shchukin, Y.; Tran, M.L.; Ballereau, S.; Levade, T.; Génisson, Y.; Dehoux, C. Concise asymmetric synthesis of new enantiomeric *C*-alkyl pyrrolidines acting as pharmacological chaperones against Gaucher disease. *Org. Biomol. Chem.* **2020**, *18*, 7852–7861. [CrossRef]
41. Theorell, H.; Yonetani, T. Studies on Liver Alcohol Dehydrogenase Complexes. IV. Spectrophotometric Observations on the Enzyme Complexes. *Arch. Biochem. Biophys.* **1964**, *106*, 252–258. [CrossRef]
42. Diot, J.D.; Garcia Moreno, I.; Twigg, G.; Ortiz Mellet, C.; Haupt, K.; Butters, T.D.; Kovensky, J.; Gouin, S.G. Amphiphilic 1-Deoxynojirimycin Derivatives through Click Strategies for Chemical Chaperoning in N370S Gaucher Cells. *J. Org. Chem.* **2011**, *76*, 7757–7768. [CrossRef]
43. Martínez-Bailén, M.; Carmona, A.T.; Patterson-Orazem, A.C.; Lieberman, R.L.; Ide, D.; Kubo, M.; Kato, A.; Robina, I.; Moreno-Vargas, A.J. Exploring substituent diversity on pyrrolidine-aryltriazole iminosugars: Structural basis of β-glucocerebrosidase inhibition. *Bioorg. Chem.* **2019**, *86*, 652–664. [CrossRef]
44. Kato, A.; Nakagome, I.; Nakagawa, S.; Koike, Y.; Nash, R.J.; Adachi, I.; Hirono, S. Docking and SAR studies of calystegines: Binding orientation and influence on pharmacological chaperone effects for Gaucher's disease. *Bioorg. Med. Chem.* **2014**, *22*, 2435–2441. [CrossRef]
45. García-Moreno, M.I.; de la Mata, M.; Sánchez-Fernández, E.M.; Benito, J.M.; Díaz-Quintana, A.; Fustero, S.; Nanba, E.; Higaki, K.; Sánchez-Alcázar, J.A.; García Fernández, J.M.; et al. Fluorinated Chaperone-β-Cyclodextrin Formulations for β-Glucocerebrosidase Activity Enhancement in Neuronopathic Gaucher Disease. *J. Med. Chem.* **2017**, *60*, 1829–1842. [CrossRef] [PubMed]
46. Gillis, E.P.; Eastman, K.J.; Hill, M.D.; Donnelly, D.J.; Meanwell, N.A. Applications of Fluorine in Medicinal Chemistry. *J. Med. Chem.* **2015**, *58*, 8315–8359. [CrossRef]
47. Bassetto, M.; Ferla, S.; Pertusati, F. Polyfluorinated groups in medicinal chemistry. *Future Med. Chem.* **2015**, *7*, 527–546. [CrossRef] [PubMed]
48. Palchevskyy, S.S.; Posokhov, Y.O.; Olivier, B.; Popot, J.-L.; Pucci, B.; Ladokhin, A.S. Chaperoning of Insertion of Membrane Proteins into Lipid Bilayers by Hemifluorinated Surfactants: Application to Diphtheria Toxin. *Biochemistry* **2006**, *45*, 2629–2635. [CrossRef]

49. Robertson, J.; Stevens, K. Pyrrolizidine alkaloids: Occurrence, biology, and chemical synthesis. *Nat. Prod. Rep.* **2017**, *34*, 62–89. [CrossRef]
50. Mena-Barragán, T.; García-Moreno, M.I.; Nanba, E.; Higaki, K.; Concia, A.L.; Clapés, P.; García Fernández, J.M.; Ortiz Mellet, C. Inhibitor versus chaperone behaviour of D-fagomine, DAB and LAB sp^2-iminosugar conjugates against glycosidases: A structure-activity relationship study in Gaucher fibroblasts. *Eur. J. Med. Chem.* **2016**, *121*, 880–891. [CrossRef]
51. Mena-Barragán, T.; García-Moreno, M.I.; Sevšek, A.; Okazaki, T.; Nanba, E.; Higaki, K.; Martin, N.I.; Pieters, R.J.; García Fernández, J.M.; Ortiz Mellet, C. Probing the Inhibitor versus Chaperone Properties of sp^2-Iminosugars towards Human β-Glucocerebrosidase: A Picomolar Chaperone for Gaucher Disease. *Molecules* **2018**, *23*, 927. [CrossRef]
52. Sevšek, A.; Čelan, M.; Erjavec, B.; Quarles van Ufford, L.; Sastre Toraño, J.; Moret, E.E.; Pieters, R.J.; Martin, N.I. Bicyclic isoureas derived from 1-deoxynojirimycin are potent inhibitors of β-glucocerebrosidase. *Org. Biomol. Chem.* **2016**, *14*, 8670–8673. [CrossRef]
53. Rodríguez-Lavado, J.; de la Mata, M.; Jiménez-Blanco, J.L.; García-Moreno, M.I.; Benito, J.M.; Díaz-Quintana, A.; Sánchez-Alcázar, J.A.; Higaki, K.; Nanba, E.; Ohno, K.; et al. Targeted delivery of pharmacological chaperones for Gaucher disease to macrophages by a mannosylated cyclodextrin carrier. *Org. Biomol. Chem.* **2014**, *12*, 2289–2301. [CrossRef]
54. Compain, P.; Bodlenner, A. The Multivalent Effect in Glycosidase Inhibition: A New, Rapidly Emerging Topic in Glycoscience. *ChemBiochem* **2014**, *15*, 1239–1251. [CrossRef]
55. Gouin, S.G. Multivalent Inhibitors for Carbohydrate-Processing Enzymes: Beyond the "Lock-and-Key" Concept. *Chem. Eur. J.* **2014**, *20*, 11616–11628. [CrossRef]
56. Zelli, R.; Longevial, J.-F.; Dumy, P.; Marra, A. Synthesis and biological properties of multivalent iminosugars. *New J. Chem.* **2015**, *39*, 5050–5074. [CrossRef]
57. Matassini, C.; Parmeggiani, C.; Cardona, F.; Goti, A. Are enzymes sensitive to the multivalent effect? Emerging evidence with glycosidases. *Tetrahedron Lett.* **2016**, *57*, 5407–5415. [CrossRef]
58. Compain, P. Multivalent Effect in Glycosidase Inhibition: The End of the Beginning. *Chem. Rec.* **2020**, *20*, 10–22. [CrossRef]
59. González-Cuesta, M.; Ortiz Mellet, C.; García Fernández, J.M. Carbohydrate supramolecular chemistry: Beyond the multivalent effect. *Chem. Commun.* **2020**, *56*, 5207–5222. [CrossRef]
60. Della Sala, P.; Vanni, C.; Talotta, C.; Di Marino, L.; Matassini, C.; Goti, A.; Neri, P.; Šesták, S.; Cardona, F.; Gaeta, C. Multivalent resorcinarene clusters decorated with DAB-1 inhitopes: Targeting Golgi α-mannosidase from *Drosoph. Melanogaster Org. Chem. Front.* **2021**, *8*, 6648–6656. [CrossRef]
61. Vanni, C.; Bodlenner, A.; Marradi, M.; Schneider, J.P.; Ramirez, M.d.l.A.; Moya, S.; Goti, A.; Cardona, F.; Compain, P.; Matassini, C. Hybrid Multivalent Jack Bean α-Mannosidase Inhibitors: The First Example of Gold Nanoparticles Decorated with Deoxynojirimycin Inhitopes. *Molecules* **2021**, *26*, 5864. [CrossRef]
62. Alvarez-Dorta, D.; King, D.T.; Legigan, T.; Ide, D.; Adachi, I.; Deniaud, D.; Désiré, J.; Kato, A.; Vocadlo, D.; Gouin, S.G.; et al. Multivalency To Inhibit and Discriminate Hexosaminidases. *Chem. Eur. J.* **2017**, *23*, 9022–9025. [CrossRef]
63. Matassini, C.; Vanni, C.; Goti, A.; Morrone, A.; Marradi, M.; Cardona, F. Multimerization of DAB-1 onto Au GNPs affords new potent and selective *N*-acetylgalactosamine-6-sulfatase (GALNS) inhibitors. *Org. Biomol. Chem.* **2018**, *16*, 8604–8612. [CrossRef]
64. D'Adamio, G.; Matassini, C.; Parmeggiani, C.; Catarzi, S.; Morrone, A.; Goti, A.; Paoli, P.; Cardona, F. Evidence for a multivalent effect in inhibition of sulfatases involved in lysosomal storage disorders (LSDs). *RSC Adv.* **2016**, *6*, 64847–64851. [CrossRef]
65. Martínez-Bailén, M.; Carmona, A.T.; Cardona, F.; Matassini, C.; Goti, A.; Kubo, M.; Kato, A.; Robina, I.; Moreno-Vargas, A.J. Synthesis of multimeric pyrrolidine iminosugar inhibitors of human β-glucocerebrosidase and α-galactosidase A: First example of a multivalent enzyme activity enhancer for Fabry disease. *Eur. J. Med. Chem.* **2020**, *192*, 112173. [CrossRef]
66. Decroocq, C.; Rodríguez-Lucena, D.; Ikeda, K.; Asano, N.; Compain, P. Cyclodextrin-Based Iminosugar Click Clusters: The First Examples of Multivalent Pharmacological Chaperones for the Treatment of Lysosomal Storage Disorders. *ChemBiochem* **2012**, *13*, 661–664. [CrossRef]
67. Joosten, A.; Decroocq, C.; de Sousa, J.; Schneider, J.P.; Etamé, E.; Bodlenner, A.; Butters, T.D.; Compain, P. A Systematic Investigation of Iminosugar Click Clusters as Pharmacological Chaperones for the Treatment of Gaucher Disease. *ChemBiochem* **2014**, *15*, 309–319. [CrossRef]
68. Laigre, E.; Hazelard, D.; Casas, J.; Serra-Vinardell, J.; Michelakakis, H.; Mavridou, I.; Aerts, J.M.F.G.; Delgado, A.; Compain, P. Investigation of original multivalent iminosugars as pharmacological chaperones for the treatment of Gaucher disease. *Carbohydr. Res.* **2016**, *429*, 98–104. [CrossRef]
69. Jing, X.; Yu, F.; Chen, L. Visualization of nitroxyl (HNO) in vivo via a lysosome-targetable near-infrared fluorescent probe. *Chem. Commun.* **2014**, *50*, 14253–14256. [CrossRef]
70. Stauffert, F.; Serra-Vinardell, J.; Gómez-Grau, M.; Michelakakis, H.; Mavridou, I.; Grinberg, D.; Vilageliu, L.; Casas, J.; Bodlenner, A.; Delgado, A.; et al. Stereodivergent synthesis of right- and left-handed iminoxylitol heterodimers and monomers. Study of their impact on β-glucocerebrosidase activity. *Org. Biomol. Chem.* **2017**, *15*, 3681–3705. [CrossRef]
71. Vanni, C.; Clemente, F.; Paoli, P.; Morrone, A.; Matassini, C.; Goti, A.; Cardona, F. 3,4,5-Trihydroxypiperidine Based Multivalent Glucocerebrosidase (GCase) Enhancers. *ChemBiochem* **2022**, *23*, e202200077. [CrossRef]
72. Castilla, J.; Rísquez, R.; Higaki, K.; Nanba, E.; Ohno, K.; Suzuki, Y.; Díaz, Y.; Ortiz Mellet, C.; García Fernández, J.M.; Castillón, S. Conformationally-locked *N*-glycosides: Exploiting long-range non-glycone interactions in the design of pharmacological chaperones for Gaucher disease. *Eur. J. Med. Chem.* **2015**, *90*, 258–266. [CrossRef]

73. Castilla, J.; Rísquez, R.; Cruz, D.; Higaki, K.; Nanba, E.; Ohno, K.; Suzuki, Y.; Díaz, Y.; Ortiz Mellet, C.; García Fernández, J.M.; et al. Conformationally-Locked *N*-Glycosides with Selective β-Glucosidase Inhibitory Activity: Identification of a New Non-Iminosugar-Type Pharmacological Chaperone for Gaucher Disease. *J. Med. Chem.* **2012**, *55*, 6857–6865. [CrossRef]
74. Whitworth, G.E.; Macauley, M.S.; Stubbs, K.A.; Dennis, R.J.; Taylor, E.J.; Davies, G.J.; Greig, I.R.; Vocadlo, D.J. Analysis of PUGNAc and NAG-thiazoline as Transition State Analogues for Human *O*-GlcNAcase: Mechanistic and Structural Insights into Inhibitor Selectivity and Transition State Poise. *J. Am. Chem. Soc.* **2007**, *129*, 635–644. [CrossRef]
75. Navo, C.D.; Corzana, F.; Sánchez-Fernández, E.M.; Busto, J.H.; Avenoza, A.; Zurbano, M.M.; Nanba, E.; Higaki, K.; Ortiz Mellet, C.; García Fernández, J.M.; et al. Conformationally-locked *C*-glycosides: Tuning aglycone interactions for optimal chaperone behaviour in Gaucher fibroblasts. *Org. Biomol. Chem.* **2016**, *14*, 1473–1484. [CrossRef]
76. Aydillo, C.; Navo, C.D.; Busto, J.H.; Corzana, F.; Zurbano, M.M.; Avenoza, A.; Peregrina, J.M. A Double Diastereoselective Michael-Type Addition as an Entry to Conformationally Restricted Tn Antigen Mimics. *J. Org. Chem.* **2013**, *78*, 10968–10977. [CrossRef]
77. Ben Bdira, F.; Kallemeijn, W.W.; Oussoren, S.V.; Scheij, S.; Bleijlevens, B.; Florea, B.I.; van Roomen, C.P.A.A.; Ottenhoff, R.; van Kooten, M.J.F.M.; Walvoort, M.T.C.; et al. Stabilization of Glucocerebrosidase by Active Site Occupancy. *ACS Chem. Biol.* **2017**, *12*, 1830–1841. [CrossRef]
78. Maegawa, G.H.B.; Tropak, M.B.; Buttner, J.D.; Rigat, B.A.; Fuller, M.; Pandit, D.; Tang, L.; Kornhaber, G.J.; Hamuro, Y.; Clarke, J.T.R.; et al. Identification and Characterization of Ambroxol as an Enzyme Enhancement Agent for Gaucher Disease. *J. Biol. Chem.* **2009**, *284*, 23502–23516. [CrossRef]
79. Chen, Y.; Sam, R.; Sharma, P.; Chen, L.; Do, J.; Sidransky, E. Glucocerebrosidase as a therapeutic target for Parkinson's disease. *Expert Opin. Ther. Targets* **2020**, *24*, 287–294. [CrossRef]
80. Bendikov-Bar, I.; Maor, G.; Filocamo, M.; Horowitz, M. Ambroxol as a pharmacological chaperone for mutant glucocerebrosidase. *Blood Cells Mol. Dis.* **2013**, *50*, 141–145. [CrossRef]
81. Luan, Z.; Li, L.; Higaki, K.; Nanba, E.; Suzuki, Y.; Ohno, K. The chaperone activity and toxicity of ambroxol on Gaucher cells and normal mice. *Brain Dev.* **2013**, *35*, 317–322. [CrossRef]
82. Zimran, A.; Altarescu, G.; Elstein, D. Pilot study using ambroxol as a pharmacological chaperone in type 1 Gaucher disease. *Blood Cells Mol. Dis.* **2013**, *50*, 134–137. [CrossRef]
83. Silveira, C.R.A.; MacKinley, J.; Coleman, K.; Li, Z.; Finger, E.; Bartha, R.; Morrow, S.A.; Wells, J.; Borrie, M.; Tirona, R.G.; et al. Ambroxol as a novel disease-modifying treatment for Parkinson's disease dementia: Protocol for a single-centre, randomized, double-blind, placebo-controlled trial. *BMC Neurol.* **2019**, *19*, 20. [CrossRef]
84. Mullin, S.; Smith, L.; Lee, K.; D'Souza, G.; Woodgate, P.; Elflein, J.; Hällqvist, J.; Toffoli, M.; Streeter, A.; Hosking, J.; et al. Ambroxol for the Treatment of Patients With Parkinson Disease With and Without Glucocerebrosidase Gene Mutations: A Nonrandomized, Noncontrolled Trial. *JAMA Neurol.* **2020**, *77*, 427–434. [CrossRef]
85. Han, T.-U.; Sam, R.; Sidransky, E. Small Molecule Chaperones for the Treatment of Gaucher Disease and *GBA1*-Associated Parkinson Disease. *Front. Cell Dev. Biol.* **2020**, *8*, 271. [CrossRef] [PubMed]
86. Goldin, E.; Zheng, W.; Motabar, O.; Southall, N.; Choi, J.H.; Marugan, J.; Austin, C.P.; Sidransky, E. High Throughput Screening for Small Molecule Therapy for Gaucher Disease Using Patient Tissue as the Source of Mutant Glucocerebrosidase. *PLoS ONE* **2012**, *7*, e29861. [CrossRef] [PubMed]
87. Aflaki, E.; Borger, D.K.; Moaven, N.; Stubblefield, B.K.; Rogers, S.A.; Patnaik, S.; Schoenen, F.J.; Westbroek, W.; Zheng, W.; Sullivan, P.; et al. A New Glucocerebrosidase Chaperone Reduces α-Synuclein and Glycolipid Levels in iPSC-Derived Dopaminergic Neurons from Patients with Gaucher Disease and Parkinsonism. *J. Neurosci.* **2016**, *36*, 7441–7452. [CrossRef] [PubMed]
88. Sun, Y.; Wang, Z. CNS-Accessible Pharmacological Chaperones for Treatment of Acid Beta-Glucosidase-Related Disease States. U.S. Patent 10,588,888, 17 March 2020.
89. Zheng, J.; Chen, L.; Skinner, O.S.; Ysselstein, D.; Remis, J.; Lansbury, P.; Skerlj, R.; Mrosek, M.; Heunisch, U.; Krapp, S.; et al. β-Glucocerebrosidase Modulators Promote Dimerization of β-Glucocerebrosidase and Reveal an Allosteric Binding Site. *J. Am. Chem. Soc.* **2018**, *140*, 5914–5924. [CrossRef]
90. Gruschus, J.M.; Jiang, Z.; Yap, T.L.; Hill, S.A.; Grishaev, A.; Piszczek, G.; Sidransky, E.; Lee, J.C. Dissociation of glucocerebrosidase dimer in solution by its co-factor, saposin C. *Biochem. Biophys. Res. Commun.* **2015**, *457*, 561–566. [CrossRef]
91. Benz, J.; Rufer, A.C.; Huber, S.; Ehler, A.; Hug, M.; Topp, A.; Guba, W.; Hofmann, E.C.; Jagasia, R.; Rodríguez Sarmiento, R.M. Novel β-Glucocerebrosidase Activators That Bind to a New Pocket at a Dimer Interface and Induce Dimerization. *Angew. Chem. Int. Ed.* **2021**, *60*, 5436–5442. [CrossRef]
92. García Collazo, A.M.; Barril Alonso, X.; Cubero Jordà, E.; Revés Vilaplana, M.; Roberts, R.S. Heteroaryl Compounds and Their Use. WO Patent 2018122775, 5 July 2018.
93. de Lau, L.M.L.; Breteler, M.M.B. Epidemiology of Parkinson's disease. *Lancet Neurol.* **2006**, *5*, 525–535. [CrossRef]
94. Kalia, L.V.; Lang, A.E. Parkinson's disease. *Lancet* **2015**, *386*, 896–912. [CrossRef]
95. García-Sanz, P.; Aerts, J.M.F.G.; Moratalla, R. The Role of Cholesterol in α-Synuclein and Lewy Body Pathology in *GBA1* Parkinson's Disease. *Mov. Disord.* **2021**, *36*, 1070–1085. [CrossRef] [PubMed]
96. Aflaki, E.; Westbroek, W.; Sidransky, E. The Complicated Relationship between Gaucher Disease and Parkinsonism: Insights from a Rare Disease. *Neuron* **2017**, *93*, 737–746. [CrossRef] [PubMed]

97. Murphy, K.E.; Gysbers, A.M.; Abbott, S.K.; Tayebi, N.; Kim, W.S.; Sidransky, E.; Cooper, A.; Garner, B.; Halliday, G.M. Reduced glucocerebrosidase is associated with increased α-synuclein in sporadic Parkinson's disease. *Brain* **2014**, *137*, 834–848. [CrossRef] [PubMed]
98. Alvarez-Erviti, L.; Rodriguez-Oroz, M.C.; Cooper, J.M.; Caballero, C.; Ferrer, I.; Obeso, J.A.; Schapira, A.H.V. Chaperone-Mediated Autophagy Markers in Parkinson Disease Brains. *Arch. Neurol.* **2010**, *67*, 1464–1472. [CrossRef]
99. Sanchez-Martinez, A.; Beavan, M.; Gegg, M.E.; Chau, K.-Y.; Whitworth, A.J.; Schapira, A.H.V. Parkinson disease-linked *GBA* mutation effects reversed by molecular chaperones in human cell and fly models. *Sci. Rep.* **2016**, *6*, 31380. [CrossRef]
100. Maor, G.; Cabasso, O.; Krivoruk, O.; Rodriguez, J.; Steller, H.; Segal, D.; Horowitz, M. The contribution of mutant *GBA* to the development of Parkinson disease in *Drosophila*. *Hum. Mol. Genet.* **2016**, *25*, 2712–2727. [CrossRef]
101. Migdalska-Richards, A.; Daly, L.; Bezard, E.; Schapira, A.H.V. Ambroxol Effects in Glucocerebrosidase and α-Synuclein Transgenic Mice. *Ann. Neurol.* **2016**, *80*, 766–775. [CrossRef]
102. Richter, F.; Fleming, S.M.; Watson, M.; Lemesre, V.; Pellegrino, L.; Ranes, B.; Zhu, C.; Mortazavi, F.; Mulligan, C.K.; Sioshansi, P.C.; et al. A GCase Chaperone Improves Motor Function in a Mouse Model of Synucleinopathy. *Neurotherapeutics* **2014**, *11*, 840–856. [CrossRef]
103. Migdalska-Richards, A.; Ko, W.K.D.; Li, Q.; Bezard, E.; Schapira, A.H.V. Oral ambroxol increases brain glucocerebrosidase activity in a nonhuman primate. *Synapse* **2017**, *71*, e21967. [CrossRef]
104. Burbulla, L.F.; Jeon, S.; Zheng, J.; Song, P.; Silverman, R.B.; Krainc, D. A modulator of wild-type glucocerebrosidase improves pathogenic phenotypes in dopaminergic neuronal models of Parkinson's disease. *Sci. Transl. Med.* **2019**, *11*, eaau6870. [CrossRef]

pharmaceuticals

Article

Bacterial Cellulose—Adaptation of a Nature-Identical Material to the Needs of Advanced Chronic Wound Care

Paul Zahel [1,2], Uwe Beekmann [1], Thomas Eberlein [3], Michael Schmitz [4], Oliver Werz [2] and Dana Kralisch [1,5,*]

1 JeNaCell GmbH—An Evonik Company, 07745 Jena, Germany; paul.zahel@evonik.com (P.Z.); uwe.beekmann@evonik.com (U.B.)
2 Department of Pharmaceutical/Medicinal Chemistry, Institute of Pharmacy, Friedrich Schiller University Jena, 07743 Jena, Germany; oliver.werz@uni-jena.de
3 Akademie-ZWM AG, 8424 Embrach, Switzerland; thomaseberlein@hotmail.com
4 MCS Medical Consulting, 56414 Oberahr, Germany; schmitzmichael@mcsoberahr.onmicrosoft.com
5 Evonik Operations GmbH, 45128 Essen, Germany
* Correspondence: dana.kralisch@evonik.com

Abstract: Modern wound treatment calls for hydroactive dressings. Among the variety of materials that have entered the field of wound care in recent years, the carbohydrate polymer bacterial cellulose (BC) represents one of the most promising candidates as the biomaterial features a high moisture-loading and donation capacity, mechanical stability, moldability, and breathability. Although BC has already gained increasing relevance in the treatment of burn wounds, its potential and clinical performance for "chronic wound" indications have not yet been sufficiently investigated. This article focuses on experimental and clinical data regarding the application of BC within the indications of chronic, non-healing wounds, especially venous and diabetic ulcers. A recent clinical observation study in a chronic wound setting clearly demonstrated its wound-cleansing properties and ability to induce healing in stalling wounds. Furthermore, the material parameters of BC dressings obtained through the static cultivation of *Komagataeibacter xylinus* were investigated for the first time in standardized tests and compared to various advanced wound-care products. Surprisingly, a free swell absorptive capacity of a BC dressing variant containing 97% moisture was found, which was higher than that of alginate or even hydrofiber dressings. We hypothesize that the fine-structured, open porous network and the resulting capillary forces are among the main reasons for this unexpected result.

Keywords: bacterial cellulose; carbohydrate polymer; advanced wound care; chronic wounds; clinical data; exudate management; moisture vapor transmission rate

Citation: Zahel, P.; Beekmann, U.; Eberlein, T.; Schmitz, M.; Werz, O.; Kralisch, D. Bacterial Cellulose—Adaptation of a Nature-Identical Material to the Needs of Advanced Chronic Wound Care. *Pharmaceuticals* **2022**, *15*, 683. https://doi.org/10.3390/ph15060683

Academic Editors: Federica Sodano and Maria Grazia Rimoli

Received: 24 April 2022
Accepted: 21 May 2022
Published: 30 May 2022

1. Introduction

Chronic wounds represent an enormous burden for patients and caregivers and have an immense economic impact on society [1,2]. Studies estimate that between 1 and 4% of total healthcare expenditure in developed countries is spent on chronic wound treatment and related interventions [1,3]. One important prerequisite for successful wound management is the selection of a suitable wound dressing [4,5].

Wounds can be described as disruptions in the epithelial integrity of the skin due to physiochemical or thermal damage or medical pathology. Consequently, essential skin barrier functions, such as thermal insulation and protection against pathogens or external mechanical damage, are impaired [6–8]. Although acute wounds usually progress through the four temporally and spatially overlapping stages of hemostasis, inflammation, proliferation, and remodeling, and close within 8–12 weeks, chronic wounds typically remain in the inflammatory stage, exceeding this period and may be associated with heavy exudation, pain, or infections, and can lead to sepsis or amputations [8–11].

Although the causes of chronic wounds are highly diverse, the majority of cases share similar underlying processes and can be classified into the categories of pressure ulcers, vascular ulcers (venous, arterial, or mixed leg ulcers) and diabetic ulcers [11,12]. Pressure ulcers often occur in patients with compromised mobility, in which case the tissue is subjected to unrelieved pressure that exceeds the capillary perfusion, leading to ischemia, hypoxia, and thus tissue necrosis [11–13]. In venous leg ulcers (VLUs), damaged or incompetent vein valves lead to venous stasis, hypertension, and oedema [13]. The increased pressure exceeds capillary perfusion pressure and blood vessel permeability resulting in macromolecules leaking into the perivascular space. Following ischemia, oedema and hypoxic conditions impede healing of minor trauma and accelerate skin breakdown, which is associated with inflammation and bacterial invasion. Arterial ulcers are much rarer than venous ulcers and result from arterial insufficiency due to atherosclerosis or arterial embolism in the extremities. A narrowed arterial vessel lumen and ischemia of the tissue are associated with hypoxia, through which healing of minor trauma is severely affected and tissue breakdown is accelerated. Lastly, diabetic (foot) ulcers differ from other chronic wound types on account of their specific pathophysiology. Diabetic peripheral polyneuropathy manifests in sensory loss, senso-motorical separation, atrophic structural changes, and thus an increased risk of ulceration from repeated mechanical stress. Impaired perfusion and metabolic derangements further disrupt wound healing and contribute to the chronification of diabetic ulcers [11–14].

Although treatment of the underlying diseases and causes is critical to successful wound management, the principles and advantages of a moist healing milieu, which were first studied by Winter in 1962 [15], are meanwhile widely accepted and backed by evidence [16,17]. A moist wound environment promotes autolytic debridement [18], the proliferation of keratinocytes and fibroblasts [19,20], and collagen synthesis [21], which leads to less pain, reduced scar formation, and faster wound closure [16,17,22].

In order to benefit from these findings in clinical praxis, a plethora of different materials for advanced (chronic) wound care have been developed over the past decades. Dependent on the wound's exudate level, one can choose from highly absorbant polyurethane foam dressings, hydrogel-forming carbohydrate polymers such as alginate and carboxymethylcellulose dressings, occlusive film and hydrocolloid dressings, or even moisture-donating products such as hydrogels and bacterial cellulose dressings [23,24]. Furthermore, many promising dressing developments based on techniques such as electrospinning or 3D bioprinting have been reported in recent years. These dressings show broad applicability and feature advantageous properties such as the controlled release of incorporated drugs or antibacterial activity [25,26]. Today, they are primarily used in regenerative medicine and tissue-engineering techniques [27].

Although there are some dressings that should be used primarily for special indications such as burns, the selection of an appropriate dressing is based less on the classification of the wound as chronic or acute and more on its characteristics such as tissue perfusion, exudate level, the occurrence of necrosis and infection, or localization of the wound [4]. Above all, ensuring a moist wound environment, protection against microorganisms, and thermal insulation are of high importance in both wound classifications [28]. Thus, the challenge of finding a material that sufficiently addresses all these properties and at the same time provides a moist wound environment has not yet been adequately overcome. Bacterial cellulose, however, may be a very promising candidate to fulfill these requirements.

The usage of polysaccharide materials has a long history in the field of wound care; from oil-soaked linen strips over traditional cotton gauze to advanced wound-care products [29]. In particular, semi-synthetic cellulose derivatives such as carboxymethylcellulose possess numerous advantageous properties and are nowadays of great value in advanced wound care [30]. However, carbohydrate-biopolymer bacterial cellulose (BC) has recently attracted greater attention as the subject of numerous works. In contrast to plant-derived cellulose, BC is biosynthesized by various bacteria strains from glucose monomers that form a three-dimensional network of interconnected nanostructured cellulose fibers [31,32].

It can be produced on a large scale by controlled fermentation [33]. In this process, planar hydrogel fleeces are formed on the surface between the nutrient medium and the air. Those dimensionally stable BC hydrogel membranes can be applied as wet wound dressings directly after purification and sterilization without further processing [34]. However, since the selected bacteria strain, as well as the biotechnological production process, can influence the key characteristics and material properties, it is important to independently investigate the suitability of each strain for the intended application [35]. For the strain *Komagataeibacter xylinus* DSM 14666 (see Methods section), an extensive basis of experimental data for the mentioned properties is already available from previous works. Dressings produced using this strain show, for example, a well-suited moisture vapor transmission rate (MVTR) of around 3000 $g/m^2/24$ h [36,37] when measured in contact with vapor. Excellent biocompatibility of the unmodified material could be shown in vitro for RAW264.7 (murine macrophage-like cells) and THP-1 (human monocytic/macrophage-like cells) in standard MTT assay [36,37] as well as for HaCaT keratinocytes in luminometric ATP assay [38]. Furthermore, ex-ovo biocompatibility using a shell-less hen's egg test on chick area vasculosa (HET-CAV) was once more confirmed, and in vitro wound healing in a HaCaT scratch assay showed no negative effects on cell monolayer closure [36,37]. All moist BC wound dressings have in common high biocompatibility, water-holding capacity, conformability, vapor permeability, and mechanical stability while providing a physical barrier against bacteria and other pathogens. Compared to many alternative products, BC also stands out for its high water-holding capacity, which helps to create a moist wound environment. This is due to the microscopic structure of the material: the interconnected fibers with a diameter of less than 100 nm result in a remarkable surface area and enclose large amounts of water, which stabilizes the cellulose network via hydrogen bonds [31,39]. With regard to these characteristics, BC dressings have been shown in several clinical observations to effectively relieve pain, absorb and retain exudate, provide an optimal moist wound environment, diminish infection rates, and last but not least, hasten re-epithelization and shorten wound healing. [34,40–44]. An overview of previously published clinical data as well as key findings can be found in the Supplementary Materials, Tables S1 and S2.

Clinical data published so far focused on clinical experiences with BC wound dressings in the treatment of burn wounds and chronic wounds alike without taking into account the high variability of the material, which allows an adaption of material properties on the wound type. Previous work furthermore focused on investigating the underlying mechanisms on a cellular level. Although the properties of BC that promote wound healing cannot yet be fully explained, some specific mechanisms have been observed in the past. After Sanchavanakit et al. first demonstrated that BC dressings support the growth, spreading, and migration of human keratinocytes in vitro [45], other groups observed accelerated angiogenesis, tissue regeneration, and collagen expression in vivo [46,47]. The high water content of the dressings has been shown to induce a local cooling effect in ex vivo burn wounds, reducing intradermal temperature and thermal damage [48]. In recent years, it has been shown that an acidic pH value is favorable for fast and successful wound healing, e.g., chronic wounds show an alkaline pH [49]. BC dressings that had an acidic pH were associated with faster wound healing than neutral or alkaline dressings, which is a promising finding for future clinical applications [50].

However, in addition to the positive effects of the unmodified material, many studies have taken the development one step further and demonstrated the strong potential of the carbohydrate-based material as a drug carrier to treat chronic wounds. To give an example, Hoff et al. showed how the controlled release of α-13′-carboxychromanol, a long-chain vitamin E metabolite, promoted wound healing and closure in a diabetic mouse model [51].

What is still missing is the thorough analysis of the material properties of BC-based wound dressings and the differences as well as similarities with respect to other advanced wound-dressing materials. Consequently, there is a lack of experience as to which parameters must be adjusted to adapt and optimize a BC-based wound dressing to the individual requirements of certain indications or stages of wound healing.

To gain further clinical experience and identify the potential for the optimization of the material with regard to chronic wounds, a post-market clinical follow-up study featuring epicite^{hydro}, a BC wound dressing already widely used in burn treatment, was carried out. The clinical observation was accompanied by comparative analyses of the material properties in standardized tests and adapting bacterial cellulose dressings to the specific requirements identified in the clinical and in vitro tests. The results of the combined studies could provide further important information for the general understanding of the unique aspects of BC in wound healing. Moreover, the study aims to shed light on how this biomaterial can be adapted to the specific requirements of treating different kinds of wounds.

2. Results and Discussion

2.1. Post-Market Clinical Follow-Up Study

The hydroactive BC wound dressing epicite^{hydro} (BC_A) has already been proven to perform very well in the treatment of burn wounds [52–55]. In order to investigate its clinical performance in chronic wound treatment, a PMCF observation study with 44 patients suffering from venous leg ulcers, mixed leg ulcers, and diabetic foot syndrome was carried out.

Selected results with respect to (a) wound area and wound-depth progression, (b) share of non-irritated wound margins, (c) share of fibrous tissue, as well as (d) exudate update and hydration are shown in Figure 1.

Figure 1a shows a significant ($p < 0.006$) reduction in the mean wound size and depth over the study period of 28 days. Both results can be explained, at least in part, by the ability of BC_A to create a moist environment that promotes the healing of even stagnating chronic wounds, proving that this ability of BC_A is not limited to burn wounds. For BC-based wound dressings in general, these results tie in well with previous clinical trials, wherein the suitability for chronic wounds was demonstrated (Supplementary Materials, Table S1). Another promising result is the consistent and significant ($p = 0.001$) reduction in fibrous tissue (c), which can be explained by the high moisture content of the wound dressing. The associated cleansing effect could be attributed to a softening of the plaques and therefore facilitation of autolytic debridement. A similar conclusion was reached by Alvarez et al. in a randomized clinical trial analyzing a wet BC dressing for the treatment of venous leg ulcers [56]. Future works should compare the wound-cleansing efficiency achieved using a dry dressing as a control group.

Since an excess of moisture at the dressing–wound interface can lead to macerations of the periwound skin, the condition of the wound margins needs to be carefully observed [57]. In the PMCF study, it was shown that the share of wound margins free of irritation significantly increased during the observed trial period (Figure 1b). These results suggest that BC_A may even have a desirable soothing effect on irritated periwound skin.

Although the aforementioned results indicate high suitability of BC_A in the treatment of chronic wounds, a survey of participating users revealed that exudate uptake was rated only as "good" to "satisfactory" and wound hydration as "good". Since exudate management plays a crucial role in successful wound treatment [24,57] and is one important factor in dressing selection, an enhancement of the exudate-handling properties of BC_A is desirable to optimize the dressing for chronic wound treatment.

In the following sections, the material properties of BC wound dressings compared to other advanced wound dressings will be discussed to provide recommendations for the adaption of the dressing for further use in exuding wound environments based on experimental evidence. The work focused on exudate-handling properties under the free-swell condition as well as compression, moisture donation capacity, and the permeability measured using the MVTR. To increase the capability of the experimental BC to absorb wound exudate, certain cultivation parameters were varied, and the dressings underwent a partial dehydration process. An overview of all evaluated parameters, tests, and results is shown in Section 2.7.

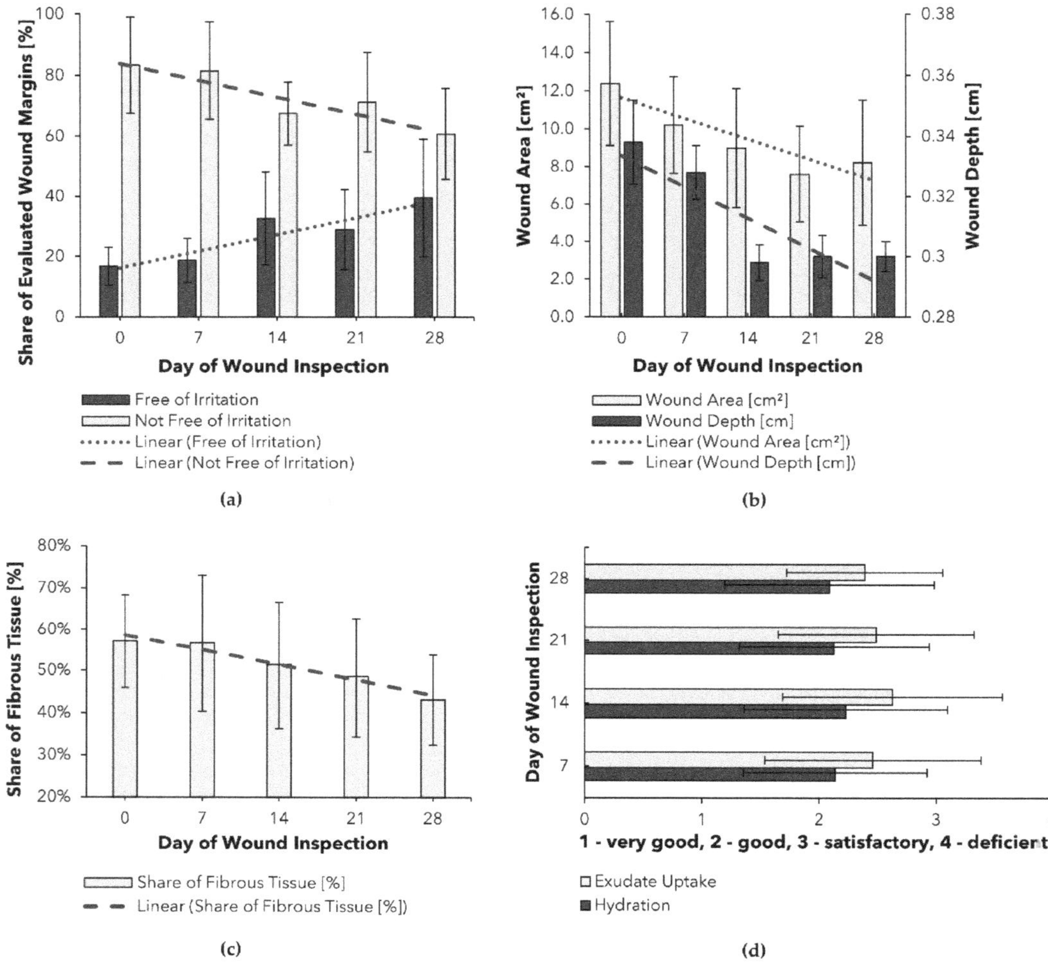

Figure 1. Selected results of a Post-Market Clinical Follow-up (PMCF) study of BC_A: (**a**) Progression of wound area and depth; (**b**) Progression of wound margin irritation; (**c**) Progression of fibrous tissue share; (**d**) Assessment of exudate management; time points of wound inspections ± 3 days.

2.2. Solid Content

To characterize the commercial and experimental BC dressings, first regarding their hydration level, the solid content (SC) of the experimentally produced wet BC wound dressings (BC_C1; BC_C2) and commercial wet BC dressings epicite^{hydro} (BC_A) and Suprasorb X (XBC) was determined (Figure 2). The solid content of a specific BC dressing describes the proportion of BC in the total mass of the sample. The parameter is of interest since moist BC wound dressings can be described as hydrogel bodies consisting of a varying amount of water and an interconnected 3D network of thin cellulose fibers [31,32]. All other commercial wound dressings investigated were excluded from this experiment since they are dry.

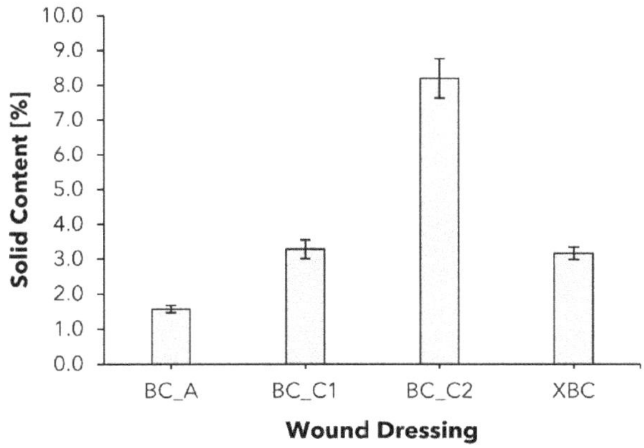

Figure 2. Solid content of wet bacterial cellulose wound dressings (mean ± SD; n = 8).

BC_A represents a moist dressing with an SC of 1.57 ± 0.10%, which is in accordance with former works [36]. The value was increased by more than two, resp. five times to 3.27 ± 0.27% for BC_C1 and 8.20 ± 0.57% for BC_C2. The commercial product XBC is comparable to BC_C1 in terms of the SC. Since the dressings tend to reswell to their original water resp. solid content when in contact with excess liquid such as wound exudate, the increased solid content represents the starting point for an improved exudate absorption.

2.3. *Free Swell Absorptive Capacity*

Although a moist healing environment is widely accepted as beneficial for successful wound healing and represents the standard for modern wound management, an excess of exudate can lead to wound complications and skin deterioration (e.g., irritation, maceration) [57,58]. In contrast to acute wounds, exudate in chronic wounds differs in its composition and has been shown to slow down cell proliferation, interfere with growth factor availability, and lead to successive degradation of the extracellular matrix [24,57]. To maintain an optimal moisture balance, a dressing must be able to absorb excess exudate and hydrate the wound at the same time. In order to compare the BC-based wound dressings, the free swell absorptive capacity according to EN 13726-1:2002 was determined for BC_A, BC_C1, BC_C2, XBC, and five different commercial wound dressings that represent a variety of commonly used options for the treatment of chronic wounds in Germany (Figure 3).

BC_A, typically used in the treatment of burns, showed a relatively low $AC_{0.5}$ of 6.7 ± 0.5 g/100 cm^2, which can be attributed to its high water content (>98%, see Figure 2). The cellulose network is already saturated with water and thus not well-suited for further fluid uptake. This property has proven to be beneficial in the treatment of burn wounds in multiple studies, mainly due to the dressing's ability to provide a moist wound environment [52,53,55]. However, in the treatment of chronic wounds, exudate absorption is of higher importance. Never-dried BC dressings feature the ability to almost completely reswell after the mechanical removal of water. In this work, the combination of increasing the thickness of the initial cellulose layer together with the intensified removal of water aimed to improve its ability to absorb aqueous fluids such as wound exudate. The determined $AC_{0.5}$ of the produced dressings BC_C1 and BC_C2 was found to be more than 3, resp. 4 times higher compared to BC_A and exceeded the $AC_{0.5}$ of XBC by 2, resp. 3 times. This finding suggests that increased solid content leads to the higher absorptive capacity of BC-based wound dressings, which could therefore be more suitable for the therapy of medium to highly exuding wounds.

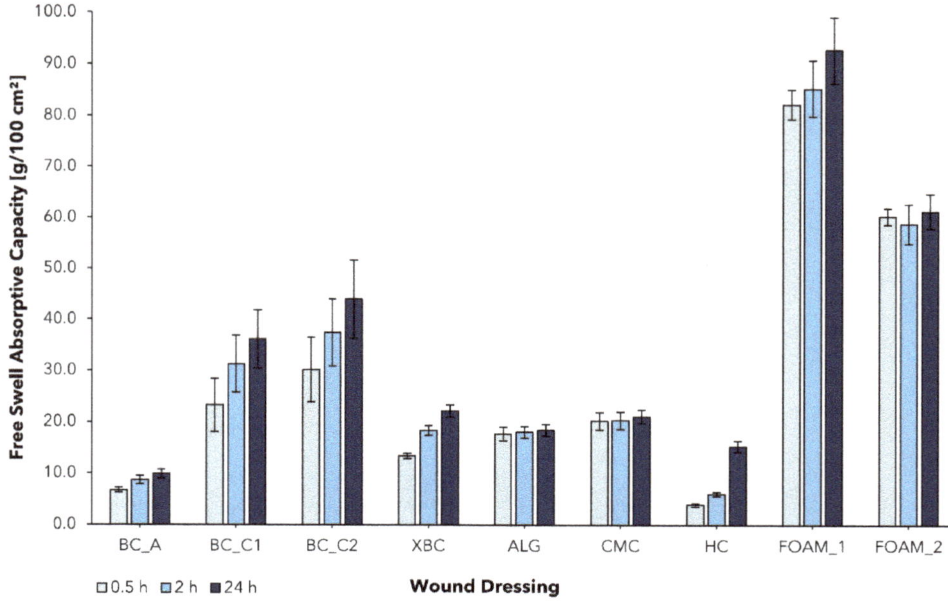

Figure 3. Free swell absorptive capacity (AC) of bacterial cellulose wound dressings and alternative wound care products (mean ± SD; n = 5). The fluid uptake was measured at three successive time points ($AC_{0.5}$; AC_2; AC_{24}) in the same experimental setting for epicite^{hydro} (BC_A); Suprasorb® X (XBC); Suprasorb® A (ALG); Aquacel® Extra™ (CMC); Hydrocoll® (HC); ALLEVYN Gentle (FOAM_1); Mepilex® (FOAM_2).

Hydrofiber and alginate dressings form gels instantly, resp. over a period of several hours when in contact with exudate, and are used for highly exuding, deep wounds [23]. In the performed test, ALG and CMC were within a range of $AC_{0.5}$ = 15–20 g/100 cm^2, which is in line with the literature [59,60]. They exceeded BC_A in terms of their absorbency but fell short of BC_C1 and BC_C2. The tested foam dressings were observed to absorb the most exudate of all dressing classes investigated and feature an $AC_{0.5}$ of ~60–85 g/100 cm^2, which compares well with previously published data on the AC of commercial foam dressings and can be explained by the open structure of the foam network [61].

It was further observed that all BC-based wound dressings were not saturated within the test period of 0.5 h and continued to absorb exudate beyond the test duration of 2 h, whereas the other tested dressings showed little difference between $AC_{0.5}$ and AC_{24}.

2.4. Absorptive Capacity under Pressure

Although the standardized measurement of the free swell absorptive capacity is a good indication for the comparison of different wound dressings, it is only partially representative for specific applications [62]. In particular for patients with chronic wounds due to oedema or venous leg ulcers, compression therapy is indicated and represents a well-established standard [63]. The applied pressure can lead to a reduction in venous pooling volume, and an improvement in muscle pump function and venous blood flow [64]. Effects such as increased ulcer healing rates and reduced healing duration under compressive bandages are supported with evidence [64]. The successful use of a wound dressing to treat chronic wounds from, e.g., venous ulcers, therefore, also depends on its capability to absorb and retain exudate under the influence of mechanical pressure. To analyze the absorption behavior in that specific case, a method based on prEN 13726 was selected with the samples subjected to a pressure of 40 mmHg corresponding to compression class 3 (CCL3, Figure 4) [65].

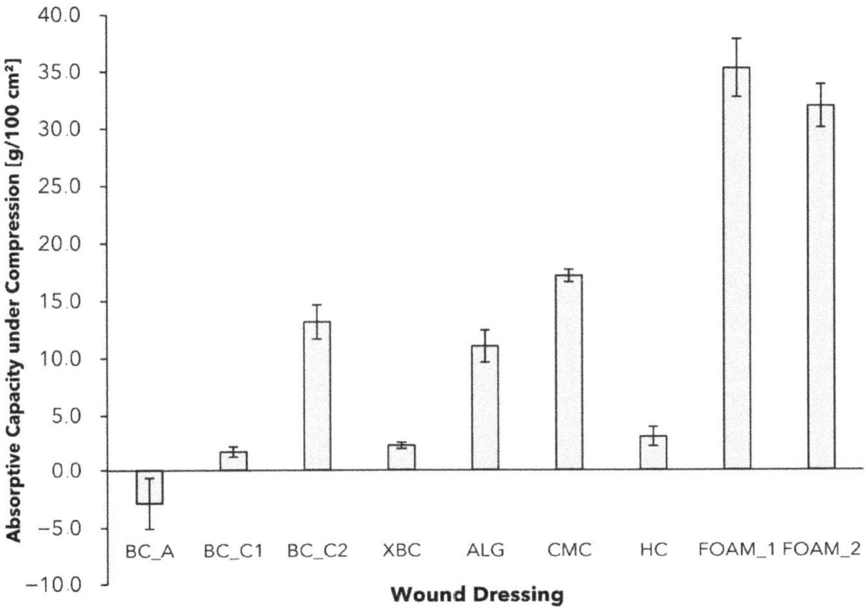

Figure 4. Absorptive capacity of different wound dressings after 0.5 h and under compression of 40 mmHg (mean ± SD; $n = 5$).

In comparison to free swell AC (Figure 3), all wound dressings exhibited reduced $AC_{0.5}$ values under the influence of pressure, with CMC dressings showing the lowest reduction of 16.7%. It could be observed that the materials react differently to the influence of mechanical pressure and vary in their ability to retain exudate. The $AC_{0.5}$ of the tested foam dressings decreased by a range of $AC_{0.5}$ = 30–40 g/100 cm^2. The differences in AC reduction can be explained by the different mechanisms of water absorption and retention. Although foam and BC-based dressings feature an open structure with free swell absorptive capacity, dressings such as ALG and CMC form hydrogels with the absorbed exudate [13,66]. Despite the fact that the absorbed water in foam dressings can be considered as "free", the fluid that is bound in a hydrogel can be retained better under compression [66].

Negative $AC_{0.5}$ in the case of BC_A can be explained by the fact that the material is already saturated with moisture, which is partly pressed out of the dressing by mechanical force. Looking at the BC-based wound dressings, BC_C2 performed best in terms of absorptive capacity, whereas BC_C1 and XBC lost most of their absorptive capacity. The reduced AC, especially of BC_A, should therefore be considered when treating chronic wounds that require compression therapy. With regard to that application, foam and hydrofiber dressings perform well in particular, but dressings such as BC_C2 with an increased solid content also show promising results.

2.5. Moisture Vapor Transmission Rate

Since the physiological skin barrier is damaged in the case of a wound, the evaporation of water represents a key factor in the regulation of a wound's moisture level. The loss of water from a wound site mainly takes place through absorption from the wound dressing and evaporation through and from the dressing. Therefore, the moisture vapor transmission rate (MVTR) of a chosen dressing is of fundamental importance in the exudate management of a wound [16,67]. To compare the evaporation of water through different wound dressings, the standardized Paddington cup method as described in EN 13726 was chosen.

As described in the standard, MVTR data was determined for two different experimental settings: dressings in contact with vapor (Supplementary Materials, Figure S1) and dressings in contact with liquid (Figure 5).

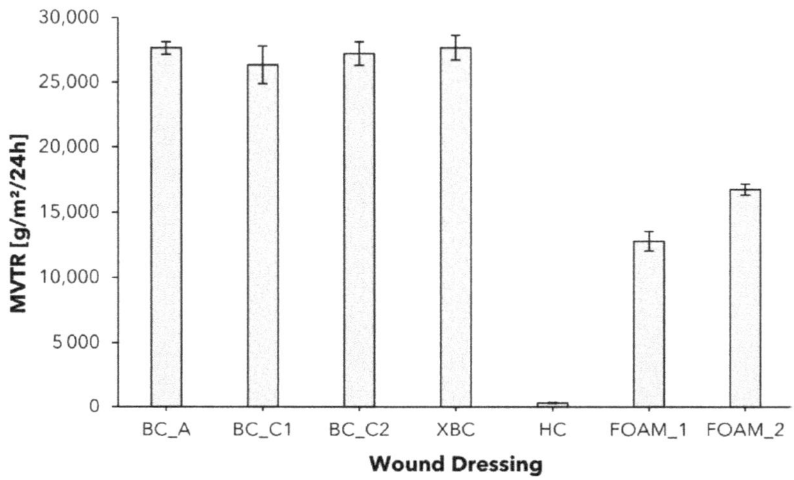

Figure 5. Moisture vapor transmission rate (MVTR) of different wound dressings determined with standard Paddington cup in contact with liquid method (mean ± SD; *n* = 5).

Although the contact with liquid method is recommended because it simulates more closely the wound environment as the dressing is in direct contact with the exudate [68], some dressings (see ALG; CMC) do not feature waterproofness and can therefore only be analyzed by the contact with vapor method (see Supplementary Materials, Figure S1).

In previous works as well as in this test, it was observed that all dressings showed relatively high MVTR values when in contact with liquid [69]. The BC dressings exceed the MVTR of foam dressings, the latter exhibiting an MVTR range close to that of the BC dressings in contact with vapor. Furthermore, the increase in water permeability when in contact with liquid instead of vapor was found to be remarkably lower for the foam dressings. Since BC dressings featuring an MVTR of >25,000 g/m^2/24 h when in contact with water exhibit lower absorptive capacity than foams, the mechanism of exudate removal is likely to be more pronounced on evaporation than on absorption. Similar results concerning the exceptional increase in water permeability of a bacterial cellulose wound dressing were published for the first time by Thomas [70]. However, they have never before been set in the context of the surprisingly high free swell absorptive capacity values. We hypothesize that the fine-structured, open-porous structure network of hydrophilic cellulose fibers supports moisture transport by capillary forces from the wound through the dressing to the surrounding air or the second dressing, facilitating a unique hydrobalance property.

In the case of the contact with vapor method, the MVTR values obtained were in a broad range. Hydrocolloid dressing (HC) showed the lowest MVTR of 98 ± 5 g/m^2/24 h, and can, therefore, be categorized as an occlusive dressing (MVTR < 300 g/m^2/24 h [5]). The low MVTR can be explained by their composition of an outer layer as being a polymeric film, which holds a layer of absorbent hydrocolloids such as sodium carboxymethyl cellulose and gelatine as well as adhesives and gelling agents [32,33]. Occlusive dressings are suited for use on low or no-exuding wounds to keep a moist wound environment, but can lead to periwound skin maceration and dressing leakage on wounds with higher levels of exudation and are not recommended to be used on infected wounds [5,28,71,72].

The MVTR in contact with vapor for the tested BC-based wound dressings was found to be in a range of 2697 ± 118 g/m^2/24 h (BC_C1) to 4039 ± 230 g/m^2/24 h (BC_A), which is in accordance with previously published data (MVTR~3000 g/m^2/24 h) [36,37] and close to the ideal MVTR range reported for maintaining the optimal moisture content of a wound (2000–2500 g/m^2/24 h, measured in contact with vapor) [73,74]. In the case of BC_A, the high MVTR is compensated by the intrinsic moisture of the dressing, which hydrates the wound and maintains a moist environment. Although the increased cellulose content and thickness of the produced dressings BC_C1 and BC_C2 clearly decreased the moisture vapor permeability, the MVTR was still found to be in the optimal range. With regard to the clinical application, this finding suggests that an increased absorptive capacity is not at the expense of an adequate permeability of moisture in vapor form.

2.6. Fluid Donation

Although the successful healing of many types of chronic wounds is likely to be affected by an excess of exudate, the moisture balance needs to be evaluated continuously. Whereas fibrous tissue and other plaque often need to be removed first by, e.g., an autolytic debridement before the healing process can be reactivated in a stagnating wound, the exudate level also changes as healing progresses to the later stages. In such cases, external hydration (e.g., via moist dressings) can ensure an optimal moisture level and thus successful wound closure [75].

Since most wound-dressing products used today in the treatment of chronic wounds come in a dry state to absorb an excess of exudate, the native hydration of BC-based wound dressings such as BC_A or XBC represents the rare property they share with only a few products, such as amorphous hydrogels or hydrogel dressings. To evaluate the influence of an increased solid content and to compare BC dressings in terms of their fluid donation (FD) capability to an artificial gelatin wound bed, a method based on EN 13726 was adapted.

In a comparison of the tested moisture-donating commercial wound dressings, BC_A showed the highest FD of 18.9 ± 1.8 g/100 cm^2, whereas the FD was remarkably lower for XBC with 9.0 ± 0.2 g/100 cm^2 (Figure 6). The difference in FD between the BC dressings is probably partly due to the differing overall water content (BC_A > 98%, XBC > 96%). The gel dressing Suprasorb® G (GEL) consists of an acrylic polymer-based hydrogel with a water content of ~70% [76] and consequently showed the lowest FD of all tested samples with 2.0 ± 0.9 g/100 cm^2. Considering the experimental dressing samples BC_C1 and BC_C2, the reduced water content of BC_C2 also correlates with a low FD capability. It stands out, however, that the modification in the case of BC_C1 did not lead to a reduction in the FD; the dressing performed similarly to BC_A in the experiment. Although this outcome seems contradictory at first, a possible rationalization could be that BC_C1 features not only an increased solid content but also a higher overall mass per area, whereas the absolute amount of water is comparable to that of BC_A. Taken together, these findings suggest that BC_C1 is as equally suited as the commercial dressing BC_A for the external hydration of a dry resp. low-exuding wound maintaining a suitable moist environment for successful wound healing, whereas the higher absorption capacity also suggests application on medium- to higher-exudation wounds.

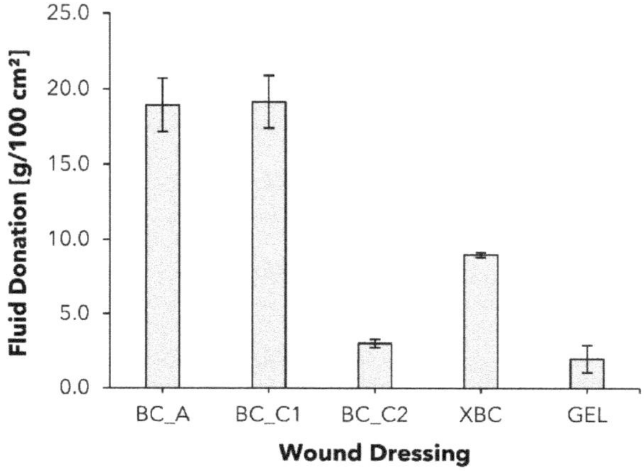

Figure 6. Fluid donation per area of five different moisture-donating wound dressings to a gelatin substrate after 48 h (mean ± SD; n = 5).

2.7. Overview of Results

The results of all performed in vitro tests are shown in Table 1.

Table 1. Overview of evaluated parameters, tests, and results.

Wound Dressing	Solid Content	Free Swell Absorptive Capacity (0.5 h; 2 h; 24 h)	Absorptive Capacity under Pressure	MVTR In Contact With Vapor	MVTR in Contact with Liquid	Fluid Donation to Gelatin
Unit	%	g/100 cm^2	g/100 cm^2	g/m^2/24 h	g/m^2/24 h	g/100 cm^2
ALG	-	17.6 ± 1.3 18.0 ± 1.1 18.4 ± 1.1	11.0 ± 1.4	6135 ± 78	-	-
CMC	-	20.1 ± 1.7 20.3 ± 1.7 21.0 ± 1.3	17.2 ± 0.5	6844 ± 282	-	-
HC	-	3.9 ± 0.3 6.0 ± 0.4 15.2 ± 1.1	3.0 ± 0.9	98 ± 5	330 ± 42	-
FOAM_1	-	82.0 ± 2.9 85.1 ± 5.4 92.5 ± 6.4	35.3 ± 2.5	3510 ± 225	12,790 ± 747	-
FOAM_2	-	60.1 ± 1.6 58.6 ± 3.9 61.1 ± 3.4	32.0 ± 1.9	2808 ± 103	16,750 ± 419	-
XBC	3.16 ± 0.18	13.3 ± 0.5 18.4 ± 0.9 22.1 ± 1.2	2.2 ± 0.3	3458 ± 198	27,663 ± 959	9.0 ± 0.2

Table 1. *Cont.*

Wound Dressing	Solid Content	Free Swell Absorptive Capacity (0.5 h; 2 h; 24 h)	Absorptive Capacity under Pressure	MVTR In Contact With Vapor	MVTR in Contact with Liquid	Fluid Donation to Gelatin
BC_A	1.57 ± 0.10	6.7 ± 0.5 8.7 ± 0.8 9.9 ± 0.9	−2.9 ± 2.2	4039 ± 230	27,648 ± 488	18.9 ± 1.8
BC_C1	3.27 ± 0.27	23.2 ± 5.1 31.2 ± 5.6 36.1 ± 5.7	1.7 ± 0.5	2697 ± 118	26,334 ± 1465	19.1 ± 1.7
BC_C2	8.20 ± 0.57	30.1 ± 6.3 37.4 ± 6.5 43.9 ± 7.7	13.1 ± 1.5	2908 ± 95	27,212 ± 913	3.0 ± 0.3
GEL	-	-	-	-	-	2.0 ± 0.9

3. Materials and Methods

3.1. Post-Market Clinical Follow-Up Study Design

Commercially available BC wound dressing epicite^{hydro} (BC_A) was evaluated in a multi-center study in 44 patients with mainly venous leg ulcers, mixed leg ulcers, and diabetic foot syndrome. This observational data collection has been performed according to Medical Devices Documents (MEDDEV) 2.12/2 Rev. 2 guidelines for post-market clinical follow-up studies in compliance with German medical devices legislation (§§ 20 ff. Medical Device Act). Therefore, formal ethical approval was not required.

The dressing was evaluated regarding criteria such as wound area and depth progression, fibrous tissue, exudation, hydration, and state of wound margins, among others. The baseline of the study is shown in Table 2, whereas diagnoses, previous local treatment, and causal therapy are provided as Supplementary Materials (Tables S2–S4).

Table 2. Baseline of BC_A Post-Market Clinical Follow-up study.

	Male	**Female**	
Sex	22	22	
	Mean	**Median**	**SD**
Patient Age [years]	66.9	71	15.94
Wound Age [weeks]	66.56	13	120.04
	Yes	**Not specified**	
Diagnosis of underlying disease	42	2	
Treatment of underlying disease	29	15	

Wound area determination was carried out by either measurement of wound length and width or calculation by wound-documentation software. Fibrous tissue share was assessed by visual inspection of wound surface. Exudation and wound hydration were evaluated through interviews with participating physicians and care specialists. To assess condition of wound margins, one or several of the following parameters: 'free from irritations, reddened, edematous, macerated, discolorated skin, necrotic, keratotic or undermined conditions', were chosen. Statistical analysis was performed using SPSS.

3.2. *In Vitro Tests*

3.2.1. Materials

The commercially available wound dressings tested in in vitro tests included BC dressing epicite^{hydro} (QRSKIN GmbH, Würzburg, Germany) (BC_A), bacterial cellulose dressing Suprasorb® X (XBC), calcium alginate dressing Suprasorb® A (ALG), and gel dressing Suprasorb® G (GEL), all by Lohmann & Rauscher GmbH & Co. KG (Neuwied, Germany); polyurethane foam dressing ALLEVYN Gentle (FOAM_1) by Smith & Nephew Medical Ltd. (Hull, UK); soft silicone foam dressing Mepilex® (FOAM_2) by Mölnlycke Health Care AB (Göteborg, Sweden); hydrocolloid dressing (HC) Hydrocoll® (PAUL HARTMANN Limited, Heywood, UK); and sodium carboxymethylcellulose dressing Aquacel® Extra™ (CMC) by ConvaTec Limited (Flintshire, UK). Sodium chloride was purchased from VWR Chemicals (Radnor, PA, USA). Calcium chloride dihydrate was obtained from Merck KgaA (Darmstadt, Germany) and gelatin 180 bloom was purchased from Carl Roth GmbH + Co KG (Karlsruhe, Germany).

3.2.2. Preparation of Native BC

Experimental, non-commercial BC dressings (BC_C1 and BC_C2) were produced using the *Komagataeibacter xylinus* strain DSM 14666, deposited at the German Collection of Microorganism and Cell Cultures (DSMZ, Braunschweig, Germany) in a static cultivation technique as previously described [36,37]. Briefly, *K. xylinus* was cultivated at 28 °C in a process-controlled pilot plant with an area of 1 m^2 (JeNaCell GmbH, Jena, Germany) using Hestrin–Schramm culture medium (HSM) [77]. In order to obtain BC dressings with high cellulose content, cultivation time and volume of HSM were adjusted empirically, so that a thickness of 8 mm was accomplished, which was chosen in order to double the thickness of most commercially available BC-based wound dressings (4 mm). Following cultivation, harvesting, and purification, BC dressings were cut into square pieces of 5 cm × 5 cm, sterilized by autoclaving (121 °C, 20 min, 2 bar), and stored at room temperature until further tests were carried out. Moisture content of BC_C was modified by flattening out the never-dried dressings by homogeneous vertical pressure until the sample weight reached an empirically predetermined weight range. BC_C1 and BC_C2 were produced with a weight range of 6.3 to 7.0 g per 5 cm × 5 cm sample, resp. 1.3 to 2.0 g per 5 cm × 5 cm sample.

3.2.3. Solid Content

Solid Content (SC) was evaluated in accordance with Beekmann et al. [37]. Briefly, moisture-adjusted dressings were cut into pieces of approximately 1 cm^2 and weighed (initial mass; m_i). After air-drying for 48 h at room temperature (RT), weighing was repeated (dry mass; m_d). SC was subsequently calculated using Equation (1):

$$SC\ [\%] = \frac{m_d}{m_i} \times 100\% \quad (1)$$

3.2.4. Free Swell Absorptive Capacity

In order to assess the capability of different wound dressing materials to absorb fluid from heavily exuding wounds, a technique based on EN 13726-1:2002 was used. Minor modifications were made to align the method with the ongoing revision published as draft prEN 13726:2021-02 [78,79]. In short, 5 cm × 5 cm sized wound dressings were measured to the nearest 0.1 cm to determine the exact sample area (A) and weighed in a dry state (m_0). Wound exudate was simulated by a test solution containing 142 mmol sodium and 2.5 mmol calcium as chloride salts (Test Solution A). Dressings were placed in a petri dish and immersed completely in the prewarmed test solution (37 °C ± 2 °C), transferred to an incubator (Binder GmbH, Tuttlingen, Germany), and allowed to soak for 0.5 h, 2 h, and 24 h ± 1 min at 37 °C ± 2 °C. At the end of each period, samples were removed from the test solution and excess fluid was allowed to drip off for 30 s ± 5 s. The dressings were reweighed ($m_{0.5\,h}$, $m_{2\,h}$, $m_{24\,h}$) and transferred back to the test solution for the remaining time until the test was completed. Each test was performed on 5 samples. Absorptive capacity (AC) was calculated for each test duration using Equation (2):

$$AC_t \left[\frac{g}{100\ cm^2}\right] = \frac{(m_t - m_0) \times 100}{A}$$
$$t = \text{test duration} = 0.5\ h \mid 2\ h \mid 24\ h \quad (2)$$

3.2.5. Absorptive Capacity under Pressure

To assess the absorptive capacity of dressing materials under pressure, the investigation was carried out in accordance with the method proposed in prEN 13726:2021 with minor modifications [78]. Briefly, the setup described in Section 3.2.4 was adapted by placing the wound dressings on a perforated plate before immersing them in Test Solution A. Once the dressing was completely submerged, a pressure of 40 mmHg was applied using a weight of 1359.5 g with the help of a rigid compression plate between the weight and the dressing for homogenous pressure distribution. After 0.5 h, the setup consisting of a perforated plate, dressing, compression plate, and weight was removed from the test solution and allowed to drain for 5 min ± 10 s while pressure was still applied. Finally, the dressing was reweighed ($m_{0.5\ h}$) and AC was calculated as reported above.

3.2.6. Moisture Vapor Transmission Rate

Moisture vapor transmission rate (MVTR) was evaluated according to EN 13726-2:2002. For ALG, CMC, HC, XBC, FOAM_1, FOAM_2, BC_A, BC_C1, and BC_C2 test variation A, 'MVTR in contact with *vapor*' was carried out. Briefly, the dry dressing materials were clamped into the upper part of a standardized Paddington cup (test area: 10 cm^2, The Surgical Materials Testing Laboratory, Bridgend, UK) filled with deionized water leaving a 5 mm gap between the water surface and the dressing. Evaporation through the dressing at 37 °C and <20% relative humidity (RH) was measured by weighing the cup at the beginning (m_1) and the end (m_2) of the 18 h to 24 h testing period (Figure 7).

A B C

Figure 7. Standard Paddington cup method used to determine moisture vapor transmission rate (MVTR) of different wound dressings. (**A**) Paddington cup with defined orifice of 10 cm^2 filled with deionized water; (**B**) Dressing sample placed over the orifice secured with clamping plate and sealed with impermeable film to prevent evaporation through the edges; (**C**) Final setup for measuring MVTR; fixation of the setup with screw clamps.

For HC, XBC, FOAM_1, FOAM_2, BC_A, BC_C1, and BC_C2 test variation B, 'MVTR in contact with *liquid*' was assessed as proposed in EN 13726-2:2002 with minor modifications. The upper part of the Paddington cup was covered with the dressing as described above. Afterward, the cup was inverted, filled with 20 mL of deionized water, and incubated at 37 °C, <20% RH for 18 h to 24 h. Determination of evaporation by weight loss was executed as described above. The Paddington cup was modified using a base plate with a fine vent hole (0.25 mm) on the opposite side of the dressing to prevent stretching of the dressing as a consequence of negative pressure inside the cup. All experiments were performed fivefold and MVTR for each variation was calculated using Equation (3):

$$\begin{gathered} \text{MVTR}\left[\frac{\text{g}}{\text{m}^2 \times 24\ \text{h}}\right] = (\text{m}_1 - \text{m}_2) \times 1000 \times \frac{24}{\text{T}} \\ \text{m}_1 = \text{weight before incubation [g]} \\ \text{m}_2 = \text{weight after incubation [g]} \\ \text{T} = \text{testing duration [h]} \end{gathered} \tag{3}$$

3.2.7. Fluid Donation

The wet dressings XBC, BC_A, BC_C, and GEL were compared regarding their fluid donation abilities using a method based on EN 13726-1:2002 part 3.4. Briefly, 35.00 g of gelatin powder was suspended in 65.00 g Test Solution A and incubated for 12 h at 60 °C until completely dissolved. An amount of 30 g of gelatin solution was poured into a petri dish of 90 mm diameter. The petri dishes were closed, sealed tightly, and incubated for 3 h at 25 °C to allow gel-forming (Figure 8).

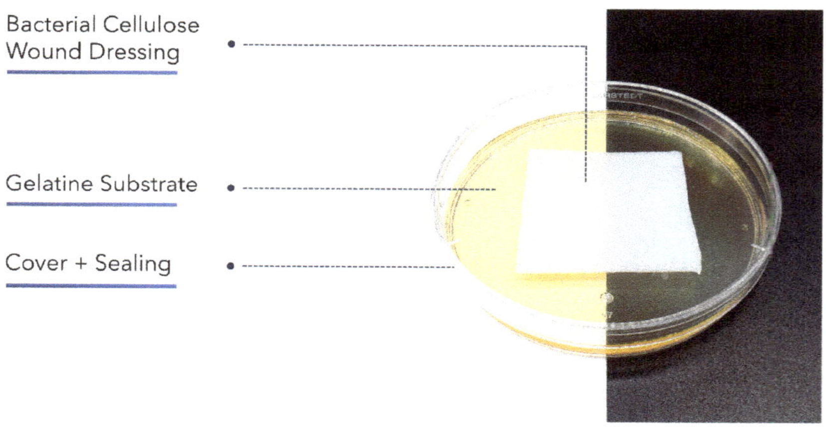

Figure 8. Setup based on EN 13726-1:2002 part 3.4 to compare moisture-donating wound dressings regarding their fluid donation. A gelatin gel with a water content < 35% was weighed before and after 48 h of contact with the dressing samples. The petri dish was covered and sealed tightly to prevent evaporation.

Petri dishes were opened and weighed (m_1), and 5 cm × 5 cm sized wound dressings were placed on the gel surface. After closing and sealing, the dishes were incubated vertically at 25 °C for 48 h. Lid was removed and petri dishes were weighed again after removal of the dressing samples (m_2). Fluid Donation (FD) was calculated using Equation (4):

$$\begin{gathered} \text{FD}\left[\frac{\text{g}}{100\ \text{cm}^2}\right] = \frac{(\text{m}_2 - \text{m}_1) \times 100}{\text{A}} \\ \text{m}_1 = \text{weight before incubation [g]} \\ \text{m}_2 = \text{weight after incubation [g]} \end{gathered} \tag{4}$$

4. Conclusions

Millions of patients worldwide suffer from chronic wounds. Although there are numerous types of wound dressings available in the market, selecting a suitable dressing can be demanding and therapy success is not self-evident. Wound dressings based on BC such as epicite^{hydro} have gained increasing relevance and show excellent results in the treatment of burn wounds. The post-market surveillance study featuring epicite^{hydro} presented herein revealed promising results regarding wound cleansing, pain reduction, a reduction in the wound area, and depths as well as fibrous tissue in chronic wounds. Nonetheless, there was still room for improvement in exudate management, where in vitro tests revealed that epicite^{hydro} fell short of higher absorptive materials such as foam dressings, alginates, or hydrofibers in terms of exudate absorption.

In this work, the exudate-handling properties of BC dressings could be altered using a readily feasiblemodification to provide a remarkably higher absorption capacity, which represents a promising enhancement concerning the therapy of medium- to higher-exuding wounds. For the first time, a potential interplay between the MVTR in contact with liquid and the free swell absorptive capacity was described, the latter being surprisingly higher for BC dressings featuring a moisture content of 92% or 97%, respectively, than for dry, advanced wound dressings made of alginates, carboxymethylcellulose (hydrofibers), or hydrocolloids. This unique property could be explained by capillary forces within the open-porous and fine network structure of BC dressings. Both observations, the increase in the free swell absorptive capacity over time (here: within 24 h), and the significantly lower free swell absorptive capacity of the moisture-saturated version (BC_A) support this hypothesis. Future studies should investigate whether these results can be replicated in alternative experimental settings and could be effective in clinical use. Biomimetic concepts such as wound-healing (scratch) assays, which have already been successfully applied to BC-based wound dressings, could also be helpful in this context [36,80].

At the same time, the proven beneficial characteristics, such as a suitable permeability and a strong ability to donate moisture, which have a clinically relevant impact on the hydrobalance of wounds, could be maintained in the new versions BC_C1 and BC_C2. Carbohydrate polymer-based wound dressings made of bacterial cellulose such as epicite^{hydro} thus represent a very promising and adaptable material for the successful treatment of both burn and chronic wounds of different origins.

Supplementary Materials: The following are available online at https://www.mdpi.com/article/10.3390/ph15060683/s1, Table S1: Clinical trials assessing different BC-based wound dressings in chronic wound care. Table S2: Clinical trials assessing different BC-based wound dressings in burn wound care. Table S3: Overview of wound diagnoses in BC_A Post-Market Clinical Follow-up Study. Table S4: Overview of local wound treatment prior to BC_A Post-Market Clinical Follow-up Study. Table S5: Overview of causal therapy in BC_A Post-Market Clinical Follow-up Study. Figure S1: Moisture vapor transmission rate (MVTR) of different wound dressings determined with standard Paddington cup in contact with vapor method (mean ± SD; $n = 5$). References [81–91] are cited in the supplementary materials.

Author Contributions: Conceptualization, P.Z., U.B. and D.K., methodology, P.Z.; formal analysis, P.Z., T.E. and M.S.; investigation, P.Z. and T.E.; resources, D.K.; data curation, P.Z., T.E. and M.S.; writing—original draft preparation, P.Z.; writing—review and editing, U.B., T.E., D.K. and O.W.; visualization, P.Z.; supervision, U.B., T.E., D.K. and O.W.; project administration, D.K.; funding acquisition, D.K. All authors have read and agreed to the published version of the manuscript.

Funding: P.Z. gratefully acknowledges the Free State of Thuringia and the European Social Fund (INP0046) for funding the study. D.K. gratefully acknowledges funding from the European Union's Horizon 2020 research and innovation programme under the Marie Skłodowska-Curie grant agreement No 955335.

Institutional Review Board Statement: Ethical review and approval were waived for this study due to the clinical data mentioned in the manuscript was collected in the context of an observational study. This observational study has been performed according to MEDDEV 2.12/2 Rev. 2 guideline for post-market clinical follow-up studies in compliance with with German medical devices legislations (§§ 20 ff. MPG). Therefore, a formal ethical approval was not required. This explanation is also included in the manuscripts Materials and Methods Section 3.1 "Post-Market Clinical Follow-up Study Design".

Informed Consent Statement: Informed consent was obtained from all subjects involved in the clinical observation study.

Data Availability Statement: The data contained within the article and supplementary.

Acknowledgments: The authors would like to thank Elena Pfaff, Elisabeth Mende, and Susanr Kugel (JeNaCell GmbH) for their excellent technical assistance as well as Ryan Heniford (Evonik Operations GmbH) for language proofreading.

Conflicts of Interest: D.K. is CEO and General Manager of JeNaCell GmbH—an Evonik company (Jena, Germany)—as well as an employee of Evonik Operations GmbH. D.K. has no other relevant affiliations or financial involvement with any organization or entity with a financial interest in or financial conflict with the subject matter or materials discussed in the publication apart from those disclosed. JeNaCell GmbH is the producer of epicite^{hydro} wound dressings, the product that is discussed in the publication. D.K. initiated the clinical observation study. T.E. and M.S. are the medical writers of the observation study presented herewith, responsible for data management, statistics, and medical writing. Both have received a consultation fee. JeNaCell GmbH had no influence on the planning and implementation of the clinical observation study and no influence on the data collection for the clinical observation study.

Abbreviations & Terms

AC	Absorptive Capacity
ALG	Calcium Alginate dressing Suprasorb® A (Lohmann & Rauscher GmbH & Co. KG, Neuwied, Germany)
BC	Bacterial Cellulose
BC_A	BC based wound dressing epicite^{hydro} (QRSKIN GmbH, Würzburg, Germany)
BC_C1	Experimental, non-commercial and modified BC-based wound dressing
BC_C2	Experimental, non-commercial, and modified BC Cellulose-based wound dressing
CCl	Compression Class
CMC	Sodium carboxymethylcellulose dressing Aquacel® Extra™ (ConvaTec Limited, Flintshire, UK)
DSMZ	Deutsche Sammlung von Mikroorganismen und Zellkulturen (German Collection of Microorganisms and Cell Cultures)
FD	Fluid Donation
FOAM_1	Polyurethan foam dressing ALLEVYN Gentle (Smith & Nephew Medical Ltd., Hull, UK)
FOAM_2	Soft silicone foam dressing Mepilex® (Mölnlycke Health Care AB, Göteborg, Sweden)
GEL	Gel dressing Suprasorb® G (Lohmann & Rauscher GmbH & Co. KG, Neuwied, Germany)
HC	Hydrocolloid dressing Hydrocoll® (PAUL HARTMANN Limited, Heywood, UK)
HET-CAV	Shell-less Hen's Egg Test on Chick Area Vasculosa
HSM	Hestrin–Schramm culture medium
K. xylinus	*Komagataeibacter xylinus*
MTT	3-(4,5-dimethylthiazol-2-yl)-2,5-diphenyltetrazolium bromide
MVTR	Moisture Vapor Transmission Rate
PMCF	Post-Market Clinical Follow-up (Study)
SC	Solid Content
VLU	Venous Leg Ulcer
XBC	BC-based wound dressing Suprasorb® X (Lohmann & Rauscher GmbH & Co. KG, Neuwied, Germany)

References

1. Olsson, M.; Järbrink, K.; Divakar, U.; Bajpai, R.; Upton, Z.; Schmidtchen, A.; Car, J. The Humanistic and Economic Burden of Chronic Wounds: A Systematic Review: The Burden of Chronic Wounds. *Wound Repair Regen.* **2019**, *27*, 114–125. [CrossRef] [PubMed]
2. Fife, C.E.; Carter, M.J. Wound Care Outcomes and Associated Cost among Patients Treated in US Outpatient Wound Centers: Data from the US Wound Registry. *Wounds Compend. Clin. Res. Pract.* **2012**, *24*, 10–17.
3. Graves, N.; Phillips, C.J.; Harding, K. A Narrative Review of the Epidemiology and Economics of Chronic Wounds. *Br. J. Dermatol.* **2021**, bjd.20692. [CrossRef] [PubMed]
4. Shi, C.; Wang, C.; Liu, H.; Li, Q.; Li, R.; Zhang, Y.; Liu, Y.; Shao, Y.; Wang, J. Selection of Appropriate Wound Dressing for Various Wounds. *Front. Bioeng. Biotechnol.* **2020**, *8*, 182. [CrossRef]
5. Seaman, S. Dressing Selection in Chronic Wound Management. *J. Am. Podiatr. Med. Assoc.* **2002**, *92*, 24–33. [CrossRef]
6. Korting, H.; Schöllmann, C.; White, R. Management of Minor Acute Cutaneous Wounds: Importance of Wound Healing in a Moist Environment: Acute Wounds and Healing. *J. Eur. Acad. Dermatol. Venereol.* **2011**, *25*, 130–137. [CrossRef]
7. Lazarus, G.S.; Cooper, D.M.; Knighton, D.R.; Margolis, D.J.; Percoraro, R.E.; Rodeheaver, G.; Robson, M.C. Definitions and Guidelines for Assessment of Wounds and Evaluation of Healing. *Wound Repair Regen.* **1994**, *2*, 165–170. [CrossRef]
8. Dai, C.; Shih, S.; Khachemoune, A. Skin Substitutes for Acute and Chronic Wound Healing: An Updated Review. *J. Dermatol. Treat.* **2020**, *31*, 639–648. [CrossRef]
9. Koehler, J.; Brandl, F.P.; Goepferich, A.M. Hydrogel Wound Dressings for Bioactive Treatment of Acute and Chronic Wounds. *Eur. Polym. J.* **2018**, *100*, 1–11. [CrossRef]
10. Dissemond, J.; Kröger, K.; Initiative Chronische Wunden, e.V. *Chronische Wunden: Diagnostik-Therapie-Versorgung*; Urban & Fischer Verlag/Elsevier GmbH: München, Germany, 2019.
11. Zhao, R.; Liang, H.; Clarke, E.; Jackson, C.; Xue, M. Inflammation in Chronic Wounds. *Int. J. Mol. Sci.* **2016**, *17*, 2085. [CrossRef] [PubMed]
12. Smith, F.; Sharp, A. Undertaking a Person-Centred Assessment of Patients with Chronic Wounds. *Nurs. Stand.* **2019**, *34*, 77–82. [CrossRef]
13. Boateng, J. Therapeutic Dressings and Wound Healing Applications. In *Advances in Pharmaceutical Technology*, 1st ed.; Wiley: Hoboken, NJ, USA, 2020; ISBN 978-1-119-43326-2.
14. Morton, L.M.; Phillips, T.J. Wound Healing and Treating Wounds. *J. Am. Acad. Dermatol.* **2016**, *74*, 589–605. [CrossRef] [PubMed]
15. Winter, G.D. Formation of the Scab and the Rate of Epithelization of Superficial Wounds in the Skin of the Young Domestic Pig. *Nature* **1962**, *193*, 293–294. [CrossRef] [PubMed]
16. Nuutila, K.; Eriksson, E. Moist Wound Healing with Commonly Available Dressings. *Adv. Wound Care* **2021**, *10*, 685–698. [CrossRef] [PubMed]
17. Benbow, M. Exploring the Concept of Moist Wound Healing and Its Application in Practice. *Br. J. Nurs.* **2008**, *17*, S4–S16. [CrossRef]
18. Constantine, B.E.; Bolton, L.L. A Wound Model for Ischemic Ulcers in the Guinea Pig. *Arch. Dermatol. Res.* **1986**, *278*, 429–431. [CrossRef]
19. Madden, M.R.; Nolan, E.; Finkelstein, J.L.; Yurt, R.W.; Smeland, J.; Goodwin, C.W.; Hefton, J.; Staiano-Coico, L. Comparison of an Occlusive and a Semi-Occlusive Dressing and the Effect of the Wound Exudate upon Keratinocyte Proliferation. *J. Trauma* **1989**, *29*, 924–930. [CrossRef]
20. Katz, M.H.; Alvarez, A.F.; Kirsner, R.S.; Eaglstein, W.H.; Falanga, V. Human Wound Fluid from Acute Wounds Stimulates Fibroblast and Endothelial Cell Growth. *J. Am. Acad. Dermatol.* **1991**, *25*, 1054–1058. [CrossRef]
21. Leipziger, L.S.; Glushko, V.; DiBernardo, B.; Shafaie, F.; Noble, J.; Nichols, J.; Alvarez, O.M. Dermal Wound Repair: Role of Collagen Matrix Implants and Synthetic Polymer Dressings. *J. Am. Acad. Dermatol.* **1985**, *12*, 409–419. [CrossRef]
22. Sharman, D. Moist Wound Healing: A Review of Evidence, Application and Outcome. *Diabet. Foot* **2003**, *6*, 5.
23. Powers, J.G.; Morton, L.M.; Phillips, T.J. Dressings for Chronic Wounds: Dressings. *Dermatol. Ther.* **2013**, *26*, 197–206. [CrossRef] [PubMed]
24. Wiegand, C.; Hipler, U.-C.; Elsner, P.; Tittelbach, J. Clinical Efficacy of Dressings for Treatment of Heavily Exuding Chronic Wounds. *Chronic Wound Care Manag. Res.* **2015**, *2*, 101. [CrossRef]
25. Si, H.; Xing, T.; Ding, Y.; Zhang, H.; Yin, R.; Zhang, W. 3D Bioprinting of the Sustained Drug Release Wound Dressing with Double-Crosslinked Hyaluronic-Acid-Based Hydrogels. *Polymers* **2019**, *11*, 1584. [CrossRef] [PubMed]
26. Jia, Y.; Zhang, H.; Yang, S.; Xi, Z.; Tang, T.; Yin, R.; Zhang, W. Electrospun PLGA Membrane Incorporated with Andrographolide-Loaded Mesoporous Silica Nanoparticles for Sustained Antibacterial Wound Dressing. *Nanomedicine* **2018**, *13*, 2881–2899. [CrossRef] [PubMed]
27. De Pieri, A.; Byerley, A.M.; Musumeci, C.R.; Salemizadehparizi, F.; Vanderhorst, M.A.; Wuertz-Kozak, K. Electrospinning and 3D Bioprinting for Intervertebral Disc Tissue Engineering. *JOR Spine* **2020**, *3*, e1117. [CrossRef]
28. Vowden, K.; Vowden, P. Wound Dressings: Principles and Practice. *Surgery* **2014**, *32*, 462–467.
29. Dhivya, S.; Padma, V.V.; Santhini, E. Wound Dressings—A Review. *BioMedicine* **2015**, *5*, 22. [CrossRef]
30. Tudoroiu, E.-E.; Dinu-Pîrvu, C.-E.; Albu Kaya, M.G.; Popa, L.; Anuța, V.; Prisada, R.M.; Ghica, M.V. An Overview of Cellulose Derivatives-Based Dressings for Wound-Healing Management. *Pharmaceuticals* **2021**, *14*, 1215. [CrossRef]

31. Klemm, D.; Kramer, F.; Moritz, S.; Lindström, T.; Ankerfors, M.; Gray, D.; Dorris, A. Nanocellulosen: Eine neue Familie naturbasierter Materialien. *Angew. Chem.* **2011**, *123*, 5550–5580. [CrossRef]
32. Klemm, D.; Petzold-Welcke, K.; Kramer, F.; Richter, T.; Raddatz, V.; Fried, W.; Nietzsche, S.; Bellmann, T.; Fischer, D. Biotech Nanocellulose: A Review on Progress in Product Design and Today's State of Technical and Medical Applications. *Carbohydr. Polym.* **2021**, *254*, 117313. [CrossRef]
33. Kralisch, D.; Hessler, N.; Klemm, D.; Erdmann, R.; Schmidt, W. White Biotechnology for Cellulose Manufacturing—The HoLiR Concept. *Biotechnol. Bioeng.* **2010**, *105*, 740–747. [CrossRef] [PubMed]
34. Ahmed, J.; Gultekinoglu, M.; Edirisinghe, M. Bacterial Cellulose Micro-Nano Fibres for Wound Healing Applications. *Biotechnol. Adv.* **2020**, *41*, 107549. [CrossRef] [PubMed]
35. Gao, H.; Sun, Q.; Han, Z.; Li, J.; Liao, B.; Hu, L.; Huang, J.; Zou, C.; Jia, C.; Huang, J.; et al. Comparison of Bacterial Nanocellulose Produced by Different Strains under Static and Agitated Culture Conditions. *Carbohydr. Polym.* **2020**, *227*, 115323. [CrossRef] [PubMed]
36. Beekmann, U.; Zahel, P.; Karl, B.; Schmölz, L.; Börner, F.; Gerstmeier, J.; Werz, O.; Lorkowski, S.; Wiegand, C.; Fischer, D.; et al. Modified Bacterial Cellulose Dressings to Treat Inflammatory Wounds. *Nanomaterials* **2020**, *10*, 2508. [CrossRef]
37. Beekmann, U.; Schmölz, L.; Lorkowski, S.; Werz, O.; Thamm, J.; Fischer, D.; Kralisch, D. Process Control and Scale-up of Modified Bacterial Cellulose Production for Tailor-Made Anti-Inflammatory Drug Delivery Systems. *Carbohydr. Polym.* **2020**, *235*, 116062. [CrossRef]
38. Wiegand, C.; Moritz, S.; Hessler, N.; Kralisch, D.; Wesarg, F.; Müller, F.A.; Fischer, D.; Hipler, U.-C. Antimicrobial Functionalization of Bacterial Nanocellulose by Loading with Polihexanide and Povidone-Iodine. *J. Mater. Sci. Mater. Med.* **2015**, *26*, 245. [CrossRef]
39. Picheth, G.F.; Pirich, C.L.; Sierakowski, M.R.; Woehl, M.A.; Sakakibara, C.N.; de Souza, C.F.; Martin, A.A.; da Silva, R.; de Freitas, R.A. Bacterial Cellulose in Biomedical Applications: A Review. *Int. J. Biol. Macromol.* **2017**, *104*, 97–106. [CrossRef]
40. Czaja, W.; Krystynowicz, A.; Bielecki, S.; Brownjr, R. Microbial Cellulose—The Natural Power to Heal Wounds. *Biomaterials* **2006**, *27*, 145–151. [CrossRef]
41. Sulaeva, I.; Henniges, U.; Rosenau, T.; Potthast, A. Bacterial Cellulose as a Material for Wound Treatment: Properties and Modifications. A Review. *Biotechnol. Adv.* **2015**, *33*, 1547–1571. [CrossRef]
42. He, W.; Wu, J.; Xu, J.; Mosselhy, D.A.; Zheng, Y.; Yang, S. Bacterial Cellulose: Functional Modification and Wound Healing Applications. *Adv. Wound Care* **2021**, *10*, 623–640. [CrossRef]
43. Portela, R.; Leal, C.R.; Almeida, P.L.; Sobral, R.G. Bacterial Cellulose: A Versatile Biopolymer for Wound Dressing Applications. *Microb. Biotechnol.* **2019**, *12*, 586–610. [CrossRef] [PubMed]
44. Pang, M.; Huang, Y.; Meng, F.; Zhuang, Y.; Liu, H.; Du, M.; Ma, Q.; Wang, Q.; Chen, Z.; Chen, L.; et al. Application of Bacterial Cellulose in Skin and Bone Tissue Engineering. *Eur. Polym. J.* **2020**, *122*, 109365. [CrossRef]
45. Sanchavanakit, N.; Sangrungraungroj, W.; Kaomongkolgit, R.; Banaprasert, T.; Pavasant, P.; Phisalaphong, M. Growth of Human Keratinocytes and Fibroblasts on Bacterial Cellulose Film. *Biotechnol. Prog.* **2006**, *22*, 1194–1199. [CrossRef] [PubMed]
46. Kwak, M.H.; Kim, J.E.; Go, J.; Koh, E.K.; Song, S.H.; Son, H.J.; Kim, H.S.; Yun, Y.H.; Jung, Y.J.; Hwang, D.Y. Bacterial Cellulose Membrane Produced by Acetobacter Sp. A10 for Burn Wound Dressing Applications. *Carbohydr. Polym.* **2015**, *122*, 387–398. [CrossRef]
47. Li, Y.; Wang, S.; Huang, R.; Huang, Z.; Hu, B.; Zheng, W.; Yang, G.; Jiang, X. Evaluation of the Effect of the Structure of Bacterial Cellulose on Full Thickness Skin Wound Repair on a Microfluidic Chip. *Biomacromolecules* **2015**, *16*, 780–789. [CrossRef] [PubMed]
48. Holzer, J.C.J.; Tiffner, K.; Kainz, S.; Reisenegger, P.; de Mattos, I.B.; Funk, M.; Lemarchand, T.; Laaff, H.; Bal, A.; Birngruber, T.; et al. A Novel Human Ex-Vivo Burn Model and the Local Cooling Effect of a Bacterial Nanocellulose-Based Wound Dressing. *Burns* **2020**, *46*, 1924–1932. [CrossRef]
49. Jones, E.M.; Cochrane, C.A.; Percival, S.L. The Effect of PH on the Extracellular Matrix and Biofilms. *Adv. Wound Care* **2015**, *4*, 431–439. [CrossRef]
50. Pourali, P.; Razavianzadeh, N.; Khojasteh, L.; Yahyaei, B. Assessment of the Cutaneous Wound Healing Efficiency of Acidic, Neutral and Alkaline Bacterial Cellulose Membrane in Rat. *J. Mater. Sci. Mater. Med.* **2018**, *29*, 90. [CrossRef]
51. Hoff, J.; Karl, B.; Gerstmeier, J.; Beekmann, U.; Schmölz, L.; Börner, F.; Kralisch, D.; Bauer, M.; Werz, O.; Fischer, D.; et al. Controlled Release of the α-Tocopherol-Derived Metabolite α-13′-Carboxychromanol from Bacterial Nanocellulose Wound Cover Improves Wound Healing. *Nanomaterials* **2021**, *11*, 1939. [CrossRef]
52. Maurer, K.; Renkert, M.; Duis, M.; Weiss, C.; Wessel, L.M.; Lange, B. Application of Bacterial Nanocellulose-Based Wound Dressings in the Management of Thermal Injuries: Experience in 92 Children. *Burns* **2021**, *48*, 608–614. [CrossRef]
53. Resch, A.; Staud, C.; Radtke, C. Nanocellulose-based Wound Dressing for Conservative Wound Management in Children with Second-degree Burns. *Int. Wound J.* **2021**, *18*, 478–486. [CrossRef] [PubMed]
54. Schiefer, J.L.; Aretz, G.F.; Fuchs, P.C.; Bagheri, M.; Funk, M.; Schulz, A.; Daniels, M. Comparison of Wound Healing and Patient Comfort in Partial-thickness Burn Wounds Treated with SUPRATHEL and Epicte ^{hydro} Wound Dressings. *Int. Wound J.* **2021**, *19*, 782–790. [CrossRef] [PubMed]
55. Cattelaens, J.; Turco, L.; Berclaz, L.M.; Huelsse, B.; Hitzl, W.; Vollkommer, T.; Bodenschatz, K.J. The Impact of a Nanocellulose-Based Wound Dressing in the Management of Thermal Injuries in Children: Results of a Retrospective Evaluation. *Life* **2020**, *10*, 212. [CrossRef] [PubMed]

56. Alvarez, O.M.; Patel, M.; Booker, J.; Markowitz, L. Effectiveness of a Biocellulose Wound Dressing for the Treatment of Chronic Venous Leg Ulcers: Results of a Single Center Randomized Study Involving 24 Patients. *Wounds Compend. Clin. Res. Pract.* **2004**, *16*, 224–233.
57. Vowden, K.; Vowden, P. Understanding Exudate Management and the Role of Exudate in the Healing Process. *Br. J. Community Nurs.* **2003**, *8*, S4–S13. [CrossRef]
58. Bishop, S.M.; Walker, M.; Rogers, A.A.; Chen, W.Y.J. Importance of Moisture Balance at the Wound-Dressing Interface. *J. Wound Care* **2003**, *12*, 125–128. [CrossRef]
59. Kim, Y.; Doh, S.J.; Lee, G.D.; Kim, C.; Im, J.N. Composite Nonwovens Based on Carboxymethyl Cellulose for Wound Dressing Materials. *Fibers Polym.* **2019**, *20*, 2048–2056. [CrossRef]
60. Thomas, S. *Technical Report: An 'In-Vitro' Comparison of the Physical Characteristics of Hydrocolloids, Hydrogels, Foams, and Alginate/CMC Fibrous Dressings;* Surgical Materials Testing Laboratory: Bridgend, UK, 2005.
61. Lee, S.M.; Park, I.K.; Kim, Y.S.; Kim, H.J.; Moon, H.; Mueller, S.; Jeong, Y.-I. Physical, Morphological, and Wound Healing Properties of a Polyurethane Foam-Film Dressing. *Biomater. Res* **2016**, *20*, 15. [CrossRef]
62. Körber, A.; Weindorf, M.; Dissemond, J. Exudate capacity of modern wound dressings during compression therapy for chronic venous leg ulcers. *Hautarzt* **2008**, *59*, 904–911. [CrossRef]
63. Dissemond, J.; Assenheimer, B.; Bültemann, A.; Gerber, V.; Gretener, S.; Kohler-von Siebenthal, E.; Koller, S.; Kröger, K.; Kurz, P.; Läuchli, S.; et al. Compression Therapy in Patients with Venous Leg Ulcers. *JDDG J. Ger. Soc. Dermatol.* **2016**, *14*, 1073–1089. [CrossRef]
64. Lurie, F.; Bittar, S.; Kasper, G. Optimal Compression Therapy and Wound Care for Venous Ulcers. *Surg. Clin. N. Am.* **2018**, *98*, 349–360. [CrossRef] [PubMed]
65. Linnitt, N.; Jones, J. Indications for Deciding Compression in the Treatment of Chronic Oedema. *Br. J. Community Nurs.* **2012**, *17*, S11–S15. [CrossRef]
66. Walker, M.; Lam, S.; Pritchard, D.; Cochrane, C.A. Biophysical Properties of a Hydrofiber® Cover Dressing. *Wounds UK* **2010**, *6*, 9.
67. Wu, P.; Nelson, E.A.; Reid, W.H.; Ruckley, C.V.; Gaylor, J.D.S. Water Vapour Transmission Rates in Burns and Chronic Leg Ulcers: Influence of Wound Dressings and Comparison with in Vitro Evaluation. *Biomaterials* **1996**, *17*, 1373–1377. [CrossRef]
68. Browning, P.; White, R.J.; Rowell, T. Comparative Evaluation of the Functional Properties of Superabsorbent Dressings and Their Effect on Exudate Management. *J. Wound Care* **2016**, *25*, 452–462. [CrossRef]
69. Zehrer, C.L.; Holm, D.; Solfest, S.E.; Walters, S.-A. A Comparison of the in Vitro Moisture Vapour Transmission Rate and in Vivo Fluid-Handling Capacity of Six Adhesive Foam Dressings to a Newly Reformulated Adhesive Foam Dressing: Comparison of Fluid-Handling Capacity of Adhesive Foam Dressings. *Int. Wound J.* **2014**, *11*, 681–690. [CrossRef]
70. Thomas, S. A Review of the Physical, Biological and Clinical Properties of a Bacterial Cellulose Wound. *J. Wound Care* **2008**, *17*, 349–352. [CrossRef]
71. Rajendran, S. *Advanced Textiles for Wound Care*, 2nd ed.; Woodhead Publishing: Duxford, UK, 2019; ISBN 978-0-08-102192-7.
72. Wu, P.; Fisher, A.; Foo, P.P.; Queen, D.; Gaylor, J.D.S. In Vitro Assessment of Water Vapour Transmission of Synthetic Wound Dressings. *Biomaterials* **1995**, *16*, 5. [CrossRef]
73. Queen, D.; Gaylor, J.D.S.; Evans, J.H.; Courtney, J.M.; Reid, W.H. The Preclinical Evaluation of the Water Vapour Transmission Rate through Burn Wound Dressings. *Biomaterials* **1987**, *8*, 367–371. [CrossRef]
74. Xu, R.; Xia, H.; He, W.; Li, Z.; Zhao, J.; Liu, B.; Wang, Y.; Lei, Q.; Kong, Y.; Bai, Y.; et al. Controlled Water Vapor Transmission Rate Promotes Wound-Healing via Wound Re-Epithelialization and Contraction Enhancement. *Sci. Rep.* **2016**, *6*, 24596. [CrossRef]
75. White, R.; Cutting, K.F. Modern Exudate Management: A Review of Wound Treatments. Available online: http://www.worldwidewounds.com/2006/september/White/Modern-Exudate-Mgt (accessed on 11 October 2021).
76. Lohmann & Rauscher GmbH & Co. KG Product Information "Suprasorb G". Available online: https://www.lohmann-rauscher.com/en/products/wound-care/modern-wound-care/suprasorb-g-gel-dressing/ (accessed on 8 November 2021).
77. Hestrin, S.; Schramm, M. Synthesis of Cellulose by Acetobacter Xylinum. II. Preparation of Freeze-Dried Cells Capable of Polymerizing Glucose to Cellulose. *Biochem. J.* **1954**, *58*, 345–352. [CrossRef] [PubMed]
78. *CEN/TC 205 Draft PrEN 13726:2021;* Test Methods for Wound Dressings—Aspects of Absorbency and Moisture Vapour Transmission, Waterproofness and Conformability. Beuth: Brussels, Belgium, 2021.
79. *CEN/TC 205 EN 13726-1:2002; EN 13726-2:2002;* Test Methods for Primary Wound Dressings. Beuth: Brussels, Belgium, 2002.
80. Jonkman, J.E.N.; Cathcart, J.A.; Xu, F.; Bartolini, M.E.; Amon, J.E.; Stevens, K.M.; Colarusso, P. An Introduction to the Wound Healing Assay Using Live-Cell Microscopy. *Cell Adhes. Migr.* **2014**, *8*, 440–451. [CrossRef] [PubMed]
81. Portal, O.; Clark, W.A.; Levinson, D.J. Microbial Cellulose Wound Dressing in the Treatment of Nonhealing Lower Extremity Ulcers. *Wounds Compend. Clin. Res. Pract.* **2009**, *2*, 1–3.
82. Solway, D.R.; Clark, W.A.; Levinson, D.J. A Parallel Open-Label Trial to Evaluate Microbial Cellulose Wound Dressing in the Treatment of Diabetic Foot Ulcers. *Int. Wound J.* **2011**, *8*, 69–73. [CrossRef] [PubMed]
83. Dini, V.; Romanelli, M.; Andriessen, A.; Barbanera, S.; Bertone, M.S.; Brilli, C.; Abel, M. Improvement of Periulcer Skin Condition in Venous Leg Ulcer Patients: Prospective, Randomized, Controlled, Single-Blinded Clinical Trial Comparing a Biosynthetic Cellulose Dressing with a Foam Dressing. *Adv. Ski. Wound Care* **2013**, *26*, 8. [CrossRef]

84. Cavalcanti, L.M.; Pinto, F.C.M.; Oliveira, G.M.D.; Lima, S.V.C.; Aguiar, J.L.D.A.; Lins, E.M. Efficacy of Bacterial Cellulose Membrane for the Treatment of Lower Limbs Chronic Varicose Ulcers: A Randomized and Controlled Trial. *Rev. Colégio Bras. Cir.* **2017**, *44*, 72–80. [CrossRef] [PubMed]
85. Colenci, R.; Miot, H.A.; Marques, M.E.A.; Schmitt, J.V.; Basmaji, P.; Jacinto, J. dos S.; Abbade, L.P.F. Cellulose Biomembrane versus Collagenase Dressing for the Treatment of Chronic Venous Ulcers: A Randomized, Controlled Clinical Trial. *Eur. J. Dermatol.* **2019**, *29*, 387–395. [CrossRef]
86. Maia, A.L.; Lins, E.M.; Aguiar, J.L.A.; Pinto, F.C.M.; Rocha, F.A.; Batista, L.L.; Fernandes, W.R. de M.A. Curativo Com Filme e Gel de Biopolímero de Celulose Bacteriana No Tratamento de Feridas Isquêmicas Após Revascularização de Membros Inferiores. *Rev. Colégio Bras. Cir.* **2019**, *46*, e20192260. [CrossRef]
87. Piatkowski, A.; Drummer, N.; Andriessen, A.; Ulrich, D.; Pallua, N. Randomized Controlled Single Center Study Comparing a Polyhexanide Containing Bio-Cellulose Dressing with Silver Sulfadiazine Cream in Partial-Thickness Dermal Burns. *Burns* **2011**, *37*, 800–804. [CrossRef]
88. Aboelnaga, A.; Elmasry, M.; Adly, O.A.; Elbadawy, M.A.; Abbas, A.H.; Abdelrahman, I.; Salah, O.; Steinvall, I. Microbial Cellulose Dressing Compared with Silver Sulphadiazine for the Treatment of Partial Thickness Burns: A Prospective, Randomised, Clinical Trial. *Burns* **2018**, *44*, 1982–1988. [CrossRef]
89. delli Santi, G.; Borgognone, A. The Use of Epiprotect®, an Advanced Wound Dressing, to Heal Paediatric Patients with Burns: A Pilot Study. *Burns Open* **2019**, *3*, 103–107. [CrossRef]
90. Karlsson, M.; Olofsson, P.; Steinvall, I.; Sjöberg, F.; Thorfinn, J.; Elmasry, M. Three Years' Experience of a Novel Biosynthetic Cellulose Dressing in Burns. *Adv. Wound Care* **2019**, *8*, 71–76. [CrossRef] [PubMed]
91. Shanks, L.A.; Cronshaw, A.; Alexander, K.S.; Davies, J.A.; O'Boyle, C.P. Evaluation of EpiProtect®Microbial Cellulose Burns Dressings in Young Children. *Scars Burns Heal.* **2020**, *6*, 205951312094050. [CrossRef] [PubMed]

pharmaceuticals

Article

Spectroscopy Approach for Highly-Efficient Screening of Lectin-Ligand Interactions in Application for Mannose Receptor and Molecular Containers for Antibacterial Drugs

Igor D. Zlotnikov and Elena V. Kudryashova *

Department of Chemistry, M.V. Lomonosov Moscow State University, Leninskie Gory, 1/11B, 119991 Moscow, Russia; zlotnikovid@my.msu.ru

* Correspondence: helena_koudriachova@hotmail.com

Abstract: Rational search of a ligand for a specific receptor is a cornerstone of a typical drug discovery process. However, to make it more "rational" one would appreciate having detailed information on the functional groups involved in ligand-receptor interaction. Typically, the 3D structure of a ligand-receptor complex can be built on the basis of time-consuming X-ray crystallography data. Here, a combination of FTIR and fluorescence methods, together with appropriate processing, yields valuable information about the functional groups of both the ligand and receptor involved in the interaction, with the simplicity of conventional spectrophotometry. We have synthesized the "molecular containers" based on cyclodextrins, polyethyleneimines (PEI) or spermine with mannose-rich side-chains of different molecular architecture (reticulated, star-shaped and branched) with variable parameters to facilitate delivery to alveolar macrophages. We have shown that synthetic mannose-rich conjugates are highly affine to the model mannose receptor ConA: $K_d \approx 10^{-5}$–10^{-7} M vs. natural ligand trimannoside (10^{-5} M). Further, it was shown that molecular containers effectively load levofloxacin (dissociation constants are $5{\cdot}10^{-4}$–$5{\cdot}10^{-6}$ M) and the eugenol adjuvant (up to 15–80 drug molecules for each conjugate molecule) by including them in the cyclodextrins cavities, as well as by interacting with polymer chains. Promising formulations of levofloxacin and its enhancer (eugenol) in star-shaped and polymer conjugates of high capacity were obtained. UV spectroscopy demonstrated a doubling of the release time of levofloxacin into the external solution from the complexes with conjugates, and the effective action time (time of 80% release) was increased from 0.5 to 20–70 h. The synergy effect of antibacterial activity of levofloxacin and its adjuvants eugenol and apiol on *Escherichia coli* was demonstrated: the minimum effective concentration of the antibiotic was approximately halved.

Keywords: targeted delivery; levofloxacin; eugenol; double inclusion complex; cyclodextrin; molecular container; synergism; allylpolyalkoxybenzenes

Citation: Zlotnikov, I.D.; Kudryashova, E.V. Spectroscopy Approach for Highly-Efficient Screening of Lectin-Ligand Interactions in Application for Mannose Receptor and Molecular Containers for Antibacterial Drugs. *Pharmaceuticals* **2022**, *15*, 625. https://doi.org/10.3390/ph15050625

Academic Editors: Federica Sodano and Maria Grazia Rimoli

Received: 9 April 2022
Accepted: 17 May 2022
Published: 19 May 2022

1. Introduction

Ligand-receptor interactions play a significant role in the functioning of many biological systems [1], especially the immune system [2]. Regulation of the complex formation processes of proteins with bioactive agents can be considered as a potential approach to targeted impact on tissues and cells. There is widespread application in medical biochemistry of linking high-affinity ligands to receptors, for example, galactosylated drug carriers to the asialoglycoprotein receptor [3], or drug carriers functionalized by monoclonal antibodies to the E-selectin [4,5], which is expressed by endothelial cells during an acute or chronic inflammatory reaction. This is due to the high therapeutic potential of these systems in the treatment of a number of diseases, including oncological, neurodegenerative, autoimmune diseases [6].

Macrophages as key cells of the immune system are a potential target for the delivery of antibacterial drugs or agents capable of regulating the activation status of macrophages, depending on the treatment objectives [7,8]. Macrophages express a large number of receptors [9], including Toll-like and scavenger receptors, integrins and the most significant in our research field, C-type lectin domains (CTLDs). Clustering of 8 CTLDs provides specific recognition of pathogens by mannose receptors, including CD206 [10]. CD206 is a transmembrane protein (175 kDa) that recognizes Man, Fuc and GlcNAc residues of oligosaccharides of the cell wall of pathogenic microorganisms [11]. Carbohydrate binding occurs due to the C-type region consisting of 8 domains; however, the greatest contribution to the ability to bind carbohydrates in isolation from other parts is shown only for the fourth carbohydrate–recognizing domain (CRD4) [12–14]. Therefore, determining the parameters of a carbohydrate ligand for high-affinity interaction with CD206 or its domains is an urgent biochemical issue. Various methods of studying receptor-ligand interactions on the example of the mannose receptor are presented in the literature: X-ray crystallography of trimmed CRD4 [13], oligosaccharide binding competition assays [13–16], NMR shifts [15,16]. However, these methods are not express and require strict validation of the experimental methodology. In addition, due to the limited availability of the CD206 receptor [13], the range of applicable methods is limited; also high throughput screening analysis of ligands is not available. Accordingly, in such cases, a model lectin ConA can be used, which is more suitable for screening purposes. The relevance of ConA as a CD206 model has been widely discussed in the literature and studied by us through computer modeling of 15 ligands [17]. We have previously shown the similarity of the structural organization of the binding sites in ConA and in CD206 (CRD4) with high correlation of the values of the free energies of complexation ($r > 0.9$).

ConA is an easily accessible lectin having a carbohydrate-binding domain structure similar to CD206 and a general similarity in carbohydrate-binding ability [8,17–21]. According to the literature, the CD206 domain shows a high similarity with mannan binding lectin A, which, in turn, is similar to ConA in terms of carbohydrate binding [22]. This is confirmed by numerous in vivo and in vitro studies [20]. Mannose-containing ligands, pre-selected by ConA assay, show high affinity (by flow cytometry) to CD206+ cells [23,24]. Similarly, mannose-containing ligands, pre-selected by ConA assay, are absorbed by macrophages via a receptor-mediated internalization mechanism [7], and are bio-distributed to lungs due to adsorption by macrophages [25]. All this data supports the relevance of using ConA as a mannose receptor model.

A wide range of methods can be applied to the study of ConA–ligand interactions [8]: isothermal titration calorimetry [21,26], fluorescence methods [18,27], FTIR spectroscopy [28], Landsteiner inhibition method [29]. In silico methods are no less valuable and highly informative [30,31].

However, the known techniques have good reliability for mono- and oligosaccharides. Moreover, in the case of polymer molecules or their conjugates, the study of receptor-ligand interactions is quite a difficult task, due to the complex 3D architecture and interpretation of data, for example, in isothermal calorimetry effect of dissolution of the polymer system itself. It is necessary to reproducibly and reliably determine the parameters of conjugate (ligand) interaction with receptors with minimal consumption of proteins and ligands. A wide range of methods are described in the literature, but not all can be applied to the systems under consideration.

The proposed spectral approaches (IR, fluorescence) make it possible to increase sensitivity to the concentrations of the receptor and ligands, which means to reduce the consumption of expensive reagents, and, more importantly, to obtain valuable information about the mechanism of complexation (microenvironment of amino acid residues, protonation of carboxyl groups, formation of hydrogen bonds and electrostatic interactions).

The methodology for investigating the effectiveness of lectin-ligand interactions has been optimized here in the application of newly developed highly-mannosylated multifunctional conjugates for targeting macrophages based on cyclodextrin with polyethylenimine or spermine and mannose. Moreover, we have been studied the drug loading and release depending on the nature of the ligand, molecular weight, degree of branching, molecular architecture—using fluorescence quenching and polarization, UV and Fourier-transform infrared spectroscopy (FTIR).

Cyclodextrins are introduced into the conjugate so that the system is able to form inclusion complexes with biologically active substances and drugs [32–35]. At the same time, cyclodextrins protect them from oxidation, hydrolysis, enzymatic destruction, excessive hygroscopicity [36], and, in addition, cause increased penetration of drugs through biological membranes, which is extremely important in the field of drug delivery [37]. To achieve an additional effect, cyclodextrins were modified with spermine molecules or attached to the polymer chain of PEI, which is often used for targeted delivery [38,39]. This brings the positive charge of the conjugate, and the ability to hold the therapeutic agent more firmly and release it in a prolonged manner due to the polymer mesh [40]. In addition, the presence of amino groups allows modification of mannose for high-affinity interaction of ligands with mannose receptors.

Currently, a promising direction in the treatment of lung infections is research related to the use of combined antibacterial drugs. Thus, the effectiveness of chemotherapy in people with severe lung infections (primarily tuberculosis) is due to the use of new generation drugs or adjuvants (enhancers) with a fundamentally different mechanism of action, for example Bedaquiline (Sirturo, Janssen Biotech, Inc., Philadelphia, PA, USA) in combination with the main antibacterial agent (such as levofloxacin) [41]. It is assumed that various adjuvants can increase the binding strength of levofloxacin to bacterial DNA gyrase, as showed for ginsenoside [42].

We chose the model fluoroquinolone [37,40,43–45], levofloxacin and the promising adjuvant eugenol (clove oil) [46–48], as expected, which showed the effect of synergism, which would reduce the dosage of the main "poisonous" drug [49–51]. For these "cargos" to function, it is necessary to dissolve them by loading cyclodextrin into the tori and deliver them to the alveolar macrophages of the lungs (for example, intranasal or inhalation forms).

Thus, in this work: (1) we synthesized promising drug delivery systems with high affinity to mannose receptor macrophages using alternative methods for obtaining mannose clusters instead of very expensive oligomannosides, (2) determined the parameters for the inclusion of levofloxacin and eugenol drugs, (3) presented a combined formulation of these drugs, and (4) developed effective spectral methods for studying receptor-ligand interactions.

2. Results and Discussion

2.1. HPCD-PEI-Man & HPCD-Spermine-Man Conjugates Synthesis and Characterization

The main goal of the work is to develop an optimal molecular container for targeted delivery of levofloxacin to mannose receptors of macrophages and to work out the spectral approach for high-throughput screening of lectin-ligand interactions using concanavalin A as a model lectin. Using different ratios of components, varying the conditions and the reaction sequence, we obtained 9 mannosylated HPCD conjugates with different molecular architecture (Table 1), with spermine as spacer for mannose clusters (high affinity for CD206) or with PEI of various molecular weights and degrees of branching and cyclodextrins' tori for drug loading—according to the schemes shown in Figures 1 and S1–S6.

Such variability of the obtained products is very promising in the field of synthesis of targeted drug delivery systems with high affinity for mannose receptors. Moreover, it is possible to adjust the capacity for loading the drug due to the amount of HPCD or polymer chains, vary the physico-chemical parameters (Table 1: particle size, hydrophilicity, charges, zeta potential, degree of mannosylation) to optimize the platform for a specific task.

All synthesis started with the activation of secondary hydroxyl groups of cyclodextrin with CDI followed by cyclodextrins' modification with the primary and secondary amines of spermine or PEI (Figures 1a,b and S1–S6) based on the reactions described in works [52–56] with author's modifications. Finally, reductive mannosylation of amino groups has been performed. The successful formation of conjugates was demonstrated by Fourier transform infrared (FTIR) spectroscopy, as shown in Figures 2 and S7–S10. Let us consider the characterization of the structure by the example of a conjugate HPCD-spermine-Man-**7**. In the HPCD spectrum (Figure 2a), characteristic strong absorption bands of valence vibrations of C–O and C–O–C bonds 1030–1150 cm^{-1} are observed After modification by spermine (Figure 2b), hydroxypropyl groups become cross-linked with carbonyl amide bond –O–C(=O)–NH–. Thus, corresponding peaks become narrower and more intense (Figure 2c—1046 and 1085 cm^{-1}) due to relative intensification of the peak of C–O bonds in the glucopyranose residues of CD themselves and also due to the C-N-C bonds of spermine molecules. In the region of 2800–3000 cm^{-1}, characteristic sharp peaks of symmetric (2855 cm^{-1}) and antisymmetric (2926 cm^{-1}) valence vibrations of CH_2 groups in spermine (Figure 2b) are superimposed on a wide absorption band of valence vibrations of charged amines NH^+, $NH_2{}^+$, $NH_3{}^+$ from spermine. The most important intense peak in the IR spectrum at 1630–1650 cm^{-1} corresponds to C=O oscillations in urethane groups –O–C(=O)–NH– formed as a result of crosslinking of OH groups of CD and NH groups of spermine with carbonyldiimidazole. The spectrum of the mannosylated conjugate (Figure 2d) is characterized by reducing the absorption of vibrations of amino groups in the region of 2700–3000 cm^{-1}, due to their modification and "fixing" the mobile hydrogen atoms; by broadening and changing the absorption form of C-O-C and C–O valence vibrations due to the introduction of mannose into the system; and by overlapping the peaks of C-N-C and C-O-C bonds. Determination of the ratio mannose:cyclodextrin was carried out by decomposing a section of the spectrum 980–1130 cm^{-1} into components, using FTIR spectra of the constituents (Supplement Figure S7). The degree of mannosylation in conjugates determined by FTIR spectroscopy is in good correlation with the data obtained by visible spectroscopy (by titration of amino groups in conjugates with TNBS) as well as by fluorimetric titration of amino groups with OPA (Figure S25).

Table 1. Physical-chemical properties of obtained β-cyclodextrin conjugates.

Sample's Designation *	Hydrodynamic Diameter (by DLS) **, nm	M_r, kDa	Molar Ratio HPCD: Spermine (or PEI): Man	Average Amount of Mannose Residues per Molecule	Degree of Mannosylation	ζ-Potential **, mV	References to Synthesis Schemes and FTIR/NMR Spectra
HPCD-spermine-Man-**1**	336 ± 68	6	7.4:1:6.2	6	95 ± 3	−2.7 ± 0.8	Supplement Figures S1, S7 and S8
HPCD-spermine-Man-**2**	523 ± 106	9	5:1:6.9	7	98 ± 2	−4.1 ± 1.9	Supplement Figures S1 and S8
HPCD-spermine-Man-**3**	781 ± 205	8.5	1.5:1:11	22	92 ± 3	−4.9 ± 0.1	Supplement Figures S2 and S8
HPCD-PEI35-Man-**4**	52.8 ± 1.3	42	1:1:34.5	35	77 ± 3	5.1 ± 1.4	Supplement Figures S3 and S9
HPCD-PEI35-Man-**5**	62.3 ± 0.7	50	8.6:1:83	83	94 ± 4	−9.4 ± 2.1	Supplement Figures S4 and S9
HPCD-PEI35-Man-**6**	59 ± 7	115	1:2:271	271	80 ± 4	19.0 ± 2.5	Figure 2h, Figures S5 and S9
HPCD-spermine-Man-**7**	152.2 ± 1.0	8	2:9:21	21	85 ± 5	−6.0 ± 0.4	Figures 1a and 2
HPCD-PEI1.8-Man-**8**	66.9 ± 0.3	20	1:6.1:45	45	83 ± 4	−2.0 ± 0.1	Figure 2f, Figures S6 and S10
HPCD-PEI10-Man-**9**	373 ± 108	28	2:1:154	154	99 ± 1	4.8 ± 1.0	Figure 1b, Figure 2e, Figure S10

* The last number in the name is the serial number of the sample, and the number after PEI is the average M_n molar mass of the polymer. ** SD for $n = 3$ is given.

The proposed conjugate's structures and molecular weights are identified on the basis of conditions and sequence of synthesis reactions, the ratio of components, IR spectra, data on the number of mannosylated amino groups (Table 1). Conjugates **1–3** (Supplement Figures S1, S2, S7 and S8) are oligomeric cyclodextrin's meshes with mannose clusters; they have low-negative zeta-potential (−3–4 mV, because charged groups are virtually absent) and a high hydrodynamic diameter from 330 to 780 nm in accordance with the increase in molecular weight from 6 to 9 kDa and most significantly—the number of mannose clusters (mannose molecules from one spermine molecule) on spermine spacers from 3–5 to 10–15. This size (780 $_{HM}$) includes a shell of water around the conjugate, so for "sugar" conjugates, the size turns out to be large (as earlier reported for mannan 530–720 nm [57]). Products **1–2** contain linear and pyranose mannose residues that have been attached to activated OH groups of cyclodextrin. Product **3** is a highly mannosylated HPCD mesh, which was synthesized using a significant excess of CDI (40 eq.), which prevented crosslinking of HPCD tori. Conjugates **4–6** (Table 1 and Supplement Figures S3–S5, S9) of PEI35-series are polymer chains grafted with cyclodextrins, compacted with medium hydrodynamic diameter of about 55 nm. Ligands **4** and **6** have a positive zeta-potential due to the amino group of

PEI35, but in conjugate **5** there are practically no free amino groups; moreover, a negative charge is possible due to the presence of COO^- groups on CD remaining after activation with carbonyldiimidazole, which has less effect on other conjugates with a smaller number of CD. In addition, we obtained star-shaped and probably highly hydrated (150 and 67 nm) conjugates **7–8** and huge branched polymer **9** (Table 1 and Figures 1 and S10): in the center there is one or more HPCD's tori with spermine (4.5 eq) attached to them, or branched PEI 1.8 and 10 kDa–6.1 eq and 0.5 eq. respectively. For conjugates **7–9**, zeta-potential close to 0 corresponds to an almost complete modification of the amino groups of PEI or spermine and effective crosslinking (there are no non-reacted COO^-).

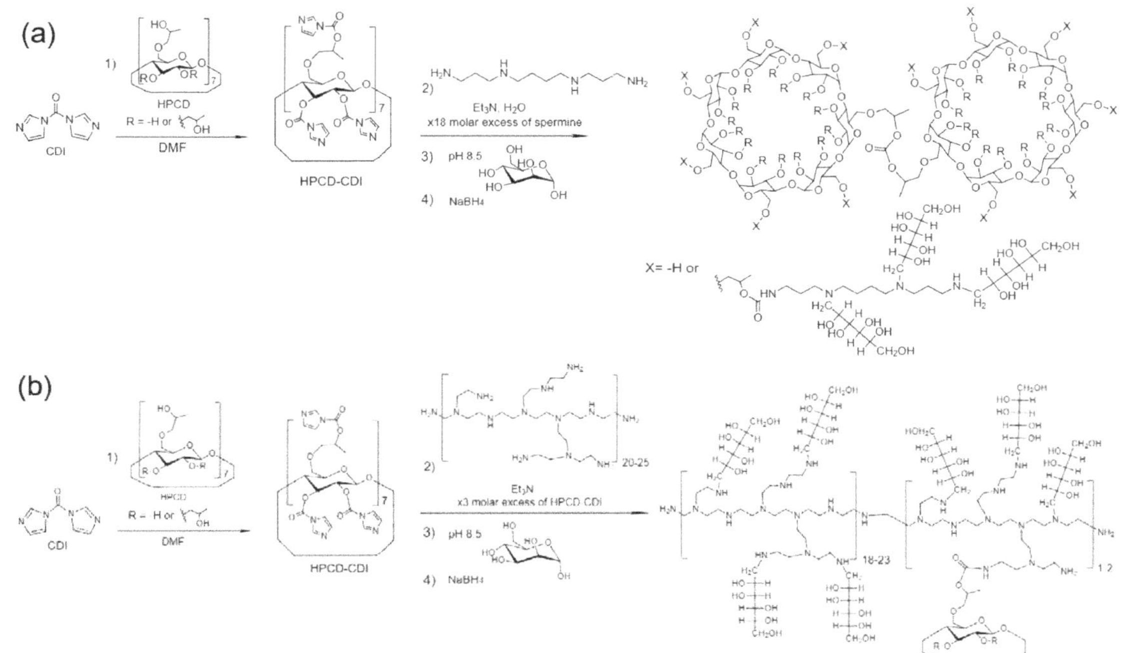

Figure 1. (**a**) The scheme of synthesis of mannosylated star-shaped HPCD-spermine-Man-**7** conjugate. The characteristics of the synthesized ligand are given in Table 1. The number of substituents X, not hydrogen atoms, is on average 9. (**b**) The scheme of synthesis of mannosylated HPCD-PEI10-Man-**9** conjugate. The characteristics of the synthesized ligand are given in Table 1. On average, one branched mannosylated chain of PEI accounts for 1–2 cyclodextrins.

The obtained data on the structure of the synthesized conjugates were confirmed by NMR spectroscopy. In the 1H NMR spectra (Figure 2e–h) of conjugates, both peaks of the initial components and new signals related to urethane bonds −N(H)−C(O) (formed as a result of the interaction of activated HPCD with spermine or PEI) were detected [58,59]. The signals in the range of 4.0–3.2 ppm and the singlet of 5.09 ppm correspond to H2–H5 and H1 of the D-glucopyranose or D-mannose link of the oligosaccharide, respectively [24]. The presence of spermine in conjugate **7** is confirmed by a 1.0–1.1 ppm doublet and 1.78 and 2.11 pm singlets in the 1H NMR spectrum (Figure 2g). The presence of PEI and the formation of urethane bonds—HFCD-CH2-O-C(O)-NH-CH2- in conjugates is confirmed by the signals in the range of 2.8–3.0 pm [40,54]. The number of urethane bonds in terms of one cyclodextrin torus is largest for conjugates **5**, **6** and **8**, and least for conjugate **9**, which is consistent with the data on the ratio of components determined by IR spectroscopy.

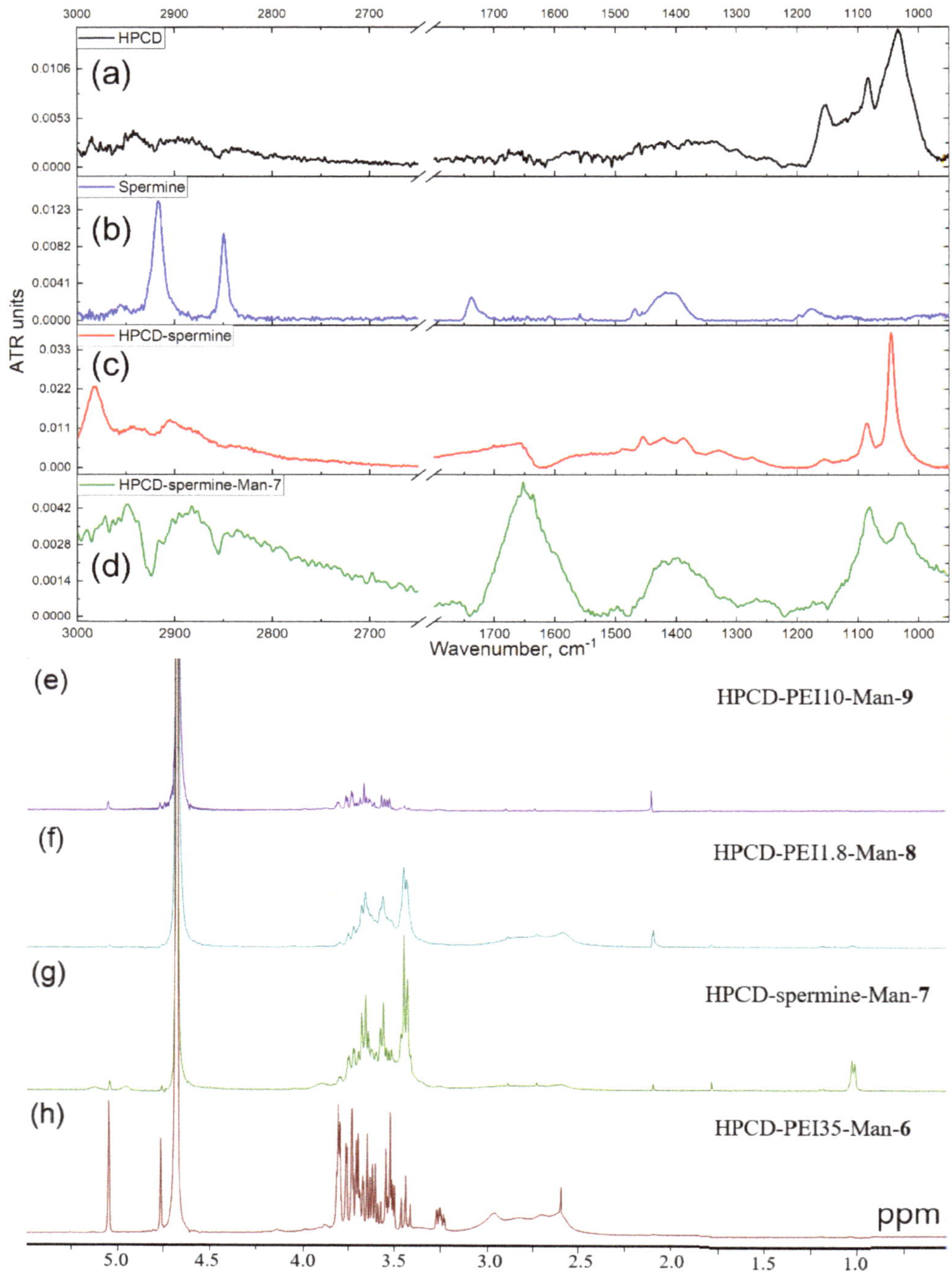

Figure 2. Fourier transform infrared spectra of (**a**) HPCD, (**b**) spermine, (**c**) HPCD-spermine conjugate, (**d**) mannosylated HPCD-spermine-Man-**7**. ^{1}H NMR spectra of conjugates: (**e**) HPCD-PEI10-Man-**9**, (**f**) HPCD-PEI1.8-Man-**8**, (**g**) HPCD-spermine-Man-**7**, (**h**) HPCD-PEI35-Man-**6**. D_2O, 12 mg/mL 400 MHz.

2.2. HPCD-PEI-Man & HPCD-Spermine-Man Conjugates Binding with a Model Mannose Receptor ConA

The requirements imposed on the drug carrier include high affinity for the mannose receptor of macrophages. Relevant data on the affinity for mannose receptors of CD206 macrophages can be obtained using model experiments with lectin concanavalin A, because of a high correlation of carbohydrate binding capacity with the fourth domain (CRD4) of CD206 as was shown by in silico methods [17]. Therefore, it is necessary to have a robust, reproducible and accessible method for studying protein-receptor binding [8,18], which is generally complicated for structurally sophisticated ligands. A small number of techniques for similar systems are described in the literature: isothermal calorimetry of titration on complexation of mannosylated multivalent carbohydrates [26] or cyclodextrins with mannose clusters [60] with ConA, with dependence of hemagglutination inhibition on neoglycopolymer length [61]. In this paper, a promising approach has been suggested to study ligand–receptor binding using Fourier infrared spectroscopy (FTIR) and Trp or FITC fluorescence quenching and anisotropy.

2.2.1. Fluorescence Quenching and Anisotropy

We proposed a technique allowing to reliably determine the efficiency of binding without the introduction of fluorescent labels into the ligand (as described in the literature [23,53]). The technique is based on the fluorescence quenching of tryptophan residues in a protein when binding to a ligand [18], as well as an increase in the anisotropy of Trp fluorescence due to an increase in the molar mass of the ConA–ligand complex in comparison with an unbound receptor. The requirement for precise and correct determination of dissociation constants is a ligand mass (for fluorescence anisotropy) of at least 10 kDa (about 20% of dimeric ConA 53 kDa). An alternative approach that increases sensitivity and removes the restriction on molecular weight is possible—the introduction of a fluorescent label (for example FITC) into a conjugate. Data of a model experiment with labeled cyclodextrin are shown—Figure S11. However, for the systems under consideration, FITC (and other fluorescent molecules) can form an inclusion complex with neighboring CD tori, which can affect the K_d values. This fact complicates the interpretation of the data significantly. Thus, the use of the protein's own fluorescence simplifies the screening of ligand-receptor affinities.

The interaction of "heavy" conjugates (PEI group with molecular weight in the range from 20 to 115 kDa) with ConA was studied by changing the fluorescence anisotropy (r), using the signal of the aromatic tryptophan residue in ConA. The adsorption curves (Figure 3a) are typical saturation lines, most accurately approximated by the Hill (3) and Boltzmann (4) models. The magnitude of the change in r correlates with the molar mass of the ligand (from 50 to 100% of initial anisotropy value of protein). The processing of sorption curves in Hill coordinates (10) leads to an S-shaped dependence (Figure 3b) [62], which can be approximated by a single straight line (short points for the blue curve) to determine the apparent dissociation constant of the ConA-ligand K_{obs} complex or to isolate three linear sections [63] and consider them separately to calculate K_d in several binding modes, to identify positive/negative cooperativeness (dash lines (– –) for red curve in Figure 3b). Thus, for the HPCD-PEI35-Man-**5** ligand, K_{obs} = (60 ± 9) μM, while the maximum affinity for the most "successful spatial configurations" of ligand molecules (for the same ligand) (in the presence of positive cooperativeness [8,18]) increases: K_{max} = (5.0 ± 0.3) μM.

To prove that mannosylated conjugates interact specifically with the model ConA receptor precisely at the mannose binding sites, a competitive complexation of HPCD-PEI35-Man-**5** was carried out in the presence of 0.5 M mannose (Figure 3a). Titration of the receptor solution with a ligand is initially characterized by a decrease in the anisotropy of tryptophan fluorescence due to changes in the microenvironment from a viscous mannose to a conjugate. 0.5 M mannose solution increases the viscosity (viscosity is about 1–1.5 mPa·s) and thereby the anisotropy of Trp in ConA from 0.08 to 0.10. Moreover, the presence of concentrated mannose changes the form of the binding curve to a gentler one, which indicates a competitive mechanism of complexation. However, ligand binding is noticeable only at a threefold molar excess: C_{lig} = 44 μM, C_{Man} = 460 mM, which indicates a high affinity binding of the mannosylated polymer, 3–4 orders of magnitude superior in affinity to mannose (K_d 0.1 μM versus 1 mM, respectively [8]).

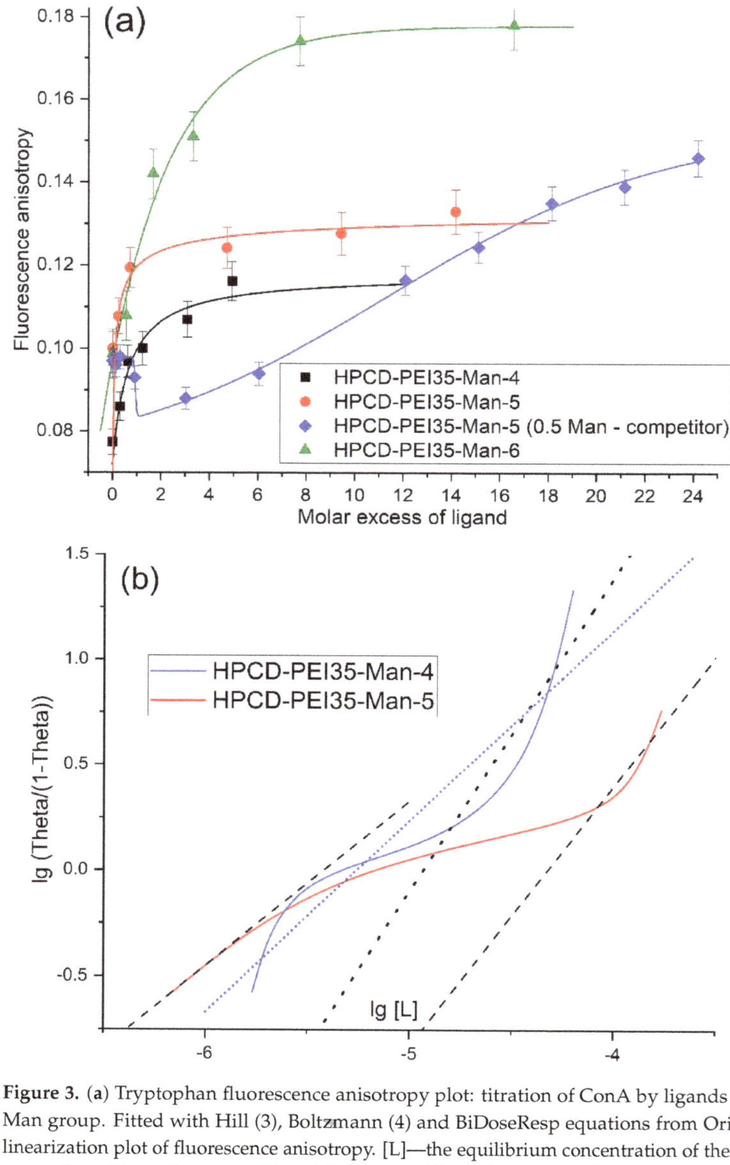

Figure 3. (**a**) Tryptophan fluorescence anisotropy plot: titration of ConA by ligands HPCD-PEI35-Man group. Fitted with Hill (3), Boltzmann (4) and BiDoseResp equations from Origin. (**b**) Hill's linearization plot of fluorescence anisotropy. [L]—the equilibrium concentration of the ligand. Theta-proportion of bound receptor. Two dash lines (– –) for red curve are a linearized model of two types of binding. Two dot lines ("') for blue curve are a linearized model of the most affinity binding and average observed binding (short dots). C_0(dimeric ConA) = 38 μM. Ligand concentration varied from 1 to 250 μM. In the case of competitor binding (Figure S12), we used mannose with [Man] = 0.5 M. Natrium-citric buffer solution (0.04 M, pH 5.5). C(Ca^{2+}) = C(Mn^{2+}) = 1 mM. T = 22 °C.

The interaction of "light" conjugates (spermine group with molecular weight in range from 6 to 9 kDa) with ConA was studied by changing the fluorescence intensity F of the aromatic tryptophan residues (the fluorescence of tyrosine residues under these conditions is significantly less: λ_{max}(Tyr) = 303 nm, λ_{max} (Trp) = 340 nm, λ_{max} (observed) = 335 nm) in ConA (Figure S13 for mannan). The analysis was carried out similarly to the recently described methods for mannosylated chitosans [18]. Despite the fact that quenching the tryptophan fluorescence in a protein is a fast affinity screening method, there is a disadvantage: at high concentrations of the ligand, dynamic factors can play a significant role. Its accounting can be carried out according to the modified Stern–Vollmer equation [18] by highlighting the static and dynamic (K_{st} and K_{din}) components: $F_0/F = 1 + (K_{din} + K_{st})[Q] + K_{din} \cdot K_{st} \cdot [Q]^2$, where F_0 and F are the initial and observed fluorescence intensities of the ConA-ligand complex, respectively, and [Q] is the equilibrium concentration of the fluorescence quencher (conjugate). According to this equation, it was possible to determine the dynamic contribution of tryptophan fluorescence quenching in ConA by mannose molecules (0.5 M) in the case of competitive binding of the HPCD-PEI35-Man-5 ligand (Figure S12): K_{st} = 3060, K_{din} = 0.06—superiority in affinity of the polymer ligand by 4–5 orders of magnitude.

So, the methods of polarization and fluorescence quenching are applicable to the studied systems to determine the affinity of the ligand to the receptor, and moreover, as will be shown later, to study the interaction of molecular containers with drugs.

2.2.2. Fourier-Transform IR Spectroscopy (FTIR)

FTIR is highly informative in relation to the studied ligand-receptor interaction, because it gives information about the spatial arrangement polymeric chains, the changes in the protein tertiary and secondary structure, the microenvironment of certain functional groups or the formation of non-covalent interactions (van der Waals, electrostatic, hydrophobic) by changing the following characteristic regions (both in proteins and polymers): amide A (valence vibrations of N–H bond atoms), amide B, amide I (contribution of valence vibrations of C=O bonds—80%; and C–N bonds—about 15%), amide II (main contribution of deformation vibrations of N–H bonds (80%) and minor contribution of valence vibrations of C–N bonds—about 15%) and amide III (the fallback option is difficult to interpret and overlaps with other peaks).

Recently we studied in silico the detailed mechanism of these interactions [17] (Figure S14), using the interaction of trimannoside with ConA as the example. The "amide" atoms of the binding site are involved in complexation-Asn14 (N)-Man (O3), Asp16 (H)-Man (O5), Arg222 (H)-Man (O2). Thus, changes in the amide II region are expected when the ligand binds to the protein. Moreover, carbonyl oxygen atoms of the Gly-peptide bond are also involved in complexation [17]. Experimental data on changes in the Amide I and Amide II regions (ConA spectrum) upon binding to the ligand HPCD-spermine-Man-**1** (Figure 4a) confirm the formation of point interactions of oxygen atoms of the carbonyl group and hydrogen or nitrogen atoms in the amide with ligand donors/acceptors. The microenvironment of the peptide bond atoms changes (predominantly in the binding site of the ligand), which is reflected in the IR spectrum: the intensity of the peaks of amides I and II decreases (Figure 4a). Computer modeling [17,64] and crystallographic methods [14,65,66] allow us to determine the point interactions of ligand and receptor atoms, the formation of which leads to a change in the energies of certain amino acid residues, which theoretically should be displayed in the IR spectrum. However, the analyzed peaks correspond to a whole set of bonds, and therefore, the target minor change (against the background of a huge number of other peptide bonds) is practically unnoticeable. Previously, it was proposed to normalize the peak of amide II to Amide I and only monitor the intensity [28], but this is not always correct. Therefore, we differentiated the spectra and obtained a more detailed picture in which energy-changing components can be distinguished. The dissociation constants were calculated based on the dependence of the peak position at 155(6–9) or 153(4–6) cm^{-1} in the second-order differential IR spectrum (Figure 4b). Differentiation allows you to select the target component of the analysis (N–H and C=O valence oscillations—1500–1575 and 1620–1650 cm^{-1}, respectively)—in a narrow energy region (1–5 cm^{-1}) and corresponding peptides.

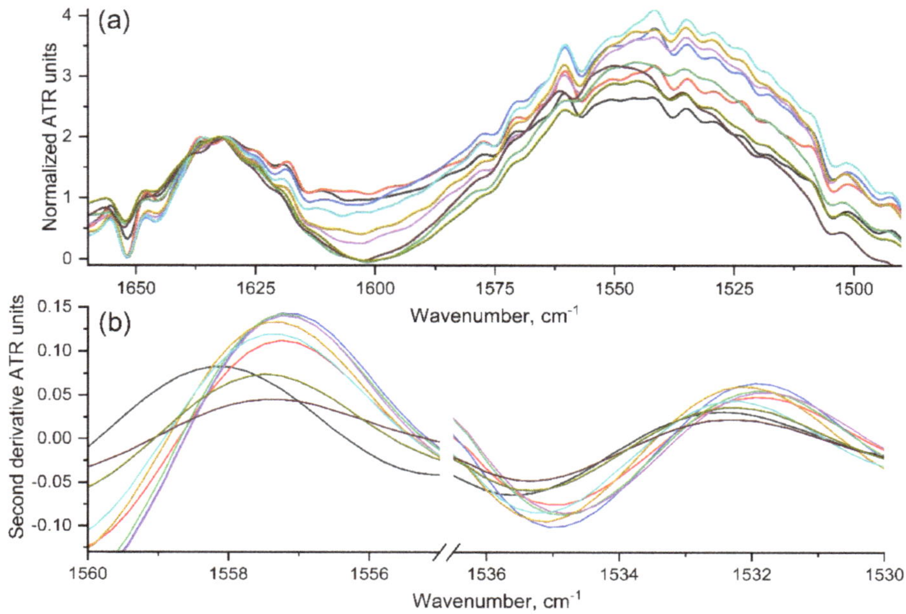

Figure 4. (**a**) Normalized to the amide I band Fourier transform infrared spectra of ConA: titration of ConA by ligand HPCD-spermine-Man-**1**. The sequential addition of the ligand corresponds to the transition from blue to red and black, i.e., from top to bottom. (**b**) Amide II areas in differential IR spectra of the second order. C_0(ConA) = 33 µM. Ligand concentration varied from 90 to 750 µM. Natrium-citric buffer solution (0.04 M, pH 5.5). C(Ca^{2+}) = C(Mn^{2+}) = 1 mM. T = 22 °C.

Thus, we were able to identify shifts in the differential IR spectrum of the second order in the Amide II region at 1558 and 1535, 1532 cm^{-1}. Changes are also observed in other peaks, but the most relevant data are believed to be obtained from the peak at the maximum position, the change in which corresponds to the energy gain during complexation with ligand. Analysis of the 2nd derivative of IR spectra leads to hyperbolic curves (Supplement Figure S15), showing the dependence of the analytical signal on the concentration of the added ligand to the receptor. Linearization is most representative in Hill's coordinates (Figure S16), which, in addition to the dissociation constant, allows us to determine the effect of cooperativeness during ligand binding.

Thus, the approaches discussed above, based on the use of IR spectroscopy and fluorescence analysis methods, make it possible to study the receptor-ligand interaction even for complex conjugates of a wide range of molecular composition, which is difficult when using classical methods such as isothermal calorimetry [21,65]. The following is a discussion of the results of experiments to determine the affinity of ligands with ConA.

2.2.3. The Dependences of Conjugates' Affinity to the Model ConA Receptor on the Parameters of the Molecular Containers

One of the tasks of the work is the selection of optimal parameters (Tables 1 and 2) for drug delivery systems to macrophages by varying the composition, molecular architecture and molar ratio of components of ligands. Therefore, we determined the dissociation constants of conjugate complexes with the ConA mannose receptor (Table 2). The "light" star-shaped conjugates HPCD-spermine-Man-(**3**,**7**) were the most affine in comparison with systems with polymer substituents PEI 1.8, 10 and 35 kDa. With an increase in the molecular weight of the polymer from 1.8 to 35 kDa, the affinity to the mannose receptor increases, provided the optimal configuration of the components (in a series of ligands **6** > **9** > **8**). It is noteworthy that conjugates with PEI are characterized by several binding modes due to the possibility of ligand binding by a low-affinity part (cyclodextrin) and a high-affinity mannose cluster on a spermine or a branch of PEI—K_{max} value. During the experiment, an average pattern is recorded—K_{obs}. Positive cooperativeness is observed for all systems, since in Hill coordinates the slope of the straight line is greater than 1. The degree of cooperativeness in Equation (3) increases with increasing conjugate mass and affinity to the receptor. In the case of a branched conjugate with a PEI 10 kDa (* in Table 2) $n > 3$ is observed, which indicates multivalent ligand binding at several sites of ConA (one dimeric protein or several globules), providing nanomolar total dissociation constants of complexes—supramolecular ensembles of protein and ligand. Let us discuss the influence of molecular architecture on affinity to mannose receptors. The HPCD-spermine-Man-**1** and **2** ligands (Figure S1) are crosslinked meshes of cyclodextrins with mannose residues, K_d of the order of $5 \cdot 10^{-5}$ M; and ligand **3** (Figure S2) of a similar composition, but with a lower degree of crosslinking of CD and mannose clusters on spermines is half an order superior to the high-affinity mannan: $5 \cdot 10^{-7}$ M vs. 1 μM. Moreover, ligand **7** (Figure 1a) with star-shaped architecture, even more affine to mannose receptor: $K_d \approx 2.5 \cdot 10^{-7}$ M. In the series of HPCD-PEI35-Man conjugates (**4–6**): the composition of ligand number **6** (1:2 CD/PEI) is the most optimal from the point of view of binding to the receptor; however, ligand **5** (8:1 CD/PEI) has a large number of CD tours, which is likely to increase loading capacity for the drug, unlike samples **4** and **6**, although it remains affine to mannose receptors: K_d of the order of $5 \cdot 10^{-5}$ M and up to $5 \cdot 10^{-6}$ M—affinity in the optimal mode. Conjugates **8** and **9** with branched PEI are highly affine to ConA and are characterized by a higher relative mass content of CD due to a decrease in the mass of the polymer. Thus, we have obtained conjugates comparable and even superior in affinity to the natural trimannoside ligand to the mannose receptor ConA ($K_d \approx 10$ μM [8,21,28]) and, moreover, superior to the high affinity polymer mannan ($K_d \approx 1$ μM [66]). Thus, the condition of high specificity of drug delivery systems to mannose receptors of macrophages is fulfilled. The physicochemical aspects of the interaction of conjugates with levofloxacin and eugenol, a promising combination of therapeutic agents, are described below.

Table 2. Dissociation constants of ConA–ligand complexes determined by fluorescence methods and FTIR spectroscopy. The conditions are given in the Methods section.

Ligand	**K_d (ConA–Ligand), μM**		
	Quenching of Trp Fluorescence		**FTIR Spectroscopy**
HPCD-spermine-Man-**1**	64 ± 11		50 ± 6
HPCD-spermine-Man-**2**	69 ± 12		60 ± 11
HPCD-spermine-Man-**3**	0.55 ± 0.19		0.49 ± 0.23
HPCD-spermine-Man-**7**	0.24 ± 0.12		0.27 ± 0.08
Ligand	Quenching of Trp fluorescence	Anisotropy of Trp fluorescence	FTIR spectroscopy
HPCD-PEI35-Man-**4**	11 ± 6	K_{obs} =17 ± 8	12 ± 3
HPCD-PEI35-Man-**5**	69 ± 21	K_{obs} = 60 ± 19 K_{max} = 5.0 ± 0.7	57 ± 17
HPCD-PEI35-Man-**6**	K_{obs} = 31 ± 12 K_{max} = 0.10 ± 0.03	K_{obs} = 1.3 ± 0.4 K_{max} = 0.97 ± 0.22	2.8 ± 0.7
HPCD-PEI1.8-Man-**8**	36 ± 13	K_{obs} = 19 ± 6 K_{max} = 2.9 ± 1.2	27 ± 10
HPCD-PEI10-Man-**9**	6.8 ± 0.4	8.1 ± 1.5	55 ± 14 *
Mannan (control)	1.9 ± 0.5	1.6 ± 0.9	1.7 ± 0.5

* Multiple site binding and positive cooperativeness. The value of the $K_{d,n}$ dissociation constant in the Hill's model is given. lg K_d = 13.71, n = 3.22. lg $K_{d,n}$ = (lg K_d)/n.

2.3. *Levofloxacin and the Adjuvant Eugenol Interactions with HPCD-PEI-Man and HPCD-Spermine-Man Conjugates*

Next, experiments are shown to study the affinity of the interaction of conjugates with fluoroquinolones and eugenol by fluorescent methods, FTIR and UV spectroscopy.

2.3.1. Levofloxacin Fluorescence Quenching and Polarization

Fluorescent methods are well suited to studying both protein-ligand interactions and the binding of aromatic molecules to molecular containers. In our case: levofloxacin interacts with mesh HPCD-spermine-Man-**2** and polymer HPCD-PEI35-Man-**4** conjugates (Figure S17). However, the sensitivity and magnitude of the change in the analytical signal increases in comparison tryptophan fluorescence: anisotropy of levofloxacin changed from about 0.02 to 0.15 units; fluorescence quenched from 650 to 250 units. If, in the case of protein, the most convenient and accurate method was polarization analysis, then the interaction of aromatic molecules with conjugates based on cyclodextrins and polyethyleneimines were described more accurately as the quenching of fluorescence.

The mesh conjugate HPCD-spermine-Man-**2** (Figure S1) interacts with 20 levofloxacin molecules maximally (determined by fitting model—see Methods), holding them into the inner cavity of the CD (tori) due to hydrophobic interactions, as well as the interactions with the polymer matrix by non-covalent interactions (hydrogen bonds, electrostatic) contributing to the stability of the complex with levofloxacin.

In the literature, loading of levofloxacin in poly(lactic-co-glycolic acid) and chitosan nanoparticles [67,68], star poly(e-caprolactone) scaffolds [69] was described. The studied conjugates are also able to effectively incorporate drugs. The dissociation constants of Lev – HPCD-spermine-Man-**2** are about 320 μM (three times better than a simple HPCD [70]) with the simultaneous increased capacity of the drug. The cyclodextrin conjugate with a heavy PEI of 35 kDa (Figure S3) retains up to 40–50 levofloxacin molecules ($K_d \approx 21$ μM—more affine than spermine's conjugate) presumably to a greater extent due to their interaction with polymer chains. Note that the quenching of fluorescence, as expected, is expressed by a Hill type dependence (Equation (3)) while, in contrast, the anisotropy increases almost linearly or exponentially (Figure S17)—due to the formation of molecular aggregates of the ligand's molecules.

2.3.2. Eugenol and Levofloxacin: Fourier-Transform IR Spectroscopy Approach

The fluorescent approach to study the interaction of fluoroquinolones with conjugates allows us to determine the dissociation constants and the loading of the drug, but does not provide information about the molecular mechanism of the interaction of an aromatic molecule with the delivery system. As a complementary method for determining the molecular details of the formation of drug inclusion complexes with the studied polymer ligands, the FTIR spectroscopy is proposed, allowing monitoring of the involvement of specific functional groups in ligand interactions on the part of the drug and drug binding sites on the part of ligands. Moreover, IR spectroscopy, unlike most other methods, makes it possible to study the interaction of multicomponent systems due to the large number of analytically significant peaks, which provides information about the state of each component.

Levofloxacin

The parameters of complexation of levofloxacin and the eugenol adjuvant with molecular containers were determined by FTIR in solutions (Figure 5). The most noticeable changes were in the microenvironment of the quinolone structure of levofloxacin (corresponding peaks of aromatic C-C bond fluctuations at 1474 and C-N bond 1398 cm^{-1}). As previously reported [44], Lev (Figure 5a, insert) exists in zwitterionic form in water solutions and consequently the molecular structure of Lev–CD complex with amino and carboxylate groups of Lev outside the CD cavity is stabilized by forming of intermolecular hydrogen bonds with amino group of spermine and hydroxyl groups of cyclodextrin. The appearance of shoulder peaks in the range of 1425–1465 cm^{-1} confirms several variants of the location of levofloxacin in relation to the cyclodextrin cavity and also corresponds to the Lev–spermine interaction (control experimental spectra—Figure S18).

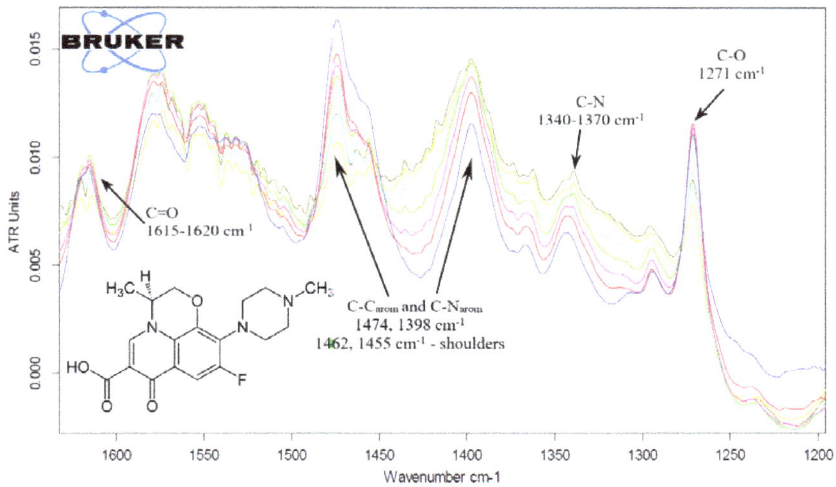

(**a**)

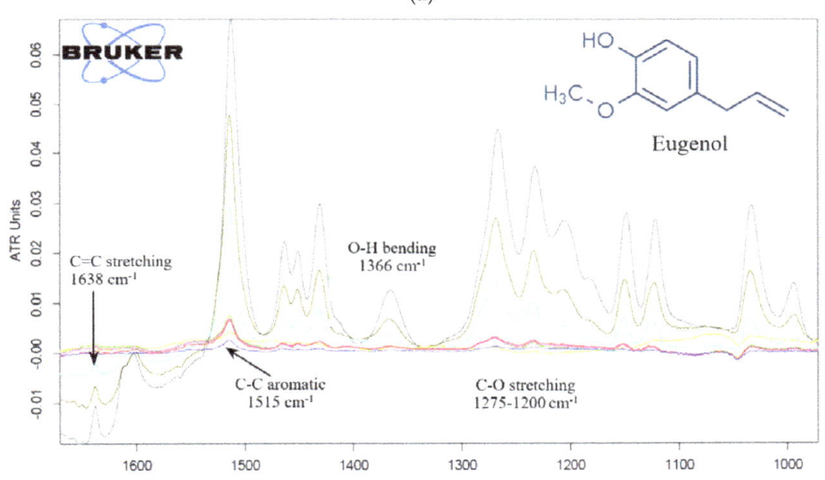

(**b**)

Figure 5. (**a**) Fourier transform infrared spectra of levofloxacin complexes with HPCD-spermine-Man-**1**. C_0(Lev) = 5 mg/mL. 1 mM HCl solution. Molar ratio HPCD tori of ligand:Lev is 0, 0.06, 0.11, 0.23, 0.4, 0.6, 0.93 respectively for gradient spectra from blue to brown. (**b**) FTIR spectra of eugenol complexes with HPCD-PEI35-Man-**5** conjugate. C_0(eugenol in water suspension) = 0.43 mg/mL. The concentration of the ligand per cyclodextrin tori ranged from 8 to 260 μM. T = 22 °C.

During complexation with the HPCD-spermine-Man-**1** molecular container, the intensity of the 1474 cm^{-1} peak decreases by about 40%, and shoulder peaks also appear (Figure 5), which indicates the inclusion of Lev in the oligomeric mesh or CD cavities. The shift of this peak to the region of low wave numbers corresponds to the interaction of Lev with spermine (Figure S18). An increase in intensity of the peak of aromatic C-N bonds at 1398 cm^{-1} is due to the effects of the polymer chains' because formation of "guest-host" complexes Lev with monomeric CD, in contrast, resulted in decreasing of intensity (Figure S19) [70,71].

The dissociation constants of the complexes of molecular containers with levofloxacin were calculated based on changes in the intensity of peaks at 1474 cm^{-1} (Figures 5a, S19 and S20, Table 3), since the most pronounced changes are observed in this area. Constants of the order of 0.5 mM are characteristic for conjugates of the «spermine» series **1–3**: at the same time, mesh ligands **1–2** include Lev more efficiently than a highly mannosylated branched ligand **3**. However, all these conjugates are capable of "loading" more than 10–15 Lev per polymer molecule. Polymer systems **5** and **6** have significantly higher parameters of the loading capacity. In this case, conjugate **5** interacts with levofloxacin 2 orders of magnitude more affine due to the optimal PEI/cyclodextrin ratio of the order of 1:10. The linear polymer can fold into a molecular tangle: therefore, the drug is fixed more firmly in the cavities of the CD: $K_d \approx 5$ μM.

Table 3. Dissociation and equilibrium constants of conjugates' complexes with levofloxacin and eugenol determined by FTIR spectroscopy. Maximum loading capacity (N) of the drug into conjugates. Entrapment efficiency (EE) and loading capacity (LC) of levofloxacin and eugenol by molecular containers. The conditions are given in the Methods section.

	Levofloxacin					
Conjugate	HPCD-spermine-Man-**1**	HPCD-spermine-Man-**2**	HPCD-spermine-Man-**3**	HPCD-spermine-Man-**7**	HPCD-PEI35-Man-**5**	HPCD-PEI35-Man-**6**
$-\lg K_d$	3.89 ± 0.24	3.75 ± 0.17	3.33 ± 0.19	4.46 ± 0.29	5.28 ± 0.41	3.15 ± 0.15
N	14	19	12	15	85	80
EE, %	97 ± 2	96 ± 3	82 ± 3	98 ± 2	98 ± 2	93 ± 3
LC, %	86 ± 5	78 ± 4	52 ± 4	69 ± 4	63 ± 5	26 ± 2
			Eugenol			
Conjugate	HPCD-PEI35-Man-**5**		HPCD-PEI35-Man-**6**	HPCD-spermine-Man-**7**		
K *	47 ± 4		7 ± 2	58 ± 5		
N	6 ± 1 (mostly cyclodextrin inclusion)		7 ± 2 (mostly polymer's interaction)	6 ± 1		
EE, %	83 ± 5		62 ± 6	86 ± 4		
LC, %	2.0 ± 0.1		1.0 ± 0.1	12 ± 1		

* K = [eugenol in complex]/C(cyclodextrin tori of conjugate).

Levofloxacin is included in conjugates with high entrapment efficiency (EE) (up to 98% of the initial amount Table 3), while the load capacity by weight of molecular containers for the drug is also high: mostly in the range from 52 to 86%. The smallest EE is found in the case of heavy polymer molecules. Conjugates **1**, **7** (oligomeric grid CD 6 kDa and star-shaped form, respectively) demonstrate the highest parameters of loading capacity and entrapment efficiency (EE and LC) for both levofloxacin and eugenol. Conjugate **7** combines the valuable properties of a drug container and a high affinity to mannose receptors due to presence of the trimannoside analogue on spermine spacer.

Eugenol

Eugenol (EG) is an oily liquid (ρ = 1.065 g/mL), practically insoluble in water, which does not allow the use of this substance for the preparation and use in aqueous forms of therapy. However, the developed molecular containers form inclusion complexes with aromatic hydrophobic molecules due to the hydrophobic cavities of cyclodextrin, significantly increase the solubility of eugenol and allow the creation of combined enhanced forms of fluoroquinolones. The interaction of eugenol with conjugates based on CD and PEI35 was studied by IR spectroscopy (Figure 5b). When in complex formation with a ligand, eugenol interacts with polymer chains or is included in the CD cavity, which is confirmed by an increase in the intensity of the corresponding characteristic peaks: C=C valence stretching (1638 cm^{-1}), C–C aromatic (1515 cm^{-1}), O–H bending (1366 cm^{-1}), alkyl aryl ether C–O stretching (1200–1275 cm^{-1})—and probably EG passes from an emulsion state to a dissolved form. The dissociation constants of complexes of molecular containers with eugenol were calculated based on changes in the intensity of peaks at 1515 cm^{-1} in the IR spectra (Figures 5b and S21, Table 3). The interaction of the oil with the HPCD-PEI35-Man-**5** conjugate is preferable due to the increased number of CD tori per polymer molecule (approximately 9 vs. 1–2 in the case of ligands **4** and **6**): the difference in equilibrium constants is half an order. However, the star-shaped cyclodextrin conjugate **7** demonstrated even superior EE in comparison with the polymer grafted conjugate **5** and, moreover, an order of magnitude increased LC due to a decrease in the molecular weight of the system. The loading capacity of eugenol is less than in the case of levofloxacin due to lower affinity of eugenol to conjugates. However, the calculated parameters characterize molecular systems as effective drug delivery systems.

Since we have successfully synthesized inclusion complexes of levofloxacin and eugenol in the tori of CD conjugates, then the task is to obtain double complexes of inclusion of both drug molecules at once for possible use as a combined drug in one vial.

2.3.3. Eugenol and Levofloxacin Double Inclusion Complexes in Molecular Containers FTIR Microscopy Approach

We synthesized complexes of inclusion of eugenol and Lev in the polymer conjugate HPCD-PEI35-Man-**5** and the star-shaped conjugate HPCD-spermine-Man-**7**. The characterization of the composition and distribution of the components was carried out by IR microscopy. The distribution of the components of the double complex was studied using IR microscopy, which allows measurements with a spatial resolution of up to 5 microns in order to analyze the inclusions and defects in polymer systems. IR spectra are recorded at several points (40) and peak intensities are integrated at specified intervals, which are displayed in a "heat" map corresponding to the quantity of components.

Thus, the HPCD-PEI35-Man-**5** conjugate uniformly includes aromatic Lev and EG molecules (pink and white colors—Figure S22). Figure 6 shows the distribution of CD (a), levofloxacin (b), eugenol (c) in a double drug complex with HPCD-spermine-Man-**7** conjugate according to the integral intensity of the characteristic peaks of the C–O–C bonds, $C–C_{arom}$ and C=O, $C–C_{arom}$ bonds (d), respectively. Cyclodextrin in the composition of the star-shaped conjugate is presented in large quantities and distributed evenly (pink and white colors in Figure 6a). Levofloxacin has areas of "medium" and "high degree of loading" (Figure 6b). Eugenol competes with levofloxacin when interacting with a conjugate. However, due to the fact that it is in a different phase, the inclusion degree of eugenol is much less (Figure 6c). However, in the regions from yellow and above to pink on the scale, double Lev–EG–conjugate complexes were obtained. Moreover, it is assumed that a very small amount of an adjuvant, for example, eugenol [46–48], presumably acting as a synergist for mechanisms that are being studied, such as blocking mitochondrial dehydrogenase, is required to enhance the antibacterial effect of fluoroquinolones. Thus, synthesized molecular containers with a combined drug are promising in the treatment of a wide range of diseases.

The interaction of levofloxacin with monomeric CD (as a control sample for comparison) and to a greater extent with its conjugates leads to significant changes in the environment of the aromatic system and functional groups (Figure S20). Shifts of characteristic peaks of Lev to the high-frequency region are observed: the peak of carbonyl C=O shifts from 1616 cm^{-1} in simple Lev to 1620 and 1622 cm^{-1} in complexes with CD and HPCD-PEI35-Man-**5** conjugate, respectively. At the same time, the maximum in the spectra of Lev shifts less—up to 1619 cm^{-1}. It is known that Lev is included in the CD's cavity from the side of the carboxyl group [43,70], but only in a protonated form, which is accompanied by the disappearance of a peak at 1580 cm^{-1} corresponding to COO^- due to protonation and immersion in the top of the CD [43,70]. We have discovered the fact that the aromatic peak (C–C 1474 cm^{-1}) in solid-phase spectra is very sensitive to the microenvironment of levofloxacin, with a separation into two components. Similar results were observed earlier for moxifloxacin– HPCD inclusion complex [72]. In our case, peaks at 1452 cm^{-1} and 1492 cm^{-1} indicate two modes of complexation process between Lev and HPCD-PEI35-Man-**5**—thus, conjugates loaded with a drug were obtained. In the IR spectra of dissolved substances (Figures 5a and S19), such splitting clearly does not occur; however, the appearance of shoulder peaks is noticeable, and deconvolution by Gaussians into the same 2 components is possible [72]—Figure S23—it is characterized by an increase in the proportion of the component at 1473 cm^{-1}. Thus, solid-phase spectra complement the spectra of solutions since they exclude hydration.

The dissolution of eugenol in a polymer conjugate proceeds much more efficiently than with a simple CD, and also leads to changes in the IR spectrum (Figure S21), especially in the aromatic peak at 1515 cm^{-1}. The presence of an alkyl-aryl ether peak at 1270 cm^{-1} confirms the presence of eugenol in the double complex.

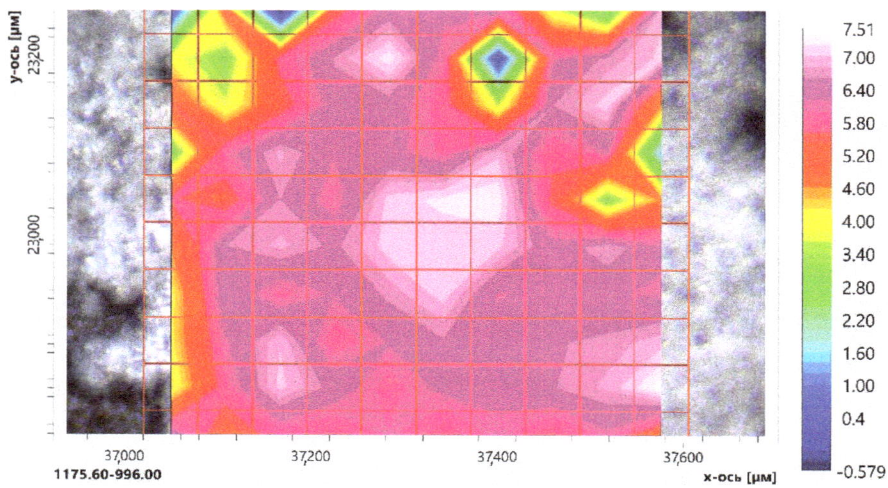

(**a**)

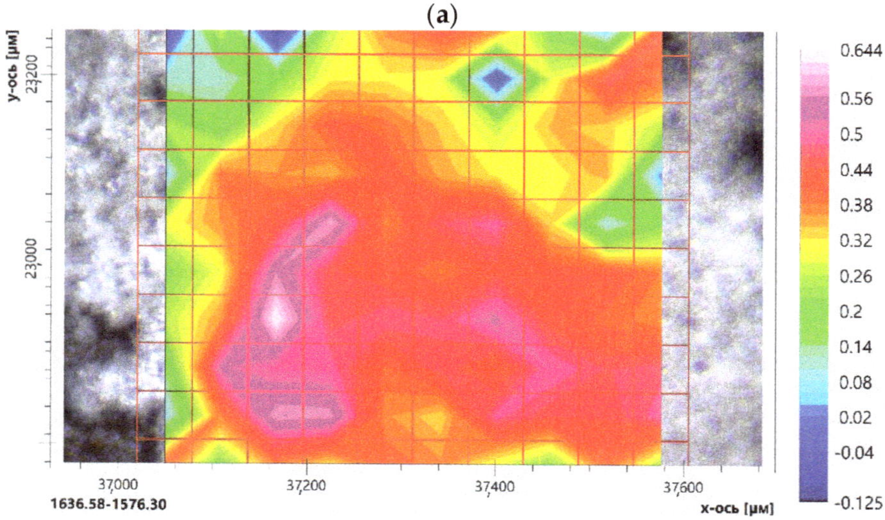

(**b**)

Figure 6. *Cont.*

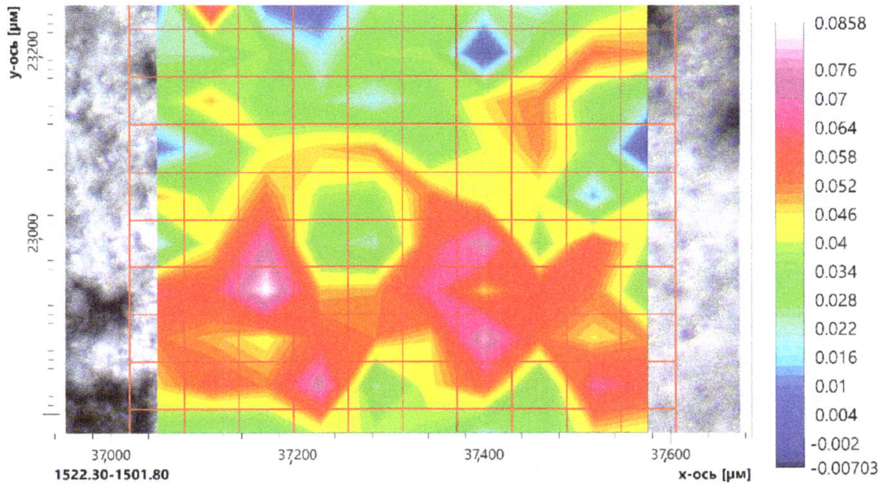

(c)

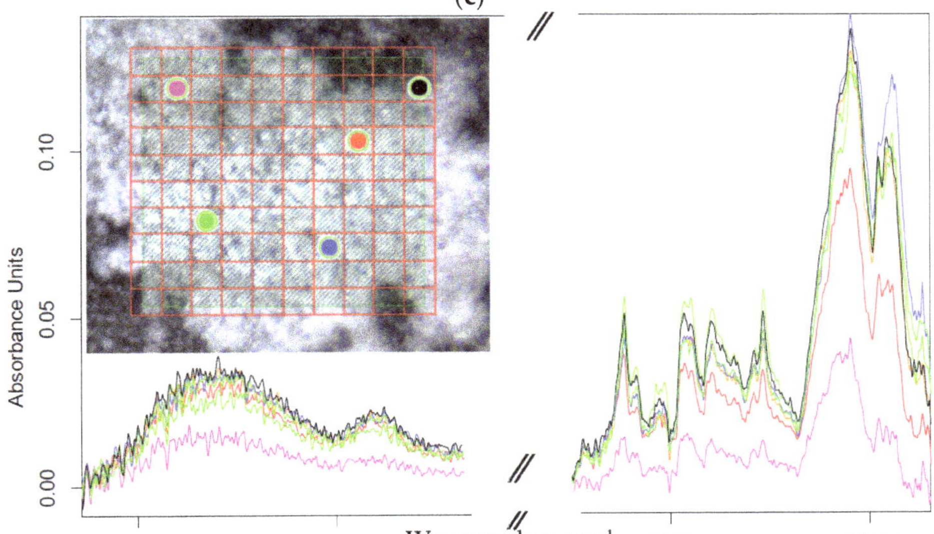

(d)

Figure 6. Integral intensity of three regions in solid phase FTIR spectra of levofloxacin (Lev) and eugenol (EG) double complex with HPCD-spermine-Man-7 conjugate, superimposed on microphotography samples of 1.5 × 0.5 microns in size. (**a**) HPCD region—996–1175 cm^{-1}. (**b**) Levofloxacin region—1576–1636 cm^{-1}. (**c**) Eugenol region—1502–1522 cm^{-1}. (**d**) Solid phase FTIR spectra in 7 points (the same substance and region).

2.3.4. Kinetics of Drug Release from Molecular Containers

Promising approaches to the treatment of diseases, where the driver or direct participants are macrophages, imply targeted and prolonged release of drugs to reduce the dose and frequency of taking the drug. Simple cyclodextrins form inclusion complexes with fluoroquinolones with constants 10^3–10^4 M [72–74], which does not lead to a noticeable effect of prolonged release [75]. Therefore, molecular systems are being developed that can retain drugs with greater affinity, for example, polymer meshes of cross-linked cyclodextrins that increase the effective release of the drug up to 1–5 days [40]. We have synthesized conjugates **1–9** of different molecular architectures and, consequently, with different properties, including drug release parameters. We used levofloxacin and eugenol to study the kinetics of dissociation of its complexes with molecular containers. Figure 7 shows the kinetic curves of the levofloxacin and EG release in free form through a dialysis membrane (12–14 kDa cut-off) into an external solution in comparison with linear conjugate **5** (PEI35 grafted with cyclodextrins), star-shaped conjugates **7–8** based on spermine or light PEI1.8 and CD, as well as CD with branched PEI10 (conjugate **9**) due to the dissociation of the complexes.

Table 4 shows the kinetic parameters of Lev and EG release under the same conditions. Molecular containers provide approximately twice orthrice the slow half-release of drugs. Using the asymptotes, we determined the approximate content of levofloxacin and eugenol in the CD tori and retained by the oligomeric/polymer matrix (Table 4). Levofloxacin, in comparison with eugenol, is included in the cyclodextrin cavity to a greater extent (higher affinity) than entangled in the matrix. Both levofloxacin and eugenol are characterized by a distribution in favor of cyclodextrin tori in the case of star-shaped conjugates, and in favor of a polymer chain for a linear grafted conjugate. Due to the time spent on dissociation of drug complexes with molecular containers, the half-release period can be increased by 2–3 times (Table 4). And the total effective time of action increases significantly from half an hour to tens of hours. The slowest release of Levofloxacin from conjugates is observed for conjugates of **7**, **9**—star CD with grafted chains of spermine or PEI10. For eugenol, the slowest release is observed for conjugates of **8–9**—star CD with grafted chains of PEI1,8 and PEI10. These data correlate with the values of dissociation constants in comparison with monomeric HPCD (10^{-3} M) (Table 3).

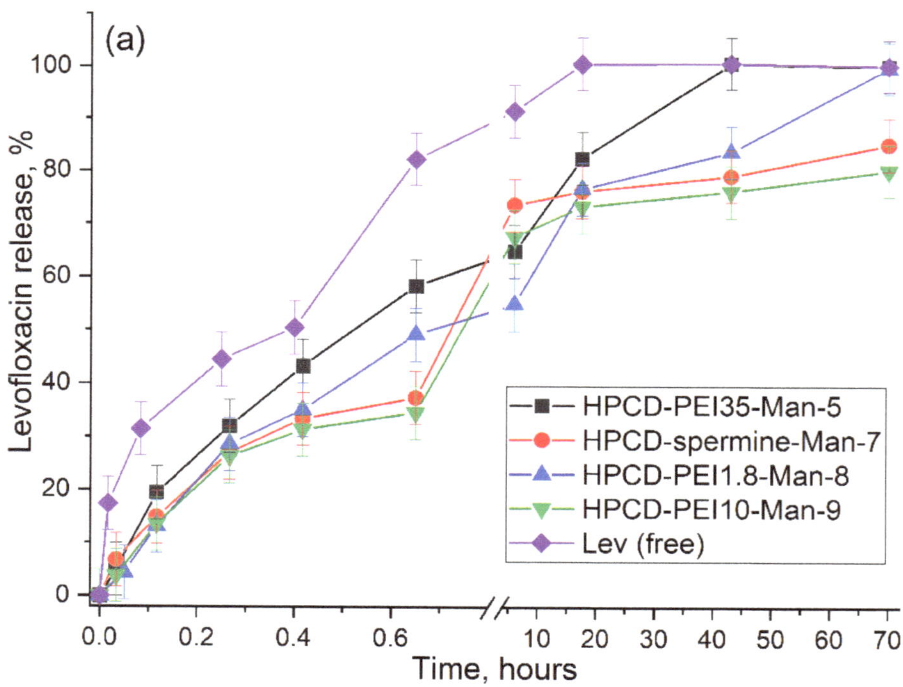

(a)

Figure 7. *Cont.*

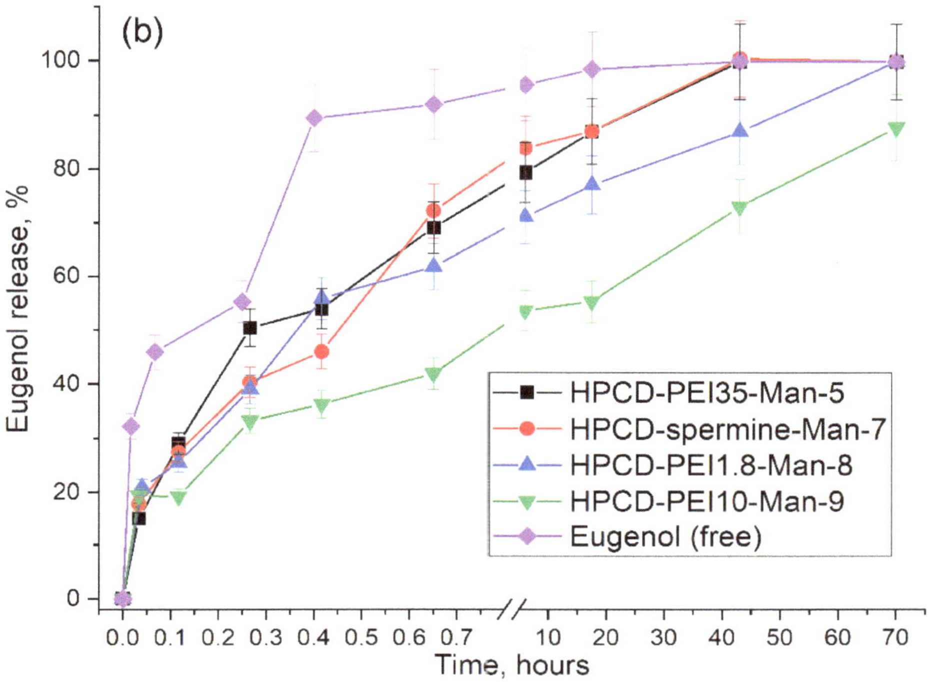

(b)

Figure 7. Release kinetics curves of (**a**) free levofloxacin and levofloxacin complexes (1:1 by mass) with molecular containers **5**, **7–9**; and (**b**) free eugenol and eugenol complexes (1:1 by mass) with molecular containers **5**, **7–9**. Dialysis membrane (12–14 kDa cut-off) into an external solution (1:5 by volumes). PBS (pH = 7.4). Levofloxacin and eugenol are detected by UV-absorption at 287 and 280 nm, correspondingly. C_0(Lev) = 0.2 mg/mL. C_0(eugenol) = 2 mg/mL. T = 37 °C. Values are presented as the mean ± SD of three experiments.

Table 4. Kinetics parameters of release curves of (a) free levofloxacin and levofloxacin complexes (1:1 by mass) with molecular containers **5**, **7–9**; and (b) free eugenol (EG) and eugenol complexes (1:1 by mass) with molecular containers **5**, **7–9**. Dialysis membrane (12–14 kDa cut-off) into an external solution (1:5 by volumes). PBS (pH = 7.4). Levofloxacin and eugenol are detected by UV-absorption at 287 and 280 nm, correspondingly. C_0(Lev) = 0.2 mg/mL. C_0(eugenol) = 2 mg/mL. T = 37 °C.

Conjugate (Sample)	Time of Semi-Release of Lev $\tau_{1/2}$, Min	Time of 80%–Release of Lev $\tau_{80\%}$, Hours	Kinetic Constants, h^{-1}: The Total Process of Dissociation of the Complex and Release through the Membrane. k_{tot}	Kinetic Constants, h^{-1}: Dissociation of the Complex. k_{diss}	The Proportion of Lev or EG Included in Cyclodextrin Tori and Retained by the Polymer Matrix, %
Levofloxacin free	21 ± 3	0.62 ± 0.06	10.4 ± 1.3	-	-
Lev in HPCD-PEI35-Man-**5**	32 ± 2	17 ± 1	1.5 ± 0.2	1.8 ± 0.2	40/60
Lev in HPCD-spermine-Man-**7**	43 ± 3	66 ± 5	2.1 ± 0.3	2.6 ± 0.3	60/40
Lev in HPCD-PEI1.8-Man-**8**	40 ± 2	30 ± 2	0.88 ± 0.07	0.96 ± 0.11	50/50

Table 4. *Cont.*

Conjugate (Sample)	Time of Semi-Release of Lev $\tau_{1/2}$, Min	Time of 80%–Release of Lev $\tau_{80\%}$, Hours	Kinetic Constants, h^{-1} The Total Process of Dissociation of the Complex and Release through the Membrane. k_{tot}	Kinetic Constants, h^{-1} Dissociation of the Complex. k_{diss}	The Proportion of Lev or EG Included in Cyclodextrin Tori and Retained by the Polymer Matrix, %
Lev in HPCD-PEI10-Man-**9**	44 ± 2	70 ± 2	1.15 ± 0.12	1.3 ± 0.2	65/35
Eugenol free	9 ± 1	0.36 ± 0.04	19.3 ± 2.4	-	-
EG in HPCD-PEI35-Man-**5**	16 ± 2	6.5 ± 0.5	4.6 ± 0.4	6.0 ± 0.5	30/70
EG in HPCD-spermine-Man-**7**	27 ± 3	4.5 ± 0.6	5.4 ± 0.4	7.5 ± 0.6	25/35
EG in HPCD-PEI1.8-Man-**8**	21 ± 3	25 ± 3	5.2 ± 0.3	7.1 ± 0.5	40/60
EG in HPCD-PEI10-Man-**9**	300 ± 20	57 ± 4	2.6 ± 0.2	3.0 ± 0.3	55/45

Thus, we have obtained molecular containers that can be adapted directly to the tasks of a specific therapeutic strategy—by regulating the composition of the conjugate (components and their ratio), as well as the molecular architecture.

2.4. *Antibacterial Activity of Levofloxacin and Adjuvants in Molecular Containers*

The antibacterial activity of free levofloxacin and in the form of inclusion complexes with molecular containers HPCD-PEI35-Man-**5** and HPCD-spermine-Man-**7** was studied by the method of diffusion in agar containing *E. coli* cells (Table 5). The drug in the composition of conjugates has an antibacterial effect slightly superior in strength to free Lev, probably due to enhanced penetration through the cell membrane [40] and due to increased adhesion of conjugates to the cell wall [37].

Table 5. Diameters of zones of antibacterial action of complex formulations of levofloxacin and adjuvants (eugenol and apiol). *E. coli* (NCIB 12210). pH 7.4 (0.01 M PBS), 37 °C, 24 h of incubation. Molar ratios are given in parentheses (conjugate by HPCD's tori). Values are presented as the mean± SD of three experiments.

Sample	Inhibition Zone Diameters (mm, ±0.5 mm) and Corresponding Concentration of Drug			
	0.5 μg/mL Lev	1 μg/mL Lev	2 μg/mL Lev	5 μg/mL Lev
Levofloxacin	13	22.5	28	31
HPCD-spermine-Man-**7** + Lev (1:1)	14	24	31	32
HPCD-spermine-Man-**7** + Lev/EG (1:5:5)	15	24.5	30.5	33
HPCD-PEI35-Man-**5** + Lev (1:1)	13	23	27	32
HPCD-PEI35-Man-**5** + Lev/EG (1:1:1)	14.5	25	27.5	34

Adjuvants (allylpolyalkoxybenzenes) can be used to reduce the concentration of the toxic main component of levofloxacin in the complex formulation due to the synergy effect, as well as to increase the therapeutic potential of the drug. Presumably, the use of complex forms of Lev in combination with allylpolyalkoxybenzenes would increase the circulation time of Lev in the bloodstream and "save" a significant part of the drug from misuse (destruction, accelerated excretion). Indeed, while eugenol has an antibacterial effect on some bacteria [73,76], it is found that eugenol [74], apiol and its derivatives [77,78], and some monoterpenes such as menthol and linalool [79] showed synergism with several antibiotics.

The introduction of the synergist eugenol into conjugates further enhances the effect of levofloxacin up to 15% (Table 5). Thus, the minimum inhibitory Lev concentration can be reduced from 0.4–0.45 μg/mL to 0.2–0.35 μg/mL due to both the molecular container and adjuvants. However, this effect is not the key.

The most significant effect is that the molecular containers provide a prolonged effect of levofloxacin (Figure 8). Bacterial growth was suppressed by all samples with Lev for 24 h. However, based on the steepness of the initial slope of curves (Figure 8b) and absorption of cell suspension (A_{600} values) after 22 h (Figure 8b), it can be concluded that EG (eugenol) and apiol enhance and accelerate the antibacterial effect of Lev in conjugates (with comparably high efficiency in both HPCD-spermine-Man-**7** and HPCD-PEI35-Man-**5**), and therefore the MIC of the antibiotic can potentially be reduced. Statistical analysis of obtained data was carried out using the Student's t-test show that the differences in the Lev efficiency in free form and in the complex with conjugates (values of both groups of experiments are presented in Table 5) were considered statistically significant, since the calculated *p*-value was less than 0.05. Comparing the values of absorption of cell suspension (correlates with the number of colony-forming units—CFU) after 1 and 6 days, we come to the conclusion that the Lev in the complex with conjugates acts more than 50% more effectively than the free form. Moreover, the effect of prolonged action is observed (the decline of curves against the plateau in the case of Lev). In this article, we have demonstrated that eugenol and apiol exhibit a synergy effect with Levofloxacin, boosting its antibacterial activity at concentrations several orders of magnitude lower than is needed for these substances to show any antibacterial activity. The ratio of Levofloxacin:eugenol was 1:1 and 1:2. Here we aimed at showing the proof-of-principle—that there is a possibility as such to obtain the double complexes of Levofloxacin and its adjuvants (allylpolyalkoxybenzenes), and to study the ligand binding constants, release kinetics and the antibacterial activity of drug molecules from both single and double complexes.

In another paper (in press), we found the ratio of Levofloxacin:eugenol = 1:500—1:1000 to be optimal, yielding antibacterial activity of Levofloxacin at concentrations several fold lower than would normally be expected [48]. This synergistic effect can be explained by the formation of defects in the bacterial membrane by an aromatic molecule, resulting the increased penetration of the antibiotic into the cell; and also, by the inhibition of bacterial outflow pumps and, as a consequence, increase of intracellular antibiotic concentration.

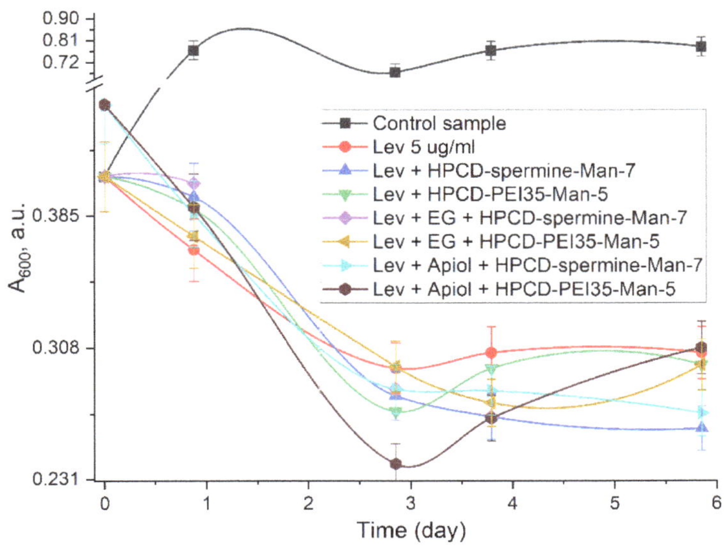

(a)

Figure 8. *Cont.*

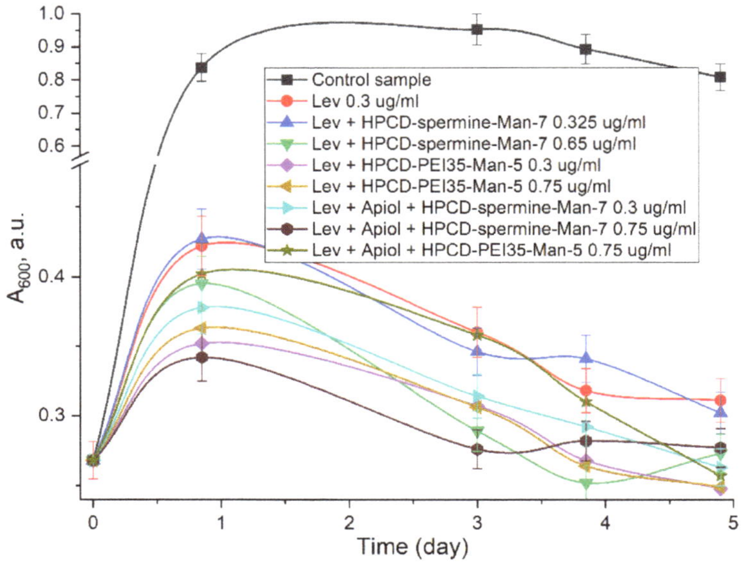

(**b**)

Figure 8. The dependence of the optical density A_{600} (correlate with CFU) on the incubation time of *E. coli* cells (NCIB 12210) with drugs. (**a**) In all samples C(Lev) = 5 µg/mL (**b**) C(Lev) = 0.3–0.75 µg/mL are indicated in the figure. Molar ratio of Lev:apiol and Lev:EG is equal to 1:1. pH 7.4 (0.01 M PBS), 37 °C, 24 h of incubation. Values are presented as the mean ± SD of three experiments.

Thus, the combined formulation of levofloxacin-adjuvant-cyclodextrin can be considered as a promising drug system for the treatment of bacterial infections in order to increase the therapeutic effect, minimize side effects on the body and ensure the absence of resistance of pathogenic microorganisms.

3. Materials and Methods

3.1. Materials

We used carbonyldiimidazole—CDI (Chemical Process, Singapore), PEI 30–40 kDa (Serva, Heidelberg, Germany), ConA (Paneco-ltd, Moscow, Russia). Spermine, D-mannose, PEI1.8 and PEI10, fluorescein isothiocyanate (FITC), DMF, DMSO, Et_3N, orthophthalic aldehyde, Et_2O, 2-mercaptoethanol and 2-hydroxipropyl-β-cyclodextrin were obtained from Sigma Aldrich (St. Louis, MI, USA), levofloxacin from Zhejiang Kangyu Pharm Co Ltd. (Zhejiang, China). Other chemicals: 2,4,6-trinitrobenzenesulfonic acid, $NaBH_4$, salts and acids—production Reakhim (Moscow, Russia).

3.2. Synthesis of HPCD Derivatives with PEI and Spermine

Adapted methodology [52–56] used for obtaining mesh and linear polymer conjugates: 6 mixtures of dry substances were prepared, consisting of HPCD sample (0.038 g, 25 mmol) and an excess of CDI (10-, 20- and 40-fold molar surpluses—two samples each). The powders are dissolved in 0.5 mL of DMF. The samples were incubated for an hour at room temperature. Then 1.5 mL of cooled to −70 °C Et_2O was added to precipitate a transparent gel (HPCD-CDI), which was separated for further synthesis. Three solutions of spermine for conjugates 1–3 and PEI (35 kDa) for conjugates 4–6 were prepared in 50% DMSO with a concentration of 0.1 g/mL. The molar excesses relative to the initial amount of HPCD are respectively 8, 15 and 30 for spermine; 0.25, 0.04 and 1 for PEI. HPCD-CDI gels and 0.1 mL Et_3N were added drop-by-drop to the indicated solutions of spermine and PEI for 90 min. The samples were incubated for 6 h at room temperature. In solutions 1–3 containing spermine, a white precipitate (22 mg) was formed, which was separated by centrifugation and dissolved in 1 mL of water. Purification of the obtained six masterbatch solutions of the samples (HPCD-PEI and HPCD-spermine conjugates) was carried out in 5–8 consecutive stages of adding 500–1000 μL of water and centrifugation with filters of 3 kDa in the case of spermine and 30 kDa in the case of PEI (5000 rpm, 30 min, Eppendorf).

Other techniques were used for obtaining star and branching polymer conjugates, as well as increasing the quantitative yields of intermediates and products, and also for narrowing the distribution by structure. The slow drop-by-drop addition of HPCD solution in water (0.4 mM) to the CDI solution in DMSO (1.5 mM) makes it possible to create a locally very high excess of the activating agent and thereby obtain a product of a strict HPCD-CDI_7 composition at the output. Activated cyclodextrin was added drop by drop to a solution of PEI or spermine. Purification from unreacted components and low molecular weight substances was carried out by dialysis (cut-off weight from 3.5 to 14 kDa, depending on the type of oligo- or polyamine). The technique does not require the use of ice ether for precipitation, and the amount of organic solvents is reduced, which simplifies the purification stages and simplifies the technological implementation of pharmaceutical production of these delivery systems at the enterprise. Schemes of synthesis are given in (Supplement Figures S1–S7).

3.3. *Mannosylation and Purification of HPCD-PEI and HPCD-Spermine Conjugates*

To samples of 200–600 μL of HPCD-spermine and HPCD-PEI conjugates with an average concentration of 25 mg/mL and 60–200 mg/mL respectively, a one-and-a-half-fold molar excess (relative to the number of NH_2-groups) of 3 M mannose water solution was added drop by drop with intensive stirring for 1 h, followed by small portions of dry Schiff base reducing agent $NaBH_4$. Purification of conjugates from low molecular weight impurities was carried out using centrifuge filters, as described above. The purity of the preparation was controlled by HPLC gel filtration in a Knauer chromatography system (Knauer, Berlin, Germany) on BioFox 17 SEC in a 15 cm × 1 cm^2 column. The eluent was 15 mM PBS (pH 7.4) containing 150 mM NaCl; the elution rate was 0.5 mL/min, 25 °C. The chromatogram of the resulting conjugate is shown in supplemental material Figure S24. The resulting conjugates were lyophilized or frozen and stored at −20 °C.

The degree of mannosylation was calculated according to spectrophotometric titration of amino groups (before and after mannosylation) with 2,4,6-trinitrobenzenesulfonic acid [28] and fluorescence analysis with orthophthalic aldehyde and 2-mercaptoethanol [80,81] (Figure S25). Mainly primary amino groups are detected.

3.4. *FTIR-Spectroscopy*

ATR-FTIR spectra of samples' solutions were recorded using a Bruker Tensor 27 spectrometer equipped with a liquid nitrogen cooled MCT (mercury cadmium telluride) detector. Samples were placed in a thermostated cell BioATR-II with ZnSe ATR element (Bruker, Germany). The IR spectrometer was purged with a constant flow of dry air (Jun-Air, USA). FTIR spectra were acquired from 900 to 3000 cm^{-1} with 1 cm^{-1} spectral resolution. For each spectrum, 50–70 scans were accumulated at 20 kHz scanning speed and averaged. ATR-FTIR spectra of solid samples placed on KBr glass were recorded using a Bruker Lumos II IR microscope in region from 700 to 4000 cm^{-1} with 1 cm^{-1} spectral resolution with scanning in the area on average 1 × 1 microns. Spectral data were processed using the Bruker software system Opus 8.2.28 (Bruker, Germany), which includes linear blank subtraction, baseline correction, differentiation (second order, 17 smoothing points), min-max normalization and atmosphere compensation [43,82]. If necessary, 5-point Savitsky–Golay smoothing was used to remove white noise. Peaks were identified by standard Bruker picking-peak procedure.

3.4.1. Complex Formation of ConA with Conjugates. FTIR

Titration of a dimeric ConA solution (C = 33 μM) in 25 μL of a sodium citrate buffer solution (C = 0.02 M, pH 5.5) containing 0.5 M NaCl, 1 mM $CaCl_2$, 1 mM $MnCl_2$ was carried out by adding aliquot ligands (HPCD-PEI-Man and HPCD-spermine-Man conjugates) 0.5–50 μL at 22 °C. The degree of complexation was determined by changes in the intensity of amide II band normalized to the amide I band, or the same changes of intensity or max-min position in Amide II areas in differential IR spectra of the second order.

3.4.2. Complex Formation of Conjugates with Levofloxacin. FTIR

Levofloxacin complexes (0.67–5 mg/mL) with HPCD-PEI-Man or HPCD-spermine-Man conjugates in 25–150 μL diluted HCl solution (C_{HCl} = 1 mM, pH = 3) were first prepared by mixing samples and one hour incubation at 37 °C. The concentration per CD ring varied in the range 10 μM–10 mM) at 37 °C. Further, the IR spectra of the samples were recorded. The proportion of bound levofloxacin was determined by the intensity of the aromatic peak at 1474 cm^{-1}, as well as the position of the minimum-maximum of the second derivative in this region.

3.4.3. Complex Formation of Conjugates with Eugenol. FTIR

Eugenol complexes (0.43–2.4 mg/mL-aqueous suspension, pre-moistened eugenol oil of 10 μL ethanol) with HPCD-PEI-Man or HPCD-spermine-Man conjugates were prepared by mixing samples and one hour incubation at 37 °C. Further—similarly as in the case of levofloxacin. Analytical aromatic peak at 1515 cm^{-1}.

3.4.4. Formation of Double Drug Inclusion Complexes of Eugenol and Levofloxacin

Solid-phase (heterogeneous) synthesis. Two samples of HPCD-PEI35-Man-5 with Lev and a star-shaped conjugate HPCD-spermine-Man-7 with Lev were obtained, moistened with eugenol oil (molar ratio CD tori: Lev: EG = 1:1:1), and a couple of drops of ethanol or acetonitrile were added. The sample is ground on glass and dried at low heat for no more than 35 °C for 30 min. Then leave at room temperature.

3.5. Fluorescence Spectroscopy

Fluorescence was measured using a Varian Cary Eclipse spectrophotometer (Agilent Technologies, Santa Clara, CA, USA) at a temperature of 22 °C in the fluorescence range of Trp residues (for nonfluorescent ligand λ_{ex} = 280 nm, λ_{em} = 337 nm) and FITC-labeled ligand (λ_{ex} = 485 nm, λ_{em} = 525 nm) [18]. Fluorescence anisotropy spectra were recorded using a manual polarizer. The fluorescence anisotropy value was calculated based on the intensity components according to the formulas:

$$r = \frac{I_{VV} - G \cdot I_{VH}}{I_{VV} + 2G \cdot I_{VH}}, \tag{1}$$

$$\text{factor } G = \frac{I_{HV}}{I_{HH}}, \tag{2}$$

3.5.1. Complex Formation of ConA with Conjugates. Fluorescence Methods

Titration of a dimeric ConA solution (C = 19 μM) in 500 μL of a sodium citrate buffer solution (C = 0.02 M, pH 5.5) containing 0.5 M NaCl, 1 mM $CaCl_2$, 1 mM $MnCl_2$ was carried out by adding aliquot ligands (HPCD-PEI-Man and HPCD-spermine-Man conjugates) 2–400 μL at 22 °C. The degree of complexation was determined by changes in the intensity and anisotropy of the fluorescence of tryptophan residues in the receptor protein.

3.5.2. Complex Formation of ConA with FITC-Labeled Ligands. Fluorescence Methods

Titration of a FITC-labeled ligand solution ($C_{FITC} \approx 0.2$ μM) in 550 μL Tris/HCl buffer solution (C = 0.01 M, pH = 7.2) containing 0.5 M NaCl, 1 mM $CaCl_2$, 1 mM $MnCl_2$ was performed by adding aliquot ConA (10–25 mg/mL) 2–200 μL at 22 °C. The degree of complexation was determined by changes in the intensity and anisotropy of FITC fluorescence.

3.5.3. Complex Formation of Conjugates with Levofloxacin. Fluorescence Methods

Titration of a levofloxacin (135 μM) in 500 μL diluted HCl solution (C_{HCl} = 1 mM, pH = 3) was performed by adding aliquot HPCD-PEI-Man or HPCD-spermine-Man conjugates (the concentration per HPCD ring varied in the range 10–20 mM) at 37 °C. The degree of Lev inclusion was determined by changes in the intensity and anisotropy of Lev fluorescence (λ_{ex} = 396 nm, λ_{em} = 480 nm).

3.6. *Dynamic Light Scattering (DLS)*

The particle sizes and zeta potentials were measured using a Zetasizer Nano S «Malvern» (England) (4mW He–Ne-laser, 633 nm, scattering angle 173°). The experiment was performed in a temperature-controlled cell at 25 °C. Autocorrelation functions of intensity fluctuations of light scattering were obtained using the correlation of the Correlator system K7032-09 «Malvern» (Worcestershire, UK). Experimental data was processed using «Zetasizer Software» (v. 8.02).

3.7. *Levofloxacin and Eugenol Release by UV-Spectroscopy*

Parameters of release kinetics of free levofloxacin (C_0(Lev) = 0.2 mg/mL), eugenol (C_0(EG) = 2 mg/mL) and eugenol/levofloxacin complexes with molecular containers **5, 7–9** through a dialysis membrane (12–14 kDa cut-off) into an external solution (1 mL: 5 mL) were measured by UV-absorption at 287 (Lev) and 280 (EG) nm. Aliquots of 400 μL were selected and replaced with a buffer solution. UV-spectroscopy was performed with Amersham Biosciences Ultraspec 2100 pro. UV-spectra were recorded 240–400 nm at 25 °C. PBS (pH = 7.4).

3.8. *NMR Spectroscopy*

12 mg of the sample was dissolved in 700 μL of D_2O. ^{1}H -spectra of the solutions were recorded on a Bruker Avance 400 spectrometer (Germany) with an operating frequency of 400 MHz. Chemical shifts (δ) in m.d. were calculated relative to the residual signals of D_2O (4.79 m.d.).

3.9. *Antibacterial Activity of Levofloxacin and Adjuvants in Molecular Containers*

The strains used in this study were *Escherichia coli* (NCIB 12210) from National Resource Center All-Russian collection of industrial microorganisms SIC "Kurchatov Institute"). The cell sample was placed in 5 mL of liquid medium Luria–Bertani (pH 7.2) in a thermostat at 37 °C. 24 h before the experiments, 150 μL of cell culture was placed in 3 mL of liquid nutrient medium Luria–Bertani (pH 7.2) and then in a thermostat at 37 °C without stirring. The study of the antibacterial effect of levofloxacin and adjuvants was carried out by agar diffusion test. 500 μL of 100-fold culture dilution was evenly distributed over a solid nutrient medium (Luria–Bertani, pH 7.2) on a Petri dish. After 20 min, agar disks with a diameter of 9 mm were removed from the cups and 50 μL of the test samples were placed in the formed wells at the required concentration in a sterile PBS buffer (see paragraph 3.3(2)). After 30 min, Petri dishes were placed in a 37 °C thermostat. After 24 h, the diameters of the growth inhibition zones were measured, and the minimum inhibitory concentrations and amounts of adjuvants necessary to enhance the main component of levofloxacin were calculated.

The study of the antibacterial effect of fluoroquinolones in liquid media was carried out using a nocturnal bacterial culture. The cell suspension was diluted twice with Luria–Bertani liquid medium (pH 7.4). 200 μL of sample was added to 1.8 mL of culture. The resulting systems were incubated at a temperature of 37 °C and stirring at 100 rpm in an ES-20 BIOSAN incubator shaker (Latvia) for 6 days. Aliquots were selected at certain intervals to measure absorption at 600 nm—to determine the number of viable cells.

3.10. Mathematical Processing of the Results

(a) Calculation of dissociation constants of ConA-ligand complexes was carried out in 3 stages:

(1) Fitting of the curves of change of the analytical signal ξ versus molar excess of the ligand x was carried out using the Hill and Boltzmann equations (Origin software):

$$\text{Hill1: } \xi = \text{start} + (\text{end}-\text{start}) \cdot x^n/(x^n + k^n) \quad (3)$$

where start and end—horizontal asymptotes;

$$\text{Boltzmann: } \xi = (A_1-A_2)/(1 + \exp((x-x_0)/dx)) + A_2 \quad (4)$$

(2) Calculation of the fraction of the bound receptor (α) through the values of the analyzed quantities (fluorescence intensity (F) or anisotropy (r), IR absorption (A)) for the receptor ξ_0, the complex ξ_∞ and the current value ξ, as described earlier [18]:

$$\alpha = \left| \frac{\xi - \xi_0}{\xi_\infty - \xi_0} \right|, \quad (5)$$

(3) Calculation of the equilibrium concentrations of the receptor [R], ligand [L] and complex [R·L] and the K_d value by following equations by following equations [63,83]:

$$\text{material balance for receptor: } C_0(R) = [R] + [R \cdot L] \quad (6)$$

$$\text{material balance for ligand: } C_0(\text{ligand}) = [L] + [R \cdot L] \quad (7)$$

$$\text{complex: } [R \cdot L] = C_0(R) \cdot \alpha \quad (8)$$

$$K_d = \frac{[R] \cdot [L]}{[R \cdot L]} \text{one} - \text{binding site model} \quad (9)$$

$$\text{Hill's linearization model: } \lg(\theta/(1-\theta)) = n \cdot \lg[L] - \lg K_d, \quad (10)$$

where $\theta = [R \cdot L]/C_0(R)$—proportion of bound receptor, n—number of binding site.

$$\text{Scatchard's linearization model: } [R \cdot L]/[L] = (C_0(R) - [R \cdot L])/K_d \quad (11)$$

Similar calculations were performed for the Lev and EG—receptor interactions.

(b) Determination of drug loading parameters:

(1) Entrapment efficiency of X [50]: EE (%) = 100·(X amount of loaded X)/(total X amount).

(2) Loading capacity of X [84]: LC(%) = 100·(mass of loaded X)/(mass of sample) = 100 · (number of drug molecules per one container molecule) · (molar mass of X)/(molar mass of conjugate).

Statistical analysis of obtained data was carried out using the Student's *t*-test and Statistica 9.0 software (StatSoft, Tulsa, OK, USA).

4. Conclusions

In this paper, the spectral approach for high-throughput screening of lectin-ligand interactions using concanavalin A as a model mannose receptor in application for molecular containers based on cyclodextrins, oligoamines and mannose for targeted delivery of levofloxacin to alveolar macrophages have been developed to increase the effectiveness of therapy for various diseases, including pneumonia and tuberculosis. By varying the conjugate structure from mesh, polymer to star-shaped and branched, it is possible to adapt the delivery system to specific treatment tasks: the time of drug half-release and effective action. The introduction of a mannose label (instead of using expensive derivatives of oligomannosides) allows targeting only activated macrophages expressing the CD206 mannose receptor, which minimizes adverse reactions and immune response. Complexes of inclusion of aromatic model fluoroquinolone levofloxacin into cyclodextrin tori and polymer matrix have been synthesized, as well as promising adjuvants eugenol and apiol—antibacterial enhancers. Conjugate **7** (star-shaped cyclodextrin with grafted spermines, respectively) demonstrate the highest parameters of loading capacity and entrapment efficiency to both levofloxacin and eugenol molecules parameters (EE and LC). These conjugates combine the valuable properties of a drug container and a high affinity ligand to mannose receptors due to presence of the trimannoside analogue on spacer spermine. Thus, the developed systems of targeted drug delivery of combined drugs due to their high specificity to macrophage mannose receptors (shown in the model ConA), presumably low immunogenicity, high drug loading capacity and prolonged release in target tissues are promising in medical biotechnology and pharmaceuticals for the creation of new dosage forms and strategies for the treatment of respiratory tract diseases.

Supplementary Materials: The following supporting information can be downloaded at: https://www.mdpi.com/article/10.3390/ph15050625/s1, Figure S1: Scheme of synthesis of mannosylated HPCD-spermine-Man-**1** and HPCD-spermine-Man-**2** conjugates. 2-hydroxypropyl-β-cyclodextrin (HPCD) diagrammatically depicted in the form of a heptagon. The characteristics of the synthesized ligand are given in Table 1; Figure S2: Scheme of synthesis of mannosylated HPCD-spermine-Man-**3** conjugate. 2-hydroxypropyl-β-cyclodextrin (HPCD) diagrammatically depicted in the form of a heptagon. The characteristics of the synthesized ligand are given in Table 1; Figure S3: Scheme of synthesis of mannosylated HPCD-PEI35-Man-**4** conjugate. On average, two hydroxypropyl groups are activated in HPCD-CDI. The characteristics of the synthesized ligand are given in Table 1; Figure S4: Scheme of synthesis of mannosylated HPCD-PEI35-Man-**5** conjugate. On average, seven hydroxypropyl groups are activated in HPCD-CDI. The characteristics of the synthesized ligand are given in Table 1; Figure S5: Scheme of synthesis of mannosylated HPCD-PEI35-Man-**6** conjugate. The characteristics of the synthesized ligand are given in Table 1; Figure S6: Scheme of synthesis of mannosylated HPCD-PEI1.8-Man-**8** conjugate. The characteristics of the synthesized ligand are given in Table 1. On average, one cyclodextrin torus accounts for 4–7 branched mannosylated chains of PEI; Figure S7: Decomposition of the IR spectrum of the HPCD-spermine-Man-**1** conjugate in the 980–1130 cm^{-1} region into HPCD and Man components. The fitted spectrum is given; Figure S8: Fourier transform infrared spectra of mannosylated spermine-series conjugates: (**a**) HPCD-spermine-Man-**1**, (**b**) HPCD-spermine-Man-**2**, (**c**) HPCD-spermine-Man-**3**; Figure S9: Fourier transform infrared spectra of mannosylated PEI35-series conjugates: (**a**) HPCD-PEI35-Man-4, (**b**) HPCD-PEI35-Man-5, (**c**) HPCD-PEI35-Man-6. (**d**) 1H NMR spectrum of HPCD-PEI35-Man-6. D2O, 12 mg/mL 400 MHz; Figure S10: FTIRspectra of mannosylated branched-PEI-series conjugates: (**a**) HPCD-PEI1.8-Man-**8**, (**b**) HPCD-PEI10-Man-**9**. ^1H NMR spectra of conjugates: (**c**) HPCD-spermine-Man-**7**, (**d**) HPCD-PEI1.8-Man-**8**, (**e**) HPCD-PEI10-Man-**9**. D_2O, 12 mg/mL, 400 MHz; Figure S11: A fluorescent approach to studying the interaction of FITC-labeled cyclodextrin with ConA. Titration of a FITC-labeled ligand solution ($C_{FITC} \approx 0.2$ μM) in 550 μL Tris/HCl buffer solution (C = 0.01 M, pH = 7.2) containing 0.5 M NaCl, 1 mM $CaCl_2$, 1 mM $MnCl_2$ was performed by adding aliquot ConA (10–25 mg/mL) 2–200 μL at 22 °C; Figure S12: Stern-Vollmer's fitting plot of fluorescence intensity in case competitor binding of HPCD-PEI35-Man-**5** conjugate with ConA in mannose solution [Man] = 0.5 M. [L]—the equilibrium concentration of the ligand. C_0(dimeric ConA) = 38 μM. Ligand concentration varied from 1 to 250 μM. Natrium-citric buffer solution (0.04 M,m). $C(Ca^{2+}) = C(Mn^{2+}) = 1$ mM. T = 22 °C; Figure S13: Tryptophan fluorescence spectra: titration of ConA by mannan. Titration of a dimeric ConA solution (C = 19 μM) in 500 μL of a sodium citrate buffer solution (C = 0.02 M, pH 5.5) containing 0.5 M NaCl, 1 mM $CaCl_2$, 1 mM $MnCl_2$ was carried out by adding aliquot of mannan at 22 °C. Concentration of mannan varied in row 0; 1.3 μM; 3.9 μM; 8 μM; 26 μM to 78 μM; Figure S14: Complex ConA–trimannoside: X-ray diffraction data of PDB (**a**) and after 50-ns molecular dynamics simulation (**b**). The green sphere is Ca^{2+}. From [17]; Figure S15: Maximum position in Amide II areas in differential IR spectra of the second order. Normalized to the amide I band Fourier transform infrared spectra of ConA: titration of ConA by ligand HPCD-spermine-

Man-**2** and **3**. C_0(ConA) = 33 μM. Natrium-citric buffer solution (0.04 M, pH 5.5). C(Ca^{2+}) = C(Mn^{2+}) = 1 mM. T = 22 °C; Figure S16: Hill's linearization of hyperbolic curves of dependences of maximum position in differential IR spectra of the second order (as in Figure S15). The conditions are similar to those specified in Figure S15. [L]—the equilibrium concentration of the ligand. Theta—percentage of occupied binding sites. lg K_d value is calculated as the ratio of the length of the segment cut off on the ordinate axis to the value of the tangent of the angle of inclination; Figure S17: A fluorescent approach to studying the interaction of levofloxacin (Lev) with mesh HPCD-spermine-Man-**2** and polymer HPCD-PEI35-Man-**4** conjugates. Quenching and anisotropy of Lev fluorescence: λ_{exci} = 396 nm, λ_{emi} = 480 nm. Titration of Lev solution (C_0 = 5 mg/mL, 1 mM HCl) by ligands. T = 22 °C; Figure S18: FTIR spectra of Lev (C0(Lev) = 5 mg/mL) in the absence of spermine (blue spectrum) and in the presence of spermine (concentration gradient from red to green—from 2 to 5 mM). 10 mM HCl solution. T = 22 °C; Figure S19: The dependence of the peak intensity at 1474 cm^{-1} in the IR spectra of levofloxacin complexes with conjugates normalized for the range of 2800–1800 cm^{-1}. The insert shows the normalized IR spectra for HPCD-spermine-Man-**2** complexes with Lev. C_0(Lev) = 5 mg/mL. 1 mM HCl solution. T = 22 °C; Figure S20: Solid phase FTIR spectra of levofloxacin (Lev), Lev–HPCD complex, and Lev complex with HPCD-PEI35-Man-**5** conjugate; Figure S21: Solid phase FTIR spectra of eugenol, eugenol–HPCD complex, and dual (Lev ang eugenol) drug inclusion complexes with HPCD-PEI35-Man-**5** conjugate; Figure S22: (**a**) Integral intensity of 1184–1680 cm^{-1} region in solid phase FTIR spectra of levofloxacin (Lev) and eugenol (EG) double complex with HPCD-PEI35-Man-**5** conjugate, superimposed on microphotography samples of 1 × 1 microns in size. (**b**) Corresponding FTIR spectra and areas on the micrograph (**c**); Figure S23: Result of deconvolution of the absorption aromatic band at 1495–1415 cm^{-1} in the FTIR spectra for the (**a**) Lev, and (**b**) Lev—HPCD-PEI35-Man-**5** complex; Figure S24: High-performance liquid chromatogram of conjugates of (**a**) HPCD-PEI1.8-Man-8 (10 mg/mL) and (**b**) HPCD-PEI35–6 (5 mg/mL). UV-absorption detector (204 nm). The eluent was 15 mM PBS (pH 7.4) containing 150 mM NaCl; the elution rate was 0.5 mL/min, 25 °C; Figure S25: (**a**) Spectrophotometric titration of amino groups (before and after mannosylation) with 2,4,6-trinitrobenzenesulfonic acid for conjugate HPCD-spermine-Man-**7**. ε(2,4,6-trinitrobenzenesulfonic acid with NH_2-group adduct) = $1.3 \cdot 10^4$ $M^{-1} \cdot cm^{-1}$. (**b**) Calibration dependence of the fluorescence intensity of the reaction product on the concentration of primary amino groups (standard—spermine). Fluorescence analysis with *ortho*-phthalic aldehyde and 2-mercaptoethanol (1 min for incubation). T = 22 °C.

Author Contributions: I.D.Z. performed the experiments, analyzed the data and wrote the manuscript; E.V.K. analyzed the data, supervised the project and edited the manuscript. All authors have read and agreed to the published version of the manuscript.

Funding: This research was funded by Russian Science Foundation, grant number 22-24-00604.

Institutional Review Board Statement: Not applicable.

Informed Consent Statement: Not applicable.

Data Availability Statement: The data presented in this study are available in the main text and Supplementary Materials.

Acknowledgments: The work was performed using equipment (FTIR spectrometer Bruker Tensor 27 and Jasco J-815 CD Spectrometer) of the program for the development of Moscow State University. NMR spectra were recorded by Rostislav Petrov (Lomonosov Moscow State University, Faculty of Chemistry, Department of Organic Chemistry, Research Institute of Physico-Chemical Methods of Analysis of the Structure of Matter, Researcher)—the authors express their sincere gratitude.

Conflicts of Interest: The authors declare no conflict of interest.

References

1. Bongrand, P. Ligand-receptor interactions. *Rep. Prog. Phys.* **1999**, *62*, 921–968. [CrossRef]
2. Rudd, P.M.; Wormald, M.R.; Dwek, R.A. Sugar-Mediated Ligand-Receptor Interactions in the Immune System. *Trends Biotechnol.* **2004**, *22*, 524–530. [CrossRef] [PubMed]
3. Managit, C.; Kawakami, S.; Yamashita, F.; Hashida, M. Effect of Galactose Density on Asialoglycoprotein Receptor-Mediated Uptake of Galactosylated Liposomes. *J. Pharm. Sci.* **2005**, *94*, 2266–2275. [CrossRef] [PubMed]
4. Kessner, S.; Krause, A.; Rothe, U.; Bendas, G. Investigation of the Cellular Uptake of E-Selectin-Targeted Immunoliposomes by Activated Human Endothelial Cells. *Biochim. Biophys. Acta-Biomembr.* **2001**, *1514*, 177–190. [CrossRef]

5. Bendas, G.; Krause, A.; Schmidt, R.; Vogel, J.; Rothe, U. Selectins as New Targets for Immunoliposome-Mediated Drug Delivery. A Potential Way of Anti-Inflammatory Therapy. *Pharm. Acta Helv.* **1998**, *73*, 19–26. [CrossRef]
6. Brekke, O.H.; Sandlie, I. Therapeutic Antibodies for Human Diseases at the Dawn of the Twenty-First Century. *Nat. Rev. Drug Discov.* **2003**, *2*, 52–62. [CrossRef] [PubMed]
7. Filatova, L.Y.; Klyachko, N.L.; Kudryashova, E.V. Targeted Delivery of Anti-Tuberculosis Drugs to Macrophages: Targeting Mannose Receptors. *Russ. Chem. Rev.* **2018**, *87*, 374–391. [CrossRef]
8. Zlotnikov, I.D.; Kudryashova, E.V. Mannose Receptors of Alveolar Macrophages as a Target for the Addressed Delivery of Medicines to the Lungs. *Russ. J. Bioorg. Chem.* **2022**, *48*, 46–75. [CrossRef]
9. Taylor, P.R.; Martinez-Pomares, L.; Stacey, M.; Lin, H.H.; Brown, G.D.; Gordon, S. Macrophage Receptors and Immune Recognition. *Annu. Rev. Immunol.* **2005**, *23*, 901–944. [CrossRef]
10. East, L.; Isacke, C.M. The Mannose Receptor Family. *Biochim. Biophys. Acta Gen. Subj.* **2002**, *1572*, 364–386. [CrossRef]
11. Stahl, P.D. The Macrophage Mannose Receptor: Current Status. *Am. J. Respir. Cell Mol. Biol.* **1990**, *2*, 317–318. [CrossRef] [PubMed]
12. Feinberg, H.; Park-Snyder, S.; Kolatkar, A.R.; Heise, C.T.; Taylor, M.E.; Weis, W.I. Structure of a C-Type Carbohydrate Recognition Domain from the Macrophage Mannose Receptor. *J. Biol. Chem.* **2000**, *275*, 21539–21548. [CrossRef] [PubMed]
13. Feinberg, H.; Jégouzo, S.A.F.; Lasanajak, Y.; Smith, D.F.; Drickamer, K.; Weis, W.I.; Taylor, M.E. Structural Analysis of Carbohydrate Binding by the Macrophage Mannose Receptor CD206. *J. Biol. Chem.* **2021**, *296*, 100368. [CrossRef] [PubMed]
14. Taylor, M.E.; Drickamer, K. Structural Requirements for High Affinity Binding of Complex Ligands by the Macrophage Mannose Receptor. *J. Biol. Chem.* **1993**, *268*, 399–404. [CrossRef]
15. Hitchen, P.G.; Mullin, N.P.; Taylor, M.E. Orientation of Sugars Bound to the Principal C-Type Carbohydrate-Recognition Domain of the Macrophage Mannose Receptor. *Biochem. J.* **1998**, *333*, 601–608. [CrossRef]
16. Mullin, N.P.; Hitchen, P.G.; Taylor, M.E. Mechanism of Ca2+ and Monosaccharide Binding to a C-Type Carbohydrate- Recognition Domain of the Macrophage Mannose Receptor. *J. Biol. Chem.* **1997**, *272*, 5668–5681. [CrossRef]
17. Zlotnikov, I.D.; Kudryashova, E.V. Computer Simulation of the Receptor–Ligand Interactions of Mannose Receptor CD206 in Comparison with the Lectin Concanavalin A Model. *Biochem.* **2022**, *87*, 54–69. [CrossRef]
18. Zlotnikov, I.D.; Vanichkin, D.A.; Kudryashova, E.V. Methods for Determining the Parameters of Receptor-Ligand Interactions on the Model of Concanavalin A and Mannosylated Chitosans Promising Carriers for Drug Delivery to Alveolar Macrophages. *Biotekhnologiya* **2021**, *37*, 28–40. [CrossRef]
19. Loontiens, F.G.; Clegg, R.M.; Van Landschoot, A. Some Physicochemical Aspects of Oligosaccharide Binding to Concanavalin A and Wheat Germ Agglutinin. *J. Biosci.* **1983**, *5*, 105–120. [CrossRef]
20. Sawyer, H.W.; Dabscheck, R.; Nott, P.R.; Selinger, B.K.; Kuntz, I.D. Hydrodynamic Changes Accompanying the Loss of Metal Ions from Concanavalin A. *Biochem. J.* **1975**, *147*, 613–615. [CrossRef]
21. Mandal, D.K.; Kishore, N.; Brewer, C.F. Thermodynamics of Lectin-Carbohydrate Interactions. Titration Microcalorimetry Measurements of the Binding of N-Linked Carbohydrates and Ovalbumin to Concanavalin A. *Biochemistry* **1994**, *33*, 1149–1156. [CrossRef] [PubMed]
22. Weis, W.I.; Drickamer, K. Structural Basis of Recognition Lectin-Carb Ohydrate. *Anna Rev. Biochen. 19%* **1996**, *65*, 441–473. [CrossRef] [PubMed]
23. Ghotbi, Z.; Haddadi, A.; Hamdy, S.; Hung, R.W.; Samuel, J.; Lavasanifar, A. Active Targeting of Dendritic Cells with Mannan-Decorated PLGA Nanoparticles. *J. Drug Target.* **2011**, *19*, 281–292. [CrossRef] [PubMed]
24. Ruan, G.X.; Chen, Y.Z.; Yao, X.L.; Du, A.; Tang, G.P.; Shen, Y.Q.; Tabata, Y.; Gao, J.Q. Macrophage Mannose Receptor-Specific Gene Delivery Vehicle for Macrophage Engineering. *Acta Biomater.* **2014**, *10*, 1847–1855. [CrossRef] [PubMed]
25. Nimje, N.; Agarwal, A.; Saraogi, G.K.; Lariya, N.; Rai, G.; Agrawal, H.; Agrawal, G.P. Mannosylated Nanoparticulate Carriers of Rifabutin for Alveolar Targeting. *J. Drug Target.* **2009**, *17*, 777–787. [CrossRef]
26. Dam, T.K.; Roy, R.; Das, S.K.; Oscarson, S.; Brewer, C.F. Binding of Multivalent Carbohydrates to Concanavalin A and Dioclea Grandiflora Lectin. Thermodynamic Analysis of the 'Multivalency Effect'. *J. Biol. Chem.* **2000**, *275*, 14223–14230. [CrossRef]
27. VAN LANDSCHOOT, A.; LOONTIENS, F.G.; DE BRUYNE, C.K. Binding of Manno-oligosaccharides to Concanavalin A: Substitution Titration with a Fluorescent-Indicator Ligand. *Eur. J. Biochem.* **1980**, *103*, 307–312. [CrossRef]
28. Le-Deygen, I.M.; Mamaeva, P.V.; Skuredina, A.A.; Kudryashova, E.V. A Spectral Approach to Study Interaction between Chitosan Modified with Mannose and Concavalin A for the Creation of Address Delivery Systems of Antituberculosis Drugs. *Moscow Univ. Chem. Bull.* **2020**, *75*, 213–217. [CrossRef]
29. So, L.L.; Goldstein, I.J. Protein-Carbohydrate Interaction. 13. The Interaction of Concanavalin A with Alpha-Mannans from a Variety of Microorganisms. *J. Biol. Chem.* **1968**, *243*, 2003–2007. [CrossRef]
30. Liu, Z.; Li, X.; Ding, X.; Yang, Y. In Silico and Experimental Studies of Concanavalin a: Insights into Its Antiproliferative Activity and Apoptotic Mechanism. *Appl. Biochem. Biotechnol.* **2010**, *162*, 134–145. [CrossRef]
31. Kaushik, S.; Mohanty, D.; Surolia, A. The Role of Metal Ions in Substrate Recognition and Stability of Concanavalin A: A Molecular Dynamics Study. *Biophys. J.* **2009**, *96*, 21–34. [CrossRef] [PubMed]
32. Saokham, P.; Muankaew, C.; Jansook, P.; Loftsson, T. Solubility of Cyclodextrins and Drug/Cyclodextrin Complexes. *Molecules* **2018**, *23*, 1161. [CrossRef] [PubMed]
33. SZEJTLI, J. ChemInform Abstract: Introduction and General Overview of Cyclodextrin Chemistry. *ChemInform* **2010**, *29*, no-no. [CrossRef]

34. Connors, K.A. The Stability of Cyclodextrin Complexes in Solution. *Chem. Rev.* **1997**, *97*, 1325–1357. [CrossRef] [PubMed]
35. Milcovich, G.; Antunes, F.E.; Grassi, M.; Asaro, F. Stabilization of Unilamellar Catanionic Vesicles Induced by β-Cyclodextrins: A Strategy for a Tunable Drug Delivery Depot. *Int. J. Pharm.* **2018**, *548*, 474–479. [CrossRef]
36. Davis, M.E.; Brewster, M.E. Cyclodextrin-Based Pharmaceutics: Past, Present and Future. *Nat. Rev. Drug Discov.* **2004**, *3*, 1023–1035. [CrossRef]
37. Skuredina, A.A.; Tychinina, A.S.; Le-Deygen, I.M.; Belogurova, N.G.; Kudryashova, E.V. Regulation of Properties of Lipid Membranes by Interaction with 2-Hydroxypropyl β-Cyclodextrin: Molecular Details. *Russ. J. Bioorganic Chem.* **2020**, *46*, 692–701. [CrossRef]
38. Lopukhov, A.V.; Yang, Z.; Haney, M.J.; Bronich, T.K.; Sokolsky-Papkov, M.; Batrakova, E.V.; Klyachko, N.L.; Kabanov, A.V. Mannosylated Cationic Copolymers for Gene Delivery to Macrophages. *Macromol. Biosci.* **2021**, *21*, 2000371. [CrossRef]
39. Sato, A.; Kawakami, S.; Yamada, M.; Yamashita, F.; Hashida, M. Enhanced Gene Transfection in Macrophages Using Mannosylated Cationic Liposome-Polyethylenimine-Plasmid DNA Complexes. *J. Drug Target.* **2001**, *9*, 201–207. [CrossRef]
40. Skuredina, A.A.; Tychinina, A.S.; Le-Deygen, I.M.; Golyshev, S.A.; Belogurova, N.G.; Kudryashova, E.V. The Formation of Quasi-Regular Polymeric Network of Cross-Linked Sulfobutyl Ether Derivative of β-Cyclodextrin Synthesized with Moxifloxacin as a Template. *React. Funct. Polym.* **2021**, *159*, 104811. [CrossRef]
41. Zheng, H.; He, W.; Jiao, W.; Xia, H.; Sun, L.; Wang, S.; Xiao, J.; Ou, X.; Zhao, Y.; Shen, A. Molecular Characterization of Multidrug-Resistant Tuberculosis against Levofloxacin, Moxifloxacin, Bedaquiline, Linezolid, Clofazimine, and Delamanid in Southwest of China. *BMC Infect. Dis.* **2021**, *21*, 330. [CrossRef] [PubMed]
42. Chen, X.Y.; Qian, F.; Wang, Y.Y.; Liu, Y.; Sun, Y.; Zha, W.B.; Hao, K.; Zhou, F.; Wang, G.J.; Zhang, J.W. Ginsenoside 20(S)-Rh2 Promotes Cellular Pharmacokinetics and Intracellular Antibacterial Activity of Levofloxacin against Staphylococcus Aureus through Drug Efflux Inhibition and Subcellular Stabilization. *Acta Pharmacol. Sin.* **2021**, *42*, 1930–1941. [CrossRef] [PubMed]
43. Deygen, I.M.; Egorov, A.M.; Kudryashova, E.V. Structure and Stability of Fluoroquinolone-(2-Hydroxypropyl)-β-Cyclodextrin Complexes as Perspective Antituberculosis Drugs. *Moscow Univ. Chem. Bull.* **2016**, *71*, 387–392. [CrossRef]
44. Jelić, R.; Tomović, M.; Stojanović, S.; Joksović, L.; Jakovljević, I.; Djurdjević, P. Study of Inclusion Complex of β-Cyclodextrin and Levofloxacin and Its Effect on the Solution Equilibria between Gadolinium(III) Ion and Levofloxacin. *Monatshefte fur Chemie* **2015**, *146*, 1621–1630. [CrossRef]
45. Tychinina, A.S.; Skuredina, A.A.; Le-Deygen, I.M.; Kudryashova, E.V. Influence of Substituents in β-Cyclodextrin on the Interaction of Levofloxacin–β-Cyclodextrin Complexes with Liposomal Membrane. *Colloid J.* **2021**, *83*, 794–801. [CrossRef]
46. Teles, A.M.; Silva-Silva, J.V.; Fernandes, J.M.P.; Abreu-Silva, A.L.; Calabrese, K.D.S.; Mendes Filho, N.E.; Mouchrek, A.N.; Almeida-Souza, F. GC-MS Characterization of Antibacterial, Antioxidant, and Antitrypanosomal Activity of Syzygium Aromaticum Essential Oil and Eugenol. *Evidence-Based Complement. Altern. Med.* **2021**, *2021*, 6663255. [CrossRef]
47. Gong, L.; Li, T.; Chen, F.; Duan, X.; Yuan, Y.; Zhang, D.; Jiang, Y. An Inclusion Complex of Eugenol into β-Cyclodextrin: Preparation, and Physicochemical and Antifungal Characterization. *Food Chem.* **2016**, *196*, 324–330. [CrossRef]
48. Wang, T.; Li, B.; Si, H.; Lin, L.; Chen, L. Release Characteristics and Antibacterial Activity of Solid State Eugenol/β-Cyclodextrin Inclusion Complex. *J. Incl. Phenom. Macrocycl. Chem.* **2011**, *71*, 207–213. [CrossRef]
49. Arana-Sánchez, A.; Estarrón-Espinosa, M.; Obledo-Vázquez, E.N.; Padilla-Camberos, E.; Silva-Vázquez, R.; Lugo-Cervantes, E. Antimicrobial and Antioxidant Activities of Mexican Oregano Essential Oils (Lippia Graveolens H. B. K.) with Different Composition When Microencapsulated Inβ-Cyclodextrin. *Lett. Appl. Microbiol.* **2010**, *50*, 585–590. [CrossRef]
50. Hill, L.E.; Gomes, C.; Taylor, T.M. Characterization of Beta-Cyclodextrin Inclusion Complexes Containing Essential Oils (Trans-Cinnamaldehyde, Eugenol, Cinnamon Bark, and Clove Bud Extracts) for Antimicrobial Delivery Applications. *LWT-Food Sci. Technol.* **2013**, *51*, 86–93. [CrossRef]
51. Weisheimer, V.; Miron, D.; Silva, C.B.; Guterres, S.S.; Schapoval, E.E.S. Microparticles Containing Lemongrass Volatile Oil: Preparation, Characterization and Thermal Stability. *Pharmazie* **2010**, *65*, 885–890. [CrossRef] [PubMed]
52. Gao, J.Q.; Zhao, Q.Q.; Lv, T.F.; Shuai, W.P.; Zhou, J.; Tang, G.P.; Liang, W.Q.; Tabata, Y.; Hu, Y.L. Gene-Carried Chitosan-Linked-PEI Induced High Gene Transfection Efficiency with Low Toxicity and Significant Tumor-Suppressive Activity. *Int. J. Pharm.* **2010**, *387*, 286–294. [CrossRef] [PubMed]
53. Raviv, L.; Jaron-Mendelson, M.; David, A. Mannosylated Polyion Complexes for in Vivo Gene Delivery into CD11c+ Dendritic Cells. *Mol. Pharm.* **2015**, *12*, 453–462. [CrossRef] [PubMed]
54. Yang, C.; Li, H.; Goh, S.H.; Li, J. Cationic Star Polymers Consisting of α-Cyclodextrin Core and Oligoethylenimine Arms as Nonviral Gene Delivery Vectors. *Biomaterials* **2007**, *28*, 3245–3254. [CrossRef] [PubMed]
55. Lu, X.; Ping, Y.; Xu, F.J.; Li, Z.H.; Wang, Q.Q.; Chen, J.H.; Yang, W.T.; Tang, G.P. Bifunctional Conjugates Comprising β-Cyclodextrin, Polyethylenimine, and 5-Fluoro-2′-Deoxyuridine for Drug Delivery and Gene Transfer. *Bioconjug. Chem.* **2010**, *21*, 1855–1863. [CrossRef]
56. Tang, G.P.; Guo, H.Y.; Alexis, F.; Wang, X.; Zeng, S.; Lim, T.M.; Ding, J.; Yang, Y.Y.; Wang, S. Low Molecular Weight Polyethylenimines Linked by β-Cyclodextrin for Gene Transfer into the Nervous System. *J. Gene Med.* **2006**, *8*, 736–744. [CrossRef]
57. Casillo, A.; Fabozzi, A.; Russo Krauss, I.; Parrilli, E.; Biggs, C.I.; Gibson, M.I.; Lanzetta, R.; Appavou, M.S.; Radulescu, A.; Tutino, M.L.; et al. Physicochemical Approach to Understanding the Structure, Conformation, and Activity of Mannan Polysaccharides. *Biomacromolecules* **2021**, *22*, 1445–1457. [CrossRef]

58. Szabó, Z.I.; Deme, R.; Mucsi, Z.; Rusu, A.; Mare, A.D.; Fiser, B.; Toma, F.; Sipos, E.; Tóth, G. Equilibrium, Structural and Antibacterial Characterization of Moxifloxacin-β-Cyclodextrin Complex. *J. Mol. Struct.* **2018**, *1166*, 228–236. [CrossRef]
59. Cameron, K.S.; Fletcher, D.; Fielding, L. An NMR Study of Cyclodextrin Complexes of the Steroidal Neuromuscular Blocker Drug Rocuronium Bromide. *Magn. Reson. Chem.* **2002**, *40*, 251–260. [CrossRef]
60. Rodríguez-Lavado, J.; De La Mata, M.; Jiménez-Blanco, J.L.; García-Moreno, M.I.; Benito, J.M.; Díaz-Quintana, A.; Sánchez-Alcázar, J.A.; Higaki, K.; Nanba, E.; Ohno, K.; et al. Targeted Delivery of Pharmacological Chaperones for Gaucher Disease to Macrophages by a Mannosylated Cyclodextrin Carrier. *Org. Biomol. Chem.* **2014**, *12*, 2289–2301. [CrossRef]
61. Kanai, M.; Mortell, K.H.; Kiessling, L.L. Varying the Size of Multivalent Ligands: The Dependence of Concanavalin A Binding on Neoglycopolymer Length. *J. Am. Chem. Soc.* **1997**, *119*, 9931–9932. [CrossRef]
62. Weatherman, R.V.; Kiessling, L.L. Fluorescence Anisotropy Assays Reveal Affinities of C- and O-Glycosides for Concanavalin A. *J. Org. Chem.* **1996**, *61*, 534–538. [CrossRef] [PubMed]
63. Stefan, M.I.; Le Novère, N. Cooperative Binding. *PLoS Comput. Biol.* **2013**, *9*, 2–7. [CrossRef] [PubMed]
64. Bryce, R.A.; Hillier, I.H.; Naismith, J.H. Carbohydrate-Protein Recognition: Molecular Dynamics Simulations and Free Energy Analysis of Oligosaccharide Binding to Concanavalin A. *Biophys. J.* **2001**, *81*, 1373–1388. [CrossRef]
65. Dani, M.; Manca, F.; Rialdi, G. Calorimetric study of concanavalin A binding to saccharides. *Biochim. et Biophys. Acta BBA Protein Struct.* **1981**, *667*, 108–117. [CrossRef]
66. Mislovičová, D.; Masárová, J.; Švitel, J.; Mendichi, R.; Šoltés, L.; Gemeiner, P.; Danielsson, B. Neoglycoconjugates of Mannan with Bovine Serum Albumin and Their Interaction with Lectin Concanavalin A. *Bioconjug. Chem.* **2002**, *13*, 136–142. [CrossRef]
67. López-López, M.; Fernández-Delgado, A.; Moyá, M.L.; Blanco-Arévalo, D.; Carrera, C.; de la Haba, R.R.; Ventosa, A.; Bernal, E.; López-Cornejo, P. Optimized Preparation of Levofloxacin Loaded Polymeric Nanoparticles. *Pharmaceutics* **2019**, *11*, 57. [CrossRef]
68. Ameeduzzafar; Imam, S.S.; Bukhari, S.N.A.; Ahmad, J.; Ali, A. Formulation and Optimization of Levofloxacin Loaded Chitosan Nanoparticle for Ocular Delivery: In-Vitro Characterization, Ocular Tolerance and Antibacterial Activity. *Int. J. Biol. Macromol.* **2018**, *108*, 650–659. [CrossRef]
69. Puppi, D.; Piras, A.M.; Pirosa, A.; Sandreschi, S.; Chiellini, F. Levofloxacin-Loaded Star Poly(ε-Caprolactone) Scaffolds by Additive Manufacturing. *J. Mater. Sci. Mater. Med.* **2016**, *27*, 44. [CrossRef]
70. Le-Deygen, I.M.; Skuredina, A.A.; Uporov, I.V.; Kudryashova, E.V. Thermodynamics and Molecular Insight in Guest–Host Complexes of Fluoroquinolones with β-Cyclodextrin Derivatives, as Revealed by ATR-FTIR Spectroscopy and Molecular Modeling Experiments. *Anal. Bioanal. Chem.* **2017**, *409*, 6451–6462. [CrossRef]
71. Skuredina, A.A.; Le-Deygen, I.M.; Belogurova, N.G.; Kudryashova, E.V. Effect of Cross-Linking on the Inclusion Complex Formation of Derivatized β-Cyclodextrins with Small-Molecule Drug Moxifloxacin. *Carbohydr. Res.* **2020**, *498*, 108183. [CrossRef] [PubMed]
72. Sukhoverkov, K.V.; Le-Deygen, I.M.; Egorov, A.M.; Kudryashova, E.V. Physicochemical Properties of the Inclusion Complex of Moxifloxacin with Hydroxypropyl-β-Cyclodextrin Synthesized by RESS. *Russ. J. Phys. Chem. B* **2018**, *12*, 1193–1204. [CrossRef]
73. Leite, A.M.; Lima, E.D.O.; De Souza, E.L.; Diniz, M.D.F.F.M.; Trajano, V.N.; De Medeiros, I.A. Inhibitory Effect of β-Pinene, α-Pinene and Eugenol on the Growth of Potential Infectious Endocarditis Causing Gram-Positive Bacteria. *Rev. Bras. Ciencias Farm. J. Pharm. Sci.* **2007**, *43*, 121–126. [CrossRef]
74. Hemaiswarya, S.; Doble, M. Synergistic Interaction of Eugenol with Antibiotics against Gram Negative Bacteria. *Phytomedicine* **2009**, *16*, 997–1005. [CrossRef] [PubMed]
75. Skuredina, A.A.; Kopnova, T.Y.; Le-Deygen, I.M.; Kudryashova, E.V. Physical and Chemical Properties of the Guest–Host Inclusion Complexes of Cyprofloxacin with β-Cyclodextrin Derivatives. *Moscow Univ. Chem. Bull.* **2020**, *75*, 218–224. [CrossRef]
76. Ali, S.M.; Khan, A.A.; Ahmed, I.; Musaddiq, M.; Ahmed, K.S.; Polasa, H.; Rao, L.V.; Habibullah, C.M.; Sechi, L.A.; Ahmed, N. Antimicrobial Activities of Eugenol and Cinnamaldehyde against the Human Gastric Pathogen Helicobacter Pylori. *Ann. Clin. Microbiol. Antimicrob.* **2005**, *4*, 20. [CrossRef]
77. Lichtenstein, E.P.; Liang, T.T.; Schulz, K.R.; Schnoes, H.K.; Carter, G.T. Insecticidal and Synergistic Components Isolated from Dill Plants. *J. Agric. Food Chem.* **1974**, *22*, 658–664. [CrossRef]
78. Samet, A.V.; Shevchenko, O.G.; Rusak, V.V.; Chartov, E.M.; Myshlyavtsev, A.B.; Rusanov, D.A.; Semenova, M.N.; Semenov, V.V. Antioxidant Activity of Natural Allylpolyalkoxybenzene Plant Essential Oil Constituents. *J. Nat. Prod.* **2019**, *82*, 1451–1458. [CrossRef]
79. Pereira de Lira, M.H.; Fernandes Queiroga Moraes, G.; Macena Santos, G.; Patrício de Andrade Júnior, F.; De Oliveira Pereira, F.; Oliveira Lima, I. Synergistic Antibacterial Activity of Monoterpenes in Combination with Conventional Antimicrobials against Gram-Positive and Gram-Negative Bacteria. *Rev. Ciências Médicas e Biológicas* **2020**, *19*, 258. [CrossRef]
80. Kudryashova, E.V.; Artemova, T.M.; Vinogradov, A.A.; Gladilin, A.K.; Mozhaev, V.V.; Levashov, A.V. Stabilization and Activation of α-Chymotrypsin in Water-Organic Solvent Systems by Complex Formation with Oligoamines. *Protein Eng.* **2003**, *16*, 303–309. [CrossRef]
81. Beketov, V.I.; Voronina, R.D.; Zorov, N.B. Fluorimetric Determination of Amino Acids and Photochemical Stability of Their Reaction Products with Ortho-Phthalic Aldehyde under Irradiation by High-Power Pulsed Laser. *Moscow Univ. Chem. Bull.* **2012**, *67*, 149–153. [CrossRef]
82. Deygen, I.M.; Kudryashova, E.V. New Versatile Approach for Analysis of PEG Content in Conjugates and Complexes with Biomacromolecules Based on FTIR Spectroscopy. *Colloids Surfaces B Biointerfaces* **2016**, *141*, 36–43. [CrossRef] [PubMed]

83. Endrenyi, L.; Kwong, F.H.F.; Fajszi, C. Evaluation of Hill Slopes and Hill Coefficients When the Saturation Binding or Velocity Is Not Known. *Eur. J. Biochem.* **1975**, *51*, 317–328. [CrossRef] [PubMed]
84. Woranuch, S.; Yoksan, R. Eugenol-Loaded Chitosan Nanoparticles: I. Thermal Stability Improvement of Eugenol through Encapsulation. *Carbohydr. Polym.* **2013**, *96*, 578–585. [CrossRef] [PubMed]

pharmaceuticals

Article

Galactosylated Prodrugs: A Strategy to Improve the Profile of Nonsteroidal Anti-Inflammatory Drugs

Federica Sodano [1,2,*], Claudia Cristiano [1], Barbara Rolando [2], Elisabetta Marini [2], Loretta Lazzarato [2], Mariarosaria Cuozzo [1], Stefania Albrizio [1], Roberto Russo [1,†] and Maria Grazia Rimoli [1,†]

1 Department of Pharmacy, "Federico II" University of Naples, 80131 Naples, Italy; claudia.cristiano@unina.it (C.C.); mariarosaria.cuozzo@unina.it (M.C.); salbrizi@unina.it (S.A.); roberto.russo@unina.it (R.R.); rimoli@unina.it (M.G.R.)

2 Department of Drug Science and Technology, University of Turin, 10125 Turin, Italy; barbara.rolando@unito.it (B.R.); elisabetta.marini@unito.it (E.M.); loretta.lazzarato@unito.it (L.L.)

* Correspondence: federica.sodano@unina.it; Tel.: +39-081-678-604

† These authors contributed equally to this work.

Abstract: Carbohydrates are one of the most abundant and important classes of biomolecules. The variety in their structures makes them valuable carriers that can improve the pharmaceutical phase, pharmacokinetics and pharmacodynamics of well-known drugs. D-galactose is a simple, naturally occurring monosaccharide sugar that has been extensively studied for use as a carrier and has proven to be valuable in this role. With the aim of validating the galactose-prodrug approach, we have investigated the galactosylated prodrugs ibuprofen, ketoprofen, flurbiprofen and indomethacin, which we have named IbuGAL, OkyGAL, FluGAL and IndoGAL, respectively. Their physicochemical profiles in terms of lipophilicity, solubility and chemical stability have been evaluated at different physiological pH values, as have human serum stability and serum protein binding. Ex vivo intestinal permeation experiments were performed to provide preliminary insights into the oral bioavailability of the galactosylated prodrugs. Finally, their anti-inflammatory, analgesic and ulcerogenic activities were investigated in vivo in mice after oral treatment. The present results, taken together with those of previous studies, undoubtedly validate the galactosylated prodrug strategy as a problem-solving technique that can overcome the disadvantages of NSAIDs.

Keywords: galactose; prodrug approach; NSAIDs; physicochemical characterization; stability; serum behavior; intestinal permeation; pain; inflammation; ulcerogenicity

Citation: Sodano, F.; Cristiano, C.; Rolando, B.; Marini, E.; Lazzarato, L.; Cuozzo, M.; Albrizio, S.; Russo, R.; Rimoli, M.G. Galactosylated Prodrugs: A Strategy to Improve the Profile of Nonsteroidal Anti-Inflammatory Drugs. *Pharmaceuticals* **2022**, *15*, 552. https://doi.org/10.3390/ph15050552

Academic Editor: Rachel Auzély

Received: 11 April 2022
Accepted: 27 April 2022
Published: 29 April 2022

1. Introduction

Prodrug design is a well-known strategy that consists of chemically modifying the structure of a drug, which is called the "parent drug". This approach allows the physicochemical and pharmacological properties of a parent drug to be optimized, resulting in an improvement in its pharmacokinetic characteristics and a decrease in its toxicity [1–3]. Prodrugs mainly fall into two categories: bioprecursors and carrier prodrugs. In the former, the molecule acts as a substrate for enzymes, which then release the active drug [4,5], whereas, in the latter, the parent drug is linked through enzymatically or chemically cleavable groups to a molecule that acts as a carrier [4]. The carrier can also be a polymer, and, in this case, they are known as macromolecular prodrugs [6,7]. Another class bears the name mutual prodrugs; here, two active compounds are linked, with each acting as a transporter of the other [4,5]. Carbohydrates are an example of successful carriers.

Carbohydrates are one of the most important and abundant classes of biomolecules [8], as they are major components of genetic material and are the main vital source of energy [9]. They play essential roles in cellular and intracellular interactions, thanks to their action as signaling molecules and their presence on cell-surface receptors [10]. These functions explain the potential of carbohydrates for use as pharmaceutical and diagnostic agents and thus reinforce the importance of the topic covered in this Special Issue.

Medicinal chemists are encouraged to design and adopt strategies to synthesize carbohydrate-based drugs [9,11]. The variety present in carbohydrate structures makes these molecules a valuable tool for the development of biologically active glycoconjugates and carrier prodrugs [12]. In particular, the presence of the carbohydrate moiety grants carrier prodrugs good hydrophilic–lipophilic balance, resulting in decreased toxicity and increased bioavailability [13].

D-galactose is a simple, naturally occurring monosaccharide sugar that has been extensively studied for use as a carrier in the prodrug approach. The use of galactosylated prodrugs has proven to be an excellent strategy, particularly when parent drugs are likely to fail therapeutically due to a lack of site-of-action specificity, a poor toxicological profile and chemical instability. It has been shown, in many studies, that the binding of a parent drug to D-galactose increases its selectivity of action and reduces its side effects. In other cases, the presence of the sugar structure ensures an improvement in the solubility and cell permeability of the parent drug, and thus in its pharmacokinetic profile [14,15], while keeping its pharmacodynamic characteristics intact [16].

Nonsteroidal anti-inflammatory drugs (NSAIDs) are among the most widely used therapeutic agents worldwide [17], making up approximately 5% of all prescribed drugs [18,19]. Despite their high therapeutic potential, long-term exposure to NSAIDs causes serious cardiovascular events, including myocardial infarction and stroke, and gastrointestinal damage, such as bleeding, ulceration and perforation, which can be fatal, as well as kidney injury, hepatotoxicity and other, more minor ailments [20–23].

There are two factors that contribute to both mild and severe gastrointestinal adverse events: firstly, the direct irritating contact of the free carboxylic function of NSAIDs with the gastrointestinal mucosa (the ion-trapping phenomenon); and secondly, the blockage of prostaglandin biosynthesis, which has more indirect effects. These adversities have severely limited the clinical use of NSAIDs and highlight the unmet need for new therapeutic strategies that aim to make their prolonged usage safer [24].

Promising results have been obtained with the prodrug approach when using D-galactose as a carrier for NSAIDs. In our previous studies, the conjugation of a galactose molecule with ketorolac (Ketogal) [25], paracetamol (PARgal) [26] and aceclofenac (ACEgal) [27] improved their physicochemical, toxicological, pharmacokinetic and pharmacodynamic profiles, resulting in remarkable safety profiles. In particular, the prolonged pharmacological action of the aforementioned prodrugs, in comparison with their parent drugs, would allow their daily intake to be reduced, while also decreasing side effects. At the same time, it may also be possible to reduce or eliminate the use of drugs that are commonly prescribed with NSAIDs, such as proton pump inhibitors (PPI), with consequent benefits in terms of health and public spending.

In light of the results obtained and with the aim of validating the galactose-prodrug approach, we have investigated the galactosylated prodrugs of ibuprofen, ketoprofen, flurbiprofen and indomethacin, which we have named IbuGAL, OkyGAL, FluGAL and IndoGAL, respectively, the syntheses of which have been described previously and the structures of which are reported in Figure 1 [28]. In this work, we evaluate their physicochemical profiles in terms of lipophilicity, solubility and chemical stability at different physiological pH values (7.4 and 1.2). The serum behavior of IbuGAL, OkyGAL, FluGAL and IndoGAL was investigated by measuring human serum stability and serum protein binding. Ex vivo intestinal permeation experiments were carried out to provide preliminary insights into the oral bioavailability of these galactosylated prodrugs. Finally, the in vivo pharmacological and toxicological profiles of galactosylated prodrugs were evaluated. Specifically, anti-inflammatory (carrageenan-induced paw edema test), analgesic (acetic acid-induced writhing test) and ulcerogenic activity were investigated in mice after oral treatment.

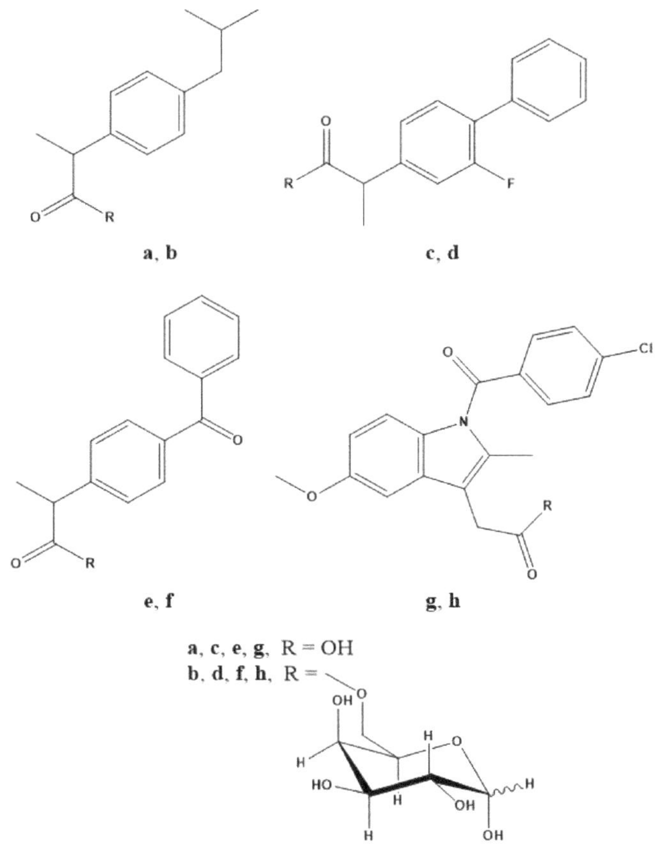

Figure 1. Chemical structures of (**a**) ibuprofen, (**b**) IbuGAL, (**c**) flurbiprofen, (**d**) FluGAL, (**e**) ketoprofen, (**f**) OkyGAL, (**g**) indomethacin and (**h**) IndoGAL.

2. Results

2.1. *Physicochemical Characterization*

2.1.1. Lipophilicity

Lipophilicity is the key physicochemical parameter that influences solubility and various other aspects of the pharmacokinetic and pharmacodynamic profile of drugs. The hydrophilic-lipophilic behavior of galactosylated prodrugs and parent drugs was investigated by measuring the distribution coefficients (log D^{pH}) in *n*-octanol-phosphate buffered saline (PBS, pH 7.4) using the shake-flask technique. The distribution coefficients of the galactosylated prodrugs were found to be pH-independent, which reflects the absence of ionizable groups, meaning that the log $D^{7.4}$ values were superimposable on the calculated log *p*-values (ClogP), as shown in Table 1. A log D value's proximity to two (as in the case of galactosylated prodrugs) suggests optimal pharmacokinetic/pharmacodynamic behavior. Ibuprofen, ketoprofen, flurbiprofen and indomethacin, which are broadly ionized at pH 7.4, exhibited lipophilicity at about one logarithmic unit lower than the related prodrugs.

Table 1. Physicochemical characterization of galactosylated prodrugs and parent drugs: lipophilicity, calculated as log P (ClogP); experimental distribution coefficient at pH 7.4 (log $D^{7.4}$); and solubility (mM) in water, SGF (pH 1.2) and PBS (pH 7.4).

Compound	Lipophilicity		Solubility at 25 °C (mM) [c]		
	ClogP [a]	logD$^{7.4}$ [b]	Water	SGF pH 1.2	PBS pH 7.4
IbuGAL	1.92	1.73 ± 0.04	1.94	2.53	2.93
Ibuprofen	3.68	0.82 ± 0.04	0.22	0.58	23.5
OkyGAL	1.00	0.92 ± 0.01	23.36	8.34	5.53
Ketoprofen	2.76	−0.22 ± 0.03	0.46	0.21	7.71
FluGAL	1.99	1.89 ± 0.09	5.21	1.44	1.01
Flurbiprofen	3.75	0.81 ± 0.01	0.07	0.02	5.43
IndoGAL	2.42	2.05 ± 0.06	0.07	0.08	0.13
Indomethacin	4.18	0.91 ± 0.01	0.07	<LOQ [d]	2.25

[a] Calculated using Bio-Loom for Windows, v. 1.5, BioByte Corporation, Claremont, CA, USA. [b] Results are expressed as mean values ± SD. [c] Results are expressed as mean values, SD < 0.01. [d] LOQ (limit of quantitation, HPLC) = 1 μM.

2.1.2. Solubility Assay

The solubilities of the prodrugs and parent drugs were evaluated at 25 °C in water, simulated gastric fluid (SGF without pepsin, pH 1.2) and PBS (pH 7.4), which are the media used for chemical stability analyses. The solubilities were quantified using RP-HPLC. As expected, the galactosylated derivatives were more soluble than the relative parent drugs in both water and SGF. Conversely, the solubility of the parent drugs at pH 7.4 increased significantly due to their ionization (Table 1).

2.1.3. Stability in SGF and PBS

The chemical stability of galactosylated prodrugs was evaluated over 48 h at 37 °C in SGF and PBS in order to simulate the gastric and physiological environments, respectively. The concentrations of the galactosylated prodrugs and the parent drugs derived from the prodrugs were quantified using RP-HPLC (Table 2). SGF was the experimental environment in which the prodrugs displayed the highest stability; the release of the parent drugs from the relative prodrugs after 24 h of incubation in SGF was lower than 4% and therefore considered negligible. The galactosylated prodrugs were shown to be stable even at physiological pH; the percentage of parent drugs released after 24 h of incubation in PBS was about 10%.

Table 2. The chemical stability in SGF (pH 1.2) and PBS (pH 7.4), enzymatic stability in human serum and serum protein binding (%bound) of galactosylated prodrugs compared to relative NSAIDs.

Compound	Stability at 37 °C % Compounds after 24 h of Incubation (Mean ± SD)			Serum Protein Binding % Bound [a]
	SGF pH 1.2 [a]	PBS pH 7.4 [a]	Human Serum [a]	
IbuGAL Ibuprofen (IbuGAL)	98.3 ± 1.0 3.4 ± 1.1	81.2 ± 2.5 9.4 ± 1.5	54.1 ± 2.2 40.8 ± 6.2	97.2 ± 0.4
Ibuprofen	stable	stable	stable	99.98 ± 0.01
OkyGAL Ketoprofen (OkyGAL)	97.9 ± 0.7 1.4 ± 0.5	91.0 ± 0.2 9.6 ± 0.8	63.0 ± 8.7 30.4 ± 7.7	95.0 ± 1.6
Ketoprofen	stable	stable	stable	99.98 ± 0.01
FluGAL Flurbiprofen (FluGAL)	98.9 ± 0.1 1.1 ± 0.4	87.3 ± 2.1 11.6 ± 0.8	57.3 ± 1.4 44.5 ± 2.4	96.1 ± 0.5

Table 2. *Cont.*

Compound	Stability at 37 °C % Compounds after 24 h of Incubation (Mean ± SD)			Serum Protein Binding % Bound [a]
	SGF pH 1.2 [a]	PBS pH 7.4 [a]	Human Serum [a]	
Flurbiprofen	stable	stable	stable	99.98 ± 0.01
IndoGAL Indomethacin (IndoGAL)	97.8 ± 0.4 2.6 ± 0.5	89.0 ± 1.6 11.5 ± 1.5	82.3 ± 2.2 17.8 ± 4.6	99.9 ± 0.1
Indomethacin	stable	stable	stable	99.98 ± 0.01

[a] Results are expressed as mean values ± SD.

2.1.4. Stability in Human Serum

Stability was also evaluated in human serum at 37 °C over 48 h (Table 2). In this esterase-rich medium, the degradation of galactosylated prodrugs into their parent drugs was more evident. For IbuGAL and FluGAL, the half-life was close to 24 h (ca 50% degradation), whereas stability increased slightly for OkyGAL (ca 30% degradation in 24 h). However, IndoGAL was the most stable compound under these conditions (ca 20% degradation in 24 h). The parent drugs were the only degradation products of the galactosylated prodrugs that were observed using RP-HPLC in these stability studies.

2.1.5. Serum Protein Binding

The serum protein binding properties of the galactosylated prodrugs and parent drugs were investigated using ultrafiltration. The compounds were incubated in human serum, and after 1 h, the concentrations of the compounds in the ultrafiltrate and the original serum were determined by means of RP-HPLC, and used to calculate the binding percentages, which are reported in Table 2. The prodrugs, with the exception of IndoGAL, showed slightly lower percentages of binding to serum proteins than their relative parent drugs, although they maintained binding percentages above 95% (Table 2).

2.2. *Ex Vivo Absorption Evaluation*

The permeability of the galactosylated prodrugs and the parent drugs through the rat small intestine was assessed ex vivo using the non-everted intestinal sac method. Despite some shortcomings (e.g., the interruption of normal blood flow, lack of a nervous system) this test is widely used to investigate the passive absorption of molecules [29] and can provide further data when equipped with a mucus layer, which includes proteins of drug transport and metabolism [30].

A suspension of the tested compounds (5 mg/Kg in PBS + methocel 3%) was syringed into intestinal sacs that consisted of different segments of rat small intestine (duodenum, jejunum, ileum). The compound-enriched tissues were incubated at 37 °C for 2 h. Samples (0.5 mL) were withdrawn every 15 min from the serosal side and the amount of analytes was determined via RP-HPLC. The results are collected in Table 3.

An analysis of the solutions withdrawn during the incubation of intestinal sacs filled with the parent drugs showed very heterogeneous behavior, with ibuprofen demonstrating high absorption (fraction absorbed, F_a, ranging from 0.84 to 0.98), flurbiprofen and ketoprofen demonstrating intermediate absorption (F_a ranging from 0.57 to 0.69 and from 0.46 to 0.53, respectively) and indomethacin demonstrating very low absorption (F_a ranging from 0.040 to 0.094). The three different intestinal segments showed no significant differences in absorption.

Table 3. Ex vivo fraction absorbed (F_a) of galactosylated prodrugs compared to relative NSAIDs.

Compound	Fraction Absorbed (F_a) ± SE		
	Duodenum	Jejunum	Ileum
IbuGAL	0.013 ± 0.006	0.026 ± 0.014	0.031 ± 0.016
Ibuprofen (IbuGAL)	0.043 ± 0.015	0.028 ± 0.011	0.036 ± 0.011
Ibuprofen	0.94 ± 0.2	0.84 ± 0.11	0.98 ± 0.10
OkyGAL	0.038 ± 0.012	0.015 ± 0.012	0.018 ± 0.006
Ketoprofen (OkyGAL)	0.10 ± 0.004	0.060 ± 0.018	0.076 ± 0.007
Ketoprofen	0.53 ± 0.02	0.48 ± 0.03	0.46 ± 0.02
FluGAL	0.060 ± 0.014	0.031 ± 0.015	0.033 ± 0.008
Flurbiprofen (FluGAL)	0.13 ± 0.01	0.075 ± 0.023	0.096 ± 0.009
Flurbiprofen	0.69 ± 0.02	0.62 ± 0.04	0.57 ± 0.04
IndoGAL Indomethacin (IndoGAL)	Not Evaluable	Not Evaluable	Not Evaluable
Indomethacin	0.040 ± 0.005	0.047 ± 0.012	0.094 ± 0.017

On the other hand, all of the galactosylated prodrugs showed similar profiles, and intestinal permeations that were significantly lower than those of their relative NSAIDs. Indeed, an analysis of the samples withdrawn revealed that very low amounts of IbuGAL, OkyGAL, FluGAL and their parent drugs had been released. IndoGAL and, consequently, indomethacin could not be evaluated in any of the three intestinal tracts (Table 3). An example of the time course of F_a in the three intestinal tracts is reported in Figure 2.

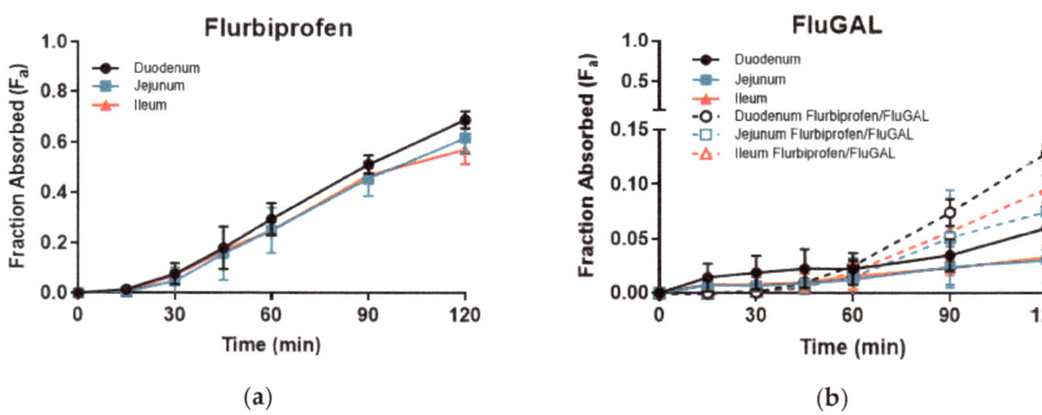

Figure 2. The time course of ex vivo fraction absorbed (F_a) of flurbiprofen (**a**) and FluGAL (**b**) in the three intestinal tracts.

In order to verify the presence of the galactosylated compounds in the intestinal tissues, the intestinal sacs were opened at the end of the incubation period, washed with fresh buffer and homogenized in ethanol and the supernatant was dried. The concentrations of the galactosylated prodrugs and the prodrug-derived parent drugs were quantified using RP-HPLC (Figure 3a–d).

As shown in Figure 3a–d, an analysis of the tissue accumulation highlighted the presence of significant amounts of the galactosylated compounds and the derived parent drugs. In particular, the total amount of prodrug plus derived parent drug was equal to the tissue accumulation of the relative NSAID alone.

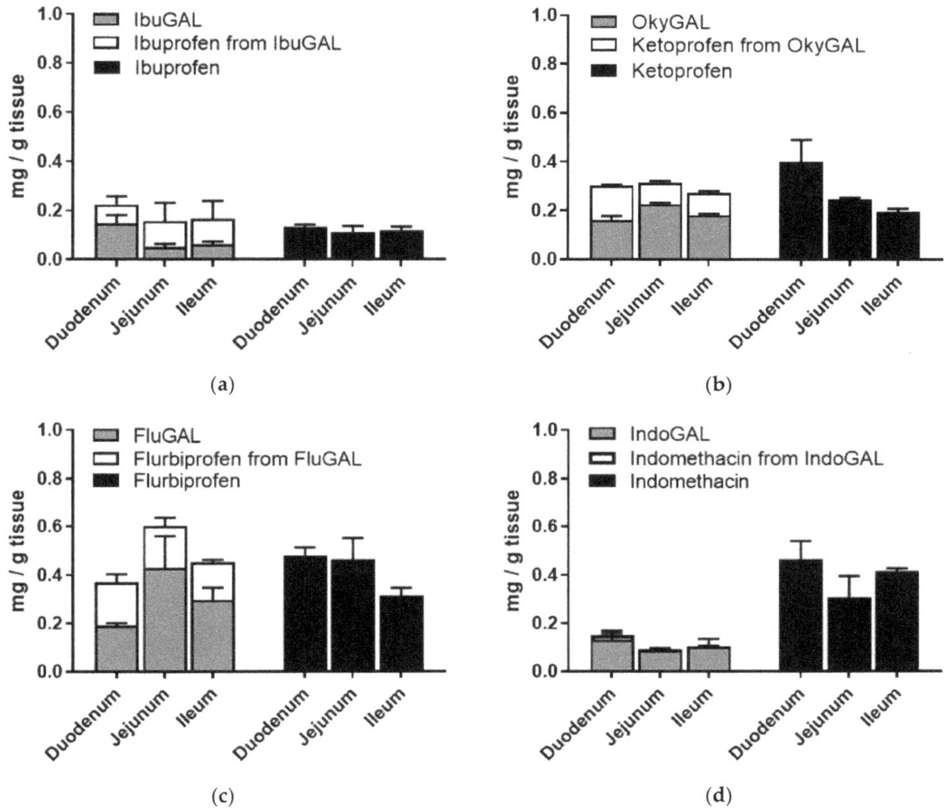

Figure 3. (**a**–**d**) Ex vivo intestinal permeation. Tissue accumulation of parent drugs (**a**) ibuprofen, (**b**) ketoprofen, (**c**) flurbiprofen and (**d**) indomethacin and the galactosylated prodrugs (**a**) IbuGAL, (**b**) OkyGAL, (**c**) FluGAL and (**d**) IndoGAL. Non-everted intestinal sacs were filled with a suspension of the compounds (5 mg/Kg, PBS + methocel 3%) and incubated in oxygenated Krebs–Ringer buffer at 37 °C with smooth shaking. At the end of the incubation period, the tissues were washed with fresh buffer, homogenized in ethanol and centrifuged and the supernatant was dried. The concentrations of the parent drugs and the galactosylated prodrugs and the parent drugs derived from the prodrugs were quantified using RP-HPLC. Tests were performed in triplicate on intestinal segments from three different rats.

2.3. *In Vivo Studies*

2.3.1. Effect of Parent Drugs and Galactosylated Prodrugs on Paw Edema in Mice

We tested the anti-inflammatory response of the parent drugs and their relative galactosylated prodrugs by subcutaneously injecting λ-carrageenan (1% (*w*/*v*) in sterile saline, 20 µL) into the hind paws of mice. Evaluation continued for 96 h after the carrageenan challenge.

Ibuprofen, indomethacin, ketoprofen and flurbiprofen were administered orally at a dose of 10 mg/kg (Figure 4a–d, respectively) and they inhibited paw edema formation during the first phase (0–6 h). The anti-inflammatory activity was significant at 2 and 4 h after carrageenan injection for ibuprofen, indomethacin and ketoprofen and up to 6 h for flurbiprofen. The antihyperalgesic effect completely disappeared after 24 h.

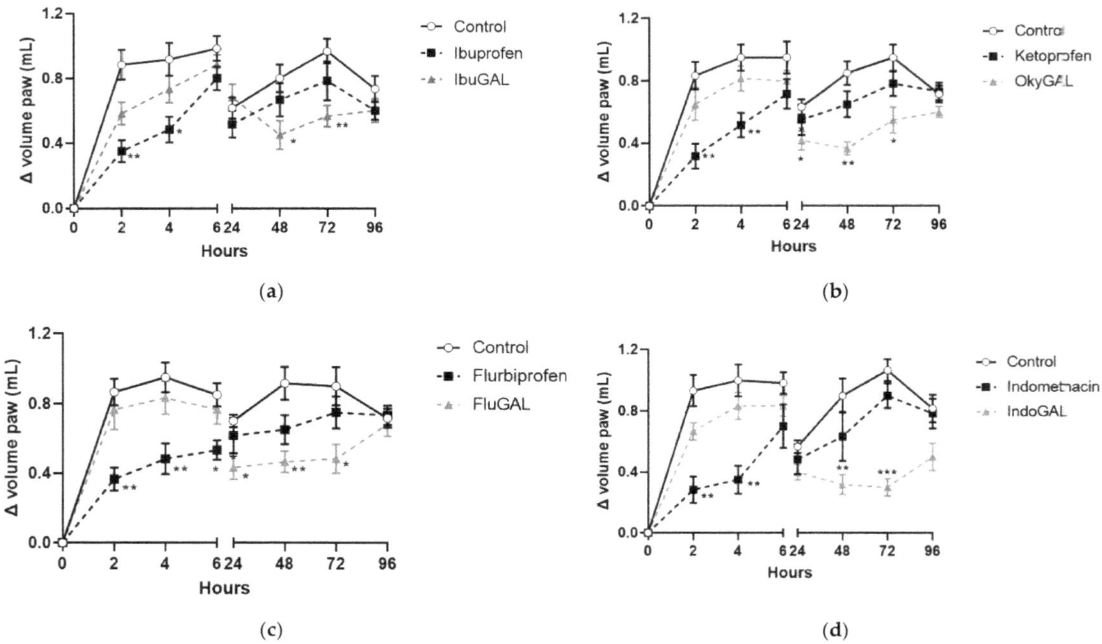

Figure 4. (**a**–**d**). Effect of the parent drugs and galactosylated prodrugs on carrageenan-induced hyperalgesia. Effect of oral administration of the parent drugs (**a**) ibuprofen, (**b**) indomethacin, (**c**) ketoprofen and (**d**) flurbiprofen and the galactosylated prodrugs (**a**) IbuGAL, (**b**) IndoGAL, (**c**) OkyGAL and (**d**) FluGAL on carrageenan-induced hyperalgesia in mouse paws evaluated at 2, 4, 6, 24, 48, 72 and 96 h after the carrageenan challenge. White circles represent control mice treated with vehicle. Black squares represent mice treated with the parent drugs (ibuprofen, indomethacin, ketoprofen, and flurbiprofen) and gray triangles represent mice treated with the galactosylated prodrugs (IbuGAL, IndoGAL, OkyGAL and FluGAL). The increase in paw volume was evaluated and is expressed as the difference in paw volume measured at each time point and the basal paw volume measured immediately before carrageenan injection. Data are expressed as means ± S.E.M. of 6 animals. *** $p < 0.001$, ** $p < 0.01$ and * $p < 0.5$ vs. Control.

By contrast, equimolecular doses of IbuGAL (17.8 mg/kg), IndoGAL (14.5 mg/kg), OkyGAL (16.4 mg/kg) and FluGAL (16.7 mg/kg) did not change the carrageenan-induced hyperalgesia during the first phase of the test, but displayed significant anti-inflammatory activity during the second phase (24–96 h). In particular, IbuGAL and IndoGAL (Figure 4a,b, respectively) showed an anti-inflammatory effect from 48 to 72 h following carrageenan administration, whereas OkyGAL and FluGAL significantly reduced paw edema from 24 to 72 h (Figure 4c,d, respectively).

2.3.2. Effect of Parent Drugs and Galactosylated Prodrugs on the Acetic-Acid-Induced Writhing Test in Mice

The number of acetic-acid-induced abdominal stretches can be evaluated in Figure 3. Control mice received an intraperitoneal injection of 0.6% acetic acid (10 mL/kg). Oral treatment was started 1 h before acetic acid administration. Ibuprofen, indomethacin, ketoprofen and flurbiprofen significantly reduced the number of writhings after 4 h, at a dose of 10 mg/kg (Figure 5a–d, respectively), compared to the control group. Conversely, IbuGAL (17.8 mg/kg), IndoGAL (14.5 mg/kg), OkyGAL (16.4 mg/kg) and FluGAL (16.7 mg/kg) (Figure 4a–d, respectively) resulted in a significant decrease, compared to the control group, in writhing responses after 48 h.

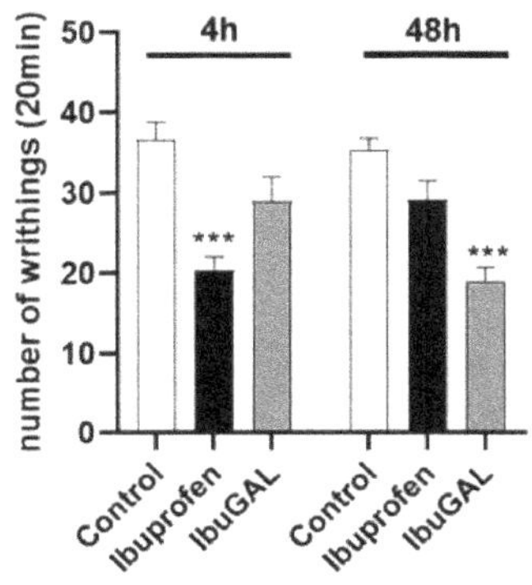

(**a**)

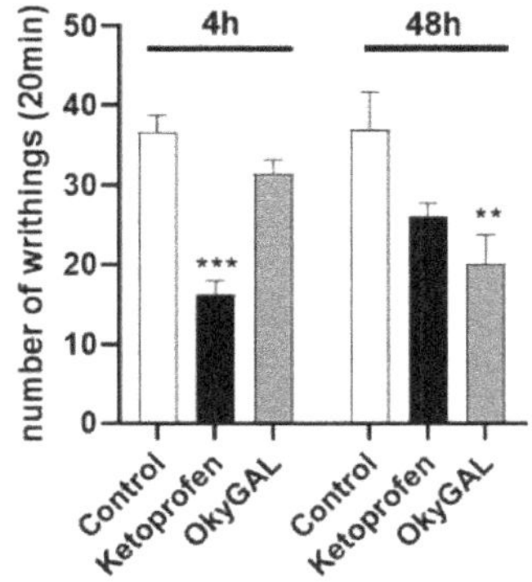

(**b**)

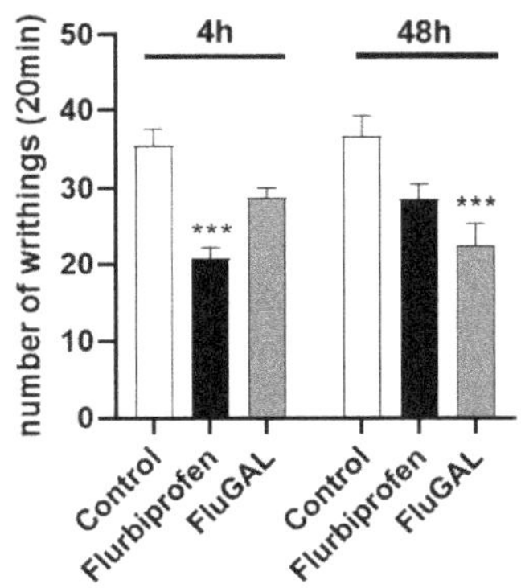

(**c**)

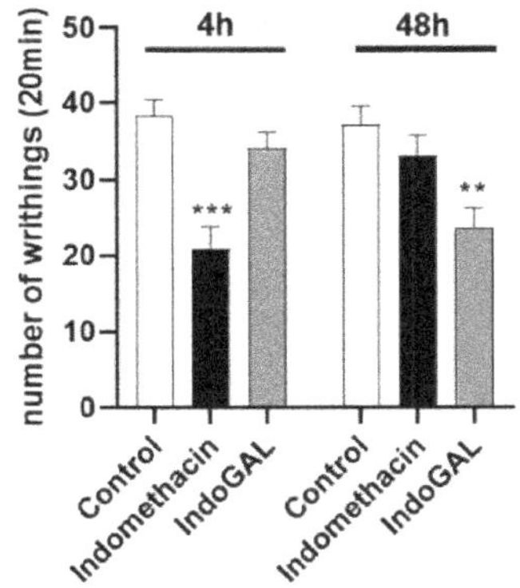

(**d**)

Figure 5. (**a**–**d**). Total number of writhing episodes within 20 min after acetic acid injection. Effect of oral administration of the parent drugs (**a**) ibuprofen, (**b**) indomethacin, (**c**) ketoprofen and (**d**) flurbiprofen and the galactosylated prodrugs (**a**) IbuGAL, (**b**) IndoGAL, (**c**) OkyGAL and (**d**) FluGAL at 4 and 48 h after administration of 0.6% acetic acid (10 mL/kg). Data are expressed as means ± S.E.M. of 6 animals. *** $p < 0.001$, ** $p < 0.01$ vs. Control.

2.3.3. Effect of Parent Drugs and Galactosylated Prodrugs on Ulcerogenic Activity

The study of ulcerogenic activity indicated that, as expected, the parent drugs (ibuprofen, indomethacin, ketoprofen and flurbiprofen), which have free acid functions, showed significantly higher ulcerogenic properties at 4 h and only indomethacin showed these increased properties at 48 h (values between 1.5–3) after oral treatment, compared with the control. However, the galactosylated prodrugs (IbuGAL, IndoGAL, OkyGAL and FluGAL) displayed less ulceration than the parent drugs at the same experimental times (values between 0–1.5). The results are reported in Table 4.

Table 4. Ulcerogenic activity of parent drugs and galactosylated prodrugs 4 and 48 h after treatment.

Compounds	Oral Dose (mg/kg)	Gastric Lesion Score (4 h) [a]	Gastric Lesion Score (48 h) [a]
Control (CMC 0.5%)	-	0	0
Ibuprofen	10	2 ##	1.5
IbuGAL	17.8	0 **	1
Ketoprofen	10	2.5 ###	1.5
OkyGAL	16.4	0 ****	1
Flurbiprofen	10	2 ##	1.5
FluGAL	16.7	0 **	0.5
Indomethacin	10	3 ####	2 ##
IndoGAL	14.5	0 ****	1

[a] A score of 0 was assigned for irritation, a score of 1 for lesions with a length between 1 and 2 mm, a score of 2 for lesions with a length between 2 and 3 mm and a score of 3 for lesions greater than 3 mm. Significance of galactosylated prodrug versus respective parent drug: **** $p < 0.0001$, ** $p < 0.01$. Significance of parent drug versus control: #### $p < 0.0001$, ### $p < 0.001$, ## $p < 0.01$.

3. Discussion

The task of a drug is to perform only its biological activity at the target site, at the lowest concentration, while avoiding side effects [31]. Several approaches have been developed to achieve this goal, including the preparation of carbohydrate-based prodrugs. The conjugation of the parent drug to a carbohydrate, such as D-galactose, is a recent strategy that still requires validation [15,32]. We applied the galactosylated prodrug strategy to a number of NSAIDs. The first successful example of this approach was Ketogal [25], with ACEgal and PARgal then following [26,27]. In each study, the galactosylated prodrug showed marked improvements, compared with its parent drug, in some or all of its physicochemical, pharmacokinetic, pharmacodynamic and toxicological parameters.

In order to confirm the suitability of D-galactose as a carrier in the development of analgesic and anti-inflammatory treatments, we examined a number of NSAIDs that had already been subjected to the galactosylated prodrug strategy in terms of design and synthesis. IbuGAL, OkyGAL, FluGAL and IndoGAL (Figure 1) were synthesized via a green method using room-temperature ionic liquids, as reported in our previous work [28].

In evaluating the physicochemical profiles of the galactosylated prodrugs, it was seen that conjugation with a D-galactose molecule improved the solubility of the starting drug (Table 1). Specifically, IbuGAL, OkyGAL and FluGAL were more soluble than ibuprofen, ketoprofen and flurbiprofen, respectively, in both water and SGF. In PBS, of course, the three starting drugs were more soluble as they are deprotonated at pH = 7.4. Surprisingly, IndoGAL was as soluble as indomethacin in water, despite the increase in the number of sugar-component hydroxyl groups. No great differences between IndoGAL and indomethacin were observed in SGF, whereas the solubility of IndoGAL at pH 7.4 was lower than that of its parent drug, as observed for the other galactosylated derivatives.

The solubility trends of the prodrugs and parent drugs in PBS reflect the values of the distribution coefficients (log $D^{7.4}$), which were measured under the same conditions (Table 1). The decrease in prodrug solubility at pH 7.4, compared to the parent drugs, is in agreement with the increase in their log D by one logarithmic unit, again compared to the parent drugs. The large difference between ClogP and log $D^{7.4}$ in the parent drugs is in accordance with their significant ionization at physiological pH (Table 1).

The galactosylated prodrug strategy ensures that the chemical stability of NSAIDs is maintained in the relative prodrugs. The percentage of unmodified IbuGAL, OkyGAL, FluGAL and IndoGAL remained greater than 97% in SGF and 81% in PBS after 24 h of incubation (see Table 2).

The galactosylated prodrugs also demonstrated stability in human serum. In fact, IndoGAL proved to be the most stable compound, followed by OkyGAL, whereas FluGAL and IbuGAL showed a half-life of around 24 h. With regards to serum protein binding, the binding percentage of ibuprofen, ketoprofen, flurbiprofen and indomethacin was 99.98%, which is in agreement with the literature (Table 2) [33]. The galactosylated prodrug strategy led to a decrease in the binding of NSAIDs to serum proteins and thus to increased free-prodrug availability in circulation.

Absorption is a primary goal in drug development and medicinal chemistry, as a drug must be absorbed before any medicinal effect can take place. In addition, the pharmacokinetic profile of a drug can be easily and significantly changed by adjusting the factors that influence absorption. In our case, the conjugation of NSAIDs to galactose led to significant changes in the absorption profile (Table 3). In fact, in the ex vivo intestinal permeation study, differences were found in the degree of absorption of the galactosylated prodrugs compared to the parent drugs, over the two-hour course of the test. This aspect may be considered disadvantageous, but, generally, the absorption and pharmacological activity of a prodrug must be delayed and prolonged over time. The permeability of IbuGAL, OkyGAL, FluGAL and IndoGAL through the rat small intestine was lower than that of the starting drugs, suggesting that the galactosylated prodrugs exhibited lower in vivo oral absorption over the first two hours. However, the small fraction that is orally absorbed would nevertheless be able to play its pharmacological role, as shown by the results of the anti-inflammatory and analgesic activity tests. The reduced binding to serum protein and the high half-life in human serum of the galactosylated prodrugs resulted in improved drug bioavailability.

No change in anti-inflammatory (Figure 3) and analgesic (Figure 4) potential was found when comparing the pharmacological profiles of the galactosylated prodrugs and their respective NSAIDs. Despite being absorbed less, IndoGAL had the same efficacy of action as indomethacin.

The anti-inflammatory activity of IbuGAL, OkyGAL, FluGAL and IndoGAL was found to be delayed and prolonged over time when assessed using the carrageenan-induced edema test. Specifically, the anti-inflammatory action of NSAIDs began 2 h after oral treatment and lasted up to 4 h, except for that of flurbiprofen, which persisted up to 6 h [34–37]. IbuGAL and IndoGAL exhibited their activity 48 h after oral administration and this lasted up to 72 h, whereas the activity of FluGAL and OkyGAL started after 24 h. The development of NSAIDs with prolonged and time-delayed activity may be advantageous as these drugs are commonly prescribed for chronic inflammatory diseases. The repeated use of NSAIDs could therefore be completely subverted by the administration of a single dose of their galactosylated analogues.

The analgesic activity of galactosylated prodrugs for a period of up to 48 h after oral administration confirmed their prolonged and delayed pharmacological profile. This was due to the masking of the free carboxylic functions of the NSAIDs. The carboxylic group is crucial for therapeutic effectiveness, meaning that no activity was observed in mice treated with galactosylated prodrugs until the enzymatic pathways cleaved the binding with D-galactose (see Figures 3 and 4).

Although, on the one hand, the carboxylic group is essential for the pharmacological activity of NSAIDs, on the other, it is also partly the cause of their main side effect, i.e., gastrointestinal toxicity. The acute administration of NSAIDs in mice has been observed to cause gastrointestinal lesions of different sizes. Indomethacin showed greater ulcerogenic properties, followed by ketoprofen and finally flurbiprofen and ibuprofen [34,35]. In our work, a slight improvement in the assessment of the damage occurred 48 h after administration, which is when tissue-repair mechanisms were already in place. When we administered the galactosylated prodrugs at the same dosage and evaluated the toxicological profile at the same time point, no damage was observed 4 h after oral administration. This may be attributed to the lack of the ion-trapping phenomenon and thus the lack of irritation of the gastric mucosa resulting from local contact with the carboxyl function.

Forty-eight hours after acute administration, mild ulcerogenic activity was observed in mice treated with the galactosylated analogues. This was probably due to complete hydrolysis of the prodrug, resulting in the release of the active drug which, in turn, led to the blockade of prostaglandin biosynthesis and mucus production in the gastrointestinal tract. It can therefore be stated that, from a toxicological point of view, the galactosylated prodrug strategy was able to alleviate the side effects of NSAIDs.

4. Materials and Methods

4.1. Chemicals and Materials

The parent drugs (ibuprofen, ketoprofen, flurbiprofen and indomethacin) were purchased from Sigma-Aldrich, Milan, Italy. The galactosylated prodrugs (IbuGAL, OkyGAL, FluGAL and IndoGAL) were synthesized as reported in the work of Magliocca [28]. The plasticware for cultures was obtained from Falcon (Becton Dickinson, Franklin Lakes, NJ, USA). Unless otherwise specified, reagents were obtained from Merck KGaA, (Darmstadt, Germany).

4.1.1. Lipophilicity

ClogP values were calculated with Bio-Loom for Windows software, v. 1.5, BioByte Corporation, Claremont, CA, USA. The distribution coefficients of the compounds between *n*-octanol and PBS at pH 7.4 (log $D^{7.4}$) were experimentally obtained using the shake-flask technique at room temperature, as previously reported [27]. The concentration of the solutes was measured in the aqueous phase on a Shimadzu UV-2501PC UV spectrophotometer, using calibration curves obtained with standard solutions ($r^2 > 0.99$). Each log D value is an average of a minimum of six measurements.

4.1.2. Solubility Assay

Solubility was assessed in deionized water, SGF without pepsin (pH 1.2) and PBS (pH 7.4, 50 mM). The test solid compounds (5 mg) were added to 1 mL of water, PBS and SGF in glass tubes and shaken for 6 h at 25 °C. The suspensions were filtered through a PTFE 0.45 μm filter (VWR), and the resulting solutions were analyzed by means of RP-HPLC (see below). Experiments were performed in triplicate for each compound. Solubility is expressed as mM concentration.

4.1.3. Stability in SGF and PBS

A 10 mM solution of each compound (the galactosylated prodrugs and their parent drugs) was added to either simulated gastric fluid (SGF-without pepsin) or phosphate-buffered saline (PBS, 50 mM). The SGF and PBS buffers were prepared as previously described [27]. The resulting solutions (100 μM) were kept at 37 °C ± 0.5 °C for 48 h; aliquots of 20 μL were withdrawn at appropriate time intervals and analyzed via RP-HPLC (as reported below). The results are expressed as percentages of unmodified galactosylated prodrugs and derived parent drugs after 24 h.

4.1.4. Stability in Human Serum

A solution of each compound (10 mM in DMSO) was added to human serum (sterile-filtered from human male AB plasma, USA origin, sterile-filtered, Sigma-Aldrich) and preheated and maintained at 37 °C ± 0.5 °C. Three-hundred microliters of the resulting solutions (200 μM) was withdrawn at appropriate time intervals and added to 300 μL of CH_3CN, which contained 0.1% TFA, in order to deproteinize the serum. Samples were sonicated, vortexed and then centrifuged for 10 min at 2150× *g*. Each clear supernatant was filtered using 0.45 μm PTFE filters (VWR International S.r.l., Milano, Italy) and analyzed via RP-HPLC as described below. The results are expressed as percentages of unmodified galactosylated prodrugs and derived parent drugs after 24 h.

4.1.5. Serum Protein Binding

Free- and protein-bound drug separation was obtained via ultrafiltration using commercially available membrane systems (Centrifree ultrafiltration devices with ultracel YM-T membrane, Merck, KGaA, Darmstadt, Germany). A solution of each compound in DMSO was added to human serum (sterile-filtered from human male AB plasma, Sigma-Aldrich) to achieve the 100 μM final concentration with 1% DMSO. One milliliter of the solution was inserted into the sample reservoir of the ultrafiltration device and gently shaken in an orbital shaker for 1 h at 37 °C. The tube was centrifuged at 1000× *g* for 15 min. The concentration of the compound both in the ultrafiltrate and filtrate was analyzed via RP-HPLC under the chromatographic conditions described above and using the calibration curves of standard solutions. The linearity of calibration curves was assessed in a concentration range of 1–100 μM ($r^2 > 0.99$). The recovery of the ultrafiltration process was calculated to observe any possible compound loss during the operations in view of the limited solubility of the tested compounds.

$$\text{Recovery} = 100 \times [(V_{\text{bound}} \times C_{\text{bound}}) + (V_{\text{unbound}} \times C_{\text{unbound}})]/V_{\text{initial serum}} \times C_{\text{initial}}$$

$\text{vol.}_{\text{bound}}$: calculated by dividing the weight of the bound fraction (difference between the weights of the empty sample reservoir and that after ultrafiltration) by its density (0.991 g/mL assessed by weighing five replicates of a known volume of bound fraction).
$\text{vol.}_{\text{unbound}}$: calculated by dividing the weight of the unbound fraction (difference between the weight of the ultrafiltrate cup before and after ultrafiltration) by its density (0.999 g/mL assessed by weighing five replicates of a known volume of unbound fraction).
$\text{conc}_{\text{bound}}$: calculated using the RP-HPLC method.
$\text{conc}_{\text{unbound}}$: calculated using the RP-HPLC method (calibration with standard additions).

The average recovery was 98% for all tested compounds.

4.1.6. HPLC Analysis

The reverse-phase HPLC procedure allowed the galactosylated prodrugs and their parent drugs to be separated and quantified. HPLC analyses were performed using a previously used chromatograph system [27]. The analytical column was a ZORBAX Eclipse XDB-C8 (150 × 4.6 mm, 5 μm; Agilent, Santa Clara, CA, USA). The mobile phase consisted of 0.1% aqueous TFA and 0.1% TFA CH_3CN 50/50 *w*/*w*% at a flow rate of 1.0 mL/min. The injection volume was 20 μL (Rheodyne, Cotati, CA, USA). The column effluent was monitored at 228 nm and 254 nm, referenced against an 800 nm wavelength. The quantitation of compounds was calculated using calibration curves and the linearity of the calibration curves was determined in a concentration range of 1–100 μM ($r^2 > 0.99$).

4.2. Permeation Measurements across Excised Rat Small Intestine

An evaluation of ex vivo absorption was carried out using permeation measurements in excised rat small intestines, as previously published [38], with some modifications. Briefly, male Wistar rats (200–250 g) were anesthetized with isoflurane, decapitated and exsanguinated. Freshly excised rat duodenum, jejunum and ileum tissue was washed with Krebs–Ringer buffer (KRB) and cut into pieces of 4–5 cm. Then, 0.2 mL of a suspension containing the compounds (5 mg/Kg, PBS + Methocel 3%) were syringed into intestinal sacs and the compound-enriched tissues were incubated in oxygenated KRB (10 mL) at 37 °C with smooth shaking. Each sample solution (0.5 mL) was taken from the serosal side at fixed time intervals up to 120 min and replaced with fresh buffer. Tests were carried out in triplicate for each compound on three intestinal segments from three different rats. The concentration in the incubation buffer was quantified via RP-HPLC, as reported above. The quantitation was performed using calibration curves of all compounds, the linearity of which was determined in a concentration range of 1–100 μM ($r^2 > 0.99$).

At the end of the incubation period, the intestinal sacs were opened, washed with fresh buffer, homogenized in ethanol (1 mL, UltraTurrax) and centrifuged at 13,000× *g* for 5 min and the supernatant was dried with nitrogen. After resolubilization in 0.5 mL of methanol, the concentrations of the galactosylated prodrugs and prodrug-derived parent drugs were quantified using RP-HPLC.

4.3. Animals

Adult male CD1 mice weighing 25–35 g, purchased from Charles River (Calco, Italy), were used for in vivo and ulcerogenicity studies. Male Wistar rats weighing 200–250 g, purchased from Charles River (Italy), were used for ex vivo absorption evaluations according to [38]. The animals were kept under standard environmental conditions of controlled temperature (22 °C ± 2 °C) and humidity with a 12/12 h light/dark cycle. Food and water were supplied ad libitum, except before the experimental days, when the animals fasted for 12 h. All experiments were approved by the Italian Ministry of Health and were in accordance with the guidelines for the use of laboratory animals. Our best efforts were made to reduce the number of animals used and to minimize their suffering.

4.3.1. Carrageenan-Induced Hyperalgesia

Carrageenan-induced paw edema was induced by injecting λ-carrageenan (1% *w*/*v* in sterile water, 20 μL) subcutaneously into the hind paws of mice using a 27 gauge needle. The drugs were administered orally by gavage 1 h earlier, and the mice were tested 2, 4, 6, 24, 48, 72 and 96 h after the λ-carrageenan injection. As previously reported [39], mouse paw edema developed in two distinct phases: an acute first phase peaking at 4 h and a second phase peaking at 72 h after carrageenan challenge. The increase in paw volume was evaluated using a plethysmometer (Ugo Basile, Milan, Italy) and reported as the difference between the paw volume measured at each time point and the basal paw volume measured immediately before carrageenan injection.

4.3.2. Acetic Acid Writhing Test

The acetic-acid-induced writhing test was performed as reported previously [40]. Briefly, the drugs were orally administered to mice 1 h before the intraperitoneal injection of 0.6% acetic acid (10 mL/kg). The number of abdominal constrictions and extensions of the trunk and hind limbs were counted for each group of mice, starting from 5 min after the injection of acetic acid up to 20 min afterwards, and expressed as the number of writhing episodes.

4.3.3. Ulcerogenicity Studies

NSAID-induced gastric damage in mice was evaluated following the procedure described by Chan et al. [41]. Solutions of either the free-acid parent drug (10 mg/kg) or equimolecular amounts of the galactosylated prodrug or vehicle alone (CMC, 0.5%) were administered orally in fasted (16–18 h) mice (n = 6 per group). Mice were euthanized either 4 h or 48 h after treatment, and the stomach was excised along its greater curvature and rinsed with normal saline. The mucosa was then examined by means of a magnifying glass for the presence of irritation or frank hemorrhagic lesions (ulcers). Irritation was assigned a score of 0, and ulcerations were scored according to their length (a score of 1 for lesions with a length between 1 and 2 mm, a score of 2 for lesions with a length between 2 and 3 mm, a score of 3 for lesions greater than 3 mm). The sum of total scores was used for comparison.

4.4. Statistical Analysis

All analyses were conducted using Graph-Pad Prism (GraphPad Prism, version 7.0 from GraphPad Software Inc., San Diego, CA, USA). For the in vivo antinociceptive tests, the significance of differences between groups was determined using two-way analysis of variance (ANOVA), followed by a Turkey's post hoc test for multiple comparisons. The level of statistical significance was * $p < 0.05$ and was evaluated vs. the control group.

5. Conclusions

By conjugating ibuprofen, ketoprofen, flurbiprofen and indomethacin to D-galactose, the physicochemical properties and bioproperties of their respective parent drugs were changed in such a way as to prolong their pharmacological activity and reduce their toxicity. If a NSAID candidate suffers from unmet requirements, its galactosylated analogue may satisfy them. In light of the results obtained in the present work and previously with ketorolac, aceclofenac and paracetamol, the galactosylated prodrug strategy can be considered a problem-solving technique that can be adopted to overcome the drawbacks of NSAIDs.

Author Contributions: Conceptualization, F.S., R.R. and M.G.R.; methodology, F.S., C.C., L.L., E.M., M.C. and B.R.; validation, M.G.R., S.A. and R.R.; formal Analysis, F.S., C.C., M.C., B.R., E.M. and L.L.; investigation, B.R., E.M. and C.C.; resources, R.R.; data curation, M.G.R. and F.S.; writing—original draft preparation, F.S., writing—review and editing, R.R., M.G.R., S.A. and L.L.; supervision, R.R. and M.G.R.; Funding acquisition, R R. All authors have read and agreed to the published version of the manuscript.

Funding: This research received no external funding.

Institutional Review Board Statement: The animal study protocol was approved by the Institutional Committee on the Ethics of Animal Experiments (CSV) by the Ministry of Health (protocol code n. 0084607,3 October 2014 and 56105.N.ZMT, 23 June 2018).

Informed Consent Statement: Not applicable.

Data Availability Statement: Data is contained within the article.

Acknowledgments: We thank Iona Christine Robinson and Dale James Matthew Lawson for English language editing.

Conflicts of Interest: The authors declare no conflict of interest.

References

1. Rautio, J.; Kumpulainen, H.; Heimbach, T.; Oliyai, R.; Oh, D.; Järvinen, T.; Savolainen, J. Prodrugs: Design and clinical applications. *Nat. Rev. Drug Discov.* **2008**, *7*, 255–270. [CrossRef] [PubMed]
2. Huttunen, K.; Raunio, H.; Rautio, J. Prodrugs: Design and clinical applications. *Pharmacol. Rev.* **2011**, *63*, 750–771. [CrossRef] [PubMed]
3. Stella, V.J.; Nti-Addae, K.W. Prodrug strategies to overcome poor water solubility. *Adv. Drug Deliv. Rev.* **2007**, *59*, 677–694. [CrossRef] [PubMed]
4. Abu-jaish, A.; Jumaa, S.; Karaman, R. Prodrug Overview. In *Prodrugs Design: A New Era*; Karaman, R., Ed.; Nova Publisher: Hauppauge, NY, USA, 2014; pp. 77–102.
5. Chung, M.C.; Silva, A.T.D.A.; Castro, L.F.; Güido, R.V.C.; Nassute, J.C.; Ferreira, E.I. Latenciação e formas avançadas de transporte de fármacos. *Rev. Bras. Ciênc. Farm.* **2005**, *41*, 155–180. [CrossRef]
6. Redasani, V.K.; Bari, S.B. *Prodrug Design: Perspectives, Approaches and Applications in Medicinal Chemistry*, 1st ed.; Academic Press: London, UK, 2015.
7. Zovko, M.; Zorc, B.; Novak, P.; Tepeš, P.; Cetina-Čižmek, B.; Horvat, M. Macromolecular prodrugs: XI. Synthesis and characterization of polymer–estradiol conjugate. *Int. J. Pharm.* **2004**, *285*, 35–41. [PubMed]
8. Fathia, M.I.; Mubark, E.O. Review: Carbohydrates chemistry. *Asian J. Sci. Technol.* **2017**, *7*, 5038–5043.
9. Mishra, S.; Upadhaya, K.; Mishra, K.B.; Shukla, A.K.; Tripathi, R.P.; Tiwari, V.K. Carbohydrate-based therapeutics: A frontier in drug discovery and development. *Stud. Nat. Prod. Chem.* **2016**, *49*, 307–361.
10. Bertozzi, C.R.; Kiessling, L.L. Chemical glycobiology. *Science* **2001**, *291*, 2357–2364. [CrossRef]
11. Wu, X.; Delbianco, M.; Anggara, K.; Michnowicz, T.; Pardo-Vargas, A.; Bharate, P.; Sen, S.; Pristl, M.; Rauschenbach, S.; Schlickum, U.; et al. Publisher Correction: Imaging single glycans. *Nature* **2020**, *582*, 375–378. [CrossRef]
12. Wang, J.; Zhang, Y.; Lu, Q.; Xing, D.; Zhang, R. Exploring Carbohydrates for Therapeutics: A Review on Future Directions. *Front. Pharmacol.* **2021**, *12*, 756724. [CrossRef]
13. Jornada, D.H.; dos Santos Fernandes, G.F.; Chiba, D.E.; de Melo, T.R.; dos Santos, J.L.; Chung, M.C. The Prodrug Approach: A Successful Tool for Improving Drug Solubility. *Molecules* **2016**, *21*, 42. [CrossRef] [PubMed]
14. Schuh, L.; Müller, P.; Torvisco, A.; Stueger, H.; Wrodnigg, T.M ; Haas, M. Synthesis of D-Galactose-Substituted Acylsilanes and Acylgermanes. Model Compounds for Visible Light Photoinitiators with Intriguing High Solubility. *Organometallics* **2021**, *40*, 1185–1189. [CrossRef] [PubMed]
15. Di Guida, F.; Pirozzi, C.; Magliocca, S.; Santoro, A.; Lama, A.; Russo, R.; Nieddu, M.; Burrai, L.; Boatto, G.; Mollica, M.P.; et al. Galactosylated Pro−Drug of Ursodeoxycholic Acid: Design, Synthesis, Characterization, and Pharmacological Effects in a Rat Model of Estrogen-Induced Cholestasis. *Mol. Pharm.* **2018**, *15*, 21–30. [CrossRef] [PubMed]

16. Perrone, M.R.; Vitale, P.; Ferorelli, S.; Boccarelli, A.; Coluccia, M.; Pannunzio, A.; Campanella, F.; Di Mauro, G.; Bonaccorso, C.; Fortuna, C.G.; et al. Effect of mofezolac-galactose distance in conjugates targeting cyclooxygenase (COX)-1 and CNS GLUT-1 carrier. *Eur. J. Med. Chem.* **2017**, *141*, 404–416. [CrossRef] [PubMed]
17. Lima, A.O.D.L.; Salamanca-Fernández, E.; Del Rey, E.J.A.; Hoces, A.M.; Vera, M.G.; Tamayo, C.B. Safety considerations during prescription of non-steroidal anti-inflammatory drugs (NSAIDs), through a review of systematic reviews. *An. Sist. Sanit. Navar.* **2021**, *44*, 261–273. [CrossRef]
18. Bindu, S.; Mazumder, S.; Bandyopadhyay, U. Non-steroidal anti-inflammatory drugs (NSAIDs) and organ damage: A current perspective. *Biochem. Pharmacol.* **2020**, *180*, 114147. [CrossRef]
19. Tsutsumi, S.; Gotoh, T.; Tomisato, W.; Mima, S.; Hoshino, T.; Hwang, H.J.; Takenaka, H.; Tsuchiya, T.; Mori, M.; Mizushima, T. Endoplasmic reticulum stress response is involved in nonsteroidal anti-inflammatory drug-induced apoptosis. *Cell Death Differ.* **2004**, *11*, 1009–1016. [CrossRef]
20. James, D.S. The multisystem adverse effects of NSAID therapy. *J. Am. Osteopath. Assoc.* **1999**, *99* (Suppl. 11), S1–S7. [CrossRef]
21. Hooper, W.P.; Gerber, H. Monte Carlo simulations of laser-generated sea surface aureole. *Appl. Opt.* **1988**, *27*, 5111–5118. [CrossRef]
22. Arfe, A.; Scotti, L.; Varas-Lorenzo, C.; Nicotra, F.; Zambon, A.; Kollhorst, B.; Schink, T.; Garbe, E.; Herings, R.; Straatman, H.; et al. Non-steroidal anti-inflammatory drugs and risk of heart failure in four European countries: Nested case-control study. *BMJ* **2016**, *354*, i4857. [CrossRef]
23. Dreischulte, T.; Morales, D.R.; Bell, S.; Guthrie, B. Combined use of nonsteroidal anti-inflammatory drugs with diuretics and/or renin-angiotensin system inhibitors in the community increases the risk of acute kidney injury. *Kidney Int.* **2015**, *88*, 396–403. [CrossRef] [PubMed]
24. Angi, A.; Lakatos, L. Adverse effects of non-steroidal anti-inflammatory drugs in the lower gastrointestinal tract. *Orv. Hetil.* **2009**, *150*, 27–34. [CrossRef] [PubMed]
25. Sodano, F.; Avallone, B.; Tizzano, M.; Fogliano, C.; Rolando, B.; Gazzano, E.; Riganti, C.; Magliocca, S.; Cuozzo, M.; Albrizio, S.; et al. Ketogal Safety Profile in Human Primary Colonic Epithelial Cells and in Mice. *Pharmaceuticals* **2021**, *14*, 1149. [CrossRef] [PubMed]
26. Sodano, F.; Lazzarato, L.; Rolando, B.; Spyrakis, F.; De Caro, C.; Magliocca, S.; Marabello, D.; Chegaev, K.; Gazzano, E.; Riganti, C.; et al. Paracetamol−Galactose Conjugate: A Novel Prodrug for an Old Analgesic Drug. *Mol. Pharm.* **2019**, *16*, 4181–4189. [CrossRef] [PubMed]
27. Magliocca, S.; De Caro, C.; Lazzarato, L.; Russo, R.; Rolando, B.; Chegaev, K.; Marini, E.; Nieddu, M.; Burrai, L.; Boatto, G ; et al. Aceclofenac−Galactose Conjugate: Design, Synthesis, Characterization, and Pharmacological and Toxicological Evaluations. *Mol. Pharm.* **2018**, *15*, 3101–3110. [CrossRef] [PubMed]
28. Magliocca, S.; Sodano, F.; Nieddu, M.; Burrai, L.; Boatto, G.; Rimoli, M.G. New Galactosylated NSAIDS prodrugs in a green context: Synthesis and stability. *IJPSR* **2017**, *8*, 1575–1581.
29. Ruan, L.P.; Chen, S.; Yu, B.Y.; Zhu, D.N.; Cordell, G.A.; Qiu, S.X. Prediction of human absorption of natural compounds by the non-everted rat intestinal sac model. *Eur. J. Med. Chem.* **2006**, *41*, 605–610. [CrossRef]
30. Luo, Z.; Liu, Y.; Zhao, B.; Tang, M.; Dong, H.; Zhang, L.; Lv, B.; Wei, L. Ex vivo and in situ approaches used to study intestinal absorption. *J. Pharmacol. Toxicol. Methods* **2013**, *68*, 208–216. [CrossRef]
31. Kapoor, V.K.; Kaur, A. Drug-glycosidation and drug development. *Mini Rev. Med. Chem.* **2013**, *13*, 584–596. [CrossRef]
32. Guerrero, A.; Guiho, R.; Herranz, N.; Uren, A.; Withers, D.J.; Martínez-Barbera, J.P.; Tietze, L.F.; Gil, J. Galactose-modified duocarmycin prodrugs as senolytics. *Aging Cell.* **2020**, *19*, e13133. [CrossRef]
33. Verbeeck, R.K.; Blackburn, J.L.; Loewen, G.R. Clinical Pharmacokinetics of Non-steroidal Anti-inflammatory Drugs. *Clin. Pharm.* **1983**, *8*, 297–331. [CrossRef] [PubMed]
34. Ashraf, Z.; Alamgeer, M.K.; Hassan, M.; Abdullah, S.; Waheed, M.; Ahsan, H.; Kim, S.J. Flurbiprofen–antioxidant mutual prodrugs as safer nonsteroidal anti-inflammatory drugs: Synthesis, pharmacological investigation, and computational molecular modeling. *Drug Des. Devel. Ther.* **2016**, *10*, 2401–2419. [PubMed]
35. Dhakane, V.D.; Thakare, V.N.; Dongare, S.B.; Bhale, P.S.; Mule, Y.B.; Bandgar, B.P.; Chavan, H.V. Preparation and Pharmacological Evaluation of Novel Orally Active Ester Prodrugs of Ketoprofen with Non-Ulcerogenic Property. *Chem. Biol. Drug Des.* **2016**, *87*, 878–884. [CrossRef] [PubMed]
36. Maciel, H.P.; Cardoso, L.G.; Ferreira, L.R.; Perazzo, F.F.; Carvalho, J.C.T. Anti-inflammatory and ulcerogenic effects of indomethacin and tenoxicam in combination with cimetidine. *Inflammopharmacology* **2004**, *2*, 203–210. [CrossRef]
37. Jarosz, M.; Szkaradek, N.; Marona, H.; Nowak, G.; Młyniec, K.; Librowski, T. Evaluation of anti-inflammatory and ulcerogenic potential of zinc–ibuprofen and zinc–naproxen complexes in rats. *Inflammopharmacology* **2017**, *25*, 653–663. [CrossRef] [PubMed]
38. Cocco, M.; Pellegrini, C.; Martínez-Banaclocha, H.; Giorgis, M.; Marini, E.; Costale, A.; Miglio, G.; Fornai, M.; Antonicli, L.; López-Castejón, G.; et al. Development of an Acrylate Derivative Targeting the NLRP3 Inflam-masome for the Treatment of Inflammatory Bowel Disease. *J. Med. Chem.* **2017**, *60*, 3656–3671. [CrossRef]
39. Posadas, I.; Bucci, M.; Roviezzo, F.; Rossi, A.; Parente, L.; Sautebin, L.; Cirino, G. Carrageenan-induced mouse paw oedema is biphasic, age-weight dependent and displays differential nitric oxide cyclooxygenase-2 expression. *Br. J. Pharmacol.* **2004**, *142*, 331–338. [CrossRef]

40. Hokanson, G. Acetic acid for analgesic screening. *J. Nat. Prod.* **1978**, *41*, 497–498.
41. Chan, C.C.; Boyce, S.; Brideau, C.; Ford-Hutchinson, A.W.; Gordon, R.; Guay, D.; Hill, R.G.; Li, C.S.; Mancini, J.; Penneton, M.; et al. Pharmacology of a selective cyclooxygenase-2 inhibitor, L-745,337: A novel nonsteroidal anti-inflammatory agent with an ulcerogenic sparing effect in rat and nonhuman primate stomach. *J. Pharmacol. Exp. Ther.* **1995**, *3*, 1531–1537.

pharmaceuticals

MDPI

Article

A Lead-Based Fragment Library Screening of the Glycosyltransferase WaaG from *Escherichia coli*

Federico Riu [1], Alessandro Ruda [2], Olof Engström [2], Claudio Muheim [3], Hani Mobarak [2], Jonas Ståhle [2], Paul Kosma [4], Antonio Carta [1], Daniel O. Daley [3] and Göran Widmalm [2,*]

1 Department of Medical, Surgical and Experimental Sciences, University of Sassari, Via Muroni, 23A, 07100 Sassari, Italy; federicoriu1@gmail.com (F.R.); acarta@uniss.it (A.C.)
2 Arrhenius Laboratory, Department of Organic Chemistry, Stockholm University, S-106 91 Stockholm, Sweden; alessandro.ruda@su.se (A.R.); j.olof.engstrom@gmail.com (O.E.); hani@gmx.se (H.M.); jstaahle@gmail.com (J.S.)
3 Arrhenius Laboratory, Department of Biochemistry and Biophysics, Stockholm University, S-106 91 Stockholm, Sweden; clamuheim@gmail.com (C.M.); daniel.daley@dbb.su.se (D.O.D.)
4 Department of Chemistry, University of Natural Resources and Life Sciences—Vienna, 1190 Vienna, Austria; paul.kosma@boku.ac.at
* Correspondence: goran.widmalm@su.se

Citation: Riu, F.; Ruda, A.; Engström, O.; Muheim, C.; Mobarak, H.; Ståhle, J.; Kosma, P.; Carta, A.; Daley, D.O.; Widmalm, G. A Lead-Based Fragment Library Screening of the Glycosyltransferase WaaG from *Escherichia coli*. *Pharmaceuticals* **2022**, *15*, 209. https://doi.org/10.3390/ph15020209

Academic Editors: Federica Sodano and Maria Grazia Rimoli

Received: 15 January 2022
Accepted: 6 February 2022
Published: 9 February 2022

Abstract: Glucosyl transferase I (WaaG) in *E. coli* catalyzes the transfer of an α-D-glucosyl group to the inner core of the lipopolysaccharide (LPS) and plays an important role in the biogenesis of the outer membrane. If its activity could be inhibited, the integrity of the outer membrane would be compromised and the bacterium would be susceptible to antibiotics that are normally prevented from entering the cell. Herein, three libraries of molecules (A, B and C) were docked in the binding pocket of WaaG, utilizing the docking binding affinity as a filter to select fragment-based compounds for further investigations. From the results of the docking procedure, a selection of compounds was investigated by molecular dynamics (MD) simulations to obtain binding free energy (BFE) and K_D values for ligands as an evaluation for the binding to WaaG. Derivatives of 1,3-thiazoles (**A7** and **A4**) from library A and 1,3,4-thiadiazole (**B33**) from library B displayed a promising profile of BFE, with K_D < mM, viz., 0.11, 0.62 and 0.04 mM, respectively. Further root-mean-square-deviation (RMSD), electrostatic/van der Waals contribution to the binding and H-bond interactions displayed a favorable profile for ligands **A4** and **B33**. Mannose and/or heptose-containing disaccharides **C1–C4**, representing sub-structures of the inner core of the LPS, were also investigated by MD simulations, and compound **C4**$^{2-}$ showed a calculated K_D = 0.4 μM. In the presence of UDP-Glc^{2-}, the best-docked pose of disaccharide **C4**$^{2-}$ is proximate to the glucose-binding site of WaaG. A study of the variation in angle and distance was performed on the different portions of WaaG (N-, the C- domains and the hinge region). The Spearman correlation coefficient between the two variables was close to unity, where both variables increase in the same way, suggesting a conformational rearrangement of the protein during the MD simulation, revealing molecular motions of the enzyme that may be part of the catalytic cycle. Selected compounds were also analyzed by Saturation Transfer Difference (STD) NMR experiments. STD effects were notable for the 1,3-thiazole derivatives **A4**, **A8** and **A15** with the apo form of the protein as well as in the presence of UDP for **A4**.

Keywords: molecular docking; molecular dynamics; binding free energy; NMR spectroscopy

1. Introduction

Pathogenic multi-drug resistant (MDR) strains of gram-negative bacteria are currently a major public health concern. They represent the causative agents of diseases such as gastroenteritis, infections of the urinary tract, the blood and the central nervous system [1]. The World Health Organization (WHO) has published a list of priority pathogens, highlighting the concern for the lack of treatment solutions for gram-negative pathogens, including

Escherichia coli [2]. Furthermore, antibiotic resistance has been repeatedly reported as an alarming global issue [3], since bacteria easily spread across countries through goods, animals and people. It also represents a significant problem for major surgery and cancer treatment, which leads to large public health expenses.

Moreover, certain strains of *E. coli* have easily and quickly spread in the community. In 2011, some countries in the WHO European Region experienced a significant *E. coli* outbreak: the increasing cases of bloody diarrhea and hemolytic syndrome (HUS) were traced to infections from verocytotoxin-producing *E. coli* O104:H4 in Germany and in 15 other countries in Europe and North America [4]. Another consequence of *E. coli* resistance is the treatment of urinary tract infections with fluoroquinolone antibiotics. In many countries, this treatment is becoming increasingly ineffective [5]. Thus, there is a pressing need to find better ways to use the currently available antibiotics [6–10].

Bacteria develop antibiotic resistance in many different ways. It can be acquired through mutation and horizontal gene transfer, or it can be intrinsic (e.g., the strain is naturally insensitive to the antibiotic). The outer membrane in gram-negative bacteria is a major contributing factor to intrinsic resistance [11–14]. It acts as a selective permeability barrier that enables the entry of essential nutrients and ions whilst simultaneously preventing the entry of many classes of antibiotics into the cell [15,16].

As a first step to overcome intrinsic resistance in *E. coli*, we have investigated the possibility to inhibit glucosyl transferase I (WaaG or RfaG), a protein that catalyzes the transfer of an α-D-glucosyl group from UDP-Glc onto L-*glycero*-D-*manno*-heptose-II during the synthesis of the core region of the lipopolysaccharide (LPS) [17]. The general structure of LPS has come from different studies on various gram-negative bacteria. From a structural point of view, lipid A is the membrane-embedded lipophilic portion of LPS consisting of glucosamine dimers (D-GlcN), which are bound to acyl chains. Lipid A is attached to the core region containing Kdo (3-deoxy-D-*manno*-2-octulosonic acid) and L-*glycero*-D-*manno*-heptose (L,D-Hep), as well as some hexosamines and hexoses (Figure 1). The O-polysaccharides (or O-antigens) are, when present, linked to the core region. Significantly, deletion of the gene encoding WaaG destabilizes the outer membrane of *E. coli* by interfering with LPS core phosphorylation and results in truncated LPS immediately after the inner core heptose residues [18]. This perturbation changes the chemical structure of the LPS and makes *E. coli* more susceptible to different classes of antibiotics—for example, rifamycins and aminoglycosides [19].

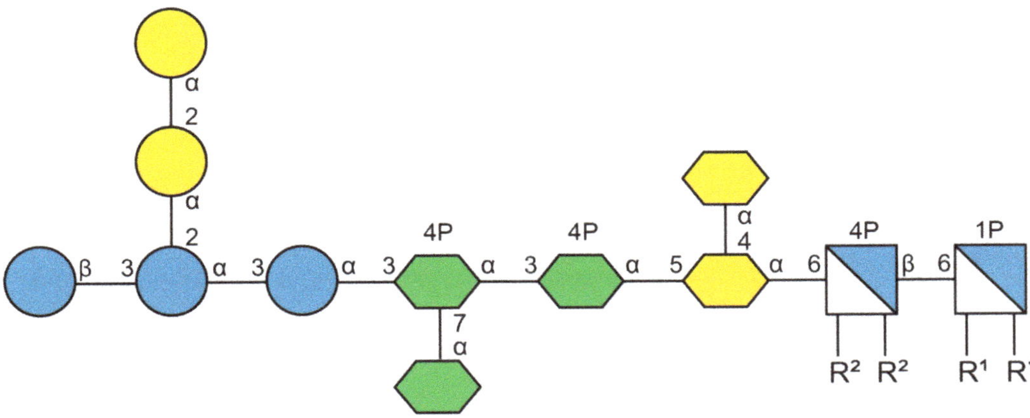

Figure 1. Structural representation of outer (to the left) and inner (middle section) portions of the R1 core and the lipid A (to the right) of LPS [20]. R^1 and R^2 represent acyl groups attached to glucosamine residues of Lipid A. The structure is reported utilizing the Symbol Nomenclature For Glycans (SFNG) representation [21].

Small molecules that bind to WaaG in vitro could represent a strategic way to make *E. coli* more sensitive to different drugs, such as rifampicin, chloramphenicol and colistin (Figure 2). Colistin-resistant bacteria are becoming significantly untreatable because of their wide spreading [5]. The small organic molecules will serve as scaffolds for the design of more effective inhibitors of WaaG and could be used as co-drugs that potentiate the action of antibiotics in gram-negative bacteria.

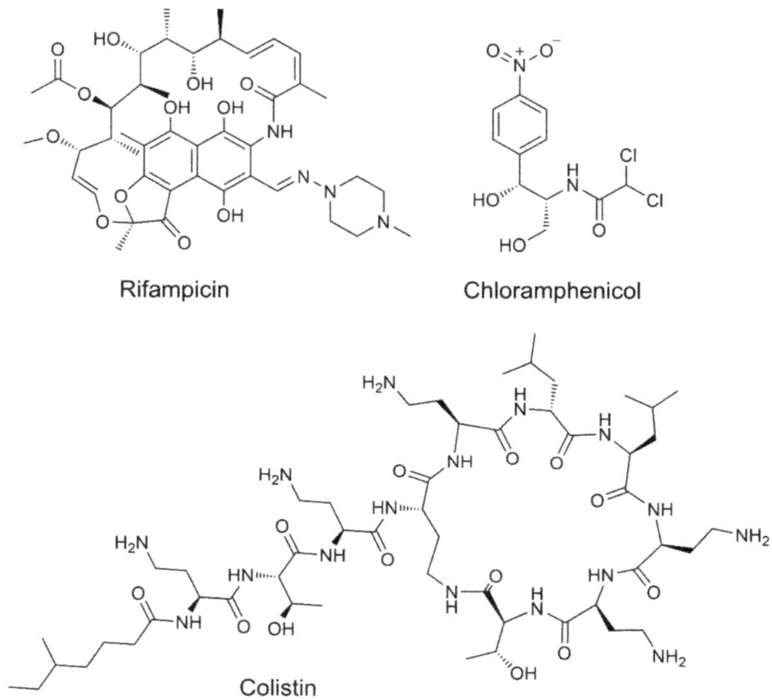

Figure 2. Structural formulas for rifampicin, chloramphenicol and colistin.

Previous molecular docking and NMR studies elucidated the binding of natural ligands to WaaG, showing 0.6 mM and 9 μM as K_D values for uridine and UDP, respectively [22]. Three compounds (in this work reported as **A1–A3**) from an Ro3 Maybridge 2006 Library compliant with the 'Rule of Three' for fragment-based lead discovery, were selected through NMR spectroscopy screening and utilized for preliminary docking studies [22]. Compound **A1** inhibited WaaG with an IC_{50} = 1.0 mM, which opened the door for new structural modifications that act as potent inhibitors [23].

In order to investigate leads to new inhibitors of WaaG, an initial fragment-based drug discovery (FBDD) process included the selection of different fragments (library A), small molecules (library B) and, for comparison, oligosaccharides (library C) (Scheme 1). Molecular docking and further molecular dynamics (MD) simulations were conducted on the best-docked poses of the most promising molecules. At a later stage, STD NMR (Saturation Transfer Difference Nuclear Magnetic Resonance) [24] spectroscopy and complementary experiments addressing inhibitory activity were carried out through in vitro biochemical assessment using libraries A and B. The STD NMR experiment is a robust and valuable method [25] for the determination of the binding of ligands to proteins [26] due to its versatility and low susceptibility to false positives. It requires only a small amount of unlabeled protein, and it is used in combination with docking studies to predict the conformational state of a ligand in a binding pocket [27].

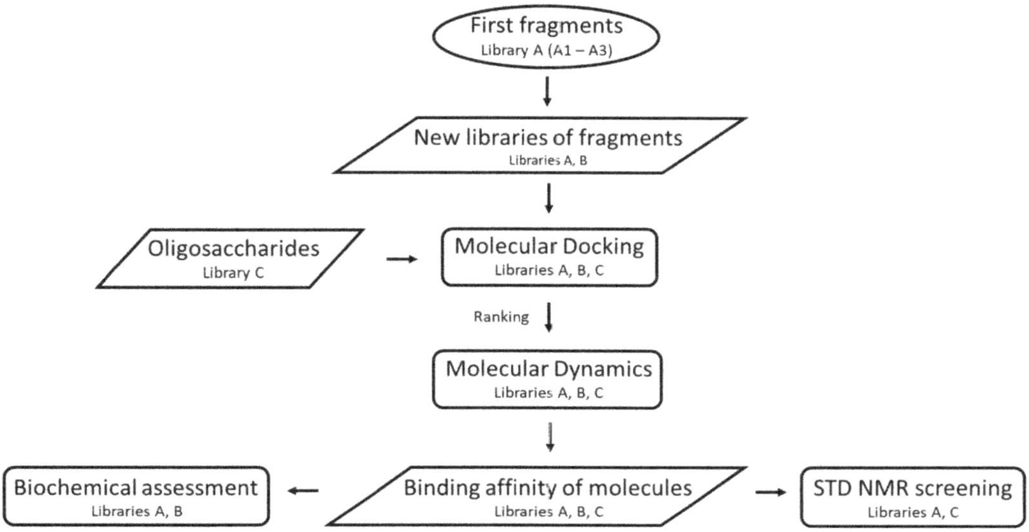

Scheme 1. Flowchart of the computational and experimental study on WaaG and its interactions with compounds from libraries A, B and C.

2. Results

2.1. Construction of Libraries

The present study considers glycosyltransferase WaaG as the target for further fragment-based studies [28]. In the RCSB protein database (PDB), there is a deposited crystal structure of WaaG bound to UDP-2-fluoro-2-deoxy-D-glucose (U2F) with the PDB code 2iw1. This protein structure was utilized for further docking and molecular dynamics investigations, as well as to generate a model of UDP-Glc^{2-} bound to the protein.

In our previous study, the drug discovery fragment-based approach involved approximately 500 compounds from the Ro3 Maybridge 2006 library and STD NMR spectroscopy. Molecular docking and MD simulations were employed to evaluate their binding affinity for the active site of WaaG [22]. Three bicyclic compounds (**A1–A3**) were selected according to their binding activity towards WaaG, competing with UDP-Glc. The docking studies showed that uridine, UDP, **A1** and **A3** bind to the same pocket. In contrast, **A2** binds to the glucosyl region in the active site [22].

In this work, we present a fragment-based approach to find a new set of compounds that may inhibit the enzymatic activity of WaaG. The above-mentioned ligands **A1–A3** led to a first ligand library consisting of 20 hetero-bicyclic compounds (**A1–A20)** with a range of molecular weights between 158 and 257 Da (library A). The scaffolds of compounds **A1–A3** (phenylthiazole, phenylpyrrole and methylpyridine-pyrrole, respectively), were modified by adding or removing polar, e.g., hydroxyl, carboxyl, amide, aldehyde or nonpolar (e.g., methyl, ethyl, phenyl, halogen atoms) functional groups. Figure 3A shows a schematic representation of the scaffolds of the library, and Figure S1 depicts the chemical structure of each ligand. Further modifications involved the phenylthiazole scaffold transformed into an indole thiazole (**A14**). Moreover, the acetate salt of a phenylthiazole derivative was included in the library (**A13**).

In order to find drug-like inhibitors, a second chemical library, library B, was generated. Library B includes 37 larger fragments (**B1–B33**) structurally expanded from library A. They are different in terms of molecular weight (M_w = 176–391 Da). Some of these fragments are hetero-bicyclic molecules bearing a variety of heterocycle rings, e.g., triazole, thiadiazole. Figure 3B presents the different heterocycles included in the structure of the ligands. In order to enhance ligand-receptor affinity, ligands with higher M_W were devised through a fragment growing process [29]. Figures S2 and S3 show the structure of each ligand.

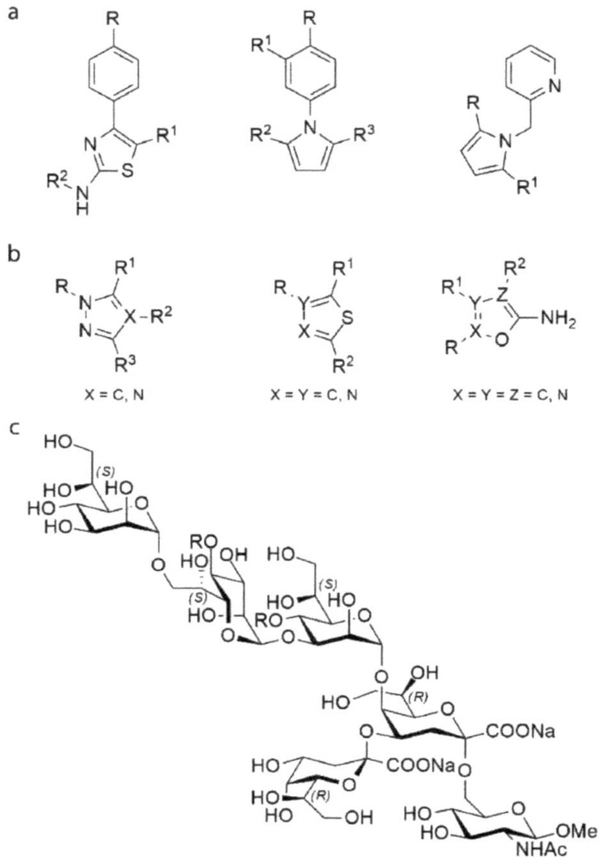

Figure 3. Chemical backbone structure of the ligands from library A (**a**) and B (**b**); in the carbohydrate-based library C (**c**) R = HPO_3Na. Figures S1–S5 report all ligands of libraries A, B and C.

In order to study the interactions between WaaG, UDP-Glc (donor) and the acceptor site of the inner core of LPS, a set of oligosaccharides were included in a third library, library C (**C1–C12**). The structures of this library mimic the outer core of LPS to different extents. A schematic representation of the oligosaccharide structures is depicted in Figure 3C. The first acceptor compound is represented by α-D-Man*p*-(1→3)-α-D-Man*p*-OMe (**C1**). Four disaccharide compounds were designed to mimic building blocks of the LPS inner core: L-α-D-Hep*p*-(1→3)-L-α-D-Hep*p*-OMe (**C2**), L-α-D-Hep*p*-(1→3)-L-α-D-Hep*p*4*P*-OMe (**C3**), L-α-D-Hep*p*4*P*-(1→3)-L-α-D-Hep*p*4*P*-OMe (**C4**) and L-α-D-Hep*p*-(1→7)-L-α-D-Hep*p*4*P*-OMe (**C5**). Focusing on the interface between the inner core and the Lipid A of LPS, the above-mentioned disaccharide structures were extended, generating additional ligands of library C: L-α-D-Hep*p*-(1→7)-L-α-D-Hep*p*4*P*-(1→3)-L-α-D-Hep*p*4*P*-(1→4)-α-Kdo*p* (**C6**), L-α-D-Hep*p*-(1→7)-L-α-D-Hep*p*-(1→3)-L-α-D-Hep*p*4*P*-(1→4)-α-Kdo*p* (**C7**), L-α-D-Hep*p*-(1→7)-L-α-D-Hep*p*4*P*-(1→3)-L-α-D-Hep*p*-(1→4)-α-Kdo*p* (**C8**), L-α-D-Hep*p*-(1→7)-L-α-D-Hep*p*-(1→3)-L-α-D-Hep*p*-(1→4)-α-Kdo*p* (**C9**). Compounds **C1–C9** are shown in Figure S4. Starting from the structure of the tetrasaccharide **C6**, the oligosaccharide structure was extended at the reducing end residue to give the following ligands: L-α-D-Hep*p*-(1→7)-L-α-

D-Hepp4P-(1→3)-L-α-D-Hepp4P-(1→5)-[α-Kdop(2→4)]-α-Kdop (**C10**), L-α-D-Hepp-(1→7)-L-α-D-Hepp4P-(1→3)-L-α-D-Hepp4P-(1→5)-[α-Kdop(2→4)]-α-Kdop-OMe (**C11**), L-α-D-Hepp-(1→7)-L-α-D-Hepp4P-(1→3)-L-α-D-Hepp4P-(1→5)-[α-Kdop(2→4)]-α-Kdop-(2→6)-β-D-GlcpNAc-OMe (**C12**). Compounds **C10–C12** are presented in Figure S5.

2.2. Molecular Docking

Four docking programs, viz., AutoDock Vina [30], LeDock [31], rDock [32] and GOLD [33], were selected based on their sampling and scoring functions [34]. The different algorithms and approaches led to similar binding poses for the ligands. However, the predicted order of poses was usually different in terms of affinity energy. Each docking program generated the top energy-ranked binding poses for ligands of libraries A, B and C. The ranked affinity energies and scores are reported in Table S2. By comparing the binding mode of the natural ligand UDP-Glc^{2-} (Figure 4) derived from the co-crystallized structure of U2F, in the 2iw1 structure and the best-predicted poses of ligands with higher affinity, it was possible to see if the ligands bind preferably in the uridine sub-pocket, in the diphosphate region or in another portion of the binding pocket. Each ligand was docked in two different environments: (i) without any donor-related structure in the binding pocket of WaaG (i.e., the *apo*-form of protein) and (ii) having UDP-Glc^{2-} in the active site. The analysis of the results for the docking studies with the *apo*-protein was made considering the interactions of each ligand with the binding pocket of WaaG.

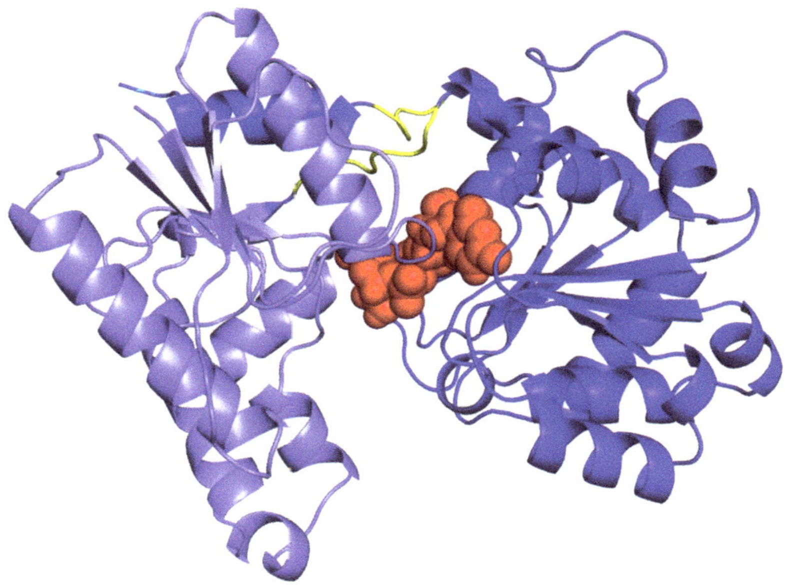

Figure 4. Conformation of UDP-Glc^{2-} in the binding pocket of WaaG, derived from U2F deposited co-crystallized structure (PDB ID 2iw1).

The first docking simulations involved the *apo*-form of WaaG and the ligands of the three libraries. The top-scoring hits of library A showed AutoDock Vina affinity energies that ranged from −6.6 to −8.7 kcal·mol^{-1}, comparable to the binding free energies of fragments investigated in the design of norovirus inhibitors [35].

Docking with WaaG apo-form: library A. In the best binding pose of **A1**, the ligand occupies the uridine sub-pocket. The HO3′ of the ribosyl residue in UDP-Glc^{2-} interacts with the carboxylic group of Glu289 present in the binding pocket of WaaG [36]. In the same way, the amino group and the nitrogen atom of the thiazole in compound **A1** make H-bonds with the same carboxyl group of Glu289. Compound **A7** sits instead in the UDP-

Glc^{2-} uridine-diphosphate portion of the binding pocket. The nitrogen of the thiazole ring in **A7** receives an H-bond from the NH_3^+ group of Lys209, while the amino group is the donor of an H-bond with the carboxylate group of Glu281. Thus, the phenyl-aminothiazole backbone is favorable for binding in the donor pocket of WaaG. The phenylthiazole scaffold of **A9** overlaps with the uridine residue of UDP-Glc^{2-}, but the thiazole ring is oriented differently. In its best-predicted pose, ligand **A9** interacts with its thiazole nitrogen with the guanidino NH_2 of Arg173, while the amino group of **A9** donates an H-bond to the carboxylate of Glu289. The top-ranked pose of compound **A14** has the ligand binding in the uridine portion of the donor pocket; in particular, the indole portion is oriented in the same way as the uracil ring, while the thiazole ring is in the ribose binding region. Compound **A16** binds in the same region as UDP-Glc^{2-}, where its carboxyl group overlaps with the α-phosphate portion of UDP-Glc^{2-} (the phosphate group linked to the ribosyl residue). The carboxyl group on the phenyl ring plays a noteworthy role in the binding: it makes H-bonds with Ile285 and Val286, known to interact with the α-phosphate group of UDP-Glc^{2-}. The hydroxyl group of Ser204 acts as a donor for a hydrogen bond to the oxygen atom of the aldehyde group on the pyrrole ring in **A16**.

Docking with WaaG apo-*form: library B*. The top-scoring ligands of library B ranked their AutoDock Vina affinity energies from −7.0 to −8.8 kcal·mol^{-1}. Compound **B3d** has a favorable binding conformation for this stereoisomer. The other high scoring ligands **B16**, **B22**, **B26** and **B33** bind in the same UDP-Glc^{2-} pocket region. The benzimidazole moiety of compound **B16** is positioned in the UDP-Glc^{2-} ribose site, and the thiadiazole ring is in the β-phosphate region, while the phenyl portion is in the UDP-Glc^{2-} glucose region. The nitrogen atoms of the thiadiazole ring in **B16** play an important role: they act as acceptors of polar interactions both with the NH_3^+ group of Lys209 and with the guanidine NH_2 group of Arg208. Compounds **B22** and **B33** bind in the uridine binding pocket, and the O-phenyl moiety of **B22** occupies the uracil portion in the binding pocket, while the ester chain overlaps with the phosphate-region of UDP-Glc^{2-}. The carbonyl group of the ester group points in the same direction as a P=O group of UDP-Glc^{2-}. Compound **B26** mainly occupies the UDP-Glc^{2-} binding region and points out of the pocket with one of the methylphenoxy groups. As for **B16**, both thiadiazole nitrogen atoms of compound **B26** interact with the donor NH_3^+ group of Lys209, while the sulfonamide group accepts two H-bonds from Gly15 and Leu16. The thiadiazole nitrogen of **B33** interacts with the amino group of Val286 in the UDP-Glc^{2-} pocket region. The lateral sulfonamide NH hydrogen atom makes H-bonds with the carboxyl group of Glu281. Each of the sulfonyl oxygens in **B33** interacts with the NH hydrogen atoms of Ala283 and Ile285.

Docking with WaaG apo-*form: library C*. By focusing on WaaG *apo*-protein, each docked pose of oligosaccharides **C1–C12** was analyzed for its interaction with the exposed amino acids within the binding pocket. The ligands of library C generally bind in the outer portion of the binding pocket of WaaG, where some overlap corresponds to the region of the β-phosphate group of UDP-Glc^{2-}. The most interesting ligands from library C were **C6** and **C10**. Tetrasaccharide **C6** docked in the binding pocket with its non-reducing end heptose (Hep-III), whereas the Kdo residue is pointing out of the pocket. Hep-III overlaps with the ribose portion of UDP-Glc^{2-}—in particular, the hydroxymethyl group of the ribose moiety and the pyranose ring of Hep-III in **C6**. Regarding the best binding poses of **C10**, the two Kdo residues point out of the pocket, while the Hep-III portion almost overlaps with UDP-Glc^{2-} ribose moiety. Furthermore, the Hep-II phosphate group binds in the region of the UDP-Glc^{2-} β-phosphate. A selection of representative ligands is reported in their best-docked poses in Figure 5a–c. Docking studies in the presence of WaaG-UDP-Glc^{2-} complex are discussed in some detail in the Supporting Information. Ligands with higher affinity for their best-predicted pose were chosen and subjected to molecular dynamics (MD) simulations.

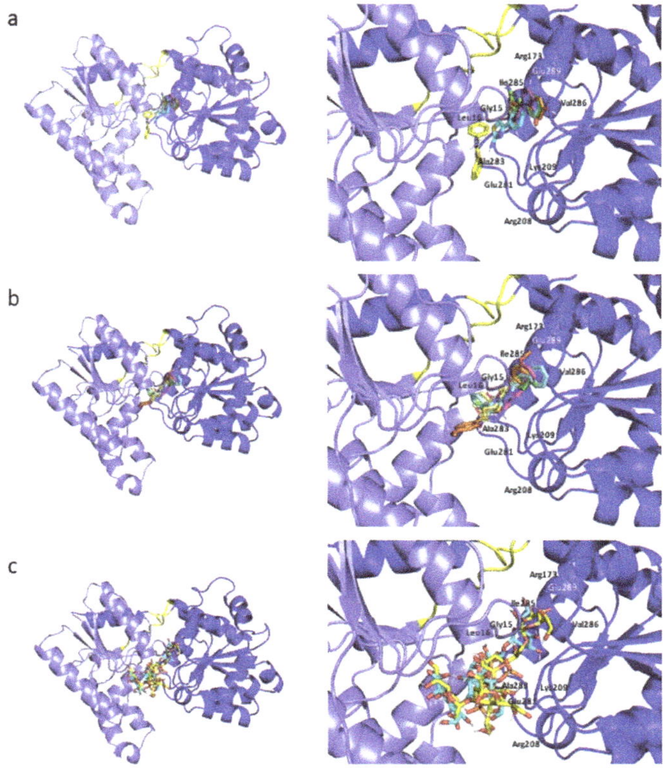

Figure 5. Docking poses of the representative ligands of libraries (**a**) A, (**b**) B and (**c**) C. On the right, a close-up of WaaG binding pocket showing the most relevant amino acids surrounding the ligands structures. (**a**) compound **A1** in green, **A7** in blue marine, **A9** in yellow, **A14** in orange and **A16** in violet. (**b**) compound **B3d** in blue marine, **B16** in light yellow, **B22** in fuchsia, **B26** in orange and **B33** in green. (**c**) compound **C6** in blue marine and **C10** in yellow.

2.3. Molecular Dynamics Simulations

In order to better understand the interactions between WaaG and its natural substrates and to further extend the results from docking studies of libraries A, B and C, 10-ns molecular dynamics (MD) simulations were carried out on both ligands and protein-ligand complexes. For library A, ligands **A4**, **A8** and **A15** were selected via STD NMR experiments (*vide infra*), exhibiting a good protein-ligand affinity profile. Docking studies (in the presence of WaaG *apo*-form or WaaG-UDP-Glc^{2-} complex) allowed a selection of ligands with the highest affinity, to be further investigated with MD simulations. Only the top-ranked ligands from each library were considered for the MD calculations. For the selected ligands, of the ten binding poses generated by each run for every ligand, the top-predicted binding pose was utilized to generate the initial coordinates for the MD trajectories. Top-ranked binding conformations of the disaccharides **C1**, **C2**, **C3** and **C4** from library C were prepared for molecular dynamics investigations, considering, in particular, their protonation states. In this regard, phosphate groups in position 4 of ligands **C3** and **C4** were considered in a protonated form (**C3** and **C4**) or a monocharged form (**C3**$^{1-}$) or were doubly charged (**C4**$^{2-}$). Explicit solvent MD calculations were carried out, solvating each ligand and protein-ligand complex in a 50 Å square water box neutralized with NaCl. The default time step was 2 fs at 310 K and 1 atm; timesteps of 1 fs or 0.1 fs were utilized in the case of unstable trajectories when a larger timestep was used.

2.4. Binding Free Energy (BFE) Calculations

Molecular dynamics output (trajectories, DCD files) of the selected ligands were first subjected to binding free energy (BFE) analysis using Linear Interaction Energy (LIE) methods [37,38]. Table 1 reports the variation in Gibbs Free Energy of Binding (ΔG_{bind}) expressed in kcal·mol^{-1} and the related dissociation constant (K_D) in mM for each WaaG/ligand complex. The results are reported in descending order, in each of the three parts of the table. For library A, ligands **A7** and **A4** have the most negative values of BFE, giving K_Ds slightly lower than mM. Compound **A9** has a K_D value in the order of mM, while ligands **A15**, **A14** and **A1** have only slightly negative values of BFE. Compounds **A8** and **A16** seem not to have a good binding profile to WaaG, having unfavorable (positive) values of BFE. Ligands from library B with bigger molecules clearly have a stronger affinity for the exposed amino acids in the binding pocket of WaaG. Ligand **B33** is predicted to have the best affinity for the active site for this library, showing a K_D value of 40 μM. Derivatives **B16** and **B22** have K_Ds in the mM range, showing similar ΔG_{bind} values, while **B3d** and **B26** both have larger K_D values, binding less tightly than the other ligands of the same library. Regarding the disaccharide ligands of library C, compound **C1** has a K_D value in the mM range, and so does **C3**. Its mono-charged form, **C3^{1-}**, seems to be a weaker binder compared to ligand **C3**. The mono phosphorylation on the reducing end heptose in **C3** seems not to play an important role for the binding with WaaG, since compound **C2** is a stronger binder, with a more negative value of ΔG_{bind} and a K_D value of 70 μM. On the other hand, the double phosphorylation on the two heptose residues in **C4** is beneficial for the binding. Both the protonated (**C4**) and the doubly charged (**C4^{2-}**) forms of the ligand have K_D values in the μM range—0.8 and 0.4 μM, respectively.

Table 1. Computed ΔG_{bind} (kcal·mol^{-1}) and calculated K_D values for ligands of libraries A, B and C. Errors in ΔG_{bind} are estimated to be 1–2 kcal·mol^{-1}, corresponding to one order of magnitude in calculated K_D values.

LIGAND	ΔG_{bind}	Calc K_D (mM)	LIGAND	ΔG_{bind}	Calc K_D (mM)	LIGAND	ΔG_{bind}	Calc K_D (mM)
A7	−5.65	0.11	**B33**	−6.29	0.04	**C4^{2-}**	−9.14	0.0004
A4	−4.58	0.62	**B16**	−3.89	1.89	**C4**	−8.70	0.0008
A9	−3.34	4.55	**B22**	−3.18	5.97	**C2**	−5.97	0.07
A15	−1.73	61.10	**B3d**	−2.19	29.09	**C3**	−4.21	1.13
A14	−0.81	269.26	**B26**	−1.03	190.48	**C1**	−4.20	1.15
A1	−0.64	354.70				**C3^{1-}**	−0.72	313.08
A8	>0	> 10^3						
A16	>0	> 10^3						

2.5. RMSD Analysis of WaaG Backbone

Analysis of the root-mean-square-deviation (RMSD) distribution of WaaG *apo*-protein backbone and the various WaaG-ligand complexes was considered as a tool to study the dynamics of each system (Figure 6). The calculation refers to WaaG backbone spatial changes during the 10-ns MD production. For the three libraries, it is evident that the flexibility of the protein backbone in the different complexes can vary in proportion to the active site adaptation to the ligand conformational state. Figure S6 depicts the RMSD as a function of the duration of the MD simulation. In addition, Table S3 in Supporting Information reports the average RMSD value and the related standard deviation (SD). The avRMSD value for the reference *apo*-form of WaaG is 1.31 ± 0.16. For the RMSD analysis in Figure 6a–c, the WaaG *apo*-form is reported as "apo" in black as a reference RMSD distribution. Regarding library A, the protein backbone for the complex WaaG-**A8** is quite flexible, defining a broad distribution of **A8** for the WaaG-binding pocket in terms of BFE and K_D values. On the other hand, compound **A16** (which was recognized as a weak binder) gives more stability to WaaG backbone compared to **A8**, even with a slightly broad main distribution. In complex with compound **A15**, the protein backbone showed three dynamic states ranging from 1.0 to 1.5 Å of RMSD variation. When bound to compounds **A7** or **A14**, WaaG backbone presents a narrow distribution of RMSD, besides a low-populated state at the initial portion of the production phase. Ligands **A1** and **A9**, when bound to WaaG, result in a comparable curve with a broad profile, whereas the complex WaaG-**A4** generates a narrow ΔRMSD distribution of WaaG backbone, with an avRMSD of 1.43 Å ± 0.21. Regarding library B, there is a trend of increased instability of WaaG backbone in comparison to the ligands of library A. In general, library B ligands generate two main ΔRMSD distributions of WaaG *apo*-protein. The more stable complex is represented by WaaG-**B33**, with a range of RMSD variation of 1.5–2.0 Å and an avRMSD of 1.29 Å ± 0.25. This is also in accordance with the best BFE value of **B33** for library B. Regarding library C, disaccharide **C1** gives two main conformations of WaaG when bound to it, but each with a very broad RMSD variation, ~2.5 Å and a total range of > 5 Å of ΔRMSD. Ligand **C2** generates two well-distinguishable and comparably populated states of the protein backbone. Disaccharide **C4**$^{2-}$ with the best affinity and best K_D value, as seen before, gives great stability to WaaG backbone, with a small range of RMSD variation (~0.5 Å), very similar to the distribution of RMSD for the protein backbone when **C3** is bound to it. The better stability of **C3** compared to **C3**$^{1-}$ is in accordance with the better affinity in terms of K_D values of **C3** compared to **C3**$^{1-}$. The charged compound **C3**$^{1-}$ generates two equally populated states of the WaaG backbone; on the other hand, WaaG has a conformational state when bound to disaccharide **C4**, presenting a quite broad RMSD distribution similar to that of the protein conformation in complex with **C2**.

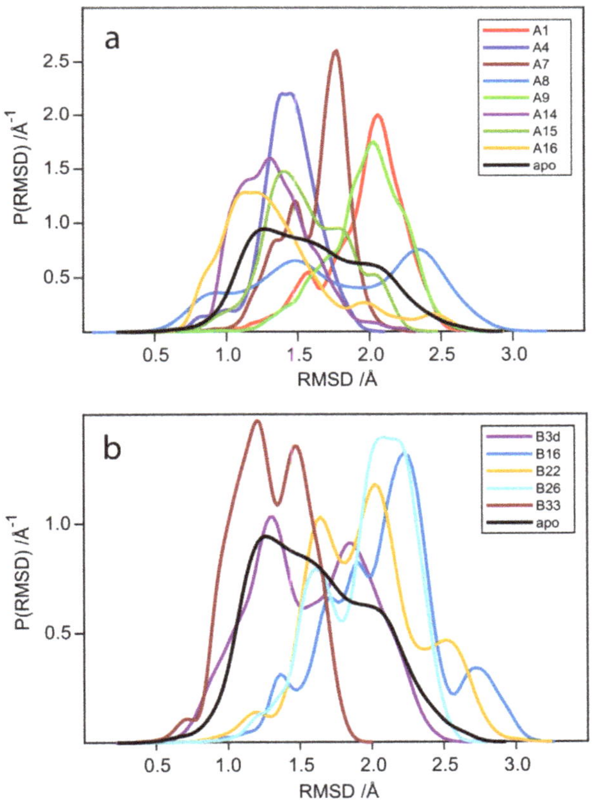

Figure 6. *Cont.*

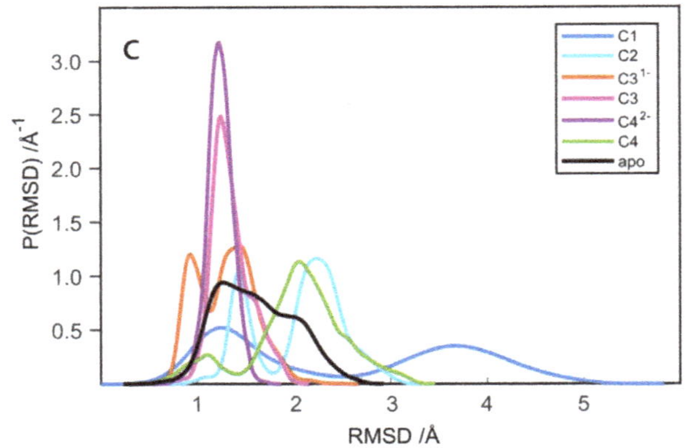

Figure 6. RMSD probability distributions obtained by a kernel density estimate using a normal distribution function. The calculation is done on the protein backbone of the *apo*-protein (from 2iw1) and the various WaaG-ligand complexes. The panels depict the RMSD distributions for WaaG in complex with ligands of (**a**) library A, (**b**) library B and (**c**) library C.

2.6. Electrostatic and van der Waals Contributions to WaaG/Ligand Binding

Figure 7a–c shows a histogram plot of the difference in polar and nonpolar interaction between each ligand and WaaG-binding cavity. In general, ligands of library A have the van der Waals (vdW) term higher than the electrostatic one for the binding affinity. Only ligands **A9** and **A14** showed a slightly higher electrostatic term compared to the vdW one. Additionally, for the ligands of library B, the vdW term is predominant, particularly for ligand **B22**. An inverse trend is displayed for the disaccharides of library C, where the electrostatic term is higher due to the presence of numerous hydroxyl groups on the disaccharides and because of their interactions with polar portions of the amino acids of the binding pocket. The electrostatic term is even more evident for the phosphate-charged disaccharides $\mathbf{C3}^{1-}$ and $\mathbf{C4}^{2-}$, similar to the importance of the electrostatic term in the interaction between negatively charged oligosaccharides and the Aβ42 peptide [39].

2.7. Influence of the H-Bonds for WaaG/Ligand Binding Affinity

Figure 8a–c presents a series of histograms with different colors representing different ligands binding the active site of WaaG. In the *x*-axis, there are reported amino acids of the binding pocket which are representative for the polar interactions of those ligands (i.e., having at least an H-bond with 30% occupancy for at least one ligand). The occupancy in the *y*-axis is reported in percentage and is estimated taking into account the presence of that particular H-bond during the 10-ns MD calculation. This analysis shows compounds **A8** and **A16** as weak binders because they do not have H-bonds with any notable occupancy during the 10-ns calculation. Phe13 has a selective H-bond for **A15** and likewise for Arg18, which has a polar contact with ligand **A4**, which, in turn, also binds with Glu289. Compound **A7** is confirmed to be a good binder, with two selective H-bonds with Asp19 and Ala99 but also an H-bond with Glu289, similarly to **A4**. Compound **A1** has two representative H-bonds with Gln236 and Arg261, while **A14** binds with Ser204 and Gln236 with a high occupancy (85–90%). As for library B, ligand **B33** has two H-bonds with Ala283 and Gly284 with a 50% occupancy. Compound **B26** has a >90% occupancy for an H-bond with Ala283. Ser204 has two H-bonds with 40–55% of occupancy with ligands **B16** and **B22**. Library C, as seen above, has a higher electrostatic term for its ligands, which is reflected in the great number of H-bonds between these disaccharides and the exposed amino acids of the binding pocket. Disaccharide **C1** has a selectivity for the hydrogen bonds with Asp172 and Arg173, with a 60–65% of occupancy. In addition, there is a marked difference between compounds **C1** and **C2** in terms of H-bonds. In support of this, disaccharide

C2 has different H-bonds with different occupancies. The most representative ones are with Asp205, Lys209, Gln280, Glu281 and Ile285, with occupancies between 70 and 90%. Disaccharide **C3**$^{1-}$ has two important H-bonds, with 65% occupancy, with Gly15 and Ala99. Compound **C3** has several H-bonds, but the most important are with Lys209 and Glu281. The doubly charged ligand **C4**$^{2-}$ has two important H-bonds with Asp100 and Arg173 and also ~10 polar bonds with lower occupancy (~30–40%). Disaccharide **C4** has a 75% occupancy for the H-bond with Arg208 and almost 100% of the MD simulation with Glu281.

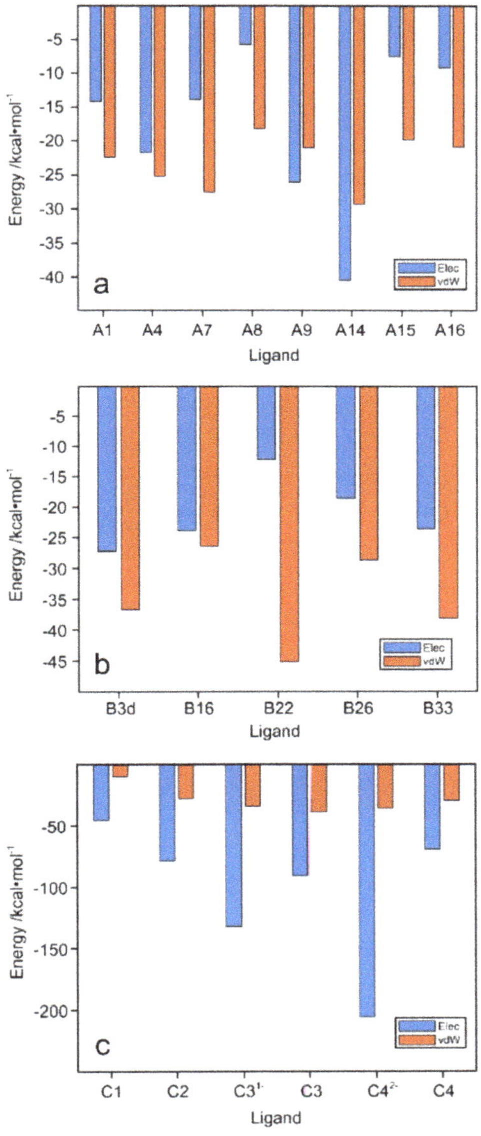

Figure 7. Contributions of electrostatic and van der Waals interactions to the binding of (**a**) library A, (**b**) library B and (**c**) library C ligands to WaaG binding pocket. Elec = electrostatic; vdW = van der Waals.

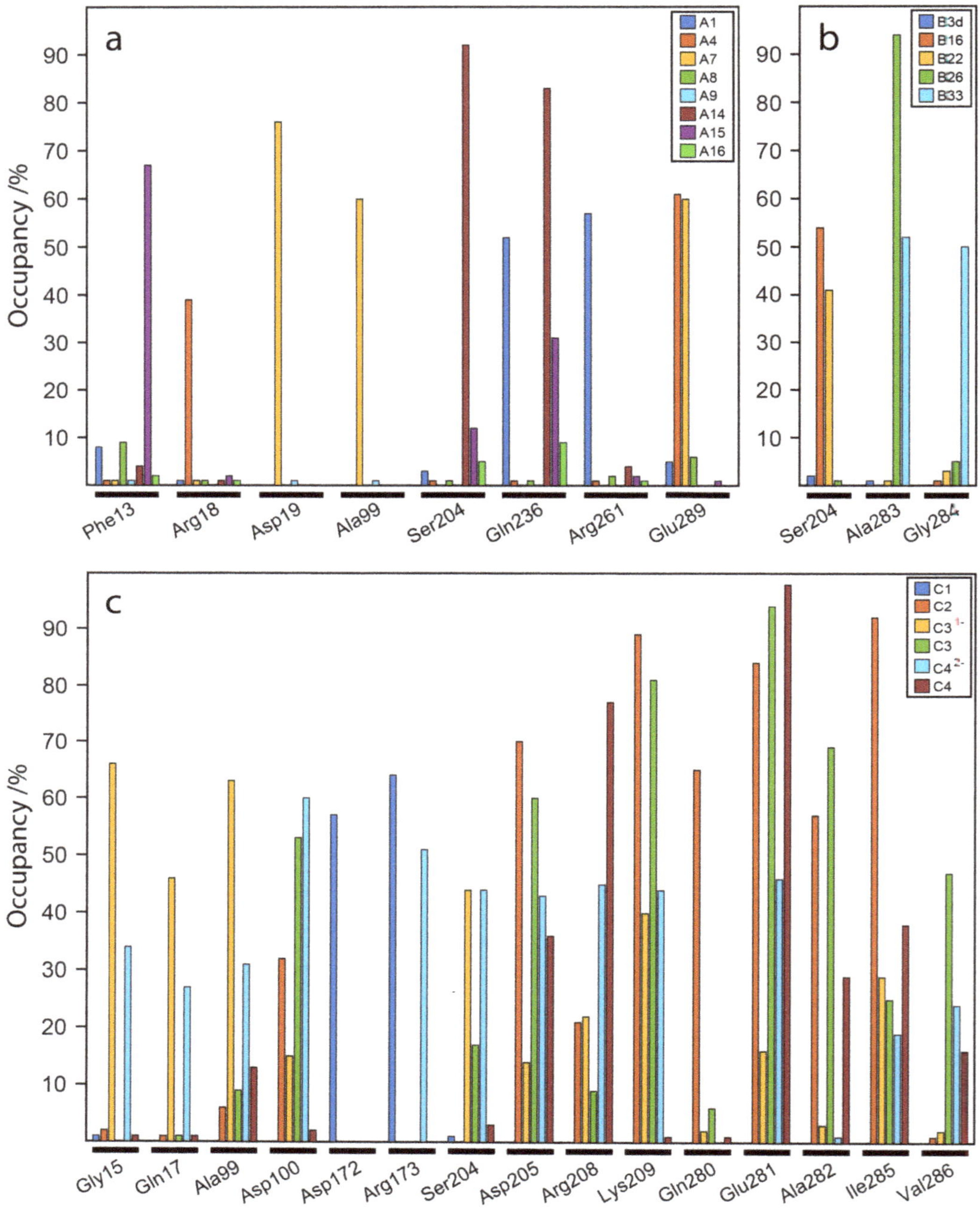

Figure 8. Hydrogen bond occupancies for ligands of (**a**) library A, (**b**) library B and (**c**) library C to the amino acids in the binding pocket of WaaG. Reported amino acids have at least 30% occupancy for one protein/ligand H-bond.

2.8. Screening by NMR Spectroscopy

Saturation Transfer Difference (STD) NMR experiments [27] were performed on different ligands of library A and disaccharides **C1**, **C2**, **C3** and **C5** of library C. Library A showed significant STD effects for three of the ligands analyzed: compounds **A4** (Figure 9), **A8** and **A15**, which was also the case for **A4** in conjunction with UDP. The STD effects were uniformly distributed for the protons throughout the **A4** molecule, and a similar pattern was observed in the binding of **A15** together with UDP and WaaG. Several other compounds from library A gave less than 1% of STD signal and were disregarded for further evaluation. Compounds of library B were not analyzed in this study via STD NMR. The oligosaccharides analyzed from library C showed only weak STD signals (less than 1%) and were therefore not analyzed further. As small saccharides generally are weak binders [40] and the STD NMR technique detects binding to proteins for ligands with K_D ranging from mM to μM or even lower [41], we conclude that the absence of notable interactions between the disaccharides and WaaG indicates that the minimal motif needed for recognition of the oligosaccharide acceptor has not been fulfilled and that a larger oligosaccharide structure is required for the glycosylation reaction to take place.

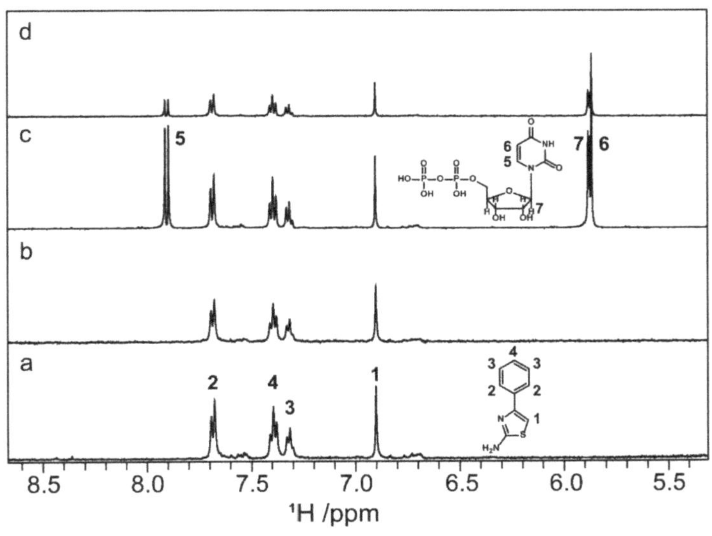

Figure 9. STD NMR spectra of ligand **A4** and WaaG carried out with and without UDP as a competitor. In sequence from the bottom to the top, 1D ^{1}H STD off-resonance (**a**) and on-resonance (**b**) irradiation of WaaG in the presence of **A4** and off-resonance (**c**) and on-resonance (**d**) irradiation with UDP added to the protein-ligand mixture. Compound **A4** shows interaction with WaaG also in the presence of UDP.

2.9. WaaG Has a Twisting-like Dynamics Involving N- and C-Domains

WaaG is a protein constituted by 371 amino acids and has two well-distinguishable domains, the N- and C-domains. The N-domain is recognizable between the amino acids Met1 and Gln163; a central part, denoted as the "hinge" region (amino acids from Ile164 to Pro171), then links the N- to the C-domain. The latter constitutes amino acids between Asp172 and Gly371, considering that there is another "hinge" portion (Gln355-Tyr358), and the C-terminus helix (Ser359-Gly371), even if the last residues are spatially in the N-domain. The starting point is the calculation of the RMSD variation for the *apo*-protein WaaG during the 10-ns production. In Figure 10a–c, the red curve represents the reference RMSD distribution of WaaG *apo*-protein. Focusing only on a particular portion of the protein, the blue, black and green curves refer to the RMSD values for the N-, C-domain and the hinge region, respectively. In Figure 10a, there is only a slight difference between the reference (*apo*) and the different domains of the protein of the RMSD distribution. In Figure 10b,c, the alignment is made on the N- and C-domain, respectively. When the alignment is done on a specific domain, it is possible to observe that the opposite domain presents a broad distribution of the backbone RMSD, which signifies the dynamic behavior of the two domains that move relative to each other.

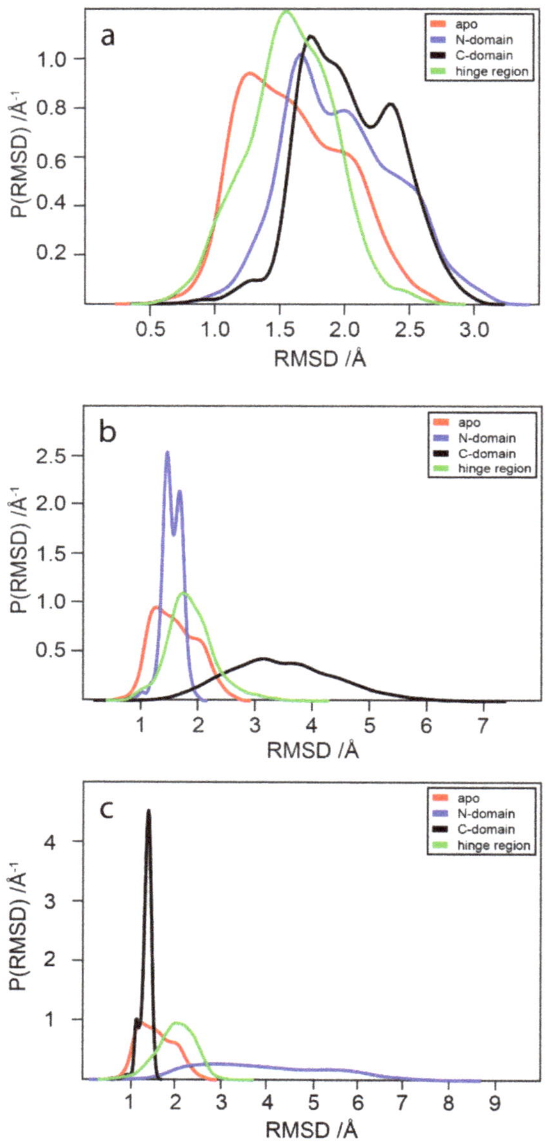

Figure 10. RMSD probability distributions obtained by a kernel density estimate using a normal distribution function. The calculation is done on the protein backbone of the *apo*-protein (from 2iw1, red) as a reference of RMSD values. Alignment of (**a**) the entire protein backbone, (**b**) N-domain, and (**c**) C-domain of *apo*, then RMSD calculation of the N- and C-domain and hinge region.

Figure 11a–c reports an additional analysis of the internal dynamics of WaaG protein over the MD trajectory. Figure 11a shows the variation over time of the angle defined between the Cα atoms of His62 in the N-domain, Lys248 in the C-domain and Gly168 in the hinge region. Figure 11b depicts the change in distance between the Cα atoms of His62 (N-domain) and Lys248 (C-domain). The variation of the angle and the distance presents a similar trend, suggesting a twisting-like movement (Figure 12). To confirm the correlation between these two variables, they were plotted together in Figure 11c, and the Spearman correlation coefficient *rho* was calculated. The correlation coefficient can have a range of values from −1 to +1, showing a perfect negative or positive correlation between the ranks of the two variables, respectively. The calculated Spearman's *rho* is 0.9028, hence indicating a good correlation between these two variables.

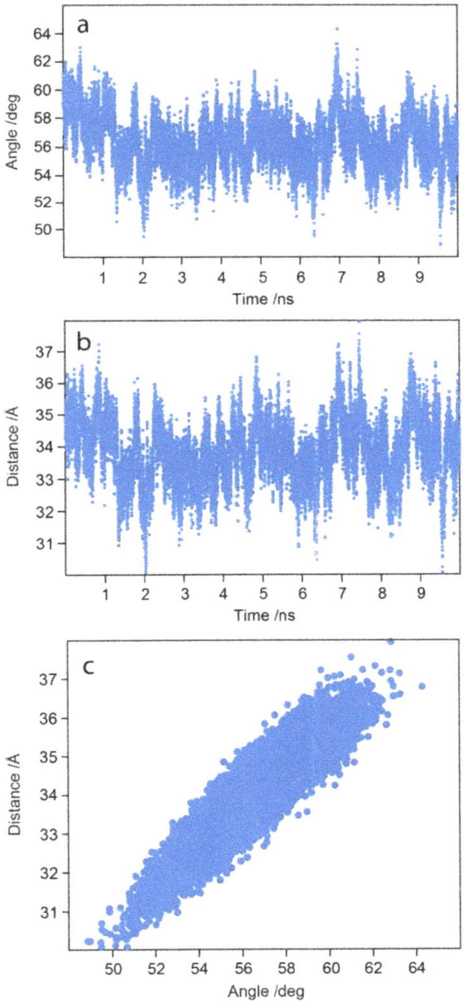

Figure 11. (**a**) Angle variation over time between the Cα atoms of His62 (N-domain), Lys248 (C-domain) and Gly168 (hinge region) and (**b**) distance fluctuations over time between the Cα atoms of His62 and Lys248 during the 10-ns MD calculation. (**c**) Scatter plot of the two variables against each other.

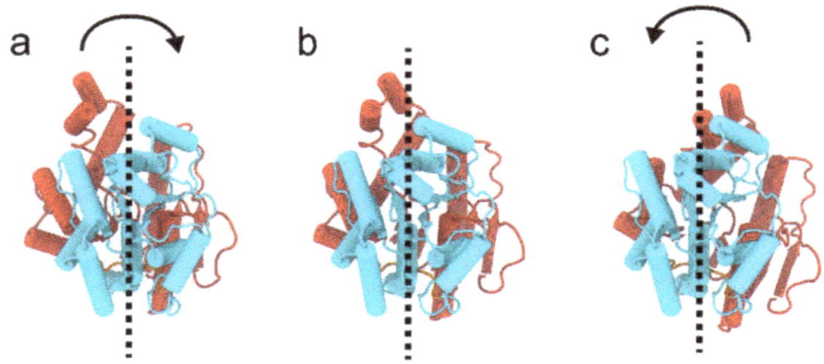

Figure 12. Graphical representation of the twist-like dynamics of WaaG. The C-domain of the protein (in cyan) is kept aligned, while the rest of the protein moves. Depicted are the two extremes (**a**,**c**) of the twist-like movements of the N-domain (in brick-red color) with respect to the C-domain and a representation of the two domains when aligned in the eclipsed state (**b**).

2.10. *In Vitro WaaG Inhibition Assay*

The WaaG activity assay utilized the previously developed protocol [23] for testing compounds of libraries A and B. However, none of the fragment-based compounds (library A) or small molecules (library B) showed any significant inhibitory effect at a lower concentration than that previously determined for **A1**, which had a half-maximal inhibitory concentration (IC_{50}) of 1 mM.

3. Discussion

The present work describes a combination of computational and experimental studies aimed at discovering new fragment molecules that could be identified as inhibitors for glucosyltransferase WaaG, but also at discovering new insights on the mechanism of action of the protein and the contact between the donor UDP-Glc^{2-} and the acceptor portion (L-*glycero*-D-*manno*-heptose-II) of LPS. Last but not least, part of this work focused on the elucidation of the dynamic behavior of the protein in the physiological environment. The background was a combination of preliminary docking, NMR spectroscopy and biochemical experiments, where compound **A1** showed some inhibitory activity towards WaaG. This evidence was the basis for further developments using a fragment-based approach to identify new molecules with the potential of increased inhibitory effects of WaaG. Library A and B were first tested with a molecular docking approach, acting as a filter and to progress only with the most interesting compounds, comparing the docking outputs with UDP-Glc^{2-} WaaG structure (obtained by the crystal structure of U2F, PDB ID 2iw1), thereby selecting for them the best-docked pose. Compounds **A1**, **A7**, **A9**, **A14**, **A16**, **B3d**, **B16**, **B22**, **B26** and **B33** from libraries A and B were selected based on docking binding affinity. Meanwhile, compounds **A4**, **A8** and **A15** were selected from an experimental study involving the STD NMR technique, where these ligands showed substantial STD effects with WaaG *apo*-protein, as well as in the presence of UDP for **A4**. The best-predicted poses of all of the above-mentioned ligands from libraries A and B were employed for further MD simulations. Analyzing the BFE and K_D values for each ligand for its binding to the active site of WaaG, compounds **A7**, **A4** and **B33** for libraries A and B displayed a promising profile of BFE, with K_D < mM (0.11, 0.62 and 0.04 mM, respectively). In order to study the effects on the protein backbone in the WaaG/ligand complex during the 10-ns MD simulation, an analysis of the variation of RMSD distribution was assessed. For library A, compound **A4** is undoubtedly the one that gives more stability to WaaG backbone, also compared to the reference RMSD analysis on the *apo*-protein, with an avRMSD of 1.43 Å ± 0.21. Regarding library B, ligand **B33** generates a good stability profile to the protein backbone, with an avRMSD of 1.29 Å ± 0.25. Investigating in more detail the binding between the ligands and the exposed amino acids in the WaaG active site, an analysis on the relative contributions of electrostatic and van der Waals interactions showed that ligand **A4** carries out a balanced binding between polar and nonpolar interactions, with an average energy of ~25 kcal·mol^{-1} for both. Library B, on the contrary, has a higher electrostatic term compared to the hydrophobic one. The hydrogen bonds between the ligands and the binding pocket were subsequently elucidated. Compound **A4** has H-bonds with Arg18 and Glu289 and compound **B33** binds to Ala283 and Gly284. Focusing on ligands **A4** and **B33**, a cluster analysis on the entire 10-ns MD simulation was explored. The different clusters of poses of a ligand were generated considering the difference in RMSD, with a cutoff value of 1 Å. The binding of the most representative pose of the clusters for compounds **A4** and **B33** is depicted in Figure 13.

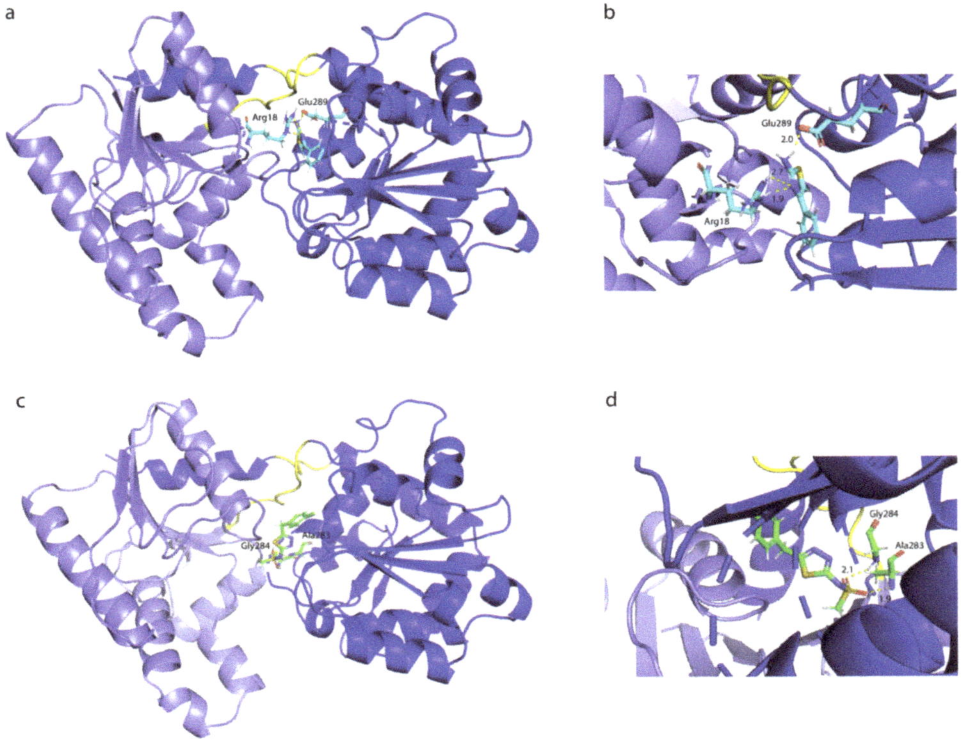

Figure 13. Molecular dynamics representation and interactions of ligands (**a**,**b**) **A4** and (**c**,**d**) **B33** with WaaG binding pocket. The structures are the most representative of the different clusters identified for each ligand.

In order to shed light on the mechanism of transfer of a glucose residue from UDP-Glc^{2-} (donor) to Hep-II (acceptor), ligands **C1–C12**, as models of different parts of the inner core of the LPS, were docked in the presence of *apo*-form, but particularly with WaaG-UDP-Glc^{2-}. Only the disaccharides from libraries C, **C1**, **C2**, **C3** and **C4** were subjected to more detailed MD calculations because of the interest in investigating Hep-II, the acceptor of the monosaccharide residue from UDP-Glc^{2-}. The phosphate-derived ligands were studied in both non-charged (**C3** and **C4**) and charged (**C3**$^{1-}$ and **C4**$^{2-}$) forms of the phosphate group(s). Doubly charged compound **C4**$^{2-}$ has a K_D value in the μM range (0.4 μM), and the most representative conformation from the 10-ns MD simulation is depicted in Figure 14a. The ligands then displayed a preferred binding to the specific region of the WaaG active site where UDP-Glc is located. In contrast, the output of the docking calculation of $\mathbf{C4}^{2-}$ in the presence of UDP-Glc^{2-} in the binding pocket of WaaG results in the disaccharide being positioned in the vicinity of the glucosyl residue (Figure 14b,c).

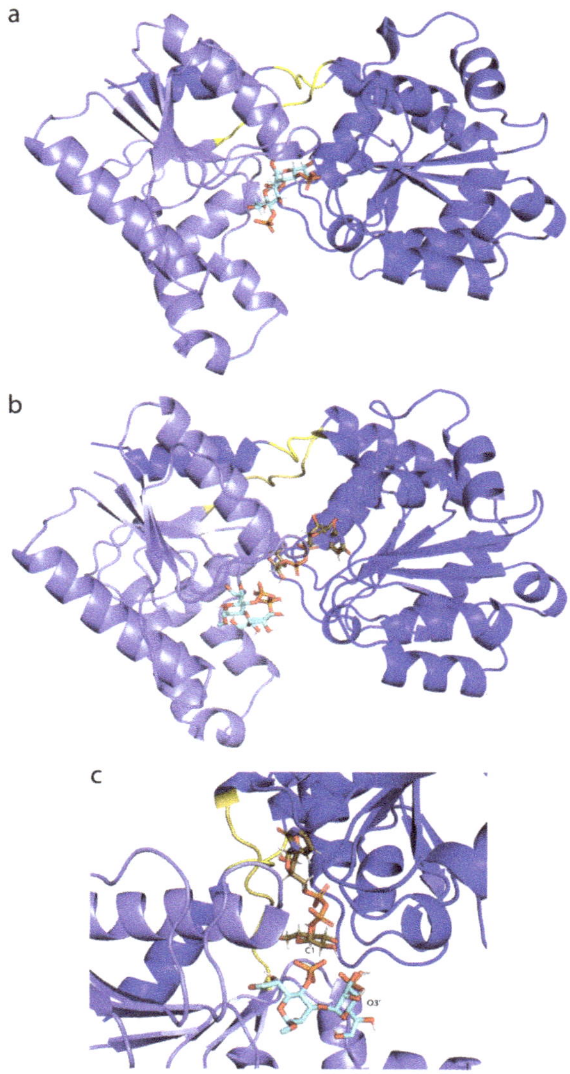

Figure 14. (**a**) Molecular dynamics representation of $\mathbf{C4}^{2-}$ and (**b**,**c**) molecular docking conformation of $\mathbf{C4}^{2-}$ with WaaG-UDP-Glc^{2-} complex.

4. Materials and Methods

4.1. General Experimental Methods

All reagents were used as delivered. The commercially available Dowex® 1×4 chloride form, 50–100 mesh, strongly basic, was used to prepare the acetate form. TLC was carried out on silica gel 60 F254 and monitored with UV light at 254 nm. NMR spectra for the characterization of compounds **A11–A13** were recorded at 25 °C on a Bruker 600 MHz spectrometer. The NMR chemical shifts are reported in ppm and are referenced to the DMSO-d_6 solvent peak at 39.52 ppm and the residual peak at 2.50 ppm for ^{13}C and ^{1}H, respectively. ^{1}H NMR chemical shifts and $J_{H,H}$ coupling constants were extracted from 1D ^{1}H NMR spectra using the NMR spin simulation software PERCH [42,43]. Disaccharides **C1**, **C3** and **C5** were available from previous studies [44,45]. A selection of protein-ligand complexes in pdb-format is given as Supplementary material.

4.1.1. 2-Acetamido-5-methyl-4-phenyl-1,3-thiazol (**A11**)

2-Amino-5-methyl-4-phenyl-1,3-thiazol (0.100 g, 0.53 mmol) and acetic anhydride (0.055 mL, 0.58 mmol) were dissolved in ethyl acetate (2 mL) and stirred overnight at rt. The formed white solid was filtered, washed with ethyl acetate and dried in vacuo to obtain the product as a white powder (100 mg, 82%). ^{1}H NMR (DMSO-d_6): δ 2.13 (s, 3H, MeCO), 2.45 (s, 3H, Me), 7.33 (dd, $J_{H\text{-ortho},H\text{-para}}$ 1.19 Hz, $J_{H\text{-meta},H\text{-para}}$ 7.41 Hz, 1H, H-para), 7.44 (dddd, $J_{H\text{-ortho},H\text{-meta}}$ 7.83 and 0.51 Hz, $J_{H\text{-meta},H\text{-meta}}$ 1.28 Hz, $J_{H\text{-meta},H\text{-para}}$ 7.41 Hz, 2H, H-meta), 7.63 (dddd, $J_{H\text{-ortho},H\text{-ortho}}$ 1.74 Hz, $J_{H\text{-ortho},H\text{-meta}}$ 7.83 and 0.51 Hz, $J_{H\text{-ortho},H\text{-para}}$ 1.19 Hz, 2H, H-ortho) (Figure S7). ^{13}C NMR (DMSO-d_6): δ 11.82 (Me), 22.40 (MeCO), 120.74 (C5), 127.17 (C-para), 127.92 (C-ortho), 128.34 (C-meta), 134.95 (C-ipso), 143.92 (C4), 153.96 (C2), 168.30 (CO) (Figure S8).

4.1.2. 2-Formamido-5-methyl-4-phenyl-1,3-thiazol (**A12**)

2-Amino-5-methyl-4-phenyl-1,3-thiazol (0.100 g, 0.53 mmol) was dissolved in 1 mL freshly prepared acetic formic anhydride and stirred overnight at rt. The solvent was evaporated and the residue was washed with ethyl acetate to obtain the product as a white powder (96 mg, 83%). ^{1}H NMR (DMSO-d_6): δ 2.47 (s, 3H, Me), 7.34 (dd, $J_{H\text{-ortho},H\text{-para}}$ 1.22 Hz, $J_{H\text{-meta},H\text{-para}}$ 7.42 Hz, 1H, H-para), 7.44 (dddd, $J_{H\text{-ortho},H\text{-meta}}$ 7.81 and 0.58 Hz, $J_{H\text{-meta},H\text{-meta}}$ 1.41 Hz, $J_{H\text{-meta},H\text{-para}}$ 7.42 Hz, 2H, H-meta), 7.63 (dddd, $J_{H\text{-ortho},H\text{-ortho}}$ 1.86 Hz, $J_{H\text{-ortho},H\text{-meta}}$ 7.81 and 0.58 Hz, $J_{H\text{-ortho},H\text{-para}}$ 1.22 Hz, 2H, H-ortho), 8.47 (s, 1H, H_{Fo}), 12.21, s, 1H, NH) (Figure S9). ^{13}C NMR (DMSO-d_6): δ 11.84 (Me), 121.55 (C5), 127.30 (C-para), 128.00 (C-ortho), 128.36 (C-meta), 134.72 (C-ipso), 144.23 (C4), 152.26 (C2), 159.43 (CO) (Figure S10).

4.1.3. 2-Amino-3,5-dimethyl-4-phenyl-1,3-thiazolium acetate (**A13**)

2-Amino-5-methyl-4-phenyl-1,3-thiazol (0.100 g, 0.53 mmol) was dissolved in acetone, methyl iodide (0.1 mL, 1.58 mmol) was added and the reaction mixture was stirred for 24 h at rt. The solvent was evaporated and the obtained white powder was dissolved in methanol and subjected to DOWEX column (AcO^- form, exchanged from Cl^- via the OH^- form) to obtain the product with acetate counter ion as a yellow syrup (60 mg, 55%). ^{1}H NMR (DMSO-d_6): δ 1.87 (s, 3H, Me), 1.88 (s, Ac), 2.91 (s, 3H, MeN^+), 7.35 (ddd, $J_{H\text{-ortho},H\text{-meta}}$ 7.63 and 0.37 Hz, $J_{H\text{-ortho},H\text{-para}}$ 1.17 Hz, 2H, H-ortho), 7.46 (dd, $J_{H\text{-ortho},H\text{-para}}$ 1.17 Hz, $J_{H\text{-meta},H\text{-para}}$ 7.52 Hz, 1H, H-para), 7.50 (ddd, $J_{H\text{-ortho},H\text{-meta}}$ 7.63 and 0.37 Hz, $J_{H\text{-meta},H\text{-para}}$ 7.52 Hz, 2H, H-meta) (Figure S11). ^{13}C NMR (DMSO-d_6): δ 12.13 (Me), 21.89 (AcO^-), 32.14 (MeN^+), 105.67 (C5), 128.77 (C-meta), 128.87 (C-para), 129.90 (C-ortho), 130.29 (C-ipso), 134.31 (C4), 161.91 (C2), 172.53 (CO) (Figure S12).

4.1.4. Methyl L-glycero-α-D-manno-heptopyranosyl-(1→3)-L-glycero-α-D-manno-heptopyranoside (**C2**)

Reported in Supporting Information (Scheme S1, Figures S13 and S14).

4.2. Molecular Docking

Docking studies were performed by using four docking programs: AutoDock Vina [30], LeDock (http://lephar.com) [31], rDock [32] and GOLD [33], selected for their established sampling and scoring functions [34].

The crystal structures of WaaG, *viz.*, co-crystallized with uridine-5′-diphosphate-2-deoxy-2-fluoro-α-D-glucose (U2F), were retrieved by the Protein Data Bank (rcsb.org, PDB ID 2iw1) [46]. Four types of docking calculations were conducted with the molecular structures of libraries A, B and C. First, U2F was removed in order to analyze the docking-pose of ligands in a free binding pocket. Moreover, additional docking studies were performed with the protein in the presence of UDP-Glc. The donor ligand UDP-Glc was obtained by modifying the crystal structure of U2F with Avogadro [47] version 1.1.1. All molecules were visualized in PyMOL 2.3.4 or VMD [48] version 1.9.3. The docking studies were conducted on a PC with an Intel® Core™ i7-9705H CPU @ 2.60 GHz with 8 GB RAM running Ubuntu 16.04 as operating system, unless otherwise stated.

The first employed docking program was AutoDock Vina 1.1.2. AutoDockTools (ADT) [49] version 1.5.6 was utilized to obtain not only the pdbqt files of both protein structures and all ligands but also to determine the different docking grid boxes (see Table S1). Docking was performed using the protein in its rigid form or by making some amino acids flexible, viz., Arg173, Arg261 and Glu289. These amino acids are known to establish important H-bonds with the uridine portion of UDP in the binding pocket of WaaG. The exhaustiveness value was set to 64 and the number of poses to 10 for each ADT run. The binding energy range was imposed to be ≤ 2 kcal·mol^{-1} above that of the best binding pose for each ligand.

LeDock was used as the second docking program. A file with all ligands in the mol2 (Tripos MOL2) format was generated from the original pdb files using Open Babel 2.3.2 [50]. LePro was employed for the pre-processing of the protein structure and to initiate the docking. Ten binding poses were generated from each docking run.

The third docking program employed was rDock. Input files in mol2 format of the protein and sdf (MDL SD) files of the ligands were utilized to carry out the docking. Using *rbcavity* to prepare the docking site, the two-sphere method was selected, and the radius of the sphere was set to 10 Å. The number of docking poses was set to 10 and the MDL SD output files reported the affinity scores for the generated poses.

GOLD was used as the last docking program. Pdb files of both protein structures and ligands were employed to execute GOLD docking predictions. The protein structure was properly protonated, and co-crystallized ligands and water molecules were removed. XYZ coordinates were edited and the binding site radius was set to 10 Å. Ligands were docked into the rigid form of the protein crystal structures, but also with some flexible side-chains, selected from the most important amino acids in the binding pocket. The amino acids were selected by checking how UDP-Glc interact with the protein and chosen in order to facilitate flexibility at the surface of the binding pocket. Ten amino acids were selected to be more flexible: Phe13, Asp100, Arg173, Arg208, Lys209, Arg261, Glu281, Ile285, Val286 and Glu289. By default, the number of dockings to be performed on each ligand was 10. Search efficiency (predictive reliability) was set to 100% (around 30.000 GA operations for each ligand). The docking studies were conducted on a PC with an Intel® Core™ i7-9700H CPU @ 3.00 GHz with 16 GB RAM on Windows 10 Pro as the operating system.

4.3. Molecular Dynamics Simulations

MD simulations were performed with NAMD v. 2.12 [51]. Both protein and ligands files were generated employing the VMD psfgen tool. The CHARMM36 force field [52–54] was utilized, and the following sections of the CHARMM-GUI website were used: *PDB R&M* (Reader and Modeler) [55,56] for protein structures, *Ligand R&M* [57] for the molecules of libraries A and B; *Glycan R&M* [58–60] was used for the oligosaccharides of library C. The protein was protonated to a physiological pH of 7.2, with an overall net charge of +1.0 e. Ligands and ligand-protein complexes were solvated in a cubic water box, 50 Å to the side, using a TIP3P water model [61]. NaCl (0.2 M) ions were subsequently added and the system neutralized with the same ions.

An initial potential energy minimization (1000 steps) was conducted on solvent molecules and ions of the systems using conjugate gradient and line search algorithms, imposing a harmonic restraining potential to ligands and complexes of 500 kcal·mol^{-1}·Å^{-2}. A subsequent minimization was performed on solvated ligands or complexes without any restraints. A gradual heating step from 1 K to 310 K in 6 ps was applied to each system, which was then equilibrated during 600 ps. Production simulations of 10 ns were run with a timestep of 2, 1, or 0.1 fs in NPT (isothermal–isobaric) ensemble with a Langevin damping coefficient set to 1 ps^{-1}. The temperature was kept constant at 310 K with a stochastic Langevin thermostat (damping coefficient of 1 ps^{-1}), while the pressure was kept stable at 1 atm (an oscillating time of 100 fs and a damping time constant of 50 fs). The coordinates were recorded every 1 fs. Hydrogens of the water molecules were kept rigid using the ShakeH algorithm, and the PME (Particle Mesh Ewald) method was employed for the calculation of non-bonded interactions, with a 1 Å grid spacing. Electrostatic and van der Waals interactions were forced to zero at the cutoff distance of 12 Å (switching function from 10 Å). For ligand **B33**, a constraint of 1.64 Å between the chlorine atom and the lone pair (LP, as generated by CHARMM *Ligand Reader & Modeler*) was imposed, and it became important to keep the contact between the halogen and the LP stable during the entire MD simulations involving the *apo*-form of WaaG protein. MD simulations were performed at the National Supercomputer Centre (NSC), being part of Linköping University and the Swedish National Infrastructure for Computing (SNIC). Computations were carried out on the Tetralith cluster using Slurm to submit jobs and ThinLinc (Cendio AB, Linköping, Sweden) as a remote desktop.

4.4. Binding Free Energy Calculation

For post-MD analysis of free binding energy, the program CaFE (Calculation of Free Energy) 1.0 [62] was used to process PSF (coordinate) and DCD (trajectory) files generated by the CHARMM/NAMD (v. 2.12) software. CaFE was employed to predict binding affinity with an end-point method, viz., LIE (Linear Interaction Energy) [37,38], which is an approximation of the linear response [37]. PSF and DCD files of both the solvated molecule and in complex with the protein are required. CaFE generates for both settings a difference in electrostatic (polar term) and van der Waals (vdW) interactions (nonpolar terms) between the ligand and the enclosing background; then, the summation of these individual energy components gives the total binding free energy [62], reported as ΔG_{calc} and expressed in kcal·mol^{-1}. The dissociation constant K_D was obtained from the free energy using the following relationship:

$$\Delta G = RT\ln(K_D/c^*)$$

where R is the ideal gas constant with a numerical value of 1.987×10^{-3} kcal·K^{-1}·mol^{-1}, T is the absolute temperature with a value of 310 K and c^* is a standard reference concentration defined as 1 M [63]. CaFE employed the entire 10-ns MD trajectories of the solvated ligand and the complex. The CaFE script implemented the required CHARMM parameter files, and the entire MD trajectory was analyzed every 5th frame by setting the stride to 5. The predefined coefficients in the LIE approach, α, β and γ, are taken as their default values, 0.18, 0.33 and 0.0, respectively.

4.5. RMSD Analysis

The root-mean-square-deviations (RMSDs) throughout the MD simulation were calculated during the 10-ns production period with the *RMSD Trajectory Tool* of VMD or each selected ligand. The reference structure is the first frame of the MD simulation. The RMSD analysis starts with a rigid-body structure alignment. Protein flexibility during the MD simulation was calculated in VMD. The stride was set to 1; therefore, each frame of the calculation was used for the RMSD analysis. RMSD values were then recorded and saved as a list of RMSDs at different frames, as a .dat file. Averaged RMSDs were calculated for each ligand, as well as the RMSD standard deviation (SD) values for each RMSD distribution. RMSD analysis was performed with MATLAB® version 7.11.0.584 (R2020b), using in-house-built scripts.

4.6. MD Post-Processing: Interactions, Angle, Distance and Spearman Correlation

The electrostatic and van der Waals contributions on the binding between WaaG and each ligand were calculated after protein backbone alignment, through an in-house tcl script utilized in VMD Tk Console, which uses NAMD Energy 1.4 as its processing program. As for the H-bonds identification, after protein backbone alignment, the DCD and PSF files of each WaaG/ligand complex are imported in VMD; then, the unique hydrogen bonds are calculated, with a donor-acceptor distance tolerance of 3.4 Å and an angle cutoff of 45°. The angle and the distance variation between strategic amino acids were calculated through VMD, and the .dat output files were used for the Spearman correlation through an in-house script in MATLAB.

4.7. Saturation Transfer Difference NMR Spectroscopy

NMR spectroscopy experiments were carried out at 5 °C on a Bruker AVANCE 500 MHz spectrometer equipped with a 5 mm PFG triple-resonance CryoProbe ($^1H/^{13}C/^{15}N$) and a Bruker AVANCE III 700 MHz equipped with a 5 mm TCI Z-Gradient Cryoprobe ($^1H/^{13}C/^{15}N$). 1D STD NMR experiments were run using a protein concentration of ~25 μM, a ligand/protein ratio of 20:1 and 600 μM UDP concentration when used. NMR tubes with an outer diameter of 3 mm (0.18 mL) were utilized for the NMR analysis. On-resonance irradiation for protein saturation was set at 0.5, 0 and −0.5 ppm, while the off-resonance control was positioned at 60 ppm using Gaussian pulses at a power level corresponding to a hard square pulse of 65 Hz and a length of 50 ms. An acquisition time of 2 s and a relaxation delay of 6 s was employed for a total number of 21,928 data points. An excitation sculpting for water suppression was employed with a squa100.1000 function of 2400 μs, while the TOCSY spinlock was set to 50 ms at 5 kHz. A total number of scans of 1024 was employed for each experiment. Blank control 1D STD experiments were run in the same way, with the sample containing only the ligand without the presence of the protein.

4.8. Sample Preparation for STD NMR Spectroscopy

Protein preparations in pure glycerol medium were kept at −80 °C and were thawed slowly to attain 0 °C when used. Three cycles of buffer exchange using Amicon Ultra centrifuge were run for 15 min each at 14,000× *g* and 4 °C. Concentration of the protein samples was obtained using a BioSpec-Nano spectrophotometer via optical density measurement at a wavelength of 280 nm, employing a 42,284 g·mol^{-1} reference value for WaaG. The samples were prepared in a phosphate-buffered saline (PBS) solution containing KH_2PO_4 (25 mM) and NaCl (150 mM) in D_2O and the pD adjusted to 8.6 with DCl and NaOD.

4.9. Biochemical In Vitro Assessment/In Vitro WaaG Activity Assay

The WaaG activity assay was performed as previously described [23], using herein ligands of libraries A and B.

5. Conclusions

Conformational aspects and the presence of dynamics in WaaG were investigated by considering the entire protein backbone, or the single parts, i.e., the N- and C-domains and the hinge region. A dynamic behavior of the two domains was observed, with movement relative to each other in a twisting-type fashion. Restricting the movements or interfering with the inherent dynamics of WaaG by using small molecule inhibitors that can act by 'putting a spanner in the works' could open a way to potentiate the action of antibiotics in pathogenic *E. coli* and other gram-negative bacteria.

Supplementary Materials: The following are available online at https://www.mdpi.com/article/10.3390/ph15020209/s1, Figure S1: Ligands of library A, Figure S2: Ligands of library B; Figure S3: Ligands of library B; Figure S4: Ligands of library C; Figure S5: Ligands of library C; Table S1: Parameters for dockings of ligands to WaaG; Table S2: Ranking of ligands in docking to WaaG; Table S3: Average RMSD on WaaG backbone during MD simulations on WaaG/ligand complexes; Figure S6: RMSDs calculated for the protein backbone during MD simulations; Figure S7: 1H NMR spectrum of compound A11; Figure S8: 13C NMR spectrum of compound A11; Figure S9: 1H NMR spectrum of compound A12; Figure S10: 13C NMR spectrum of compound A12; Figure S11: 1H NMR spectrum of compound A13; Figure S12: 13C NMR spectrum of compound A13; Scheme S1:

Synthesis of heptobioside C2; Figure S13: 1H NMR spectrum of compound C2; Figure S14: 13C NMR spectrum of compound C2; ligand-protein complexes in pdb format (9 files).

Author Contributions: Conceptualization, D.O.D. and G.W.; Data curation, F.R., A.R., O.E., C.M., H.M., J.S., D.O.D. and G.W.; Formal analysis, F.R., A.R., O.E., C.M., H.M., J.S., D.O.D. and G.W.; Funding acquisition, A.C., D.O.D. and G.W.; Investigation, F.R., A.R., O.E., C.M., H.M., J.S., P.K., D.O.D. and G.W.; Methodology, F.R., A.R., C.M., H.M., J.S., P.K., D.O.D. and G.W.; Project administration, D.O.D. and G.W.; Resources, A.C., D.O.D. and G.W.; Supervision, A.C., D.O.D. and G.W.; Writing—original draft, F.R., A.R. and G.W.; Writing—review & editing, F.R., A.R., O.E., P.K., D.O.D. and G.W. All authors have read and agreed to the published version of the manuscript.

Funding: This work was funded by grants from the Swedish Research Council (no. 2017-03703), the Knut and Alice Wallenberg Foundation and the Novo Nordisk Foundation.

Institutional Review Board Statement: Not applicable.

Informed Consent Statement: Not applicable.

Data Availability Statement: Data is contained within the article.

Acknowledgments: Compounds of library B and plating thereof were provided by the Chemical Biology Consortium Sweden (CBCS). The National Supercomputer Centre (NSC) at Linköping University and the Swedish National Infrastructure for Computing (SNIC) are thanked for providing computational resources.

Conflicts of Interest: The authors declare no conflict of interest.

References

1. Croxen, M.A.; Finlay, B.B. Molecular mechanisms of Escherichia coli pathogenicity. *Nat. Rev. Microbiol.* **2010**, *8*, 26–38. [CrossRef]
2. World Health Organization. Prioritization of Pathogens to Guide Discovery, Research and Development of New Antibiotics for Drug Resistant Bacterial Infections, Including Tuberculosis. 2017. Available online: https://apps.who.int/iris/handle/10665/311820 (accessed on 14 January 2022).
3. Levin, S. The crisis in antibiotic resistance. *Infect. Dis. Clin. Pract.* **1993**, *2*, 53. [CrossRef]
4. World Health Organization. Outbreaks of *E. coli* O104:H4 Infection. 2011. Available online: https://www.euro.who.int/en/countries/germany/outbreaks-of-e.-coli-o104h4-infection (accessed on 14 January 2022).
5. World Health Organization. Antimicrobial Resistance. 2020. Available online: https://www.who.int/health-topics/antimicrobial-resistance (accessed on 14 January 2022).
6. Bernal, P.; Molina-Santiago, C.; Daddaoua, A.; Llamas, M.A. Antibiotic adjuvants: Identification and clinical use. *Microb. Biotechnol.* **2013**, *6*, 445–449. [CrossRef] [PubMed]
7. Ejim, L.; Farha, M.A.; Falconer, S.B.; Wildenhain, J.; Coombes, B.K.; Tyers, M.; Brown, E.D.; Wright, G.D. Combinations of antibiotics and nonantibiotic drugs enhance antimicrobial efficacy. *Nat. Chem. Biol.* **2011**, *7*, 348–350. [CrossRef] [PubMed]
8. Taylor, P.L.; Rossi, L.; de Pascale, G.; Wright, G.D. A forward chemical screen identifies antibiotic adjuvants in *Escherichia coli*. *ACS Chem. Biol.* **2012**, *7*, 1547–1555. [CrossRef] [PubMed]
9. Harris, T.L.; Worthington, R.J.; Hittle, L.E.; Zurawski, D.V.; Ernst, R.K.; Melander, C. Small molecule downregulation of PmrAB reverses lipid a modification and breaks colistin resistance. *ACS Chem. Biol.* **2014**, *9*, 122–127. [CrossRef]
10. Worthington, R.J.; Melander, C. Combination approaches to combat multidrug-resistant bacteria. *Trends Biotechnol.* **2013**, *31*, 177–184. [CrossRef]
11. Delcour, A.H. Outer membrane permeability and antibiotic resistance. *Biochim. Biophys. Acta-Proteins Proteom.* **2009**, *1794*, 808–816. [CrossRef]
12. Pagès, J.M.; James, C.E.; Winterhalter, M. The porin and the permeating antibiotic: A selective diffusion barrier in Gram-negative bacteria. *Nat. Rev. Microbiol.* **2008**, *6*, 893–903. [CrossRef]
13. Cox, G.; Wright, G.D. Intrinsic antibiotic resistance: Mechanisms, origins, challenges and solutions. *Int. J. Med. Microbiol.* **2013**, *303*, 287–292. [CrossRef]
14. Bolla, J.M.; Alibert-Franco, S.; Handzlik, J.; Chevalier, J.; Mahamoud, A.; Boyer, G.; Kieć-Kononowicz, K.; Pags, J.M. Strategies for bypassing the membrane barrier in multidrug resistant Gram-negative bacteria. *FEBS Lett.* **2011**, *585*, 1682–1690. [CrossRef] [PubMed]
15. Nikaido, H. Molecular basis of bacteria outer membrane permeability revisited. *Microbiol. Mol. Biol. Rev.* **2003**, *67*, 593–656. [CrossRef] [PubMed]
16. Ruiz, N.; Wu, T.; Kahne, D.; Silhavy, T.J. Probing the barrier function of the outer membrane with chemical conditionality. *ACS Chem. Biol.* **2006**, *1*, 385–395. [CrossRef] [PubMed]
17. Raetz, C.R.H.; Whitfield, C. Lipopolysaccharide endotoxins. *Annu. Rev. Biochem.* **2002**, *71*, 635–700. [CrossRef] [PubMed]

18. Yethon, J.A.; Vinogradov, E.; Perry, M.B.; Whitfield, C. Mutation of the lipopolysaccharide core glycosyltransferase encoded by waaG destabilizes the outer membrane of *Escherichia coli* by interfering with core phosphorylation. *J. Bacteriol.* **2000**, *182*, 5620–5623. [CrossRef] [PubMed]
19. Liu, A.; Tran, L.; Becket, E.; Lee, K.; Chinn, L.; Park, E.; Tran, K.; Miller, J.H. Antibiotic sensitivity profiles determined with an *Escherichia coli* gene knockout collection: Generating an antibiotic bar code. *Antimicrob. Agents Chemother.* **2010**, *54*, 1393–1403. [CrossRef]
20. Vinogradov, E.V.; van der Drift, K.; Thomas-Oates, J.E.; Meshkov, S.; Brade, H.; Holst, O. The structures of the carbohydrate backbones of the lipopolysaccharides from *Escherichia coli* rough mutants F470 (R1 core type) and F576 (R2 core type). *Eur. J. Biochem.* **1999**, *261*, 629–639. [CrossRef]
21. Neelamegham, S.; Aoki-Kinoshita, K.; Bolton, E.; Frank, M.; Lisacek, F.; Lütteke, T.; O'Boyle, N.; Packer, N.H.; Stanley, P.; Toukach, P.; et al. Updates to the Symbol Nomenclature for Glycans guidelines. *Glycobiology* **2019**, *29*, 620–624. [CrossRef]
22. Landström, J.; Persson, K.; Rademacher, C.; Lundborg, M.; Wakarchuk, W.; Peters, T.; Widmalm, G. Small molecules containing hetero-bicyclic ring systems compete with UDP-Glc for binding to WaaG glycosyltransferase. *Glycoconj. J.* **2012**, *29*, 491–502. [CrossRef]
23. Muheim, C.; Bakali, A.; Engström, O.; Wieslander, Å.; Daley, D.O.; Widmalm, G. Identification of a fragment-based scaffold that inhibits the glycosyltransferase WaaG from *Escherichia coli*. *Antibiotics* **2016**, *5*, 10. [CrossRef]
24. Wagstaff, J.L.; Taylor, S.L.; Howard, M.J. Recent developments and applications of saturation transfer difference nuclear magnetic resonance (STD NMR) spectroscopy. *Mol. Biosyst.* **2013**, *9*, 571–577. [CrossRef] [PubMed]
25. Davis, B.J. Fragment Screening by NMR. In *Protein-Ligand Interactions Methods and Applications*; Daviter, T., Johnson, C.M., McLaughlin, S.H., Williams, M.A., Eds.; Methods in Molecular Biology; Springer Nature: Cham, Switzerland, 2021; pp. 247–270.
26. Meyer, B.; Peters, T. NMR spectroscopy techniques for screening and identifying ligand binding to protein receptors *Angew. Chem.-Int. Ed.* **2003**, *42*, 864–890. [CrossRef] [PubMed]
27. Walpole, S.; Monaco, S.; Nepravishta, R.; Angulo, J. STD NMR as a Technique for Ligand Screening and Structural Studies. *Methods Enzymol.* **2019**, *615*, 423–451. [CrossRef]
28. Price, A.J.; Howard, S.; Cons, B.D. Fragment-based drug discovery and its application to challenging drug targets. *Essays Biochem.* **2017**, *61*, 475–484. [CrossRef] [PubMed]
29. Bian, Y.; Xie, X.Q.S. Computational Fragment-Based Drug Design: Current Trends, Strategies, and Applications. *AAPS J.* **2018**, *20*, 59. [CrossRef]
30. Trott, O.; Olson, A.J. AutoDock Vina: Improving the Speed and Accuracy of Docking with a New Scoring Function, Efficient Optimization, and Multithreading. *J. Comput. Chem.* **2010**, *31*, 455–461. [CrossRef]
31. Liu, N.; Xu, Z. Using LeDock as a docking tool for computational drug design. *IOP Conf. Ser. Earth Environ. Sci.* **2019**, *218*, 012143. [CrossRef]
32. Ruiz-Carmona, S.; Alvarez-Garcia, D.; Foloppe, N.; Garmendia-Doval, A.B.; Juhos, S.; Schmidtke, P.; Barril, X.; Hubbard, R.E.; Morley, S.D. rDock: A Fast, Versatile and Open Source Program for Docking Ligands to Proteins and Nucleic Acids. *PLoS Comput. Biol.* **2014**, *10*, e1003571. [CrossRef]
33. Liebeschuetz, J.W.; Cole, J.C.; Korb, O. Pose prediction and virtual screening performance of GOLD scoring functions in a standardized test. *J. Comput. Aided. Mol. Des.* **2012**, *26*, 737–748. [CrossRef]
34. Wang, Z.; Sun, H.; Yao, X.; Li, D.; Xu, L.; Li, Y.; Tian, S.; Hou, T. Comprehensive evaluation of ten docking programs on a diverse set of protein-ligand complexes: The prediction accuracy of sampling power and scoring power. *Phys. Chem. Chem. Phys.* **2016**, *18*, 12964–12975. [CrossRef]
35. Lundborg, M.; Ali, E.; Widmalm, G. An in silico virtual screening study for the design of norovirus inhibitors: Fragment-based molecular docking and binding free energy calculations. *Carbohydr. Res.* **2013**, *378*, 133–138. [CrossRef] [PubMed]
36. Martinez-Fleites, C.; Proctor, M.; Roberts, S.; Bolam, D.N.; Gilbert, H.J.; Davies, G.J. Insights into the Synthesis of Lipopolysaccharide and Antibiotics through the Structures of Two Retaining Glycosyltransferases from Family GT4. *Chem. Biol.* **2006**, *13*, 1143–1152. [CrossRef] [PubMed]
37. Åqvist, J.; Medina, C.; Samuelsson, J.E. A new method for predicting binding affinity in computer-aided drug design. *Protein Eng. Des. Sel.* **1994**, *7*, 385–391. [CrossRef]
38. Rifai, E.A.; van Dijk, M.; Geerke, D.P. Recent Developments in Linear Interaction Energy Based Binding Free Energy Calculations. *Front. Mol. Biosci.* **2020**, *7*, 114. [CrossRef]
39. Jiang, L.; Sun, Q.; Li, L.; Lu, F.; Liu, F. Molecular Insights into the Inhibitory Effect of GV971 Components Derived from Marine Acidic Oligosaccharides against the Conformational Transition of Aβ42 Monomers. *ACS Chem. Neurosci.* **2021**, *12*, 3772–3784. [CrossRef] [PubMed]
40. Landström, J.; Bergström, M.; Hamark, C.; Ohlson, S.; Widmalm, G. Combining weak affinity chromatography, NMR spectroscopy and molecular simulations in carbohydrate-lysozyme interaction studies. *Org. Biomol. Chem.* **2012**, *10*, 3019–3032. [CrossRef]
41. Mayer, M.; Meyer, B. Characterization of ligand binding by saturation transfer difference NMR spectroscopy. *Angew. Chem.-Int. Ed.* **1999**, *38*, 1784–1788. [CrossRef]
42. Laatikainen, R.; Niemitz, M.; Weber, U.; Sundeun, J.; Hassinen, T.; Vepsäläinen, J. General strategies for total-lineshape-type spectral analysis of NMR spectra using integral-transform iterator. *J. Magn. Reson.-Ser. A* **1996**, *120*, 1–10. [CrossRef]

43. Rönnols, J.; Pendrill, R.; Fontana, C.; Hamark, C.; D'Ortoli, T.A.; Engström, O.; Ståhle, J.; Zaccheus, M.V.; Säwén, E.; Hahn, L.E.; et al. Complete 1H and 13C NMR chemical shift assignments of mono- to tetrasaccharides as basis for NMR chemical shift predictions of oligosaccharides using the computer program CASPER. *Carbohydr. Res.* **2013**, *380*, 156–166. [CrossRef]
44. Jansson, P.; Widmalm, G. NMR Studies of some 1,2- and 1,3-Linked Disaccharides. *J. Chem. Soc.* **1992**, *2*, 1085–1090. [CrossRef]
45. Stanetty, C.; Walter, M.; Kosma, P. Convergent Synthesis of 4-O-Phosphorylated L-glycero-D-manno-Heptosyl Lipopolysaccharide Core Oligosaccharides Based on Regioselective Cleavage of a 6,7-O-Tetraisopropyldisiloxane-1,3-diyl Protecting Group. *J. Org. Chem.* **2014**, *79*, 582–598. [CrossRef] [PubMed]
46. Berman, H.M.; Battistuz, T.; Bhat, T.N.; Bluhm, W.F.; Bourne, P.E.; Burkhardt, K.; Feng, Z.; Gilliland, G.L.; Iype, L.; Jain, S.; et al. The protein data bank. *Acta Crystallogr. Sect. D Biol. Crystallogr.* **2002**, *58*, 899–907. [CrossRef]
47. Hanwell, M.D.; Curtis, D.E.; Lonie, D.C.; Vandermeersch, T.; Zurek, E.; Hutchison, G.R. Avogadro: An advanced semantic chemical editor, visualization, and analysis platform. *J. Cheminform.* **2012**, *4*, 17. [CrossRef] [PubMed]
48. Humphrey, W.; Dalke, A.; Schulten, K. VMD: Visual Molecular Dynamics. *J. Mol. Graph.* **1996**, *14*, 33–38. [CrossRef]
49. Morris, G.M.; Huey, R.; Lindstrom, W.; Sanner, M.F.; Belew, R.K.; Goodsell, D.S.; Olson, A.J. AutoDock4 and AutoDockTools4: Automated Docking with Selective Receptor Flexibility. *J. Comput. Chem.* **2009**, *30*, 2785–2791. [CrossRef] [PubMed]
50. O'Boyle, N.M.; Banck, M.; James, C.A.; Morley, C.; Vandermeersch, T.; Hutchison, G.R. Open Babel. *J. Cheminform.* **2011**, *3*, 33. [CrossRef]
51. Phillips, J.C.; Braun, R.; Wang, W.; Gumbart, J.; Tajkhorshid, E.; Villa, E.; Chipot, C.; Skeel, R.D.; Kalé, L.; Schulten, K. Scalable molecular dynamics with NAMD. *J. Comput. Chem.* **2005**, *26*, 1781–1802. [CrossRef]
52. Huang, J.; Rauscher, S.; Nawrocki, G.; Ran, T.; Feig, M.; de Groot, B.L.; Grubmüller, H.; MacKerell, A.D., Jr. CHARMM36m: An Improved Force Field for Folded and Intrinsically Disordered Proteins. *Nat. Method* **2017**, *14*, 71–73. [CrossRef]
53. Best, R.B.; Zhu, X.; Shim, J.; Lopes, P.E.M.; Mittal, J.; Feig, M.; MacKerell, A.D. Optimization of the additive CHARMM all-atom protein force field targeting improved sampling of the backbone φ, ψ and side-chain χ1 and χ2 Dihedral Angles. *J. Chem. Theory Comput.* **2012**, *8*, 3257–3273. [CrossRef]
54. Brooks, B.R.; Brooks, C.L., III; Mackerell, A.D.; Nilsson, L.; Petrella, R.J.; Roux, B.; Won, Y.; Archontis, G.; Bartels, C.; Boresch, S.; et al. CHARMM: The biomolecular simulation program. *J. Comput. Chem.* **2009**, *30*, 1545–1614. [CrossRef]
55. Jo, S.; Kim, T.; Iyer, V.G.; Im, W. CHARMM-GUI: A Web-based Graphical User Interface for CHARMM. *J. Chem. Theory Comput.* **2008**, *29*, 1859–1865. [CrossRef] [PubMed]
56. Jo, S.; Cheng, X.; Islam, S.M.; Huang, L.; Rui, H.; Zhu, A.; Lee, H.S.; Qi, Y.; Han, W.; Vanommeslaeghe, K.; et al. CHARMM-GUI PDB Manipulator for Advanced Modeling and Simulations of Proteins Containing Non-standard Residues. *Adv. Protein Chem. Struct. Biol.* **2014**, *96*, 235–265. [CrossRef] [PubMed]
57. Kim, S.; Lee, J.; Jo, S.; Brooks, C.L., III; Lee, S.; Im, W. CHARMM-GUI Ligand Reader and Modeler for CHARMM Force Field Generation of Small Molecules. *J. Comput. Chem.* **2017**, *38*, 1879–1886. [CrossRef] [PubMed]
58. Jo, S.; Song, K.C.; Desaire, H.; MacKerell, J.A.D.; Im, W. Glycan Reader: Automated Sugar Identification and Simulation Preparation for Carbohydrates and Glycoproteins. *J. Comput. Chem.* **2011**, *32*, 3135–3141. [CrossRef]
59. Park, S.-J.; Lee, J.; Patel, D.S.; Ma, H.; Lee, H.S.; Jo, S.; Im, W. Glycan Reader is Improved to Recognize Most Sugar Types and Chemical Modifications in the Protein Data Bank. *Bioinformatics* **2017**, *33*, 3051–3057. [CrossRef] [PubMed]
60. Park, S.-J.; Lee, J.; Qi, Y.; Kern, N.R.; Lee, H.S.; Jo, S.; Joung, I.; Joo, K.; Lee, J.; Im, W. CHARMM-GUI Glycan Modeler for Modeling and Simulation of Carbohydrates and Glycoconjugates. *Glycobiology* **2019**, *29*, 320–331. [CrossRef]
61. Durell, S.R.; Brooks, B.R.; Ben-Naim, A. Solvent-induced forces between two hydrophilic groups. *J. Phys. Chem.* **1994**, *98*, 2198–2202. [CrossRef]
62. Liu, H.; Hou, T. CaFE: A tool for binding affinity prediction using end-point free energy methods. *Bioinformatics* **2016**, *32*, 2216–2218. [CrossRef]
63. Lecca, P.; Laurenzi, I.; Jordan, F. Deterministic chemical kinetics. *Determ. Versus Stoch. Model. Biochem. Syst. Biol.* **2013**, 1–34. [CrossRef]

pharmaceuticals

Review

An Overview of Cellulose Derivatives-Based Dressings for Wound-Healing Management

Elena-Emilia Tudoroiu [1], Cristina-Elena Dinu-Pîrvu [1,*], Mădălina Georgiana Albu Kaya [2,*], Lăcrămioara Popa [1], Valentina Anuța [1], Răzvan Mihai Prisada [1] and Mihaela Violeta Ghica [1]

1 Department of Physical and Colloidal Chemistry, Faculty of Pharmacy, Carol Davila University of Medicine and Pharmacy Bucharest, 6 Traian Vuia Str., 020956 Bucharest, Romania; elena-emilia.tudoroiu@drd.umfcd.ro (E.-E.T.); lacramioara.popa@umfcd.ro (L.P.); valentina.anuta@umfcd.ro (V.A.); razvan.prisada@umfcd.ro (R.M.P.); mihaela.ghica@umfcd.ro (M.V.G.)

2 Department of Collagen, Division Leather and Footwear Research Institute, National Research and Development Institute for Textile and Leather, 93 Ion Minulescu Str., 031215 Bucharest, Romania

* Correspondence: cristina.dinu@umfcd.ro (C.-E.D.-P.); albu_mada@yahoo.com (M.G.A.K.); Tel.: +40-72-230-4894 (C.-E.D.-P.); Tel.: +40-72-339-5108 (M.G.A.K.)

Abstract: Presently, notwithstanding the progress regarding wound-healing management, the treatment of the majority of skin lesions still represents a serious challenge for biomedical and pharmaceutical industries. Thus, the attention of the researchers has turned to the development of novel materials based on cellulose derivatives. Cellulose derivatives are semi-synthetic biopolymers, which exhibit high solubility in water and represent an advantageous alternative to water-insoluble cellulose. These biopolymers possess excellent properties, such as biocompatibility, biodegradability, sustainability, non-toxicity, non-immunogenicity, thermo-gelling behavior, mechanical strength, abundance, low costs, antibacterial effect, and high hydrophilicity. They have an efficient ability to absorb and retain a large quantity of wound exudates in the interstitial sites of their networks and can maintain optimal local moisture. Cellulose derivatives also represent a proper scaffold to incorporate various bioactive agents with beneficial therapeutic effects on skin tissue restoration. Due to these suitable and versatile characteristics, cellulose derivatives are attractive and captivating materials for wound-healing applications. This review presents an extensive overview of recent research regarding promising cellulose derivatives-based materials for the development of multiple biomedical and pharmaceutical applications, such as wound dressings, drug delivery devices, and tissue engineering.

Keywords: cellulose derivatives; wound dressings; wound management; bioactive agents delivery systems

Citation: Tudoroiu, E.-E.; Dinu-Pîrvu, C.-E.; Albu Kaya, M.G.; Popa, L.; Anuța, V.; Prisada, R.M.; Ghica, M.V. An Overview of Cellulose Derivatives-Based Dressings for Wound-Healing Management. *Pharmaceuticals* **2021**, *14*, 1215. https://doi.org/10.3390/ph14121215

Academic Editor: Federica Sodano

Received: 1 November 2021
Accepted: 18 November 2021
Published: 24 November 2021

1. Introduction

Skin is the largest and the main organ that forms the body covering, with a complex structure of tissues, and creates an exterior defense barrier, which protects the internal organs from mechanical impairments, radiation, chemicals, and foreign invaders (bacteria and viruses) [1]. More than being a passive barrier, the skin defends the body against contamination, infection, and surrounding environment influence [2]. Skin is also a sensory organ and contains a large category of sensory neuron subtypes (thermoreceptors, nociceptors, pruriceptors, and low-threshold mechanoreceptors), that take over and transfer to the brain information about the environment [3]. Moreover, the skin has an important role in homeostasis, elimination of toxins, sustaining regular hydration levels, prevention of electrolytes loss [4], and in control of body temperature and blood pressure. The skin is made up of three particular layers: epidermis, dermis, and hypodermis or subcutaneous layer [5].

The epidermis is a physical protective barrier against the external factors, which does not contain blood vessels. It is comprised of two main categories of cells: dendritic cells and keratinocytes (keratin synthesis), along with Langerhans cells (engaged in the immune reaction), Merkel cells (sensory corpuscles), and melanocytes (melanin synthesis) [6,7]. The dermis is situated between the epidermis and hypodermis, and it is responsible for skin thickness. The dermis structure is principally fibrous because it contains collagen and elastic fibers [8]. Moreover, this layer also includes hair follicles, sebaceous glands, sweat glands, nerves, and blood vessels. The fundamental component of the dermis is collagen and the most abundant are type I and type III (approximately 95%) [9,10]. The dermis plays an essential role in thermoregulation, skin protection, preservation of skin support, and perception of sensation [11]. The hypodermis (subcutaneous tissue), the widest and the thickest part of the skin, is located between the dermis and muscles or bones beneath it and it is made up of elastin and loose connective tissue [12]. The principal roles of subcutaneous tissue are thermal insulation, energy resource, nutritional reserve, and mechanical conservation [13,14].

There are many factors and systemic diseases that can alter skin functions, for example, pathogens, pollution, radiation exposure, smoking, malnutrition, obesity, diabetes mellitus, peripheral vascular diseases, pressure ulcer, inflammatory, bleeding, or metabolic conditions and immunosuppression [15,16]. In many cases, all these factors can delay wound healing, with harmful risks for patients, such as oxidative stress, chronic inflammation, infection, increased tissue loss, and necrosis [17]. Thus, skin tissue disorders are a public health problem worldwide, with a higher incidence from year to year. For example, in 2005 there were estimated ~5 million skin and subcutaneous conditions, in 2015 ~6.1 million [18], and in 2018 there were ~8.2 million people, who suffered from wounds. Total costs of treatments for wound healing vary between $28.1 billion and $96.8 billion [19]. With a prevalence of 1–2% in the global population, chronic wounds have the largest frequency from all types of skin tissue injuries, especially surgical wounds, and leg/foot ulcers (pressure ulcer and diabetic foot ulcer) [20]. An injury involves physical impact, such as pain, inflammation, mobility limitation, disturbance of sleep, alterations of skin appearance, and restriction of daily activities; consequently, these effects have a negative impact on the patient quality of life, affecting emotional, social, and physical states [21]. To restore the impaired tissue and to rebalance the quality of life for patients with wounds, optimal and multidisciplinary wound management has an essential role. Its main purpose is to obtain a proper functional, structural, and cosmetic result [22]. Frequently, to alleviate the pain and inflammation, which accompany the wounds, analgesic and anti-inflammatory drugs are prescribed. Because of their side effects on the gastrointestinal system when they are administered orally [23], a more advantageous and simple treatment is to apply on the lesion site a wound dressing [24]. The main purpose of wound management is to reduce the period of wound healing through the prevention of infection, alleviation of inflammation and pain, and diminishing the scars [25].

In this review, we mainly present studies from the last 10 years regarding the cellulose derivatives-based wound dressings with various bioactive agents to accelerate the regeneration of skin tissue injuries. Cellulose derivatives have proper and optimal applicability to develop novel wound dressings that can enhance wound healing, obtained by substituting the hydroxyl groups from cellulose molecule with different alkyl groups [26]. Thus, they exhibit high solubility in water and represent a feasible alternative to water-insoluble cellulose. Moreover, these semi-synthetic biopolymers possess other multiple characteristics, such as biocompatibility, biodegradability, proper physicochemical properties, low toxicity and costs, poor immunogenicity, adequate absorption of wound exudates, thermo-gelling power, and antibacterial effect [27,28]. Besides wound-dressings development, cellulose derivatives are promising biomaterials for tissue engineering, drug delivery, hemodialysis, osseointegration, and biosensors [29].

2. Wound Classification

A wound represents a lesion, break, tear, or damage of skin structure and function, produced by physical, mechanical (surgery), thermal, chemical, and electrical (burns) factors; an injury can also be the result of an underlying medical or physiological disorder (diabetes and malignancies) [30–32]. The Wound Healing Society define a wound as a consequence of 'disruption of normal anatomic structure and function' [33]. The National Institutes of Health from the United States assesses that approximately 3% of people over the age of 65 can suffer at any one time from a cutaneous lesion [34].

Wounds are classified according to many parameters:

- etiology: surgical, traumatic, radiation and malignant wound, chemical or thermal injury, pressure ulcer, diabetic foot ulcer, vascular ulcer, or atypical injury;
- healing time (duration) and nature of the injury restoration process: acute or chronic wounds;
- depth of injury or number of skin layers affected: superficial, partial thickness (deep dermal) or full-thickness wounds;
- complexity: simple, complex, or complicated wounds;
- contamination and postoperative infection risk: clean wound (class I), clean/contaminated wound (class II), contaminated wound (class III) or dirty wound (class IV);
- mode of lesion: abrasion, ulceration, incision, laceration or degloving;
- tissue loss: without tissue loss (surgical wounds) and with tissue loss (burns, traumatic wounds, diabetic foot ulcers, and iatrogenic wounds);
- appearance: necrotic, sloughy, infected, malodorous, granulating, and epithelializing wound;
- injured tissue coloration: black, green, yellow, white, brown, purple, beefy red, or pale pink wounds [35–37].

From all of the classification criteria, the most significant and decisive criterion for selection of an adequate dressing and for optimal wound-healing management is the healing time (duration) and the nature of the injury restoration process. Hence, an acute injury heals totally, without external support, with minimum scarring, and usually demands a period for healing from 8 to 12 weeks [38]. An acute lesion can be simple or complex, but it depends on the affected anatomical parts and on the dimension and depth of this lesion. In this category are found mechanical injuries, burns, and chemical wounds [33,39]. In contrast to the acute wound, a chronic injury heals slowly, requires a long time for healing, more than 12 weeks, usually reoccurs, and leaves severe scars; mostly, a chronic wound does not have any time limitation for the repair process [40]. The main chronic wounds are venous ulcer, ischemic injuries (especially of atherosclerotic origin), diabetic foot ulcer, pressure ulcer, and malignant wounds [41,42].

3. Wound-Healing Process

Wound healing is a sophisticated and well-coordinated process [43], which involves a variety of cellular and biochemical reactions that need a complex and dynamic cascade of biological processes [44,45] for the reestablishment of skin layers, growth and tissue regeneration, anatomical continuity, and skin functions [46]. Damaged skin tissue has the capacity to repair itself to form a new epithelium that closes the wound and repairs the barrier function, through an intricate process [47].

3.1. Wound-Healing Stages

The wound repair process consists of four different, overlapping, and exactly programmed stages: hemostasis, inflammation, proliferation, and remodeling (maturation) phases [48], illustrated in Figure 1.

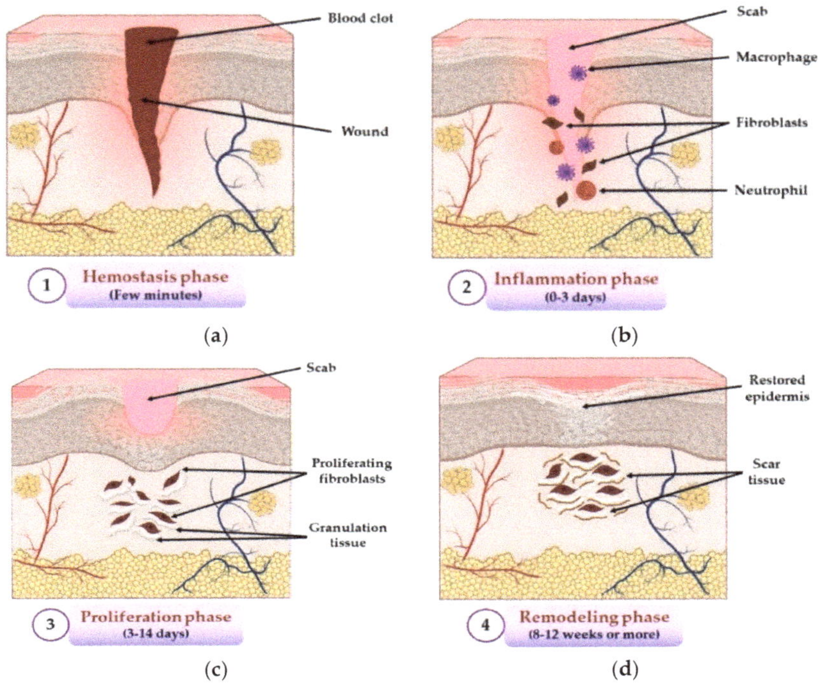

Figure 1. The four stages of wound healing process: (**a**) hemostasis phase; (**b**) inflammation phase; (**c**) proliferation phase; (**d**) remodeling phase. All illustrations have been created with BioRender.com, Inkscape, and PowerPoint.

The first and the shortest stage (5–10 min) of the wound repair process is hemostasis, an instantaneous reaction towards lesion [49], whose main purpose is to stop the bleeding through vasoconstriction, primary hemostasis (thrombocytes aggregation with thrombocyte plug formation) and secondary hemostasis (fibrin clot formation) [50]. The inflammation stage happens approximately at the same time as the hemostatic stage and consists of enrollment of neutrophils and macrophages, cytokines secretion, destruction and elimination of bacteria and formation of a wound bed [51]. Inflammation induces accumulation of leukocytes at the lesion's level, activating different mediators and chemotactic factors in 1–2 days after injury and lasts for about 3 days [47]. The proliferation stage begins on day 3 and can have a duration up to 14 days after tissue damage [52]. This phase represents a complex process that includes the next events: neoangiogenesis, production of granulation tissue through fibroblasts proliferation and collagen deposition, synthesis of extracellular matrix, re-epithelialization, and injury retirement, all these happening simultaneously [53]. The final phase of the wound repair process is the remodeling (maturation) stage, and its major purpose is the production of cellular connective tissue and hardening of the new epithelium that establishes the final scar nature [54]. It is the longest stage of all five and can last from weeks to 1–2 years or more. The main event of this phase is the remodeling of granulation tissue, where collagen type I will take the place of collagen type III because type I is more stable [55].

3.2. Factors Affecting Wound-Healing Process

Many factors can interfere with the wound-healing phases, the consequence being an improper or damaged wound repair process. In general, these factors can be categorized as local and systemic. Local factors affect features of the lesion itself, and systemic factors represent the general health or condition states of one person, which influence the capacity to heal [56]. The main factors that affect the wound-healing process are presented in Figure 2.

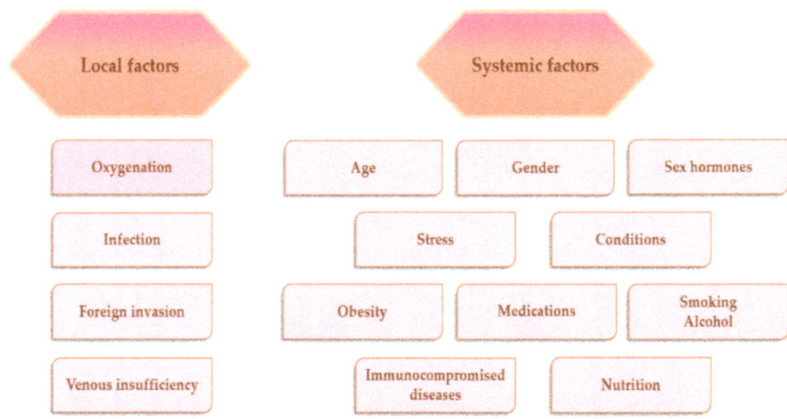

Figure 2. Factors affecting wound-healing process [57,58].

3.2.1. Local Factors That Affect Wound-Healing Process

In injured tissues, vascular disruption generates depletion of oxygen, causing hypoxia, and, consequently, impaired wound healing. Temporary hypoxia stimulates the lesion repair process, but persistent hypoxia prolongs this process [59]. Hypoxia is characterized by high levels of reactive oxygen species (ROS) in cells and the impact on tissue healing is deleterious [31,57]. Oxygen has many roles in the injured tissue: avoids infection, activates the angiogenesis, enhances keratinocytes differentiation, movement, and re-epithelialization, increases fibroblast proliferation and biosynthesis of collagen, and stimulates lesion contraction [60]. When skin tissues are injured, the physical protective barrier against foreign invaders is damaged, these germs easily invade the lesion and contaminate or colonize it, causing local infection, and in severe cases, when the injury is not treated properly, they cause systemic infection [61]. Moreover, bacteria and endotoxins may induce the extended elevation of matrix metalloproteinases and pro-inflammatory cytokines (IL-1 and TNF-α), prolonging the inflammatory phase [56].

3.2.2. Systemic Factors That Affect Wound-Healing Process

A main risk factor for the damaged lesion repair process is increased age due to multiple comorbidities [62]. Acute injuries have a prolonged healing time for elderly males compared to elderly females. This fact can be explained through sex hormones, which have an essential role in the wound repair process [63]. Alongside them, stress causes the decrease of pro-inflammatory cytokine levels and the reduction of chemo-attractants expression, which are involved in the inflammation stage of wound healing [64]. Regarding the conditions, the major disease, which strongly and negatively influences the wound-healing process, is diabetes mellitus, because of the diabetic foot ulcer, which causes hypoxia, inhibition of the expansion of macrophages and neutrophils and reduction of fibroblasts proliferation [65]. Obesity represents another major factor that affects the normal repair process, because it is characterized by an augmented workload of the heart to provide oxygenated blood to skin tissues, it cannot perfuse them, the outcome being the onset of ischemia and a higher risk to develop infections [66]. Among medications, steroids and chemotherapeutic drugs can lead to delayed healing. Mechanisms through steroids that affect the wound healing are the inhibition of lesion contraction, and fibroblasts proliferation, the decrease of tensile strength, and collagen production [67]. Chemotherapeutic drugs disturb the proliferative stage through slowing cells' movement to the lesion, angiogenesis inhibition, reduction of biosynthesis of collagen and decrease of fibrin deposition [68]. The quality and rate of the normal repair process can be also altered by smoking and alcohol because they lead to lesion infection and dehiscence, reduction of neutrophils, lessening of angiogenesis, inhibition of epithelial reconstruction and lesion contraction, and in severe cases, to necrosis of tissues [69]. Poor nutrition slows the lesion repair process through inflammation extension, inhibition of fibroblasts functions, decline of angiogenesis, and reduction of collagen biosynthesis and deposition [70].

4. Wound Dressings: Properties and Classification

In past years, due to the technology's noteworthy progress, various wound dressings were formulated worldwide to cure all types of tissue lesion. Dressings play a fundamental role in wound-healing management because these protect tissue lesions from external invasion (wound dressings are permeable for oxygen and moisture and function as physical barriers) [71], preventing the infection on the wound site [72]. Moreover, dressings contribute to the regeneration and restoration of epidermis and dermis layers [73,74].

4.1. Wound Dressing Properties

For the development of dressings, which allow rapid healing, with minimal scars on the body surface, it is necessary to develop new biopolymeric materials that accomplish some properties to create the ideal wound dressing that are reviewed in Figure 3.

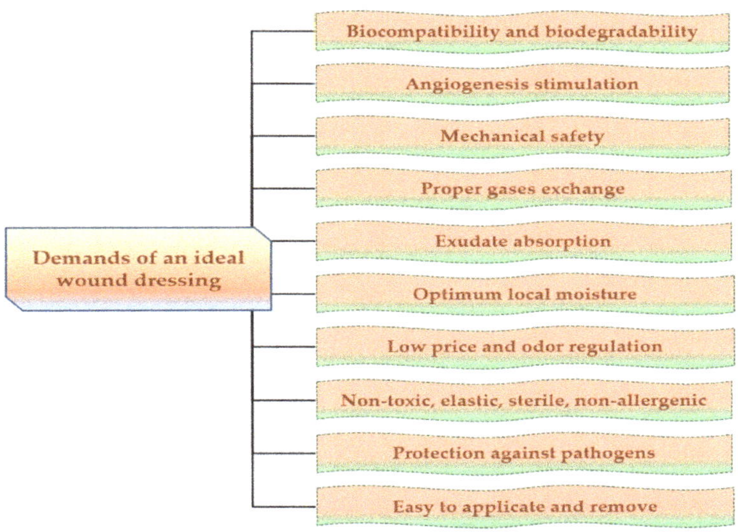

Figure 3. Major demands of an ideal wound dressing.

The ideal wound dressing preferably presents the following features: biocompatibility, biodegradability, non-toxicity, chemical inertness [75], to be applied effortlessly, to have the capacity to keep local moisture, to ensure a suitable exchange of gases (O_2 and CO_2), to absorb exudates that form on the lesion site [76], to stimulate the angiogenesis, to protect against extraneous pathogens, to clear the injured tissue, to eliminate nonviable tissues, to reduce the exposed area [77], to be able to be removed and replaced without difficulty [78], to adjust the odor, to sustain an adequate temperature to the lesion bed, to promote the blood circulation, and to stimulate cell expansion, to ensure mechanical safety [79]. Also, wound dressings materials must be elastic, sterile, non-adherent, non-allergenic [80], to have an acceptable price and to provide thermal insulation [81].

4.2. Wound-Dressing Classification

A potential classification of wound dressings comprises passive dressings and active dressings, depending on the presence or absence of one or more pharmacologically active substances or natural substances [82], which can act to the site of the lesion, with local or systemic action, conditioned by the depth of the wound. Moreover, the progress of manufacturing led to the evolution of wound dressings from traditional dressings to modern (advanced) dressings [83].

Passive dressings can be considered dry traditional dressings, which are fundamental for a faster wound-healing process. There is a wide simple range of passive dressings for several types of skin lesion: cotton wool, lint, gauze, natural, and synthetic bandages–they work as primary dressing or secondary dressing [79,84]. Active dressings contain a large variety of pharmacologically active substances (antibiotics or other antimicrobials, non-steroidal anti-inflammatory, analgesic, antifungal, and local anesthetics drugs) or natural substances (plant extracts) with anti-inflammatory, astringent, emollient, epithelializing, antioxidant, demulcent, and antimicrobial properties [73].

Modern or advanced dressings were designed to cover tissue lesions and in this category are included the hydrogels, hydrocolloids, semi-permeable films, semi-permeable foams, and alginate dressings [52,85]. The biggest difference between traditional and modern dressings is local moisture maintenance. Thus, traditional dressings have a lower capacity to maintain the local moisture on the wound site [83], and modern dressings sustain excellent local moisture to enhance wound healing [86]. The classification of wound dressings is illustrated in Figure 4.

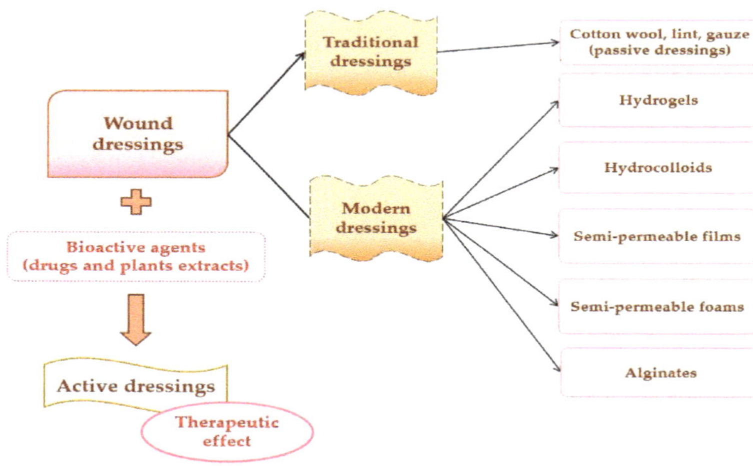

Figure 4. Wound dressings classification.

The main materials underlying the modern wound dressings are polymers, which can be natural (collagen, gelatin, cellulose, hemicellulose, chitin, chitosan, pectins, gums, chondroitin sulfate, alginic acid, alginates, agar, dextran, carrageenan, elastin, hyaluronic acid, silk fibroin, fibrinogen, and fibrin) [87,88], semi-synthetic (cellulose derivatives) [89] or synthetic (poly(α-ester)s, polyanhydrides, polycarbonates, poly(amide), poly(esteramide)s, polyphosphazenes, polyurethanes, pseudo poly(amino acids), polyacetals) [90–92].

The first class of modern wound dressings includes hydrogels, also called hungry or smart networks, which can be defined as three-dimensional networks, consisting of cross-linked polymeric materials [93], with a significant capacity to absorb inside their structure a massive volume of water or body fluids, without dissolution in these liquids [94]. Hydrogels are transparent dressings, and this fact allows the tissue lesion to be observed and controlled without the dressing being eliminated [95]. Hydrogels dressings have a remarkable application in many domains due to their high water content (up to 96%) [96], such as biomedical and pharmaceutical sciences (wound dressings, drug delivery systems, diagnostics, tissue engineering, contact lenses, regenerative medicines) [97], agriculture, food industry, biotechnology, separation technology (cells and biomolecules), electroconductive hydrogels and biosensors, oil recovery, the cosmetic industry, and hygienic products [98–100]. Hydrogels, which stimulate autolytic debridement, are used as wound dressings in pressure ulcers, thermal injuries, and lesions caused by surgery [101].

Hydrocolloids are another class of modern wound dressings, which are based on a combination between elastomers, alginates, and colloidal materials. They present the ability to take in a small or medium quantity of exudates, have good biocompatibility, biodegradability, and adhere to the skin [102]. These dressings are occlusive, so they do not allow microorganisms to penetrate tissue lesions, do not afford gases exchanges, and are water-resistant [103].

The third class of modern wound dressing includes semi-permeable films. They are flexible and elastic sheets, made from transparent polyurethane. For a good attachment to the skin, polyurethane films present an acrylic adhesive on one part [104,105]. Films are impermeable to pathogens and water but allow the exchange of oxygen and water vapor. Films are used in surgical injuries or wounds with a reduced volume of exudates [106]. They cannot be applied on tissue lesions with necrosis or infection, on sensitive skin (newborns and elder people), and also on wounds, which have a substantial amount of exudates because films offer a poor capacity to absorb wound fluids [41,96].

The following category of modern dressing includes semi-permeable foams, which showed a vast improvement and favorable biocompatibility [107]. They have a considerable capacity to swallow a great quantity of liquids formed at the injury site, due to their content of hydrophilic polyurethane and silicone, so they can be recommended for tissue lesions with a medium to high status of wound fluids [108]. The disadvantages of foams consist in the limitation of use for dry wounds (foams have a dehydration effect) and in the impossibility to follow the evolution of wound healing because foams are totally opaque [109].

The last class of modern wound dressings is represented by alginates, a category of polysaccharides extracted from brown algae and kelp, with remarkable absorption properties [109]. Alginates are the result of the alginic acid reaction with calcium and sodium; therefore, alginates are called salts [50,110]. The formation of alginate gel is based on the exchange of calcium ions, which are inside the dressing, with sodium ions, which are in lesion exudates. Thus, the alginate gel presents an exceptional power to absorb a large volume of wound fluids, especially in the case of foot ulcers [111]. Another advantage of alginates dressings is that they have a hemostatic effect due to calcium ions (known as clotting factor IV, which plays an important role in blood coagulation); accordingly, they can be used if an injury bleeds [112].

5. Cellulose Derivatives as Wound Dressings

During the last decades, cellulose derivatives, also known as cellulosics, have become extensively used in many fields, from food, cosmetics, biomedical, and pharmaceutical industry [113] to biofuels and oilfield industry (petrochemicals) [114]. These semi-synthetic biopolymers present many advantageous characteristics, such as biocompatibility, biodegradability, non-toxicity, sustainability, abundance, and a suitable price; therefore, cellulose derivatives represent the first option for wound dressings development [115,116].

5.1. Cellulose Derivatives Classification

Cellulose, discovered by Anselme Payen in the 19th century, is a natural polymer, an organic polysaccharide from plant origin, non-toxic, with a structural role, being the most plentiful and renewable biopolymer on Earth [117]. Structurally, cellulose is a linear macromolecule composed of many molecules of D-glucose (the number of the glucose units can reach more than ten thousand), which are bound through 1-4-β-glycosidic linkages and its chemical formula is $(C_6H_{10}O_5)_n$ [118]. The chemical structure of cellulose shows the presence of free hydroxyl groups at C_2, C_3, and C_6 of each molecule of glucose, which have a good capacity to form powerful inter- and intramolecular hydrogen bonds [119]. As a result of this property, cellulose has a crystalline and stiff structure and, consequently, it is insoluble in water and the majority of the organic solvents; moreover, this natural biopolymer cannot be digested by the human digestive system [120,121]. Cellulose has good stability to pH fluctuations and temperature [122].

To improve the solubility problems of cellulose and to extend its applications, the chemical structure of this polymer can suffer several changes to obtain the cellulose derivatives, which have suitable physicochemical properties to be used in many fields, especially in the pharmaceutical and biomedical industry [123]. The modifications in the cellulose molecule can be chemical, physical, or biological [114], but the most used and significant of the three is the chemical modification. Targeted by this method are the hydroxyl groups, which suffer an esterification or an etherification reaction [124]. Therefore, the cellulose derivatives can be classified in two major classes: cellulose esters derivatives and cellulose ethers derivatives, which have particular mechanical and physicochemical characteristics [125]. The chemical structures of cellulose and cellulose derivatives are presented in Figure 5.

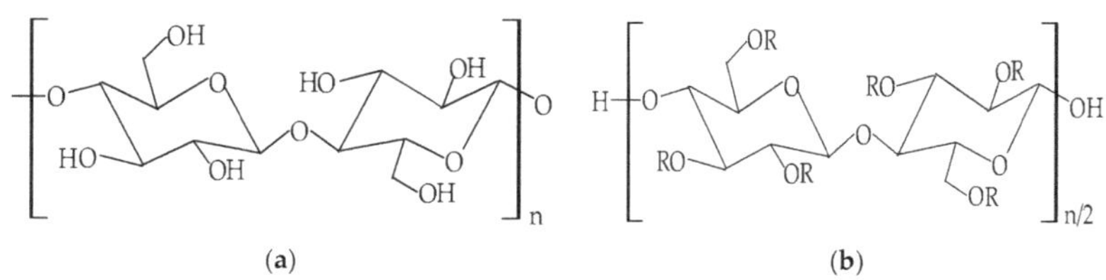

Figure 5. Chemical structures of: (**a**) cellulose; (**b**) cellulose derivatives.

Cellulose ethers derivatives are characterized by high molecular weight and the greatest applicability in the pharmaceutical domain of all these derivatives are: sodium carboxymethylcellulose (NaCMC), hydroxypropylmethylcellulose (HPMC), methylcellulose (MC), hydroxyethylcellulose (HEC), ethylcellulose (EC), hydroxypropylcellulose (HPC), hydroxyethylmethylcellulose (HEMC) and benzylcellulose (BC) [126,127]. The cellulose ethers are illustrated in Table 1.

Table 1. Main cellulose ether derivatives according to R groups [128].

Cellulose Ethers	R Groups
Methylcellulose	H, CH_3
Ethylcellulose	H, CH_2CH_3
Benzylcellulose	H, $C_6H_5CH_2$
Sodium carboxymethylcellulose	H, CH_2COONa
Hydroxyethylcellulose	H, $[CH_2CH_2O]_nH$
Hydroxypropylcellulose	H, $[CH_2CH(CH3)O]_nH$
Hydroxyethylmethylcellulose	H, CH_3, $[CH2CH2O]_nH$
Hydroxypropylmethylcellulose	H, CH_3, $[CH_2CH(CH_3)O]_nH$

Cellulose esters derivatives are extensively used in the pharmaceutical industry as enteric coated drug delivery devices, and they also have excellent properties to form films. There are two categories of cellulose esters: organic and inorganic, but the most common in the pharmaceutical practice are organic esters [129]. Among them are cellulose acetate (CA), cellulose acetate butyrate (CAB), cellulose acetate phthalate (CAP), cellulose acetate trimelitate (CAT), hydroxypropylmethylcellulose phthalate (HPMCP), and hydroxypropylmethylcellulose acetate succinate (HPMCAS). With fewer applications in the pharmaceutical industry are inorganic esters, such as cellulose nitrate (CN) and cellulose sulphate (CS) [130]. The cellulose esters are illustrated in Table 2.

Table 2. Main cellulose ester derivatives according to R groups [128].

Cellulose Esters	R Groups
Acetate	H, I
Acetate trimelliate	H, I, II
Acetate phthalate	I, III
Hydroxypropylmthylphthalate	H, CH_3, $CH_2CH(OH)CH_3$, III, IV
Hydroxypropylmthylphthalate acetate succinate	H, CH_3, $CH_2CH(OH)CH_3$, III, V

Another classification of cellulose derivatives depends on the water solubility of these polymers; thus, there are described in water-soluble cellulose derivatives and water-insoluble cellulose derivatives. In the first category are included the majority of cellulose ethers (methylcellulose (MC), sodium carboxymethylcellulose (NaCMC), hydroxyethylcellulose (HEC), hydroxypropylcellulose (HPC), hydroxyethylmethylcellulose (HEMC) and hydroxypropylmethylcellulose (HPMC)) [131], while the other cellulose ethers (ethylcellulose and benzylcellulose) and cellulose esters are included in the category of water-insoluble cellulose derivatives. Between the two categories, water-soluble cellulose derivatives are the most used biopolymers in the pharmaceutical and biomedical industry [115,132], because they present several favorable features, such as solubility, surface activity, viscosity in solution, similar properties to thermoplastic film, and proper stability to oxidative and hydrolytic reactions, heat and biodegeneration [130,133].

Due to the general properties of wound dressings presented in Section 4.1, but also the particular properties, such as hydrophilicity, mechanical toughness, pH stability, and rheological characteristics, cellulose and cellulose derivatives have multiple applications in many fields [134]. Areas of the applicability of all these biopolymers involve: biomedical and pharmaceutical industries, where they can act as drug-delivery devices, wound dressings, muco- and bioadhesive drugs, excipients for drug formulations, and support for tissue engineering [29]; also, they can be used for cosmetic and hygienic products, in the textile area, in the food industry and agriculture [128,135]. The representation of cellulose derivatives-based wound dressing on an open wound is illustrated in Figure 6.

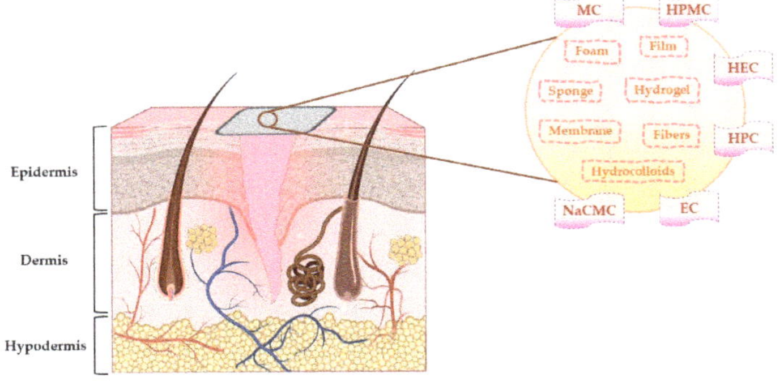

Figure 6. The representation of cellulose derivatives-based wound dressing on an open wound. This illustration has been created with BioRender.com, Inkscape, and PowerPoint.

Cellulose ethers derivatives (Table 1) are the most used biopolymers for tailoring of new wound dressings, compared to cellulose esters derivatives. Therefore, we will further describe them, and we will present their main different types of wound dressings for an optimal wound management, from gels to foams, as we summarized in Section 4.2.

5.2. Sodium Carboxymethylcellulose-Based Wound Dressings

Carboxymethylcellulose (CMC), also known as carmellose, is a semi-synthetic and hydrophilic polymer, a water-soluble cellulose ether derivative, and one of the polymers with the lowest price [136]. Sodium carboxymethylcellulose (NaCMC) is the sodium salt of CMC, an anionic polymer, with a great solubility in water [137]. NaCMC was the first compound from the group of cellulose derivatives; therefore, all the researchers' attention was focused on it because, compared to other cellulose derivatives, NaCMC can be synthesized through simple methods with low-cost materials [138]. It results from the etherification reaction of the cellulose with sodium monochloroacetate in an alkaline solution (NaOH) [139]. In the cellulose molecule, three hydroxyl groups (from 2, 3, and 6 positions) are substituted by carboxymethyl groups [140], resulting in different values of substitution degree from 0.4 to 1.5 and different molecular weights of NaCMC, varying from 90,000 to 2,000,000 g/mol [51]. The optimal substitution degree to be used in the pharmaceutical industry is from 0.60 to 1.00 [139]. The chemical structure of NaCMC is shown in Figure 7.

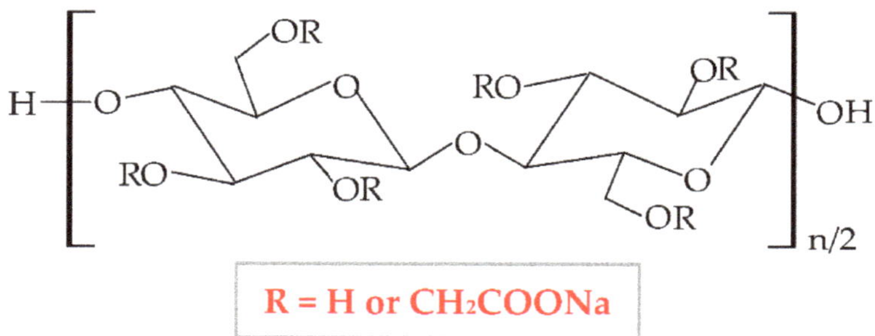

Figure 7. Chemical structure of sodium carboxymethylcellulose (NaCMC).

The NaCMC network illustrates a thixotropic behavior to generate 3D structures through intermolecular attraction. Its thixotropy is influenced by concentration and degree of substitution [141]. NaCMC presents excellent physicochemical and mechanical properties [142], optimal biocompatibility and biodegradability, proper capacity to absorb the water and to swell, high gelation behavior, non-toxicity, and low-immunogenicity [143]. It is the most used cellulose derivative in the pharmaceutical industry, mainly for the development of new wound dressings because it has the capacity to absorb heavy exudates [144,145], to ensure excellent moisture at the lesion site, and to avoid skin tissues water loss and tissues necrosis. Moreover, an optimal local humidity can impede dehydration, facilitate the synergy between target cells and growth factors, promote angiogenesis advancement, the mitigation of the ache, and the disruption of the fibrin network [146]. NaCMC is also used as a drug-delivery device and excipient for drug formulations (used as an emulsifier, thickener, stabilizer, and film-maker) [147]. Besides its applicability in the pharmaceutical area, this biopolymer possesses different usefulness in the food (E466 food additive) industry [148], in paper, textile and cosmetics domains [51,149], for tissue culture and dental medicine field [150].

NaCMC can be combined with other polymers to enhance its properties and to develop its applicability. Thus, it is more advantageous to blend two or more polymers for the development of a new material comparative to the chemical industrial development of that material. Moreover, the new material obtained by mixing other well-known polymers presents all the properties or is more favorable than the component polymers [75]. Furthermore, the blend of polymers can be realized to compensate for their drawbacks. Hence, Liu et al., combined NaCMC with HEC by electrostatic complexing and obtained a sponge and a membrane with a porous network, enhanced viscoelastic properties, and high swelling behavior [151]. Hu et al., mixed NaCMC with PVA and quaternized chitosan and designed a new composite with enhanced flexibility, water absorption rate, mechanical strength, swelling ratio, and humidity permeability [152]. A novel NaCMC/PVA-based composite was formulated, with higher properties than two polymers: improved swelling capacity, elasticity, water solubility, porosity, water vapor transmission rate, bioavailability, and biodegradability for the tissue repair process; this formulation also presented an extension of its applicability, such as agriculture, biomedical field as drug delivery systems and food packaging [153,154]. NaCMC was blended with PEG through a photo-click reaction based on thiol-norbornene. It formed a pH-sensitive hydrogel with an augmented swelling ratio [150]. Zhang et al., designed a novel hydrogel based on NaCMC and sodium alginate. In a ratio of 1:4, the hydrogel exhibited high biocompatibility, mechanical characteristics, degradation rate, and local humidity [155]. Shin et al., blended NaCMC with PVA and PEG 400 through cyclic freezing/thawing method and obtained a hydrogel with improved properties: the swelling rate, the compressive strength, and cytocompatibility [156]. NaCMC can also be blended with diverse biopolymers to develop a new potential wound dressing with better properties to accelerate the wound healing process. All these combinations are illustrated in Table 3.

Table 3. Recent studies on the use of sodium carboxymethylcellulose as a wound dressing.

Biopolymer/-s	Active Pharmaceutical Ingredient (Natural or Synthetic Substances)	Type of Wound Dressing	Main Findings	References
NaCMC	-	Gel	Anticoagulant activity at second-degree burn injuries; this action was influenced by concentration, substitution degree, and molecular weight of the biopolymer.	[157]
		Hydrocolloid	Appropriate management of humidity, maintained self-adhesiveness, and increase of the surface energy.	[85]
		Film	Enhanced tensile strength, absorption capacity, and decreased percent erosion.	[147]
		Foam	Suitable material for wound dressing, because it can maintain optimal local moisture.	[158]
	Merremia mammosa	Gel	Good gel properties, non-toxicity, speeding the wound healing process in diabetic ulcer (increases collagen synthesis and improves angiogenesis).	[159]
	Chestnut honey	Hydrogel	Enhanced moisture retention, non-adherence, and excellent antibacterial action to treat diabetic ulcer injuries.	[160]
	Reduced graphene oxide	Hydrogel	The inhibition of bacteria biofilm from infected wounds.	[161]
	Neomycin trisulphate, sulphacetamide sodium, and silver nitrate	Hydrogel	Proper elasticity, sponginess, homogenous texture, suitable smell, high drugs delivery, and antimicrobial effect.	[162]
	Tetracycline–UiO-66	Hydrogel	Increased mechanical properties, antibiotic delivery, and good antibacterial activity.	[163]
	ZnO-MCM-41 and tetracycline	Hydrogel	Improved tensile strength, permeability for gases, and swelling ratio; optimal antibacterial effect.	[164]
	Lidocaine hydrochloride	Film	Optimal physicochemical and mechanical properties; high drug release (90%) in the first 15 min.	[165]
	AgNPs	Membrane	Good morphology and superior antibacterial activity.	[166]
NaCMC/Chitosan	-	Sponge	High porosity, air permeability, and proper antibacterial effect; in the presence of the cyanine dye, the sponge indicated the pH of the wound site.	[167]
NaCMC/Pullulan	-	Hydrogel	An important decrease of the hydrogel adherence at the postoperative lesions.	[168]
NaCMC/Sericin	-	Hydrogel	Increased mechanical stability and the strength to hydrolysis and enzymatic degeneration; the molecular weight of CMC influenced the sericin release.	[169]
		Film	Higher mechanical properties, hydrophilicity, swelling power, stability, compatibility with cells and blood, and stimulation of cell proliferation.	[170]
NaCMC/ε-Polylysine	-	Hydrogel	Optimal rheological properties, degradability, compression modulus, and high antibacterial activity.	[171]

Table 3. *Cont.*

Biopolymer/-s	Active Pharmaceutical Ingredient (Natural or Synthetic Substances)	Type of Wound Dressing	Main Findings	References
NaCMC/Pectin/Cellulose	-	Film	NaCMC and cellulose enhanced the mechanical properties of pectin; the new composite accelerated the tissue repair through re-epithelialization and collagen deposition.	[172]
NaCMC/Gelatin/Pectin	-	Film	Excellent capacity to absorb biological fluid, elongation at break, mechanical properties, and water vapor transmission rate; ideal to cure infected moist injuries.	[173]
NaCMC/Chitosan/Sodium alginate	-	Hydrogel	Optimal water vapor permeability, anti-adhesion capacity to treat second-degree burn injuries; reduced the TNF-α and IL-6 levels.	[174]
NaCMC/Gelatin/PEG	-	Hydrogel	CMC increased the mechanical properties, enzymatic stability, and sanguineous compatibility; the new hydrogel had 3D and porous structure, absorbed the exudates, and maintained proper moisture.	[175]
NaCMC/PEG	-	Hydrogel/Film	Superabsorbent hydrogel with various swelling degrees from 100% to 5000%, appropriate physicochemical and morphological properties to regenerate chronic wounds.	[45]
NaCMC/PE/PP	-	Fibers	A serious increase of wet strength; future perspective as a material for the treatment of wounds with mild exudate.	[176]
NaCMC/Collagen	Mefenamic acid	Hydrogel	Good ability to swell, optimal morphological structure and excellent release model to treat burns.	[177]
	AgNPs	Membrane	Suitable antimicrobial action against pathogens from infected burn wounds.	[178]
NaCMC/Gelatin	Ciprofloxacin	Film	In a ratio of 1:1 and 1:2, the film showed good thickness, sorption capacity, blending endurance, and proper antibacterial effect.	[16]
NaCMC/Keratin	Clindamycin	Sponge	The keratin enhanced the water stability and the water vapor transmission rate; the sponge impeded the bacterial multiplication.	[2]
NaCMC/PEO	*Achillea millefolium*, *Calendula officinalis*, *Matricaria chamomilla*, *Echinacea purpurea* and *Hypericum perforatum*	Fibers	Antibacterial and antioxidant activity for acute injuries treatment.	[179]
	Lidocaine	Nanofibers	Similar characteristics as extracellular matrix; ~50% of lidocaine has been delivered in the first 10 min, relieving the pain quickly.	[180]
	AgNPs	Nanofibers	Smooth surface and limited size distribution; excellent antimicrobial effect and photoluminescent act, used as anti-adhesion composite and wound dressing.	[181]

Table 3. *Cont.*

Biopolymer/-s	Active Pharmaceutical Ingredient (Natural or Synthetic Substances)	Type of Wound Dressing	Main Findings	References
NaCMC/PVA	Propolis	Hydrogel	High swelling power and antimicrobial effect for second-degree burn injuries treatment.	[182]
	ZnO and heparin	Hydrogel	Favorable mechanical properties, water vapor transmission rate, swelling ratio, and excellent antibacterial effect.	[183]
NaCMC/Soluplus®	18β-glycyrrhetinic acid (licorice)	Hydrogel	Soluplus® augmented the solubility in water of the bioactive agent; at the lesion site, and in situ hydrogel formed, which had high swelling power and was easy to eliminate by washing.	[184]
NaCMC/Sodium alginate/PET/Viscose	Diclofenac sodium	Membrane	Optimal drug release and efficient anti-inflammatory activity.	[185]
NaCMC/Sodium alginate	Diclofenac	Film	Homogeneous film with proper water vapor transmission rate; faster alleviation of pain.	[186]

Abbreviations: AgNPs—Silver (Ag) nanoparticles, MCM-41—Mesoporous silica, PE—Polyethylene, PEG—Polyethylene glycol, PEO—Polyethylene oxide, PET—Polyethylene terephthalate, PP—polypropylene, PVA—Polyvinyl alcohol, Soluplus®—Polyvinyl caprolactam-polyvinyl acetate-polyethylene glycol graft copolymer (PCL-PVAc-PEG), UiO-66—The University of Oslo, a metal-organic framework based on Zr, ZnO—Zinc oxide.

5.3. Hydroxypropylmethylcellulose-Based Wound Dressings

Hydroxypropylmethylcellulose (HPMC), hypromellose [187], is a semi-synthetic hydrophilic polymer, a nonionic cellulose ether derivative [188], with higher stability at a lower pH. In terms of physical properties, HPMC is a white, fibrous, or granular powder, whose particles are not cohesive, and it does not have a taste and odor [189]. This biopolymer results from hydroxyl groups substitution from cellulose molecule with methyl and hydroxypropyl groups. The chemical structure of HPMC is illustrated in Figure 8.

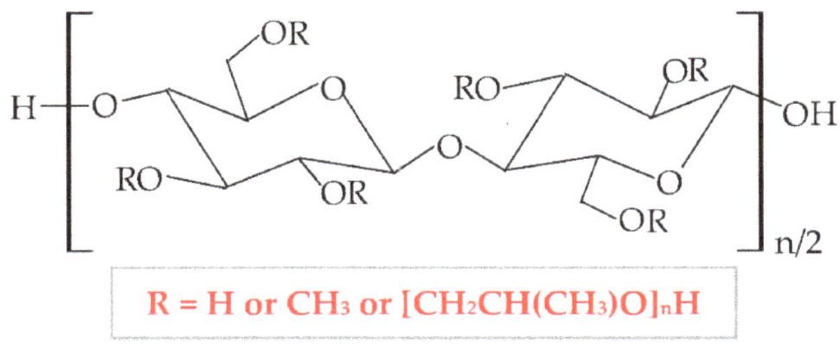

Figure 8. Chemical structure of hydroxypropylmethylcellulose (HPMC).

Therefore, HPMC presents many degrees of substitution, that give to this biopolymer different molecular weight and physicochemical features (rheological properties and crystalline nature) [190,191]. The hydrophilic or hydrophobic nature is related to the values of the degree of substitution (DS) and the molar substitution (MS). Thus, the HPMC molecule with decreased values of DS and MS is more hydrophilic and the HPMC molecule with increased values of DS and MS is more hydrophobic [192]. Following this chemical substitution, HPMC gets both polar (hydroxypropyl) and non-polar (methyl) character; consequently, it can form hydrophobic, intermolecular, and intramolecular linkages with many other materials [190]. The non-ionic character leads to a limited adhesive capacity [193]. At high temperature, the biopolymer can suffer a thermoreversible phase transition from sol to gel, with a temperature of gelation over 60 °C, superior to the temperature of the body (37 °C) [194]. HPMC-based hydrogels are temperature-responsive [195].

According to United States Pharmacopeia (USP), there are four distinct forms of HPMC, which are categorized by the content of methoxy, respectively hydroxypropoxy groups in: HPMC 1828, HPMC 2208, HPMC 2906, and HPMC 2910 [196]. This biopolymer has been approved as a food additive, E464 [197], by the American Institute, Food and Drug Administration (FDA), by the European Institution, European Parliament, and Council Directive, and by the Joint Expert Committee on Food Additives [198].

HPMC has a proper solubility in water, and it is one of the most used cellulose derivatives in many industries. It is widely used in the biotechnological field (construction, food, cosmetics, biomedical, and pharmaceutical industry), due to its excellent characteristics, such as biocompatibility, biodegradability, superior stability, large availability, excellent swelling, high surface activity. and mechanical properties [199], remarkable ability to form films and poor toxicity [200]. Regarding the applicability of HPMC in biomedical and pharmaceutical domains, it is used as a drug-delivery device, with a large practice for wound dressings development and it can also have remarkable applicability in tissue engineering [201]. HPMC can also be used as an excipient because it possesses proper abilities of emulsification, stabilization, suspension, and thickening [202,203].

HPMC can be combined with other polymers to enhance its properties and to develop its applicability [204]. To improve the physicochemical properties of a new composite, HPMC has been blended with several natural, semi-synthetic, or synthetic polymers [205]. In this way, to improve the thermal stability, HPMC has been blended with collagen [206,207], gelatin [204], chitosan [208], chitosan, and xanthan gum [209]; to improve the mechanical properties (tensile strength and ultimate elongation), HPMC has been mixed with chitosan [210], collagen [207], poloxamer 407 [211], silk fibroin [212], PVA and PVP [213], chitosan and xanthan gum [209]; to increase the swelling rate, HPMC has been combined with methylcellulose [214], κ-carrageenan [215], chitosan and hyaluronic acid [216], chitosan and xanthan gum [209]. HPMC can also be blended with diverse biopolymers and multiple bioactive agents (plants extracts, organic or inorganic substances, and chemical drugs) to develop new potential wound dressing to accelerate the wound-healing process. All these studies are summarized in Table 4.

Table 4. Recent studies on the use of hydroxypropylmethylcellulose as a wound dressing.

Biopolymer/-s	Active Pharmaceutical Ingredient (Natural or Synthetic Substances)	Type of Wound Dressing	Main Findings	References
HPMC	*Lawsonia inermis* (henna) and *Matricaria chamomilla*	Gel	Adequate stability, extrudability, viscosity, homogeneity, good herbal extract release, and antibacterial effect for burn infection cure.	[217]
	Liposomal farnesol	Gel	The combination of HPMC:farnesol in a ratio of 1:2 and 2:1 had an excellent effect on tissue regeneration at third-degree burns.	[218]
	Bacteriophage	Gel	The gel containing 3% HPMC presented good stability at 37 °C and had an effective antibacterial action against *Klebsiella pneumoniae* in wound infection treatment.	[219]
	CuNPs-licorice and phenytoin	Gel	Effect on acute lesions by the suppression of the inflammatory JAK3 and the synthesis of the procollagen type I.	[220]
	Honey and Aloe vera	Hydrogel	At 3% concentration, the hydrogel presented an adequate viscosity to be applied on the burn lesions and showed a proper antibacterial activity on infection with *Klebsiella pneumoniae*.	[197]
	Cefotaxime sodium	Hydrogel	The hydrogel containing 3% HPMC 400 exhibited a high spreadability and released all the drug content after 4 h.	[221]
	Haruan/fusidic acid	Film	Films with 1% plasticizers, respectively 2% plasticizers presented a suitable elongation at break and water vapor permeability.	[222]
	Epigallocatechin-3-gallate	Film	Enhanced tensile strength and water vapor barrier trait.	[223]
	Cooper nanoparticles	Film	Appropriate antibacterial activity.	[190]
HPMC/Silk fibroin	-	Nanofibers	The mixture of HPMC:silk fibroin in a ratio of 7:1 exhibited suitable biocompatibility, mechanical properties, hydrophilicity, and porosity for skin tissue engineering.	[201]
HPMC/Collagen/Polyurethane	-	Hydrogel	Increased proteolytic and thermal degradation.	[224]
HPMC/Polyacrylate/Tri-isocyanate crosslinked polyurethane	-	Hydrocolloid	The newly designed wound dressing containing 1% cross-linking promoter showed proper humidity (the capacity of the water uptake was 5.1% after 1 h).	[199]
HPMC succinate/Chitosan	Gentamycin sulfate	Film	High mechanical properties and antibacterial activity.	[225]
HPMC/Chitosan	Simvastatin	Gel	High viscosity and bioadhesive strength.	[17]
	Toluidine blue O	Hydrogel	High viscosity, hardness, bioadhesion, and bactericidal effect to alleviate the burn wounds caused by light irradiation.	[226]
	Pioglitazone hydrochloride	Hydrogel	The formulation containing HPMC E5:chitosan (1:2) showed the best physicochemical properties and the highest drug release.	[193]
	Silver sulphadiazine	Film	In a ratio of 1:1, the film showed the best physicochemical properties and the highest drug release; the period of wound healing was of 8 days in comparison with the marketed cream (14 days).	[227]

Table 4. *Cont.*

Biopolymer/-s	Active Pharmaceutical Ingredient (Natural or Synthetic Substances)	Type of Wound Dressing	Main Findings	References
HPMC K100/Collagen	Curcumin	Nanogel	In vivo studies illustrated a substantial rate of lesion contraction (95.42 ± 12.20%) on the 20th day.	[228]
HPMC/Hydroxyapatite	AgNPs	Hydrogel	High porous 3D structure, with excellent mechanical properties and antibacterial activity; after 16 days, wounds healed considerably (94.5 ± 1.4%).	[229]
HPMC/Hydroxypropyl-β-cyclodextrin	Coumestrol	Hydrogel	50% of wounds healed in a shorter time compared to the commercial product, with a suitable tissue re-epithelialization after 12 days.	[230]
	Gallic acid	Hydrogel	High physicochemical properties and antibacterial effect to prevent wound infection.	[231]
HPMC/PVA/PAA/PVP/PEG	Aloe vera (1%, 6%, 10%)	Nanofibers	The sample with 10 wt% aloe vera showed higher porosity and durability, faster healing of burns, and nonlinear structure.	[200]
HPMC/PVA/PVP-I/PEG	Aloe vera (2%, 4%, 6%)	Fibers	Fibers with 6% aloe vera were thinner without any beading, which led to a higher porosity of the fibers.	[232]
HPMC K15M/Tara gum	Lawsone	Gel	Suitable homogeneity, uniformity, and drug release; in vivo studies indicated total epithelialization of the excision lesion.	[233]
HPMC K15M/Xanthan gum	Nano calcium oxide	Hydrocolloid	Great mechanical strength, superior flexibility, homogeneous thickness, and higher lesion contraction compared to the commercial product.	[234]
HPMC/Xyloglucan	Gentamicin sulfate	Film	The best formulation of these polymers was in a ratio of 50:50 and it presented reliable physicochemical properties, good drug release, and favorable antibacterial effect.	[235]
HPMC/Sodium alginate	Gatifloxacin	Hydrogel	Advantageous physicochemical properties regarding the tensile strength, the swelling capacity, the elongation, and the drug release.	[236]
HPMC/Pluronic® 127	AuNPs	Gel	Good bioavailability, skin permeation, anti-inflammatory, and antibacterial effect, with an excellent drug release of 98.03% after 6 h.	[237]
HPMC/Polyacrylamide	AgNPs	Hydrogel	Superporous hydrogel with high porosity (91.4%), which allows a quicker wound healing, with minimal scar formation.	[238]

Table 4. *Cont.*

Biopolymer/-s	Active Pharmaceutical Ingredient (Natural or Synthetic Substances)	Type of Wound Dressing	Main Findings	References
HPMC/PEO	Beta-glucan	Nanofibers	At the lesion site, the nanofibers produced a hydrogel in situ.	[239]
HPMC/Poly(lactic acid)	Tetracycline hydrochloride	Nanofibers	High water sorption rate and antimicrobial activity.	[240]
HPMC/Polyurethane	Silver and asiaticoside	Foam	Increased absorption capacity and compressive strength development; adequate antimicrobial activity.	[241]
HPMC/Chondroitin sulfate/Sodium hyaluronate	Silver sulphadiazine	Sponge	Proper elasticity, softness, flexibility, and bioadhesive properties, alongside antibacterial activity.	[46]
HPMC K100M/Gum Odina/Gelatin	Fluconazole and ofloxacin	Sponge	The formulation with gum Odina-HPMC K100M:gelatin (1:1) showed excellent physicochemical properties and antimicrobial activity to cure chronic wounds.	[242]
HPMC/HA/Methyl-β-cyclodextrin	Curcumin	Film	Excellent antimicrobial effect.	[243]
HPMC/Hydroxypropyl-β-cyclodextrin/Chitosan	Caffeic acid	Hydrogel	Drug delivery system with superabsorbent capacity, higher swelling property at pH 7, and good antimicrobial effect to prevent wound infection.	[244]
HPMC/Chitosan/Sodium alginate	Lidocaine chloride and polymyxin B sulphate	Biomembrane	Proper mechanical properties (elasticity, tension, stiffness) and thickness; in vivo: high antimicrobial effect for tissue regeneration.	[245]
HPMC/Polyglycolic acid/Vicryl®/Catgut	Ofloxacin	Hydrogel	Proper physicochemical properties, which ensures a high lesion size contraction after 14 days (95%) and large collagen deposition on the 21st day.	[246]

Abbreviations: AgNPs—Silver (Ag) nanoparticles, AuNPs—Gold (Au) nanoparticles, CuNPs—Copper (Cu) nanoparticles.

5.4. Methylcellulose-Based Wound Dressings

Methylcellulose (MC) is a semi-synthetic and non-ionic polymer, a cellulose ether derivative with high solubility in water, which is influenced by temperature [247]. It forms through the etherification of cellulose molecule with methyl chloride or dimethyl sulfate in basic solution [27] when the hydroxyl groups from the mother molecule are substituted with methyl groups, which leads to a diminishing of crystallinity [248]. The chemical structure of MC is presented in Figure 9.

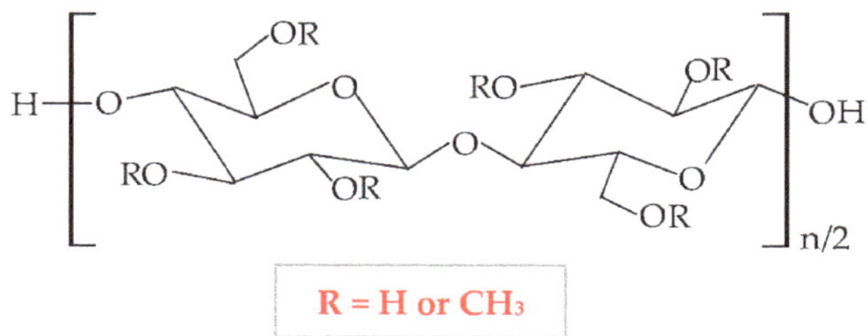

Figure 9. Chemical structure of methylcellulose (MC).

At a variation of temperature, MC has a thermo-sensitive behavior with a reversible sol-gel transition in an aqueous solution [249]. At a lower temperature than lower critical solution temperature, it realizes the hydration of the MC network in solution, with the formation of hydrogen bonds. At a higher temperature than lower critical solution temperature, the MC aqueous solution takes in the heat, with the disintegration of hydrogen bonds [195]. Thus, MC presents increased viscosity at higher temperatures, and at lower temperatures it exhibits a reduced viscosity [250].

The degree of substitution for commercial MC varies from 1.7 to 2.2 when it results in a semiflexible biopolymer because the inter-and intra- hydrogen bonds from cellulose molecule break off [251]. There are many substances, which influence the gelation behavior of MC, such as inorganics salts, ethanol, propylene glycol, polyethylene glycol 400, sucrose, glycerin, sorbitol, and different surfactants (sodium dodecyl sulfate and cetyltrimethylammonium bromide) [252]. MC is extensively used in biomedical, pharmaceutical, cosmetic, and food industries as a thickening, binding, and film-forming agent because it possesses excellent biocompatibility, biodegradability, and reduced toxicity [253–255].

To improve the characteristics of MC, it can be blended with other polymers in different ratios to enhance the physicochemical, morphological, and structural properties of both polymers and of the resulting composite [255]. Abu et al., illustrated that a higher concentration of MC led to increased hydrophilicity and porosity of the MC-chitosan scaffold due to the hydroxyl groups from the MC molecule, which can attract water molecules. The higher wettability has been described by the suitable results of the water uptake capacity [256]. Another combination of MC and chitosan was studied by Tan et al., They illustrated that an augmented concentration of MC led to improved tensile strength, moisture content, whitish index, and elongation at break [257]. El-Naggar et al., mixed MC with PVA and doxycycline hyclate (drug model) to develop a new drug delivery device, which showed a proper swelling capacity and a high drug release at basic medium [253]. The combination between MC and poly(acrylic acid) presented optimal mechanical properties and thermal stability [258]. The novel composite resulting by blending MC and tragacanth gum exhibited a higher capacity to form a gel and adequate mechanical and rheological properties [259]. MC can also be blended with diverse biopolymers and multiple bioactive agents (plants extracts, organic or inorganic substances, and chemical drugs) to develop a new potential wound dressing to accelerate the wound-healing process. All these mixtures are presented in Table 5.

Table 5. Recent studies on the use of methylcellulose as a wound dressing.

Biopolymer/-s	Active Pharmaceutical Ingredient (Natural or Synthetic Substances)	Type of Wound Dressing	Main Findings	References
MC	Zinc oxide and Silver	Ointment	MC 3% ointment showed an elastic behavior and thixotropy, which can be used for injured skin with complex relief.	[260]
	Allium hirtifolium	Gel	In a ratio of 1:1, the newly developed gel accelerated the open wounds healing through the tissue re-epithelialization.	[261]
	Cryopreserved human culture of fibroblasts and AuNPs	Gel	High capacity to heal the third-degree burns because it restored the composition of type I and III collagen on 21st day of the treatment.	[262]
	Vitamins (C, B_1, and B_6)	Hydrogel	Enhanced gelation rate and mechanical strength, with suitable applicability for wound treatment.	[263]
	Gallic acid and doxycycline	Hydrogel	At body temperature, the novel formulation formed an in situ gel and released the bioactive agents to heal the deep tissue injuries.	[252]
	Neomycin trisulphate, sulphacetamide sodium, and Silver nitrate	Hydrogel	New formulation exhibited elasticity, sponginess, homogenous texture, proper smell, and white color.	[162]
	Silver oxide nanoparticles	Hydrogel	Thermo-responsive hydrogel, which led to wound burn regeneration due to its superior antimicrobial effect.	[264]
	Pedilanthus tithymaloides	Film	A concentration of 0.5% of the plant extract accelerated the re-epithelialization of the wounded skin.	[265]
	AgNPSs	Film	Increased water absorption capacity and contact angle value; the temperature influenced the release of AgNPs.	[266]
	Borate bioactive glass and Manuka honey	Foam	The foam exhibited high porosity, better wettability, mechanical properties, and antibacterial effect.	[248]
MC/α-Chitin nanocrystals	-	Hydrogel	The novel nanocomposite presented high mechanical strength and gelation rate, being a promising dressing for tissue engineering.	[267]
MC/Fucoidan	-	Film	The newly designed formulation could be a promising dressing for wounds with a smaller production of exudates because it possessed a lower capacity to swell.	[268]
MC/Nano hyaluronic acid	-	Hydrogel	The mixture of these polymers led to an in situ hydrogel development, which enhanced tissue regeneration.	[269]
MC/Mucin	-	Hydrogel	MC enhanced the mechanical properties of the mucin and formed a thermoresponsive gel, which can be used for different wounds treatment.	[270]
MC/2-Methacryloyloxy ethyl phosphorylcholine	-	Hydrogel	It reduced the postoperative adhesion effect by inhibition of collagen proteins.	[271]
MC/Gelatin/Gellan gum/PVP	-	Film	The blend of all these polymers led to a novel wound dressing with suitable physicochemical properties to restore minor lesions.	[272]
MC/Chitosan	Adenosine and vitamin C	Hydrogel	Thermo-responsive and self-healing ability and a higher release of adenosine.	[273]
	Exosomes	Hydrogel	Suitable mechanical properties, good gelation time, and excellent self-healing for treatment of severe tissue injuries.	[274]

Table 5. *Cont.*

Biopolymer/-s	Active Pharmaceutical Ingredient (Natural or Synthetic Substances)	Type of Wound Dressing	Main Findings	References
MC/Hyaluronic acid	-	Gel	It was shown that a low molecular weight of hyaluronic acid increased the biocompatibility and the thermogelation of this new formulated composite.	[275]
	AgNPs	Hydrogel	Excellent morphological, swelling, and spectral properties, together with the high antibacterial effect to treat burn wounds in children (99.6%).	[276]
MC/sECM	Stem cells	Hydrogel	At the wound site, the hydrogel had a thermosensitive sol-gel transition, which accelerate the wound healing through neovascularization and re-epithelialization.	[277]
MC/Sericin	Tranilast	Ointment	The combination of the two polymers and the antiallergic drug exhibited excellent results to augment wound healing and to decrease the redness in diabetic rats.	[278]
MC/Silk fibroin	5-Aminosalicylic acid	Hydrogel	The gelation time of the novel composite has been enhanced by adding MC; thus, this hydrogel presents important uses as a wound dressing and drug release system.	[279]
MC/Sodium alginate	Manuka honey, aloe vera, and eucalyptus essential oil	Hydrogel	The formulation presented proper swelling capacity, biocompatibility, and suitable antimicrobial and antibiofilm action.	[280]
	Gallium (Ga^{3+})	Hydrogel	It was illustrated that this novel hydrogel is a promising dressing for wounds infection prevention due to its large antibacterial action (99.99%) and cytocompatibility.	[281]
	Montmorillonite	Film	A high content of montmorillonite led to higher tensile strength and antibacterial activity of the newly designed film.	[282]
MC/Sodium alginate/Poly(*N*-isopropylacrylamide)	Octenisept®	Hydrogel	The novel hydrogel had a large viscosity, an expressed shear-thinning nature, high antibacterial action, a homogenous and microporous structure.	[283]
MC/Pluronic F-127	GT/siMMP9	Hydrogel	It was demonstrated that this new dressing had thermosensitive traits by forming in situ and it reduced the MMP-9 level in diabetic chronic injuries.	[284]
MC/Polyacrylamide	*Aloe barbadensis*	Hydrogel	At a 2% concentration of *Aloe barbadensis*, the hydrogel showed suitable thermal stability; it also presented an antibacterial effect to cure chronic cutaneous wounds.	[285]
MC/Poly(ε-caprolactone)	Bioactive glass and Manuka honey	Fibers	The blend of these two polymers led to higher mechanical properties and wettability.	[286]
MC/PVA	Asiaticoside	Film	This newly formulated dressing exhibited significant mechanical properties, flexibility, transparency, and a shorter healing time of skin wounds.	[287]
	-	Nanofibers	Adequate physicochemical features and antibacterial activity to be used as a drug-delivery system for injuries management.	[288]

Abbreviations: AgNPs—Silver nanoparticles, AuNPs—Gold nanoparticles, GT/siMMP9—Glycogen triethylenetetramine/small interfering RNA matrix metalloproteinase 9, PVA—Polyvinyl alcohol, PVP—Polyvinylpyrrolidone, sECM—Soluble extracellular matrix.

5.5. Hydroxyethylcellulose-Based Wound Dressings

Hydroxyethylcellulose (HEC) is a semi-synthetic, nonionic, and inert polymer, a water-soluble cellulose ether derivative [289]. It forms through etherification of alkaline cellulose with chlorohydrin or ethylene oxide, when hydroxyl groups from cellulose molecule are substituted with hydroxyethyl groups [290]. The chemical structure of HEC is illustrated in Figure 10.

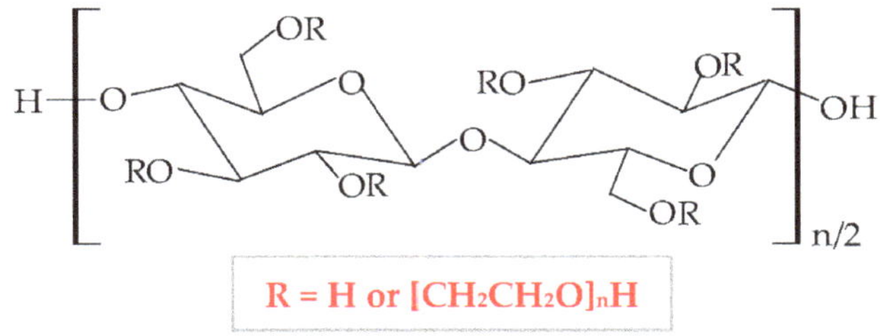

Figure 10. Chemical structure of hydroxyethylcellulose (HEC).

It has a low price, without taste and smell, with no color to light yellowish [291]; presents optimal stability at pH values between 2 and 12 [292]. HEC exhibits a proper capacity to scavenge free radicals and to form hydrogen and electrostatic bonds [293]. HEC is regarded as a hydrogel-like material, with two important characteristics: liquid-like and solid-like. Due to its polysaccharide structure, this hydrophilic biopolymer exhibits a high capacity to absorb and hold a large quantity of water or wound exudates. The elastic strength of its structure leads to an expansion of the molecule dimensions, without the modification of the structural stability and the gel form [294]. HEC possesses excellent physicochemical properties: rheological, hydrodynamic, and thermodynamic [295]. HEC also presents adequate biocompatibility, biodegradability, insignificant toxicity, immunogenicity, and cementing properties [296]. Due to its nonionic behavior, HEC exhibits the ability to coexist with a large field of other polymers, which have an appropriate solubility in water, salts, or surfactants. Therefore, HEC presents optimal toughness in a dielectric solution with a large concentration [297]. This biopolymer presents the largest commercial availability from all cellulose derivatives [298]; therefore, HEC is a noticeable biopolymer, which can be used successfully as an emulsifier, film-coating, stabilizer, suspender, and thickener agent in biomedical, pharmaceutical (wound dressing development) [299], cosmetic, food, adhesive, and textile industries [291,300,301]. The most predictive method for hydrogels synthesis is the crosslinking of free radicals generated by irradiation (electron beam and gamma-radiation) [302].

To enhance its properties, HEC can be blended with other polymers. For example, Zia et al., mixed HEC with poly(lactic acid) and polyurethane. They obtained a new composite with higher thermal stability and mechanical (tensile strength and elongation) properties compared to other polymers [303,304]. Moreover, HEC has been blended with polyvinyl alcohol (PVA), resulting in suitable electrical conductibility, viscoelasticity, stretchability, and thermosensitivity [305]. Guo et al., combined HEC with poly(caprolactone) by trimethylsilyl group technology and the result was the formation of a new copolymer with enhanced thermal properties [306]. HEC was also blended with chitosan to obtain a copolymer with improved physicochemical and mechanical characteristics [307], with gelatin to obtain a superparamagnetic composite [308], with sodium alginate to form a copolymer with enhanced swelling efficacy and drug delivery profile. HEC can also be combined with diverse biopolymers to formulate novel wound dressing, which can stimulate the wound healing process and restore the damaged skin. The main combinations are summarized in Table 6.

Table 6. Recent studies on the use of hydroxyethylcellulose as a wound dressing.

Biopolymer/-s	Active Pharmaceutical Ingredient (Natural or Synthetic Substances)	Type of Wound Dressing	Main Findings	References
HEC	-	Hydrogel	Good antibacterial effect; at a concentration between 12.5% and 15% of HEC, the water absorption rate was the highest.	[309]
HEC	Curcuma longa	Hydrogel	Suitable porous network, mechanical and release properties; besides therapeutic effect, the new hydrogel can illustrate the pH of the lesion site.	[310]
	WO_3	Hydrogel	The hydrogel with 0.04% WO_3 showed the highest capacity to heal the injuries due to optimal antibacterial and anti-inflammatory action.	[300]
	Mesocellular silica foam	Sponge	Adequate cytocompatibility, antibacterial and hemostatic action to heal full-thickness wounds.	[311]
	Graphene oxide	Film	Improved mechanical and thermal properties.	[312]
	AgNPs	Membrane	Limited degradation rate, excellent porosity, and water absorption value.	[313]
HEC/Collagen	-	Film	High swelling ratio, mechanical and thermal characteristics, degradation, and adequate biocompatibility.	[314]
HEC/Collagen/PVA	-	Nanofibers	Suitable water absorption rate and degradation behavior, a promising material for skin tissue engineering.	[315]
HEC/Carboxymethyl chitosan	-	Hydrogel	High biocompatibility, gelation time, water evaporation rate, and swelling power.	[290]
HEC/Sodium alginate/Hydroxyapatite	-	Membrane	High porous network, improved mechanical strength, and rigidity gradient.	[316]
HEC/Hyaluronic acid	-	Hydrogel	Optimal biocompatibility and blood compatibility, gelation time, water evaporation rate, water retention ability, and swelling power.	[295]
	Isoliquiritigenin	Hydrogel	pH-sensitive hydrogel with suitable adhesion, rheological properties, and antibacterial efficacy.	[317]

Abbreviations: AgNPs—Silver nanoparticles, WO_3—Tungsten oxide.

5.6. *Ethylcellulose-Based Wound Dressings*

Ethylcellulose (EC) is a nonionic semi-synthetic polymer, a cellulose ether derivative insoluble in water [318]. It forms through the etherification of alkali cellulose with ethyl chloride when the hydroxyl groups from cellulose molecule are substituted with ethyl groups [27]. The chemical structure of MC is presented in Figure 11.

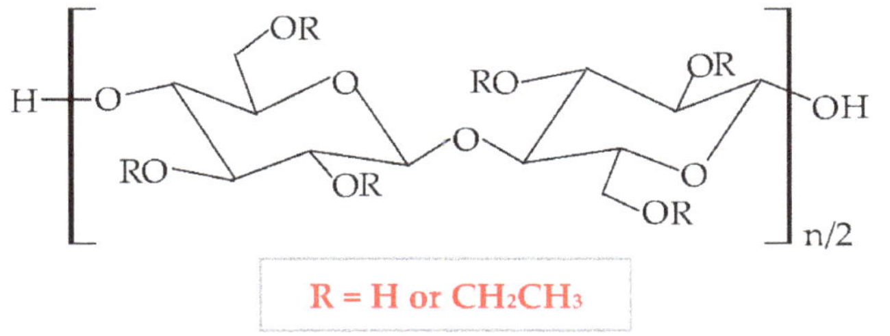

Figure 11. Chemical structure of ethylcellulose (EC).

This biopolymer presents numerous advantageous characteristics, such as mechanical properties, biodegradability, flexibility, low toxicity, hydrophobicity, gelling capacity [319], light, moisture, oxygen resistance, thermoplasticity [320], and low price, which make EC an excellent material for use in different industries (pharmaceutical, cosmetic and food) [321]. Moreover, this biopolymer has several particular features in addition to the other cellulose derivatives: high film-forming capacity, suitable chemical strength, and optimal mechanical properties [322]. EC represents the most extensively analyzed biopolymer due to its capacity to form film for coating solid pharmaceutical forms (tablets, microcapsules, and microspheres) and formulation of new topical forms [323]. EC is a promising material to be used for encapsulation due to its optimal optical transparency, processing temperature, and electronic insulation [324]. It also presents a good capacity to bind, preserve and dissolve [325], and possesses a proper control of drug delivery [326]. Films based on EC are brittle because of the stiffness of hydrogen linkages from its molecule. This biopolymer has high stability to chemical substances and can be associated with different plasticizers to design heavy and impermeable films [327].

EC can be mixed with various polymers to enhance the physicochemical and mechanical properties and thus, its applicability. To develop a novel drug-delivery device, Li et al., blended EC by electrospinning method with poly(di(ethylene glycol) methyl ether methacrylate), a thermosensitive polymer. The new formulation showed normal morphology, a large porosity, and an increased wettability at a higher temperature, which led to more hydrophobic behavior, causing an extended release of the drug [328]. EC was mixed with poly (ethylene-co-vinyl acetate) and resulted in a new composite with higher mechanical properties [329]. Chen et al., mixed EC and poly(β-hydroxybutyrate) when EC acted as a thickening agent because it increased the viscosity of the new composite. In a concentration of 1%, EC augmented the tensile strength [330]. Li et al., blended EC with konjac glucomannan to formulate a novel composite with higher mechanical properties, moisture resistance, permeability of oxygen, and stability at a high temperature [331]. EC was also associated with another cellulose derivative, HPC, and obtained a scaffold with enhanced mechanical properties and 3D printing capacity [332].

EC can also be combined with other polymers to develop new wound dressings, with enhanced physicochemical and mechanical properties that can accelerate the wound-healing process. Principal blends are presented in Table 7.

Table 7. Recent studies on the use of ethylcellulose as a wound dressing.

Biopolymer/-s	Active Pharmaceutical Ingredient (Natural or Synthetic Substances)	Type of Wound Dressing	Main Findings	References
EC/Gum tragacanth	Honey	Nanofibers	The ratio of 85:15 with 20% honey showed the best mechanical, biological, and antibacterial properties to be used as an efficient wound dressing.	[321]
EC/PVP	Naproxen	Nanofibers	The ratio of 4:1 had the fastest release of the drug; thus, it can be used as a dressing to reduce the inflammation and the pain in acute wounds.	[333]
	AgNPs and ciprofloxacin	Nanofibers	Uniform and cylindrical morphology, with over 90% release of ciprofloxacin and a high antibacterial effect.	[334]
EC/Poly(3-hydroxybutyrate)	p-4-hydroxybenzoic acid and ferulic acid	Film	Potent bacteriostatic and bactericidal action to heal wounds infection facilitating the skin restoration.	[335]
EC/Polylactic acid/Collagen	Silver sulfadiazine	Nanofibers	Good mechanical properties, proper antimicrobial effect, and enhanced cell proliferation.	[336]
EC	Ciprofloxacin	Nanofibers	Cylindrical and homogenous aspect; high antibacterial action.	[318]
EC/Zein	Photosensitizer protoporphyrin and vaccarin	Membrane	Adequate flexibility, hygroscopicity, nanonetwork structure, and mechanical properties; in vivo study showed increased angiogenesis of the injured tissue.	[337]
EC/PVA	Luliconazole	Nanosponge	Optimal viscosity, spreadability, retention time, permeation rate, and also high antifungal effect against dermatophytes.	[338]

Abbreviations: AgNPs—Silver nanoparticles, PVA—Polyvinyl alcohol, PVP—Polyvinylpyrrolidone.

5.7. Hydroxypropylcellulose-Based Wound Dressings

Hydroxypropyl cellulose (HPC) is a semi-synthetic hydrophilic polymer, a cellulose ether derivative, with proper solubility in water and organic solvents [339]. Its solubility depends on the degree of substitution. At values smaller than 12%, HPC is water-soluble and at values higher than 12%, HPC is ethanol-soluble [340]. This biopolymer results from the etherification reaction of alkali cellulose with 1,2-propylene oxide. Thus, the 2,3,6-hydroxyl groups from the cellulose molecule are replaced with hydroxypropyl groups [27]. The chemical structure of HPC is presented in Figure 12.

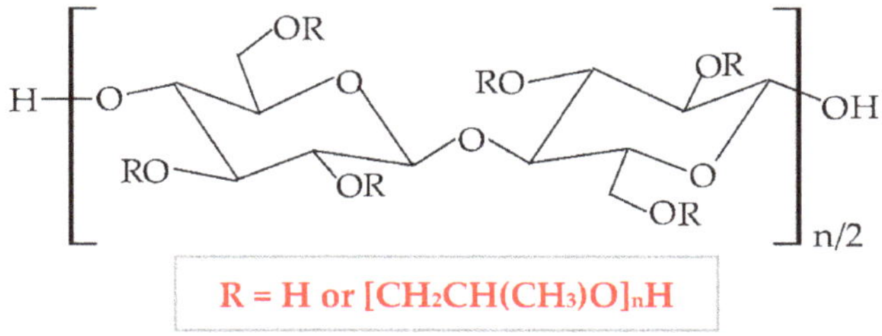

Figure 12. Chemical structure of hydroxypropylcellulose (HPC).

It has numerous advantageous properties, such as amphiphilicity, low price, electrical neutrality, biocompatibility, biodegradability, non-toxicity, high power of swelling the wounds exudate [341,342], adequate chemical strength, and film-forming efficiency [236]. At a high temperature and in a concentrated aqueous solution, HPC generates a cholesteric liquid crystalline network, depending on its concentration [343]. HPC exhibits a thermoplastic behavior and develops temperature-responsive hydrogels [195,344]. Regarding the HPC-based films, these are defined by high flexibility, good impermeability for oil and fat, and a low value of T_g (glass transition temperature) at excessive humidity. The LCST (lower critical solution temperature) water value is about 41 °C. At a slightly higher temperature than LCST, HPC presents a phase change because the water solution of this biopolymer generates metastable nanosphere aggregates [345]. Moreover, the solubility of HPC is influenced by LCST values. At a lower temperature than LCST, HPC dissolves easily in water and at a higher temperature than LCST, HPC does not dissolve [346]. Thus, this cellulose derivative is an optimal material to be used in biomedical and pharmaceutical fields as a binding, disintegrating, emulsifying, thickening, filler, and coating agent [347,348] and in the construction domain [349]. It can also be used in the food industry because the United States Food and Drug Administration (FDA) authorized HPC as a safe food additive [350].

HPC can be blended with other polymers to improve the physicochemical and mechanical properties and thus, to extend its applicability. For instance, Veerapur et al., combined HPC and chitosan, and the new formulated composite presented higher hydrophilicity, swelling capacity, and permeation rate [351]. By mixing HPC with cellulose acetate phthalate resulted a composite with higher properties than compounds: increased pseudoplasticity and viscoelastic behavior [352]. Gan et al., prepared a high-performance hydrogel with enhanced tensile strength, toughness, biocompatibility, wear resistance, and low friction coefficient from HPC, sodium alginate, and poly(vinyl alcohol); these excellent characteristics extend the area of use to biosensors and nerve replacement [353]. Lu et al., blended HPC with poly(vinyl alcohol) to obtain a new scaffold with augmented toughness, elasticity, conductivity, and mechanical strength that is a promising material for the development of biosensors and interaction between humans and machines [354]. HPC can also be mixed with other polymers to develop novel wound dressing, with higher physicochemical traits that can restore the impaired skin tissue. The main combinations are summarized in Table 8.

Table 8. Recent studies on the use of hydroxypropylcellulose as a wound dressing.

Biopolymer/-s	Active Pharmaceutical Ingredient (Natural or Synthetic Substances)	Type of Wound Dressing	Main Findings	References
HPC	Peptide PXL150	Gel	Due to the antimicrobial activity, the gel healed the infection of third-degree burns injuries and surgical wounds.	[355]
	-	Hydrogel	Suitable water absorption; the hydrogel can maintain an adequate moisture balance at the lesion site.	[339]
	Graphene oxide, Isophorone diisocyanate, and Ag/ZnO	Film	Higher mechanical, anti-ultraviolet properties, and antimicrobial activity.	[356]
HPC/Gelatin	Chloramphenicol	Hydrogel	Adequate mechanical strength, water vapor permeability, light transmittance, and excellent antibacterial effect.	[357]
HPC/Chitosan/PEO	Graphene	Membrane	Proper mechanical properties, hydrophilicity, water vapor transmission rate, and high antibacterial effect; the membrane can impede the bacterial adhesion.	[341]
HPC/Sodium alginate	Gatifloxacin	Hydrogel	Advantageous physicochemical properties regarding the tensile strength, the swelling capacity, the elongation, and the drug release.	[236]
	Pioglitazone	Hydrogel	By adding HPC, the gelation behavior of the hydrogel increased, with an extended release of the drug up to 5 days; thus, the novel formulation can be used for skin ulcer treatment.	[358]

Abbreviations: PEO—polyethylene oxide, Ag—the chemical symbol of silver, ZnO—zinc oxide.

5.8. Combinations of Cellulose Derivatives-Based Wound Dressings

One or more cellulose derivatives may combine with other cellulose derivatives to formulate novel wound dressings, with enhanced properties that can accelerate the wound-healing process and alleviate the pain, inflammation, and stress caused by damaged skin tissue. Moreover, they can be combined to counteract their drawbacks [75]. The main combinations are summarized in Table 9.

Table 9. Recent studies on the use of combinations of cellulose derivatives as wound dressing.

Biopolymer/-s	Active Pharmaceutical Ingredient (Natural or Synthetic Substances)	Type of Wound Dressing	Main Findings	References
EC/HPMC	Paromomycin and Gentamicin	Film	Optimum drugs release and inhibition of *Leishmania tropica growth.*	[359]
	Aloe vera	Nanofibers	Nanofibers with 10% Aloe vera showed suitable mechanical properties, biocompatibility, bioadhesion, and suitable antibacterial activity.	[360]
NaCMC/HPMC	Grapefruit seed extract	Film	Suitable elongation at break, stability in water, and proper antibacterial action.	[361]
	CuO	Film	Good biocompatibility and antibacterial effect.	[362]
	Tetracycline/Methylene blue	Film	Nanoporous network, increased T_g and elongation at break, sustained drug release for 72 h and high antibacterial effect.	[363]
	ZnO NPs	Film	Biocompatibility and optimum antibacterial action.	[364]
NaCMC/HPMC/CAB	Resveratrol	Membrane	Excellent adhesive capacity, hydration efficiency, and higher porous structure; in vivo studies showed accelerated wound healing.	[365]
NaCMC/MC	Simvastatin	Membrane	In a ratio of 2:1, the membrane exhibited appropriate flexibility, viscosity, stability, and sponginess; optimal drug delivery for suppurating injuries.	[366]

Abbreviations: CAB—Cellulose acetate butyrate, CuO—Copper oxide, T_g—Glass transition temperature, ZnO NPs—Zinc oxide nanoparticles.

6. Conclusions and Future Perspectives

This review has focused on different types of wound dressing (gel, hydrogel, sponge, hydrocolloid, film, membrane, foam, and nanofibers) based on cellulose derivatives as biopolymeric scaffolds, and various bioactive agents, from plant extracts to chemical drugs. We have considered the cellulose ethers derivatives (NaCMC, HPMC, MC, HEC, EC, and HPC). It has been illustrated that cellulose derivatives can manifest a therapeutic effect on wound healing, alone or in combination with other natural, semi-synthetic, and synthetic polymers. The major advantage of mixing two or more biopolymers is, besides the beneficial action on damaged tissue, the improvement of physicochemical properties of the novel dressing. Cellulose derivatives have a particular chemical structure, obtained by etherification of hydroxyl groups from cellulose molecule with different alkyl groups, the consequence being the improvement of water solubility. Therefore, these biopolymers can be successfully used as a base for diverse formulations, due to their high gelation properties. Cellulose derivatives exhibit an efficient capacity to absorb the exudates on the site of the lesion, retain them, and swell. Consequently, the newly formulated wound dressings show an excellent ability to maintain relevant moisture on the wound bed and allow gas exchanges with the environment. Due to their high biocompatibility, biodegradability, physicochemical properties, eco-friendliness, and low cost, cellulose derivatives are promising materials for biomedical and pharmaceutical domains (electrochemical biosensors for medical diagnosis, bone tissue engineering, hemodialysis, drug delivery and 3D printing), for oilfields, carbon capture and the food industry.

Author Contributions: The authors had equal contribution. All authors have read and agreed to the published version of the manuscript.

Funding: This paper was financially supported by the Carol Davila University of Medicine and Pharmacy Bucharest through Contract No. CNFIS-FDI-2021-0300 and RDI Capability consolidation at the Institutional level of the multidisciplinary research teams involved in the sustainability of UMFCD priority research directions.

Institutional Review Board Statement: Not applicable.

Informed Consent Statement: Not applicable.

Data Availability Statement: No new data were created or analyzed in this study. Data sharing is not applicable to this article.

Conflicts of Interest: The authors declare no conflict of interest.

References

1. Kim, B.S.; Gao, G.; Kim, J.Y.; Cho, D.W. 3D Cell Printing of Perfusable Vascularized Human Skin Equivalent Composed of Epidermis, Dermis, and Hypodermis for Better Structural Recapitulation of Native Skin. *Adv. Healthc. Mater.* **2019**, *8*, e1801019. [CrossRef] [PubMed]
2. Sadeghi, S.; Nourmohammadi, J.; Ghaee, A.; Soleimani, N. Carboxymethyl cellulose-human hair keratin hydrogel with controlled clindamycin release as antibacterial wound dressing. *Int. J. Biol. Macromol.* **2020**, *147*, 1239–1247. [CrossRef] [PubMed]
3. Zimmerman, A.; Bai, L.; Ginty, D.D. The gentle touch receptors of mammalian skin. *Science* **2014**, *346*, 950–954. [CrossRef] [PubMed]
4. Zhao, S.; Li, L.; Wang, H.; Zhang, Y.; Cheng, X.; Zhou, N.; Rahaman, M.N.; Liu, Z.; Huang, W.; Zhang, C. Wound dressings composed of copper-doped borate bioactive glass microfibers stimulate angiogenesis and heal full-thickness skin defects in a rodent model. *Biomaterials* **2015**, *53*, 379–391. [CrossRef]
5. Kilic Bektas, C.; Kimiz, I.; Sendemir, A.; Hasirci, V.; Hasirci, N. A bilayer scaffold prepared from collagen and carboxymethyl cellulose for skin tissue engineering applications. *J. Biomater. Sci. Polym. Ed.* **2018**, *29*, 1764–1784. [CrossRef]
6. Kolarsick, P.A.; Kolarsick, M.A.; Goodwin, C. Anatomy and physiology of the skin. *J. Dermatol. Nurses Assoc.* **2011**, *3*, 203–213. [CrossRef]
7. McLafferty, E.; Hendry, C.; Alistair, F. The integumentary system: Anatomy, physiology and function of skin. *Nurs. Stand.* **2012**, *27*, 35–42. [CrossRef]
8. Joodaki, H.; Panzer, M.B. Skin mechanical properties and modeling: A review. *Proc. Inst. Mech. Eng. Part H J. Eng. Med.* **2018**, *232*, 323–343. [CrossRef]
9. Lai-Cheong, J.E.; McGrath, J.A. Structure and function of skin, hair and nails. *Medicine* **2021**, *49*, 337–342. [CrossRef]
10. Brown, T.M.; Krishnamurthy, K. Histology, Dermis. In *StatPearls*; StatPearls Publishing LLC.: Treasure Island, FL, USA, 2020.
11. Hashmi, S.; Marinkovich, M.P. Molecular organization of the basement membrane zone. *Clin. Dermatol.* **2011**, *29*, 398–411. [CrossRef]
12. Zimoch, J.; Zielinska, D.; Michalak-Micka, K.; Rütsche, D.; Böni, R.; Biedermann, T.; Klar, A.S. Bioengineering a prevascularized human tri-layered skin substitute containing a hypodermis. *Acta Biomater.* **2021**, *134*, 215–227. [CrossRef]
13. Wong, R.; Geyer, S.; Weninger, W.; Guimberteau, J.C.; Wong, J.K. The dynamic anatomy and patterning of skin. *Exp. Dermatol.* **2016**, *25*, 92–98. [CrossRef] [PubMed]
14. Bragazzi, N.L.; Sellami, M.; Salem, I.; Conic, R.; Kimak, M.; Pigatto, P.D.M.; Damiani, G. Fasting and Its Impact on Skin Anatomy, Physiology, and Physiopathology: A Comprehensive Review of the Literature. *Nutrients* **2019**, *11*, 249. [CrossRef] [PubMed]
15. Rodrigues, C.; de Assis, A.M.; Moura, D.J.; Halmenschlager, G.; Saffi, J.; Xavier, L.L.; Fernandes Mda, C.; Wink, M.R. New therapy of skin repair combining adipose-derived mesenchymal stem cells with sodium carboxymethylcellulose scaffold in a pre-clinical rat model. *PLoS ONE* **2014**, *9*, e96241. [CrossRef]
16. Okoye, E.; Okolie, T. Development and in vitro characterization of ciprofloxacin loaded polymeric films for wound dressing. *Int. J. Health Allied Sci.* **2015**, *4*, 234. [CrossRef]
17. Varshosaz, J.; Taymouri, S.; Minaiyan, M.; Rastegarnasab, F.; Baradaran, A. Development and in vitro/in vivo evaluation of HPMC/chitosan gel containing simvastatin loaded self-assembled nanomicelles as a potent wound healing agent. *Drug Dev. Ind. Pharm.* **2018**, *44*, 276–288. [CrossRef]
18. Martinengo, L.; Olsson, M.; Bajpai, R.; Soljak, M.; Upton, Z.; Schmidtchen, A.; Car, J.; Jarbrink, K. Prevalence of chronic wounds in the general population: Systematic review and meta-analysis of observational studies. *Ann. Epidemiol.* **2019**, *29*, 8–15. [CrossRef]
19. Sen, C.K. Human Wound and Its Burden: Updated 2020 Compendium of Estimates. *Adv. Wound Care* **2021**, *10*, 281–292. [CrossRef]
20. Walker, J.; Cullen, M.; Chambers, H.; Mitchell, E.; Steers, N.; Khalil, H. Identifying wound prevalence using the Mobile Wound Care program. *Int. Wound J.* **2014**, *11*, 319–325. [CrossRef]
21. Kapp, S.; Miller, C.; Santamaria, N. The quality of life of people who have chronic wounds and who self-treat. *J. Clin. Nurs.* **2018**, *27*, 182–192. [CrossRef]

22. Moreira, M.E.; Markovchick, V.J. Wound Management. *Crit. Care Nurs. Clin. N. Am.* **2012**, *24*, 215–237. [CrossRef]
23. Ghica, M.V.; Kaya, M.G.A.; Dinu-Pirvu, C.E.; Lupuleasa, D.; Udeanu, D.I. Development, Optimization and In Vitro/In Vivo Characterization of Collagen-Dextran Spongious Wound Dressings Loaded with Flufenamic Acid. *Molecules* **2017**, *22*, 1552. [CrossRef] [PubMed]
24. Clark, M. Alginates in Dressings and Wound Management. In *Alginates and Their Biomedical Applications*; Rehm, B.H.A., Moradali, M.F., Eds.; Springer Series in Biomaterials Science and Engineering; Springer: Berlin/Heidelberg, Germany, 2018; Volume 11, pp. 213–222.
25. Aramwit, P. Introduction to biomaterials for wound healing. In *Wound Healing Biomaterials, Vol. 2: Functional Biomaterials*; Agren, M.S., Ed.; Woodhead Publishing Series in Biomaterials; Woodhead Publ. Ltd.: Cambridge, UK, 2016; Volume 115, pp. 3–38.
26. Lee, D.; Yoo, B. Cellulose derivatives agglomerated in a fluidized bed: Physical, rheological, and structural properties. *Int. J. Biol. Macromol.* **2021**, *181*, 232–240. [CrossRef]
27. Seddiqi, H.; Oliaei, E.; Honarkar, H.; Jin, J.F.; Geonzon, L.C.; Bacabac, R.G.; Klein-Nulend, J. Cellulose and its derivatives: Towards biomedical applications. *Cellulose* **2021**, *28*, 1893–1931. [CrossRef]
28. Jia, B.; Zhou, J.; Zhang, L. Electrospun nano-fiber mats containing cationic cellulose derivatives and poly (vinyl alcohol) with antibacterial activity. *Carbohydr. Res.* **2011**, *346*, 1337–1341. [CrossRef]
29. Oprea, M.; Voicu, S.I. Recent advances in composites based on cellulose derivatives for biomedical applications. *Carbohydr. Polym.* **2020**, *247*, 116683. [CrossRef] [PubMed]
30. Agrawal, P.; Soni, S.; Mittal, G.; Bhatnagar, A. Role of polymeric biomaterials as wound healing agents. *Int. J. Low Extrem. Wounds* **2014**, *13*, 180–190. [CrossRef]
31. Mir, M.; Ali, M.N.; Barakullah, A.; Gulzar, A.; Arshad, M.; Fatima, S.; Asad, M. Synthetic polymeric biomaterials for wound healing: A review. *Prog Biomater* **2018**, *7*, 1–21. [CrossRef] [PubMed]
32. Shende, P.; Gupta, H. Formulation and comparative characterization of nanoparticles of curcumin using natural, synthetic and semi-synthetic polymers for wound healing. *Life Sci.* **2020**, *253*, 117588. [CrossRef] [PubMed]
33. Boateng, J.S.; Matthews, K.H.; Stevens, H.N.; Eccleston, G.M. Wound healing dressings and drug delivery systems: A review. *J. Pharm. Sci.* **2008**, *97*, 2892–2923. [CrossRef]
34. Serena, T.E. A Global Perspective on Wound Care. *Adv. Wound Care* **2014**, *3*, 548–552. [CrossRef]
35. Ambekar, R.S.; Kandasubramanian, B. Advancements in nanofibers for wound dressing: A review. *Eur. Polym. J.* **2019**, *117*, 304–336. [CrossRef]
36. Kordestani, S.S. Chapter 6—Wound Classification. In *Atlas of Wound Healing*; Kordestani, S.S., Ed.; Elsevier: Amsterdam, The Netherlands, 2019; pp. 49–50.
37. Kumar, K.S.; Reddy, B.E.J.S.; Processing, I. Wound image analysis classifier for efficient tracking of wound healing status. *Signal Image Process.* **2014**, *5*, 15–27.
38. Wang, M.; Huang, X.; Zheng, H.; Tang, Y.; Zeng, K.; Shao, L.; Li, L. Nanomaterials applied in wound healing: Mechanisms, limitations and perspectives. *J. Control. Release* **2021**, *337*, 236–247. [CrossRef] [PubMed]
39. Low, J.S.; Mak, K.-K.; Zhang, S.; Pichika, M.R.; Marappan, P.; Mohandas, K.; Balijepalli, M.K. In vitro methods used for discovering plant derived products as wound healing agents—An update on the cell types and rationale. *Fitoterapia* **2021**, *154*, 105026. [CrossRef] [PubMed]
40. Nosrati, H.; Khodaei, M.; Alizadeh, Z.; Banitalebi-Dehkordi, M. Cationic, anionic and neutral polysaccharides for skin tissue engineering and wound healing applications. *Int. J. Biol. Macromol.* **2021**, *192*, 298–322. [CrossRef] [PubMed]
41. Skorkowska-Telichowska, K.; Czemplik, M.; Kulma, A.; Szopa, J. The local treatment and available dressings designed for chronic wounds. *J. Am. Acad. Dermatol.* **2013**, *68*, e117–e126. [CrossRef] [PubMed]
42. Thomas, S. *Surgical Dressings and Wound Management*; Dr Stephen Thomas: Joplin, MI, USA, 2010.
43. Hachem, R.; Parikh, U.M.; Reitzel, R.; Rosenblatt, J.; Kaul, A.; Vargas-Cruz, N.; Hill, L.; Moore, L.; Meyer, J.; Chaftari, A.M.; et al. Novel antimicrobial ointment for infected wound healing in an in vitro and in vivo porcine model. *Wound Repair Regen.* **2021**, *29*, 830–842. [CrossRef]
44. Albu, M.G.; Ghica, M.V.; Stefanescu, G.A.; Hodorogea, M.; Marin, M.M.; Marin, S.; Danila, E.; Voicu, S.; Simonca, A.G.; Popa, L. Design and Characterization of Collagen-Sodium Carboxymethylcellulose-Lidocaine 3D Composites for Wound Management. *Key Eng. Mater.* **2016**, *695*, 309–316. [CrossRef]
45. Capanema, N.S.V.; Mansur, A.A.P.; de Jesus, A.C.; Carvalho, S.M.; de Oliveira, L.C.; Mansur, H.S. Superabsorbent crosslinked carboxymethyl cellulose-PEG hydrogels for potential wound dressing applications. *Int. J. Biol. Macromol.* **2018**, *106*, 1218–1234. [CrossRef]
46. Sandri, G.; Bonferoni, M.C.; D'Autilia, F.; Rossi, S.; Ferrari, F.; Grisoli, P.; Sorrenti, M.; Catenacci, L.; Del Fante, C.; Perotti, C.; et al. Wound dressings based on silver sulfadiazine solid lipid nanoparticles for tissue repairing. *Eur. J. Pharm. Biopharm.* **2013**, *84*, 84–90. [CrossRef] [PubMed]
47. Cañedo-Dorantes, L.; Cañedo-Ayala, M. Skin Acute Wound Healing: A Comprehensive Review. *Int. J. Inflamm.* **2019**, *2019*, 3706315. [CrossRef] [PubMed]
48. Verma, Y.K.; Verma, R.; Sing, A.K.; Gangenahalli, G. LiCl Containing Thermosensitive Formulation Improves Hemostasis, Wound Healing, and Hair Regrowth. *Regen. Eng. Transl. Med.* **2020**, *17*, 362–378. [CrossRef]
49. Kenet, G.; Barg, A.A.; Nowak-Gottl, U. Hemostasis in the Very Young. *Semin. Thromb. Hemost.* **2018**, *44*, 617–623. [CrossRef] [PubMed]

50. Iacob, A.T.; Dragan, M.; Ionescu, O.M.; Profire, L.; Ficai, A.; Andronescu, E.; Confederat, L.G.; Lupascu, D. An Overview of Biopolymeric Electrospun Nanofibers Based on Polysaccharides for Wound Healing Management. *Pharmaceutics* **2020**, *12*, 983. [CrossRef]
51. Kanikireddy, V.; Varaprasad, K.; Jayaramudu, T.; Karthikeyan, C.; Sadiku, R. Carboxymethyl cellulose-based materials for infection control and wound healing: A review. *Int. J. Biol. Macromol.* **2020**, *164*, 963–975. [CrossRef]
52. Pawar, H.V.; Tetteh, J.; Boateng, J.S. Preparation, optimisation and characterisation of novel wound healing film dressings loaded with streptomycin and diclofenac. *Colloid Surf. B Biointerfaces* **2013**, *102*, 102–110. [CrossRef]
53. Harper, D.; Young, A.; McNaught, C.-E.J.S. The physiology of wound healing. *Adv. Ski. Wound Care* **2014**, *32*, 445–450. [CrossRef]
54. Yang, L.; Liang, F.; Zhang, X.; Jiang, Y.; Duan, F.; Li, L.; Ren, F. Remodeling microenvironment based on MOFs-Hydrogel hybrid system for improving diabetic wound healing. *Chem. Eng. J.* **2022**, *427*, 131506. [CrossRef]
55. Schreml, S.; Szeimies, R.; Prantl, L.; Karrer, S.; Landthaler, M.; Babilas, P.J. Oxygen in acute and chronic wound healing. *Br. J. Dermatol.* **2010**, *163*, 257–268. [CrossRef]
56. El-Ashram, S.; El-Samad, L.M.; Basha, A.A.; El Wakil, A. Naturally-derived targeted therapy for wound healing: Beyond classical strategies. *Pharmacol. Res.* **2021**, *170*, 105749. [CrossRef] [PubMed]
57. Guo, S.; Dipietro, L.A. Factors affecting wound healing. *J. Dent. Res.* **2010**, *89*, 219–229. [CrossRef]
58. Khalil, H.; Cullen, M.; Chambers, H.; Carroll, M.; Walker, J. Elements affecting wound healing time: An evidence based analysis. *Wound Repair Regen.* **2015**, *23*, 550–556. [CrossRef]
59. Desmet, C.M.; Preat, V.; Gallez, B. Nanomedicines and gene therapy for the delivery of growth factors to improve perfusion and oxygenation in wound healing. *Adv. Drug Deliv. Rev.* **2018**, *129*, 262–284. [CrossRef]
60. Gueldner, J.; Zhang, F.; Zechmann, B.; Bruce, E.D. Evaluating a novel oxygenating therapeutic for its potential use in the advancement of wound healing. *Toxicol. In Vitro* **2017**, *43*, 62–68. [CrossRef] [PubMed]
61. Paunica-Panea, G.; Ficai, A.; Marin, M.M.; Marin, S.; Albu, M.G.; Constantin, V.D.; Dinu-Pirvu, C.; Vuluga, Z.; Corobea, M.C.; Ghica, M.V. New Collagen-Dextran-Zinc Oxide Composites for Wound Dressing. *J. Nanomater.* **2016**, *2016*, 5805034. [CrossRef]
62. Powers, J.G.; Higham, C.; Broussard, K.; Phillips, T.J. Wound healing and treating wounds Chronic wound care and management. *J. Am. Acad. Dermatol.* **2016**, *74*, 607–625. [CrossRef]
63. Emmerson, E.; Hardman, M.J. The role of estrogen deficiency in skin ageing and wound healing. *Biogerontology* **2012**, *13*, 3–20. [CrossRef]
64. Gouin, J.P.; Kiecolt-Glaser, J.K. The Impact of Psychological Stress on Wound Healing: Methods and Mechanisms. *Crit. Care Nurs. Clin. N. Am.* **2012**, *24*, 201–213. [CrossRef]
65. Okonkwo, U.A.; DiPietro, L.A. Diabetes and Wound Angiogenesis. *Int. J. Mol. Sci.* **2017**, *18*, 1419. [CrossRef]
66. Goodarzi, P.; Alavi-Moghadam, S.; Sarvari, M.; Beik, A.T.; Falahzadeh, K.; Aghayan, H.; Payab, M.; Larijani, B.; Gilany, K.; Rahim, F.; et al. Adipose Tissue-Derived Stromal Cells for Wound Healing. In *Cell Biology and Translational Medicine, Vol. 4: Stem Cells and Cell Based Strategies in Regeneration*; Turksen, K., Ed.; Advances in Experimental Medicine and Biology; Springer International Publishing AG: Cham, Switzerland, 2018; Volume 1119, pp. 133–149.
67. Serra, M.B.; Barroso, W.A.; Silva, N.N.; Silva, S.d.N.; Borges, A.C.R.; Abreu, I.C.; Borges, M.O.d.R. From Inflammation to Current and Alternative Therapies Involved in Wound Healing. *Int. J. Inflamm.* **2017**, *2017*, 3406215. [CrossRef] [PubMed]
68. Beyene, R.T.; Derryberry, S.L.; Barbul, A. The Effect of Comorbidities on Wound Healing. *Surg. Clin. N. Am.* **2020**, *100*, 695–705. [CrossRef] [PubMed]
69. McDaniel, J.C.; Browning, K.K. Smoking, chronic wound healing, and implications for evidence-based practice. *J. Wound Ostomy Cont. Nurs.* **2014**, *41*, 415–423. [CrossRef] [PubMed]
70. Sen, C.K. Human Wounds and Its Burden: An Updated Compendium of Estimates. *Adv. Wound Care* **2019**, *8*, 39–48. [CrossRef] [PubMed]
71. Moura, L.I.; Dias, A.M.; Carvalho, E.; de Sousa, H.C. Recent advances on the development of wound dressings for diabetic foot ulcer treatment—A review. *Acta Biomater.* **2013**, *9*, 7093–7114. [CrossRef]
72. Kus, K.J.B.; Ruiz, E.S. Wound Dressings—A Practical Review. *Curr. Dermatol. Rep.* **2020**, *9*, 298–308. [CrossRef]
73. Mogoşanu, G.D.; Grumezescu, A.M. Natural and synthetic polymers for wounds and burns dressing. *Int. J. Pharm.* **2014**, *463*, 127–136. [CrossRef]
74. Namazi, H.; Rakhshaei, R.; Hamishehkar, H.; Kafil, H.S. Antibiotic loaded carboxymethylcellulose/MCM-41 nanocomposite hydrogel films as potential wound dressing. *Int. J. Biol. Macromol.* **2016**, *85*, 327–334. [CrossRef]
75. Trevisol, T.C.; Fritz, A.R.M.; de Souza, S.M.A.G.U.; Bierhalz, A.C.K.; Valle, J.A.B. Alginate and carboxymethyl cellulose in monolayer and bilayer films as wound dressings: Effect of the polymer ratio. *J. Appl. Polym. Sci.* **2019**, *136*, 46941. [CrossRef]
76. Mishra, S.K.; Mary, D.S.; Kannan, S. Copper incorporated microporous chitosan-polyethylene glycol hydrogels loaded with naproxen for effective drug release and anti-infection wound dressing. *Int. J. Biol. Macromol.* **2017**, *95*, 928–937. [CrossRef]
77. Kamoun, E.A.; Chen, X.; Mohy Eldin, M.S.; Kenawy, E.-R.S. Crosslinked poly(vinyl alcohol) hydrogels for wound dressing applications: A review of remarkably blended polymers. *Arab. J. Chem.* **2015**, *8*, 1–14. [CrossRef]
78. Patil, P.P.; Reagan, M.R.; Bohara, R.A. Silk fibroin and silk-based biomaterial derivatives for ideal wound dressings. *Int. J. Biol. Macromol.* **2020**, *164*, 4613–4627. [CrossRef] [PubMed]
79. Dhivya, S.; Padma, V.V.; Santhini, E. Wound dressings—A review. *Biomedicine* **2015**, *5*, 22. [CrossRef] [PubMed]

80. Marin, S.; Albu Kaya, M.G.; Ghica, M.V.; Dinu-Pirvu, C.; Popa, L.; Udeanu, D.I.; Mihai, G.; Enachescu, M. Collagen-Polyvinyl Alcohol-Indomethacin Biohybrid Matrices as Wound Dressings. *Pharmaceutics* **2018**, *10*, 224. [CrossRef] [PubMed]
81. Akrami-Hasan-Kohal, M.; Tayebi, L.; Ghorbani, M. Curcumin-loaded naturally-based nanofibers as active wound dressing mats: Morphology, drug release, cell proliferation, and cell adhesion studies. *New J. Chem.* **2020**, *44*, 10343–10351. [CrossRef]
82. Gruppuso, M.; Turco, G.; Marsich, E.; Porrelli, D. Polymeric wound dressings, an insight into polysaccharide-based electrospun membranes. *Appl. Mater. Today* **2021**, *24*, 101148. [CrossRef]
83. Farahani, M.; Shafiee, A. Wound Healing: From Passive to Smart Dressings. *Adv. Healthc. Mater.* **2021**, *10*, 2100477. [CrossRef]
84. Sharma, G.; Lee, S.W.; Atanacio, O.; Parvizi, J.; Kim, T.K. In search of the optimal wound dressing material following total hip and knee arthroplasty: A systematic review and meta-analysis. *Int. Orthop.* **2017**, *41*, 1295–1305. [CrossRef]
85. Kong, D.; Zhang, Q.; You, J.; Cheng, Y.; Hong, C.; Chen, Z.; Jiang, T.; Hao, T. Adhesion loss mechanism based on carboxymethyl cellulose-filled hydrocolloid dressings in physiological wounds environment. *Carbohydr. Polym.* **2020**, *235*, 115953. [CrossRef]
86. Yadav, V.; Mittal, A.; Bansal, P.; Singh, S.K. Regulatory approval process for advanced dressings in India: An overview of rules. *J. Wound Care* **2019**, *28*, S32–S42. [CrossRef]
87. Olatunji, O. Classification of Natural Polymers. In *Natural Polymers*; Olatunji, O., Ed.; Springer: Cham, Switzerland, 2016; pp. 1–17.
88. Kulkarni Vishakha, S.; Butte Kishor, D.; Rathod Sudha, S. Natural polymers–A comprehensive review. *Int. J. Res. Pharm. Biomed. Sci.* **2012**, *3*, 1597–1613.
89. Alves, T.F.; Morsink, M.; Batain, F.; Chaud, M.V.; Almeida, T.; Fernandes, D.A.; da Silva, C.F.; Souto, E.B.; Severino, P.J.C. Applications of natural, semi-synthetic, and synthetic polymers in cosmetic formulations. *Cosmetics* **2020**, *7*, 75. [CrossRef]
90. Bhatia, S. Natural Polymers vs. Synthetic Polymer. In *Natural Polymer Drug Delivery Systems*; Springer: Cham, Switzerland, 2016; pp. 95–118.
91. Maitz, M.F. Applications of synthetic polymers in clinical medicine. *Biosurf. Biotribol.* **2015**, *1*, 161–176. [CrossRef]
92. Gunatillake, P.; Mayadunne, R.; Adhikari, R. Recent developments in biodegradable synthetic polymers. *Biotechnol. Annu. Rev.* **2006**, *12*, 301–347.
93. Tavakoli, S.; Klar, A.S. Advanced Hydrogels as Wound Dressings. *Biomolecules* **2020**, *10*, 1169. [CrossRef] [PubMed]
94. Brumberg, V.; Astrelina, T.; Malivanova, T.; Samoilov, A. Modern Wound Dressings: Hydrogel Dressings. *Biomedicines* **2021**, *9*, 1235. [CrossRef]
95. Sklenar, Z.; Vitkova, Z.; Herdova, P.; Horackova, K.; Simunkova, V. Formulation and release of alaptide from cellulose-based hydrogels. *Acta Vet. Brno* **2012**, *81*, 301–306. [CrossRef]
96. Keogh, S.J.; Nelson, A.; Webster, J.; Jolly, J.; Ullman, A.J.; Chaboyer, W.P. Hydrocolloid dressings for treating pressure ulcers. *Cochrane Database Syst. Rev.* **2018**, *2018*, CD010364. [CrossRef]
97. Khorasani, M.T.; Joorabloo, A.; Moghaddam, A.; Shamsi, H.; MansooriMoghadam, Z. Incorporation of ZnO nanoparticles into heparinised polyvinyl alcohol/chitosan hydrogels for wound dressing application. *Int. J. Biol. Macromol.* **2018**, *114*, 1203–1215. [CrossRef]
98. Ullah, F.; Othman, M.B.; Javed, F.; Ahmad, Z.; Md Akil, H. Classification, processing and application of hydrogels: A review. *Mater. Sci. Eng. C Mater. Biol. Appl.* **2015**, *57*, 414–433. [CrossRef]
99. Ahmed, E.M. Hydrogel: Preparation, characterization, and applications: A review. *J. Adv. Res.* **2015**, *6*, 105–121. [CrossRef] [PubMed]
100. Mahinroosta, M.; Jomeh Farsangi, Z.; Allahverdi, A.; Shakoori, Z. Hydrogels as intelligent materials: A brief review of synthesis, properties and applications. *Mater. Today Chem.* **2018**, *8*, 42–55. [CrossRef]
101. Vowden, K.; Vowden, P. Wound dressings: Principles and practice. *Surgery* **2017**, *35*, 489–494. [CrossRef]
102. Borda, L.J.; Macquhae, F.E.; Kirsner, R.S. Wound Dressings: A Comprehensive Review. *Curr. Dermatol. Rep.* **2016**, *5*, 287–297. [CrossRef]
103. Aljghami, M.E.; Saboor, S.; Amini-Nik, S. Emerging Innovative Wound Dressings. *Ann. Biomed. Eng.* **2019**, *47*, 659–675. [CrossRef]
104. Meuleneire, F. A vapour-permeable film dressing used on superficial wounds. *Br. J. Nurs.* **2014**, *23*, S36–S43. [CrossRef] [PubMed]
105. Samprasit, W.; Chamsai, B.; Settharaksa, S.; Opanasopit, P. Synergistic antibacterial activity of alpha mangostin and resveratrol loaded polymer-based films against bacteria infected wound. *J. Drug Deliv. Sci. Technol.* **2020**, *57*, 101629. [CrossRef]
106. Shi, C.Y.; Wang, C.Y.; Liu, H.; Li, Q.J.; Li, R.H.; Zhang, Y.; Liu, Y.Z.; Shao, Y.; Wang, J.C. Selection of Appropriate Wound Dressing for Various Wounds. *Front. Bioeng. Biotechnol.* **2020**, *8*, 182. [CrossRef]
107. Lundin, J.G.; McGann, C.L.; Daniels, G.C.; Streifel, B.C.; Wynne, J.H. Hemostatic kaolin-polyurethane foam composites for multifunctional wound dressing applications. *Mater. Sci. Eng. C Mater. Biol. Appl.* **2017**, *79*, 702–709. [CrossRef]
108. Dong, R.; Guo, B. Smart wound dressings for wound healing. *Nano Today* **2021**, *41*, 101290. [CrossRef]
109. Powers, J.G.; Morton, L.M.; Phillips, T.J. Dressings for chronic wounds. *Dermatol. Ther.* **2013**, *26*, 197–206. [CrossRef]
110. Dumville, J.C.; Keogh, S.J.; Liu, Z.; Stubbs, N.; Walker, R.M.; Fortnam, M. Alginate dressings for treating pressure ulcers. *Cochrane Database Syst. Rev.* **2015**, *5*, CD011277. [CrossRef]
111. Weller, C.D.; Team, V.; Sussman, G. First-Line Interactive Wound Dressing Update: A Comprehensive Review of the Evidence. *Front. Pharmacol.* **2020**, *11*, 155. [CrossRef]
112. Aderibigbe, B.A.; Buyana, B. Alginate in Wound Dressings. *Pharmaceutics* **2018**, *10*, 42. [CrossRef]
113. Crabbe-Mann, M.; Tsaoulidis, D.; Parhizkar, M.; Edirisinghe, M. Ethyl cellulose, cellulose acetate and carboxymethyl cellulose microstructures prepared using electrohydrodynamics and green solvents. *Cellulose* **2018**, *25*, 1687–1703. [CrossRef]

114. Liu, K.; Du, H.; Zheng, T.; Liu, H.; Zhang, M.; Zhang, R.; Li, H.; Xie, H.; Zhang, X.; Ma, M.; et al. Recent advances in cellulose and its derivatives for oilfield applications. *Carbohydr. Polym.* **2021**, *259*, 117740. [CrossRef] [PubMed]
115. Vlaia, L.; Coneac, G.; Olariu, I.; Vlaia, V.; Lupuleasa, D. Cellulose-Derivatives-Based Hydrogels as Vehicles for Dermal and Transdermal Drug Delivery. In *Emerging Concepts in Analysis and Applications of Hydrogels*; Majee, S.B., Ed.; IntechOpen: London, UK, 2016; Volume 2, p. 64.
116. Teixeira, M.A.; Paiva, M.C.; Amorim, M.T.P.; Felgueiras, A.H.P. Electrospun Nanocomposites Containing Cellulose and Its Derivatives Modified with Specialized Biomolecules for an Enhanced Wound Healing. *Nanomaterials* **2020**, *10*, 557. [CrossRef] [PubMed]
117. Heinze, T. Cellulose: Structure and Properties. In *Cellulose Chemistry and Properties: Fibers, Nanocelluloses and Advanced Materials*; Rojas, O.J., Ed.; Advances in Polymer Science; Springer: Berlin/Heidelberg, Germany, 2016; Volume 271, pp. 1–52.
118. Miao, J.J.; Pangule, R.C.; Paskaleva, E.E.; Hwang, E.E.; Kane, R.S.; Linhardt, R.J.; Dordick, J.S. Lysostaphin-functionalized cellulose fibers with antistaphylococcal activity for wound healing applications. *Biomaterials* **2011**, *32*, 9557–9567. [CrossRef]
119. Wang, Y.G.; Wang, X.J.; Xie, Y.J.; Zhang, K. Functional nanomaterials through esterification of cellulose: A review of chemistry and application. *Cellulose* **2018**, *25*, 3703–3731. [CrossRef]
120. Dmour, I.; Taha, M.O. Chapter 2—Natural and semisynthetic polymers in pharmaceutical nanotechnology. In *Organic Materials as Smart Nanocarriers for Drug Delivery*; Grumezescu, A.M., Ed.; William Andrew Publishing: Amsterdam, The Netherlands, 2018; pp. 35–100.
121. Haldar, D.; Purkait, M.K. Micro and nanocrystalline cellulose derivatives of lignocellulosic biomass: A review on synthesis, applications and advancements. *Carbohydr. Polym.* **2020**, *250*, 116937. [CrossRef]
122. Raucci, M.; Alvarez-Perez, M.; Demitri, C.; Giugliano, D.; De Benedictis, V.; Sannino, A.; Ambrosio, L.J. Effect of citric acid crosslinking cellulose-based hydrogels on osteogenic differentiation. *J. Biomed. Mater. Res. Part A* **2015**, *103*, 2045–2056. [CrossRef]
123. Yu, J.; Wang, C.P.; Wang, J.F.; Chu, F.X. Synthesis and Characterization of Ethyl Cellulose Based Acrylate. In Proceedings of the 4th International Conference on Manufacturing Science and Engineering (ICMSE 2013), Dalian, China, 30–31 March 2013; pp. 124–130.
124. Fekete, T.; Borsa, J.; Takács, E.; Wojnárovits, L.J.C. Synthesis of cellulose derivative based superabsorbent hydrogels by radiation induced crosslinking. *Cellulose* **2014**, *21*, 4157–4165. [CrossRef]
125. Shaghaleh, H.; Xu, X.; Wang, S.F. Current progress in production of biopolymeric materials based on cellulose, cellulose nanofibers, and cellulose derivatives. *RSC Adv.* **2018**, *8*, 825–842. [CrossRef]
126. Abdelhak, M.J. A Review: Application of Biopolymers in the Pharmaceutical Formulation. *J. Adv. Bio-Pharm. Pharmacovigil.* **2019**, *1*, 15–25. [CrossRef]
127. Goncalves, C.; Favre, C.; Feuardant, P.; Klein, S.; Vaca-Garcia, C.; Cecutti, C.; Thiebaud-Roux, S.; Vedrenne, E. Synthesis of new cellulose ethers using Suzuki-Miyaura reactions. *Carbohydr. Polym.* **2015**, *116*, 51–59. [CrossRef] [PubMed]
128. Kabir, S.M.F.; Sikdar, P.P.; Haque, B.; Bhuiyan, M.A.R.; Ali, A.; Islam, M.N. Cellulose-based hydrogel materials: Chemistry, properties and their prospective applications. *Prog. Biomater.* **2018**, *7*, 153–174. [CrossRef]
129. Gao, C.Z.; Liu, S.; Edgar, K.J. Regioselective chlorination of cellulose esters by methanesulfonyl chloride. *Carbohydr. Polym.* **2018**, *193*, 108–118. [CrossRef]
130. Shokri, J.; Adibki, K. Application of Cellulose and Cellulose Derivatives in Pharmaceutical Industries. In *Cellulose—Medical, Pharmaceutical and Electronic Applications*; van de Ven, T., Godbout, L., Eds.; IntechOpen: London, UK, 2013.
131. Karlsson, K.; Schuster, E.; Stading, M.; Rigdahl, M. Foaming behavior of water-soluble cellulose derivatives: Hydroxypropyl methylcellulose and ethyl hydroxyethyl cellulose. *Cellulose* **2015**, *22*, 2651–2664. [CrossRef]
132. Ambrosio, L.; Demitri, C.; Sannino, A. Superabsorbent cellulose-based hydrogels for biomedical applications. In *Biomedical Hydrogels*; Rimmer, S., Ed.; Woodhead Publishing: Cambridge, UK, 2011; pp. 25–50.
133. Sannino, A.; Demitri, C.; Madaghiele, M. Biodegradable Cellulose-based Hydrogels: Design and Applications. *Materials* **2009**, *2*, 353–373. [CrossRef]
134. Mao, J.; Li, S.; Huang, J.; Meng, K.; Chen, G.; Lai, Y. Recent Advances of Multifunctional Cellulose-Based Hydrogels. In *Cellulose-Based Superabsorbent Hydrogels*; Mondal, M., Ed.; Springer: Cham, Switzerland, 2018; pp. 37–64.
135. Kamel, S.; Khattab, T.A. Recent Advances in Cellulose-Based Biosensors for Medical Diagnosis. *Biosensors* **2020**, *10*, 67. [CrossRef]
136. Burduşel, A.-C.; Stancu, I.; Marin, I.B.M.Ş.; Chelaru, C.; Serafim, A.; Drăguşin, D.; Kaya, M.G.A.; Coară, G. Development and characterization of collagen–carboxymethylcellulose materials for lenses. In Proceedings of the International Conference on Advanced Materials and Systems (ICAMS), Qingdao, China, 26–27 March 2016; pp. 215–220.
137. Vinklarkova, L.; Masteikova, R.; Vetchy, D.; Dolezel, P.; Bernatoniene, J. Formulation of Novel Layered Sodium Carboxymethylcellulose Film Wound Dressings with Ibuprofen for Alleviating Wound Pain. *Biomed. Res. Int.* **2015**, *2015*, 892671. [CrossRef] [PubMed]
138. Zennifer, A.; Senthilvelan, P.; Sethuraman, S.; Sundaramurthi, D. Key advances of carboxymethyl cellulose in tissue engineering & 3D bioprinting applications. *Carbohydr. Polym.* **2021**, *256*, 117561. [CrossRef] [PubMed]
139. Ramli, N.A.; Wong, T.W. Sodium carboxymethylcellulose scaffolds and their physicochemical effects on partial thickness wound healing. *Int. J. Pharm.* **2011**, *403*, 73–82. [CrossRef] [PubMed]
140. Fan, L.H.; Peng, M.; Zhou, X.Y.; Wu, H.; Hu, J.; Xie, W.G.; Liu, S.H. Modification of carboxymethyl cellulose grafted with collagen peptide and its antioxidant activity. *Carbohydr. Polym.* **2014**, *112*, 32–38. [CrossRef]

141. Rahman, M.S.; Hasan, M.S.; Nitai, A.S.; Nam, S.; Karmakar, A.K.; Ahsan, M.S.; Shiddiky, M.J.A.; Ahmed, M.B. Recent Developments of Carboxymethyl Cellulose. *Polymers* **2021**, *13*, 1345. [CrossRef] [PubMed]
142. Mallakpour, S.; Tukhani, M.; Hussain, C.M. Recent advancements in 3D bioprinting technology of carboxymethyl cellulose-based hydrogels: Utilization in tissue engineering. *Adv. Colloid Interface Sci.* **2021**, *292*, 102415. [CrossRef]
143. Udeanu, D.I.; Kaya, M.G.A.; Ghica, M.V.; Marin, S.; Marin, M.M.; Kaya, D.A.; Popa, L.; Dinu-Pirvu, C. Anti-inflammatory drug-loaded biopolymeric spongious matrices with therapeutic perspectives in burns treatment. *Farmacia* **2018**, *66*, 783–790. [CrossRef]
144. Miraftab, M.; Qiao, Q.; Kennedy, J.F.; Knill, C.J.; Groocock, M.R. Advanced Wound-care Materials: Ultra High Absorbing Fibres made from Alginates Containing Branan Ferulate and Carboxymethyl Cellulose. *J. Text. Inst.* **2010**, *95*, 341–348. [CrossRef]
145. Alavi, M.; Nokhodchi, A. Antimicrobial and Wound Treatment Aspects of Micro- and Nanoformulations of Carboxymethyl, Dialdehyde, and TEMPO-Oxidized Derivatives of Cellulose: Recent Advances. *Macromol. Biosci.* **2020**, *20*, 1900362. [CrossRef]
146. Wong, T.W.; Ramli, N.A. Carboxymethylcellulose film for bacterial wound infection control and healing. *Carbohydr. Polym.* **2014**, *112*, 367–375. [CrossRef]
147. Ali, M.; Khan, N.R.; Basit, H.M.; Mahmood, S. Physico-chemical based mechanistic insight into surfactant modulated sodium Carboxymethylcellulose film for skin tissue regeneration applications. *J. Polym. Res.* **2019**, *27*, 20. [CrossRef]
148. Krizova, H.; Wiener, J. Development of carboxymethyl cellulose/polyphenols gels for textile applications. *Autex Res. J.* **2013**, *13*, 33–36. [CrossRef]
149. Fukuyama, Y.; Maruo, T.; Nishiyama, Y.; Nemoto, Y.; Murayama, K.; Kayanuma, H.; Kawarai, S. Application of a novel carboxymethyl cellulose-based Mohs sol-gel on malignant wounds in three dogs. *J. Vet. Med. Sci.* **2021**, *83*, 385–389. [CrossRef]
150. Lee, S.; Park, Y.H.; Ki, C.S. Fabrication of PEG-carboxymethylcellulose hydrogel by thiol-norbornene photo-click chemistry. *Int. J. Biol. Macromol.* **2016**, *83*, 1–8. [CrossRef] [PubMed]
151. Liu, Y.; Chen, Y.; Zhao, Y.; Tong, Z.R.; Chen, S.S. Superabsorbent Sponge and Membrane Prepared by Polyelectrolyte Complexation of Carboxymethyl Cellulose/Hydroxyethyl Cellulose-Al^{3+}. *BioResources* **2015**, *10*, 6479–6495. [CrossRef]
152. Hu, D.; Qiang, T.; Wang, L. Quaternized chitosan/polyvinyl alcohol/sodium carboxymethylcellulose blend film for potential wound dressing application. *Wound Med.* **2017**, *16*, 15–21. [CrossRef]
153. Nargesi Khoramabadi, H.; Arefian, M.; Hojjati, M.; Tajzad, I.; Mokhtarzade, A.; Mazhar, M.; Jamavari, A. A review of Polyvinyl alcohol/Carboxymethyl cellulose (PVA/CMC) composites for various applications. *J. Compos. Compd.* **2020**, *2*, 69–76. [CrossRef]
154. Lim, S.J.; Lee, J.H.; Piao, M.G.; Lee, M.K.; Oh, D.H.; Hwang, D.H.; Quan, Q.Z.; Yong, C.S.; Choi, H.G. Effect of sodium carboxymethylcellulose and fucidic acid on the gel characterization of polyvinylalcohol-based wound dressing. *Arch. Pharm. Res.* **2010**, *33*, 1073–1081. [CrossRef]
155. Zhang, K.; Wang, Y.N.; Wei, Q.H.; Li, X.P.; Guo, Y.; Zhang, S. Design and Fabrication of Sodium Alginate/Carboxymethyl Cellulose Sodium Blend Hydrogel for Artificial Skin. *Gels* **2021**, *7*, 115. [CrossRef]
156. Shin, J.Y.; Lee, D.Y.; Kim, B.Y.; Yoon, J.I. Effect of polyethylene glycol molecular weight on cell growth behavior of polyvinyl alcohol/carboxymethyl cellulose/polyethylene glycol hydrogel. *J. Appl. Polym. Sci.* **2020**, *137*, 49568. [CrossRef]
157. Fan, L.H.; Zhou, X.Y.; Wu, P.H.; Xie, W.G.; Zheng, H.; Tan, W.; Liu, S.H.; Li, Q.Y. Preparation of carboxymethyl cellulose sulfates and its application as anticoagulant and wound dressing. *Int. J. Biol. Macromol.* **2014**, *66*, 245–253. [CrossRef] [PubMed]
158. Lee, J.Y.; Im, J.N.; Kim, T.H.; Chung, D.J.; Doh, S.J. Structure and Liquid Handling Properties of Water-insoluble Carboxymethyl Cellulose Foam. *Fiber. Polym.* **2015**, *16*, 726–734. [CrossRef]
159. Marchianti, A.C.N.; Sakinah, E.N.; Elfiah, U.; Putri, N.K.S.; Wahyuliswari, D.I.; Maulana, M.; Ulfa, E.U. Gel formulations of Merremia mammosa (Lour.) accelerated wound healing of the wound in diabetic rats. *J. Tradit. Complement. Med.* **2021**, *11*, 38–45. [CrossRef] [PubMed]
160. Park, J.S.; An, S.J.; Jeong, S.I.; Gwon, H.J.; Lim, Y.M.; Nho, Y.C. Chestnut Honey Impregnated Carboxymethyl Cellulose Hydrogel for Diabetic Ulcer Healing. *Polymers* **2017**, *9*, 248. [CrossRef] [PubMed]
161. Ali, N.H.; Amin, M.; Ng, S.F. Sodium carboxymethyl cellulose hydrogels containing reduced graphene oxide (rGO) as a functional antibiofilm wound dressing. *J. Biomater. Sci.-Polym. Ed.* **2019**, *30*, 629–645. [CrossRef]
162. Ng, S.F.; Jumaat, N. Carboxymethyl cellulose wafers containing antimicrobials: A modern drug delivery system for wound infections. *Eur. J. Pharm. Sci.* **2014**, *51*, 173–179. [CrossRef] [PubMed]
163. Javanbakht, S.; Nabi, M.; Shadi, M.; Amini, M.M.; Shaabani, A. Carboxymethyl cellulose/tetracycline@UiO-66 nanocomposite hydrogel films as a potential antibacterial wound dressing. *Int. J. Biol. Macromol.* **2021**, *188*, 811–819. [CrossRef]
164. Rakhshaei, R.; Namazi, H. A potential bioactive wound dressing based on carboxymethyl cellulose/ZnO impregnated MCM-41 nanocomposite hydrogel. *Mater. Sci. Eng. C* **2017**, *73*, 456–464. [CrossRef]
165. Vinklarkova, L.; Masteikova, R.; Foltynova, G.; Muselik, J.; Pavlokova, S.; Bernatoniene, J.; Vetchy, D. Film wound dressing with local anesthetic based on insoluble carboxymethycellulose matrix. *J. Appl. Biomed.* **2017**, *15*, 313–320. [CrossRef]
166. Shi, D.D.; Wang, F.J.; Lan, T.; Zhang, Y.H.; Shao, Z.Q. Convenient fabrication of carboxymethyl cellulose electrospun nanofibers functionalized with silver nanoparticles. *Cellulose* **2016**, *23*, 1899–1909. [CrossRef]
167. Tu, C.; Zhang, R.D.; Yan, C.; Guo, Y.; Cui, L. A pH indicating carboxymethyl cellulose/chitosan sponge for visual monitoring of wound healing. *Cellulose* **2019**, *26*, 4541–4552. [CrossRef]
168. Bang, S.; Ko, Y.G.; Kim, W.I.; Cho, D.; Park, W.H.; Kwon, O.H. Preventing postoperative tissue adhesion using injectable carboxymethyl cellulose-pullulan hydrogels. *Int. J. Biol. Macromol.* **2017**, *105*, 886–893. [CrossRef]

169. Siritientong, T.; Aramwit, P. Characteristics of carboxymethyl cellulose/sericin hydrogels and the influence of molecular weight of carboxymethyl cellulose. *Macromol. Res.* **2015**, *23*, 861–866. [CrossRef]
170. Wang, P.; He, H.W.; Cai, R.; Tao, G.; Yang, M.R.; Zuo, H.; Umar, A.; Wang, Y.J. Cross-linking of dialdehyde carboxymethyl cellulose with silk sericin to reinforce sericin film for potential biomedical application. *Carbohydr. Polym.* **2019**, *212*, 403–411. [CrossRef]
171. Wang, X.; Qi, J.; Zhang, W.; Pu, Y.; Yang, R.; Wang, P.; Liu, S.; Tan, X.; Chi, B. 3D-printed antioxidant antibacterial carboxymethyl cellulose/ε-polylysine hydrogel promoted skin wound repair. *Int. J. Biol. Macromol.* **2021**, *187*, 91–104. [CrossRef]
172. Ninan, N.; Muthiah, M.; Park, I.K.; Kalarikkal, N.; Elain, A.; Wong, T.W.; Thomas, S.; Grohens, Y. Wound healing analysis of pectin/carboxymethyl cellulose/microfibrillated cellulose based composite scaffolds. *Mater. Lett.* **2014**, *132*, 34–37. [CrossRef]
173. Jantrawut, P.; Bunrueangtha, J.; Suerthong, J.; Kantrong, N. Fabrication and Characterization of Low Methoxyl Pectin/Gelatin/Carboxymethyl Cellulose Absorbent Hydrogel Film for Wound Dressing Applications. *Materials* **2019**, *12*, 1628. [CrossRef]
174. Liang, M.; Chen, Z.M.; Wang, F.P.; Liu, L.; Wei, R.N.; Zhang, M. Preparation of self-regulating/anti-adhesive hydrogels and their ability to promote healing in burn wounds. *J. Biomed. Mater. Res. Part B* **2019**, *107*, 1471–1482. [CrossRef]
175. Li, D.F.; Ye, Y.X.; Li, D.R.; Li, X.Y.; Mu, C.D. Biological properties of dialdehyde carboxymethyl cellulose crosslinked gelatin-PEG composite hydrogel fibers for wound dressings. *Carbohydr. Polym.* **2016**, *137*, 508–514. [CrossRef]
176. Doh, S.J.; Lee, J.Y.; Lim, D.Y.; Im, J.N. Manufacturing and analyses of wet-laid nonwoven consisting of carboxymethyl cellulose fibers. *Fibers Polym.* **2013**, *14*, 2176–2184. [CrossRef]
177. Nitipir, C.; Marin, S.; Marin, M.M.; Kaya, M.A.; Ghica, M.V.; Mederle, N. Hybrid Collagen-NaCMC Matrices Loaded with Mefenamic Acid for Wound Healing. *Rev. Chim.* **2017**, *68*, 2605–2609. [CrossRef]
178. Leau, S.A.; Marin, S.; Coara, G.; Albu, L.; Constantinescu, R.R.; Kaya, M.A.; Neacsu, I.A. Study of wound-dressing materials based on collagen, sodium carboxymethylcellulose and silver nanoparticles used for their antibacterial activity in burn injuries. In Proceedings of the 7th International Conference on Advanced Materials and Systems, Bucharest, Romania, 18–20 October 2018; pp. 123–128.
179. Maver, T.; Kurecic, M.; Pivec, T.; Maver, U.; Gradisnik, L.; Gasparic, P.; Kaker, B.; Bratusa, A.; Hribernik, S.; Kleinschek, K.S. Needleless electrospun carboxymethyl cellulose/polyethylene oxide mats with medicinal plant extracts for advanced wound care applications. *Cellulose* **2020**, *27*, 4487–4508. [CrossRef]
180. Maver, T.; Kurecic, M.; Smrke, D.M.; Kleinschek, K.S.; Maver, U. Electrospun nanofibrous CMC/PEO as a part of an effective pain-relieving wound dressing. *J. Sol-Gel Sci. Technol.* **2016**, *79*, 475–486. [CrossRef]
181. Wang, F.J.; Shi, D.D.; Lan, T.; Zhang, Y.H.; Shao, Z.Q. One-step preparation of carboxymethyl cellulose/polyoxyethylene nanofiber mats containing silver nanoparticles. *Integr. Ferroelectr.* **2016**, *169*, 50–57. [CrossRef]
182. Oliveira, R.N.; Moreira, A.P.D.; Thire, R.; Quilty, B.; Passos, T.M.; Simon, P.; Mancini, M.C.; McGuinness, G.B. Absorbent Polyvinyl Alcohol-Sodium Carboxymethyl Cellulose Hydrogels for Propolis Delivery in Wound Healing Applications. *Polym. Eng. Sci.* **2017**, *57*, 1224–1233. [CrossRef]
183. Joorabloo, A.; Khorasani, M.T.; Adeli, H.; Mansoori-Moghadam, Z.; Moghaddam, A. Fabrication of heparinized nano ZnO/poly(vinylalcohol)/carboxymethyl cellulose bionanocomposite hydrogels using artificial neural network for wound dressing application. *J. Ind. Eng. Chem.* **2019**, *70*, 253–263. [CrossRef]
184. Pagano, C.; Calarco, P.; Di Michele, A.; Ceccarini, M.R.; Beccari, T.; Primavilla, S.; Scuota, S.; Marmottini, F.; Ramella, D.; Ricci, M.; et al. Development of sodium carboxymethyl cellulose based polymeric microparticles for in situ hydrogel wound dressing formation. *Int. J. Pharm.* **2021**, *602*, 120606. [CrossRef]
185. Maver, T.; Hribernik, S.; Mohan, T.; Smrke, D.M.; Maver, U.; Stana-Kleinschek, K. Functional wound dressing materials with highly tunable drug release properties. *RSC Adv.* **2015**, *5*, 77873–77884. [CrossRef]
186. Trevisol, T.C.; Scartazzini, L.; Valerio, A.; de Souza, S.; Bierhalz, A.C.K.; Valle, J.A.B. Diclofenac release from alginate/carboxymethyl cellulose mono and bilayer films for wound dressing applications. *Cellulose* **2020**, *27*, 6629–6642. [CrossRef]
187. Birsan, M.; Bibire, N.; Vieriu, M.; Panainte, A.D.; Cojocaru, I. Influence of Hydroxypropyl Methylcellulose on Flowing and Swelling Parameters in Biomucoadhesive Tablets with Miconazole Nitrate. *Rev. Chim.* **2017**, *68*, 2346–2349. [CrossRef]
188. Zhang, L.; Lu, Y.Q.; Qian, J.Y.; Yue, L.N.; Li, Q.; Xiao, L.X.; Ding, X.L.; Guan, C.R. Microstructures, physical and sustained antioxidant properties of hydroxypropyl methylcellulose based microporous photophobic films. *Int. J. Biol. Macromol.* **2020**, *152*, 1002–1009. [CrossRef]
189. Rao, B.L.; Shivananda, C.S.; Shetty, G.R.; Harish, K.V.; Madhukumar, R.; Sangappa, Y. Influence of UV Irradiation on Hydroxypropyl Methylcellulose Polymer Films. In Proceedings of the 2nd International Conference on Condensed Matter and Applied Physics (ICC), Bikaner, India, 24–25 November 2017.
190. Jayaramudu, T.; Varaprasad, K.; Pyarasani, R.D.; Reddy, K.K.; Akbari-Fakhrabadi, A.; Carrasco-Sanchez, V.; Amalraj, J. Hydroxypropyl methylcellulose-copper nanoparticle and its nanocomposite hydrogel films for antibacterial application. *Carbohydr. Polym.* **2021**, *254*, 117302. [CrossRef]
191. Akinosho, H.; Hawkins, S.; Wicker, L. Hydroxypropyl methylcellulose substituent analysis and rheological properties. *Carbohydr. Polym.* **2013**, *98*, 276–281. [CrossRef]
192. Tundisi, L.L.; Mostaco, G.B.; Carricondo, P.C.; Petri, D.F.S. Hydroxypropyl methylcellulose: Physicochemical properties and ocular drug delivery formulations. *Eur. J. Pharm. Sci.* **2021**, *159*, 105736. [CrossRef]

193. Kamel, R.; El-batanony, R.; Salama, A. Pioglitazone-loaded three-dimensional composite polymeric scaffolds: A proof of concept study in wounded diabetic rats. *Int. J. Pharm.* **2019**, *570*, 118667. [CrossRef]
194. Wang, T.; Chen, L.M.; Shen, T.T.; Wu, D.Y. Preparation and properties of a novel thermo-sensitive hydrogel based on chitosan/hydroxypropyl methylcellulose/glycerol. *Int. J. Biol. Macromol.* **2016**, *93*, 775–782. [CrossRef]
195. Bonetti, L.; De Nardo, L.; Fare, S. Thermo-Responsive Methylcellulose Hydrogels: From Design to Applications as Smart Biomaterials. *Tissue Eng. Part B Rev.* **2020**, *27*, 486–513. [CrossRef]
196. Siepmann, J.; Peppas, N.A. Modeling of drug release from delivery systems based on hydroxypropyl methylcellulose (HPMC). *Adv. Drug Deliv. Rev.* **2012**, *64*, 163–174. [CrossRef]
197. Kumari, S.; Harjai, K.; Chhibber, S. Topical treatment of Klebsiella pneumoniae B5055 induced burn wound infection in mice using natural products. *J. Infect. Dev. Ctries.* **2010**, *4*, 367–377. [CrossRef]
198. Ghadermazi, R.; Hamdipour, S.; Sadeghi, K.; Ghadermazi, R.; Asl, A.K. Effect of various additives on the properties of the films and coatings derived from hydroxypropyl methylcellulose—A review. *Food Sci. Nutr.* **2019**, *7*, 3363–3377. [CrossRef]
199. Liu, H.W.; Chaw, J.R.; Shih, Y.C.; Huang, C.C. Designed hydrocolloid interpenetrating polymeric networks for clinical applications of novel drug-carrying matrix systems using Tris (6-isocyanatohexyl) isocyanurate and hydroxypropylmethylcellulose. *Bio-Med. Mater. Eng.* **2014**, *24*, 2065–2072. [CrossRef]
200. Song, J.; Feng, H.; Wu, M.; Chen, L.; Xia, W.; Zhang, W. Preparation and characterization of arginine-modified chitosan/hydroxypropyl methylcellose antibacterial film. *Int. J. Biol. Macromol.* **2020**, *145*, 750–758. [CrossRef]
201. Yin, J.; Fang, Y.; Xu, L.; Ahmed, A. High-throughput fabrication of silk fibroin/hydroxypropyl methylcellulose (SF/HPMC) nanofibrous scaffolds for skin tissue engineering. *Int. J. Biol. Macromol.* **2021**, *183*, 1210–1221. [CrossRef]
202. Uslu, I.; Aytimur, A.; Serincay, H. Preparation of PVA/PAA/PEG/PVP Nanofibers with HPMC and Aloe Vera. *Curr. Nanosci.* **2013**, *9*, 489–493. [CrossRef]
203. Zhang, L.; Yue, L.N.; Qian, J.Y.; Ding, X.L. Effect of Curdlan on the Rheological Properties of Hydroxypropyl Methylcellulose. *Foods* **2021**, *10*, 34. [CrossRef]
204. Liu, X.X.; Ji, Z.L.; Peng, W.W.; Chen, M.; Yu, L.; Zhu, F. Chemical mapping analysis of compatibility in gelatin and hydroxypropyl methylcellulose blend films. *Food Hydrocoll.* **2020**, *104*, 105734. [CrossRef]
205. Wang, Y.F.; Yu, L.; Xie, F.W.; Li, S.; Sun, Q.J.; Liu, H.S.; Chen, L. On the investigation of thermal/cooling-gel biphasic systems based on hydroxypropyl methylcellulose and hydroxypropyl starch. *Ind. Crop. Prod.* **2018**, *124*, 418–428. [CrossRef]
206. Ding, C.C.; Zhang, M.; Li, G.Y. Rheological Properties of Collagen/Hydroxypropyl Methylcellulose (COL/HPMC) Blended Solutions. *J. Appl. Polym. Sci.* **2014**, *131*, 40042. [CrossRef]
207. Shao, X.R.; Sun, H.T.; Zhou, R.; Zhao, B.B.; Shi, J.F.; Jiang, R.P.; Dong, Y. Effect of bovine bone collagen and nano-TiO_2 on the properties of hydroxypropyl methylcellulose films. *Int. J. Biol. Macromol.* **2020**, *158*, 937–944. [CrossRef]
208. Barros, S.C.; da Silva, A.A.; Costa, D.B.; Costa, C.M.; Lanceros-Mendez, S.; Maciavello, M.N.T.; Ribelles, J.L.G.; Sentanin, F.; Pawlicka, A.; Silva, M.M. Thermal-mechanical behaviour of chitosan-cellulose derivative thermoreversible hydrogel films. *Cellulose* **2015**, *22*, 1911–1929. [CrossRef]
209. Ngwabebhoh, F.A.; Zandraa, O.; Patwa, R.; Saha, N.; Capakova, Z.; Saha, P. Self-crosslinked chitosan/dialdehyde xanthan gum blended hypromellose hydrogel for the controlled delivery of ampicillin, minocycline and rifampicin. *Int. J. Biol. Macromol.* **2021**, *167*, 1468–1478. [CrossRef]
210. da Silva, M.N.; Fonseca, J.D.; Feldhaus, H.K.; Soares, L.S.; Valencia, G.A.; de Campos, C.E.M.; Di Luccio, M.; Monteiro, A.R. Physical and morphological properties of hydroxypropyl methylcellulose films with curcumin polymorphs. *Food Hydrocoll.* **2019**, *97*, 105217. [CrossRef]
211. da Silva, J.B.; Cook, M.T.; Bruschi, M.L. Thermoresponsive systems composed of poloxamer 407 and HPMC or NaCMC: Mechanical, rheological and sol-gel transition analysis. *Carbohydr. Polym.* **2020**, *240*, 116268. [CrossRef]
212. Dong, T.; Mi, R.X.; Wu, M.; Zhong, N.P.; Zhao, X.; Chen, X.; Shao, Z.Z. The regenerated silk fibroin hydrogel with designed architecture bioprinted by its microhydrogel. *J. Mat. Chem. B* **2019**, *7*, 4328–4337. [CrossRef]
213. Tekko, I.A.; Chen, G.Y.; Dominguez-Robles, J.; Thakur, R.R.S.; Hamdan, I.M.N.; Vora, L.; Larraneta, E.; McElnay, J.C.; McCarthy, H.O.; Rooney, M.; et al. Development and characterisation of novel poly (vinyl alcohol)/poly (vinyl pyrrolidone)-based hydrogel-forming microneedle arrays for enhanced and sustained transdermal delivery of methotrexate. *Int. J. Pharm.* **2020**, *586*, 119580. [CrossRef]
214. Chen, Y.C.; Chen, Y.H. Thermo and pH-responsive methylcellulose and hydroxypropyl methylcellulose hydrogels containing K_2SO_4 for water retention and a controlled-release water-soluble fertilizer. *Sci. Total Environ.* **2019**, *655*, 958–967. [CrossRef]
215. Sun, G.H.; Liang, T.Q.; Tan, W.Y.; Wang, L.J. Rheological behaviors and physical properties of plasticized hydrogel films developed from kappa-carrageenan incorporating hydroxypropyl methylcellulose. *Food Hydrocoll.* **2018**, *85*, 61–68. [CrossRef]
216. Hu, M.; Yang, J.L.; Xu, J.H. Structural and biological investigation of chitosan/hyaluronic acid with silanized-hydroxypropyl methylcellulose as an injectable reinforced interpenetrating network hydrogel for cartilage tissue engineering. *Drug Deliv.* **2021**, *28*, 607–619. [CrossRef]
217. Bagheri, M.; Shokoohinia, Y.; Pourmanouchehri, Z.; Jalilian, F.; Khaledian, S.; Mirzaie, S.; Behbood, L. Formulation and evaluation of the novel herbal antibacterial gel to the treatment of cutaneous burn infections. *J. Rep. Pharm. Sci.* **2021**, *10*, 93–100. [CrossRef]
218. Wu, Y.C.; Wu, G.X.; Huang, H.H.; Kuo, S.M. Liposome-encapsulated farnesol accelerated tissue repair in third-degree burns on a rat model. *Burns* **2019**, *45*, 1139–1151. [CrossRef]

219. Chang, R.Y.K.; Morales, S.; Okamoto, Y.; Chan, H.-K. Topical application of bacteriophages for treatment of wound infections. *Transl. Res.* **2020**, *220*, 153–166. [CrossRef]
220. Saddik, M.S.; Alsharif, F.M.; El-Mokhtar, M.A.; Al-Hakkani, M.F.; El-Mahdy, M.M.; Farghaly, H.S.; Abou-Taleb, H.A. Biosynthesis, Characterization, and Wound-Healing Activity of Phenytoin-Loaded Copper Nanoparticles. *AAPS PharmSciTech* **2020**, *21*, 175. [CrossRef]
221. Zakaria, A.S.; Afifi, S.A.; Elkhodairy, K.A. Newly Developed Topical Cefotaxime Sodium Hydrogels: Antibacterial Activity and In Vivo Evaluation. *BioMed Res. Int.* **2016**, *2016*, 6525163. [CrossRef]
222. Febriyenti; Noor, A.M.; Baie, S.B. Mechanical properties and water vapour permeability of film from haruan (Channa striatus) and fusidic acid spray for wound dressing and wound healing. *Pak. J. Pharm. Sci.* **2010**, *23*, 155–159.
223. Huang, T.W.; Lu, H.T.; Ho, Y.C.; Lu, K.Y.; Wang, P.; Mi, F.L. A smart and active film with tunable drug release and color change abilities for detection and inhibition of bacterial growth. *Mater. Sci. Eng. C Mater. Biol. Appl.* **2021**, *118*, 111396. [CrossRef]
224. Claudio-Rizo, J.A.; Hernandez-Hernandez, N.G.; Cano-Salazar, L.F.; Flores-Guia, T.E.; de la Cruz-Duran, F.N.; Cabrera-Munguia, D.A.; Becerra-Rodriguez, J.J. Novel semi-interpenetrated networks based on collagen-polyurethane-polysaccharides in hydrogel state for biomedical applications. *J. Appl. Polym. Sci.* **2021**, *138*, 49739. [CrossRef]
225. Jiang, Q.; Zhou, W.; Wang, J.; Tang, R.P.; Zhang, D.; Wang, X. Hypromellose succinate-crosslinked chitosan hydrogel films for potential wound dressing. *Int. J. Biol. Macromol.* **2016**, *91*, 85–91. [CrossRef]
226. Chen, C.P.; Hsieh, C.M.; Tsai, T.; Yang, J.C.; Chen, C.T. Optimization and Evaluation of a Chitosan/Hydroxypropyl Methylcellulose Hydrogel Containing Toluidine Blue O for Antimicrobial Photodynamic Inactivation. *Int. J. Mol. Sci.* **2015**, *16*, 20859–20872. [CrossRef]
227. Anuroop, U.P.; Nair, S.S.; Chacko, A.J. Development and evaluation of medicated biodegradable film for wounds and burns. *Int. J. Pharm.l Sci. Res.* **2019**, *10*, 5664–5672. [CrossRef]
228. Pathan, I.B.; Munde, S.J.; Shelke, S.; Ambekar, W.; Setty, C.M. Curcumin loaded fish scale collagen-HPMC nanogel for wound healing application: Ex-vivo and In-vivo evaluation. *Int. J. Polym. Mater. Polym. Biomat.* **2019**, *68*, 165–174. [CrossRef]
229. Qiu, Y.M.; Sun, X.X.; Lin, X.L.; Yi, W.Y.; Jiang, J.Y. An injectable metal nanoparticle containing cellulose derivative-based hydrogels: Evaluation of antibacterial and in vitro-vivo wound healing activity in children with burn injuries. *Int. Wound J.* **2021**, *13*, 13664. [CrossRef] [PubMed]
230. Bianchi, S.E.; Machado, B.E.K.; da Silva, M.G.C.; da Silva, M.M.A.; Dal Bosco, L.; Marques, M.S.; Horn, A.P.; Persich, L.; Geller, F.C.; Argenta, D.; et al. Coumestrol/hydroxypropyl-beta-cyclodextrin association incorporated in hydroxypropyl methylcellulose hydrogel exhibits wound healing effect: In vitro and in vivo study. *Eur. J. Pharm. Sci.* **2018**, *119*, 179–188. [CrossRef] [PubMed]
231. Pinho, E.; Henriques, M.; Soares, G. Cyclodextrin/cellulose hydrogel with gallic acid to prevent wound infection. *Cellulose* **2014**, *21*, 4519–4530. [CrossRef]
232. Uslu, I.; Aytimur, A. Production and characterization of poly(vinyl alcohol)/poly(vinylpyrrolidone) iodine/poly(ethylene glycol) electrospun fibers with (hydroxypropyl)methyl cellulose and aloe vera as promising material for wound dressing. *J. Appl. Polym. Sci.* **2012**, *124*, 3520–3524. [CrossRef]
233. Lakshmi, P.K.; Thangellapalli, N.; Chennuri, A.; Prasanthi, D.; Veeresh, B. Wound healing activity of topical lawsone gel on rat model. *Int. J. Pharm. Sci. Res.* **2017**, *8*, 3162–3169. [CrossRef]
234. Abraham, S.; Harsha, G.G.S.; Desai, K.; Furtado, S.; Srinivasan, B. Nano Calcium Oxide Incorporated Hydrocolloid Dressings for Wound Care. *J. Pharm. Innov.* **2020**, 12. [CrossRef]
235. Kondaveeti, S.; Damato, T.C.; Carmona-Ribeiro, A.M.; Sierakowski, M.R.; Petri, D.F.S. Sustainable hydroxypropyl methylcellulose/xyloglucan/gentamicin films with antimicrobial properties. *Carbohydr. Polym.* **2017**, *165*, 285–293. [CrossRef]
236. Prabu, D.; Majdalawieh, A.F.; Abu-Yousef, I.A.; Inbasekaran, K.; Balasubramaniam, T.; Nallaperumal, N.; Gunasekar, C.J. Preparation and characterization of gatifloxacin-loaded sodium alginate hydrogel membranes supplemented with hydroxypropyl methylcellulose and hydroxypropyl cellulose polymers for wound dressing. *Int. J. Pharm. Investig.* **2016**, *6*, 86–95. [CrossRef]
237. Arafa, M.G.; El-Kased, R.F.; Elmazar, M.M. Thermoresponsive gels containing gold nanoparticles as smart antibacterial and wound healing agents. *Sci. Rep.* **2018**, *8*, 13674. [CrossRef]
238. Wu, Z.; Hong, Y.L. Combination of the Silver-Ethylene Interaction and 3D Printing to Develop Antibacterial Superporous Hydrogels for Wound Management. *ACS Appl. Mater. Interfaces* **2019**, *11*, 33734–33747. [CrossRef]
239. Grip, J.; Engstad, R.E.; Skjaeveland, I.; Skalko-Basnet, N.; Isaksson, J.; Basnet, P.; Holsaeter, A.M. Beta-glucan-loaded nanofiber dressing improves wound healing in diabetic mice. *Eur. J. Pharm. Sci.* **2018**, *121*, 269–280. [CrossRef] [PubMed]
240. Bilbao-Sainz, C.; Chiou, B.-S.; Valenzuela-Medina, D.; Du, W.-X.; Gregorski, K.S.; Williams, T.G.; Wood, D.F.; Glenn, G.M.; Orts, W.J. Solution blow spun poly(lactic acid)/hydroxypropyl methylcellulose nanofibers with antimicrobial properties. *Eur. Polym. J.* **2014**, *54*, 1–10. [CrossRef]
241. Namuiriyachote, N.; Lipipun, V.; Althhatuattananglzul, Y.; Charoonrut, P.; Ritthidej, G.C. Development of polyurethane foam dressing containing silver and asiaticoside for healing of dermal wound. *Asian J. Pharm. Sci.* **2019**, *14*, 63–77. [CrossRef] [PubMed]
242. Das, S.; De, A.; Das, B.; Mukherjee, B.; Samanta, A. Development of gum odina-gelatin based antimicrobial loaded biodegradable spongy scaffold: A promising wound care tool. *J. Appl. Polym. Sci.* **2021**, *138*, 50057. [CrossRef]
243. Wikene, K.O.; Hegge, A.B.; Bruzell, E.; Tonnesen, H.H. Formulation and characterization of lyophilized curcumin solid dispersions for antimicrobial photodynamic therapy (aPDT): Studies on curcumin and curcuminoids LII. *Drug Dev. Ind. Pharm.* **2015**, *41*, 969–977. [CrossRef]

244. Pinho, E.; Machado, S.; Soares, G. Smart Hydrogel for the pH-Selective Drug Delivery of Antimicrobial Compounds. In Proceedings of the 24th Polymer-Networks-Group Conference/82nd Prague Meeting on Macromolecules, Prague, Czech Republic, 17–21 June 2018.
245. Oliveira, D.M.L.; Rezende, P.S.; Barbosa, T.C.; Andrade, L.N.; Bani, C.; Tavares, D.S.; da Silva, C.F.; Chaud, M.V.; Padilha, F.; Cano, A.; et al. Double membrane based on lidocaine-coated polymyxin-alginate nanoparticles for wound healing: In vitro characterization and in vivo tissue repair. *Int. J. Pharm.* **2020**, *591*, 120001. [CrossRef]
246. Agubata, C.O.; Okereke, C.; Nzekwe, I.T.; Onoja, R.I.; Obitte, N.C. Development and evaluation of wound healing hydrogels based on a quinolone, hydroxypropyl methylcellulose and biodegradable microfibres. *Eur. J. Pharm. Sci.* **2016**, *89*, 1–10. [CrossRef]
247. Pišlová, M.; Kolářová, K.; Vokatá, B.; Brož, A.; Ulbrich, P.; Bačáková, L.; Kolská, Z.; Švorčík, V. A new way to prepare gold nanoparticles by sputtering—Sterilization, stability and other properties. *Mater. Sci. Eng. C* **2020**, *115*, 111087. [CrossRef]
248. Schuhladen, K.; Mukoo, P.; Liverani, L.; Nescakova, Z.; Boccaccini, A.R. Manuka honey and bioactive glass impart methylcellulose foams with antibacterial effects for wound-healing applications. *Biomed. Mater.* **2020**, *15*, 065002. [CrossRef]
249. Fattahpour, S.; Shamanian, M.; Tavakoli, N.; Fathi, M.; Sadeghi-Aliabadi, H.; Sheykhi, S.R.; Fesharaki, M.; Fattahpour, S. An injectable carboxymethyl chitosan-methylcellulose-pluronic hydrogel for the encapsulation of meloxicam loaded nanoparticles. *Int. J. Biol. Macromol.* **2020**, *151*, 220–229. [CrossRef]
250. Kar, A.K.; Shil, A.; Kar, B.; Dey, S. Formulation development and statistical optimization of zingiberol incorporated sodium alginate-methyl cellulose blend microspheres. *Int. J. Biol. Macromol.* **2020**, *162*, 1578–1586. [CrossRef] [PubMed]
251. Coughlin, M.L.; Liberman, L.; Ertem, S.P.; Edmund, J.; Bates, F.S.; Lodge, T.P. Methyl cellulose solutions and gels: Fibril formation and gelation properties. *Prog. Polym. Sci.* **2021**, *112*, 101324. [CrossRef]
252. Sangfai, T.; Tantishaiyakul, V.; Hirun, N.; Li, L. Microphase Separation and Gelation of Methylcellulose in the Presence of Gallic Acid and NaCl as an In Situ Gel-Forming Drug Delivery System. *AAPS PharmSciTech* **2017**, *18*, 605–616. [CrossRef] [PubMed]
253. El-Naggar, A.W.M.; Senna, M.M.; Mostafa, T.A.; Helal, R.H. Radiation synthesis and drug delivery properties of interpenetrating networks (IPNs) based on poly(vinyl alcohol)/methylcellulose blend hydrogels. *Int. J. Biol. Macromol.* **2017**, *102*, 1045–1051. [CrossRef] [PubMed]
254. Sharma, K.; Bullock, A.; Ralston, D.; MacNeil, S. Development of a one-step approach for the reconstruction of full thickness skin defects using minced split thickness skin grafts and biodegradable synthetic scaffolds as a dermal substitute. *Burns* **2014**, *40*, 957–965. [CrossRef] [PubMed]
255. Synytsya, A.; Grafova, M.; Slepicka, P.; Gedeon, O.; Synytsya, A. Modification of chitosan-methylcellulose composite films with meso-tetrakis(4-sulfonatophenyl)porphyrin. *Biomacromolecules* **2012**, *13*, 489–498. [CrossRef]
256. Abu, N.; Kasim, S.H.; Hisham, S.F.; Shamsudin, S.; Noorsal, K.; Mastor, A. Effect of Methylcellulose on the Hydrophilicity of Chitosan 3D-Porous Scaffold. *Adv. Mater. Res.* **2016**, *1133*, 55–59. [CrossRef]
257. Tan, W.; Zhang, J.; Zhao, X.; Li, Q.; Dong, F.; Guo, Z. Preparation and physicochemical properties of antioxidant chitosan ascorbate/methylcellulose composite films. *Int. J. Biol. Macromol.* **2020**, *146*, 53–61. [CrossRef]
258. Negim, E.S.M.; Nurpeissova, Z.A.; Mangazbayeva, R.A.; Khatib, J.M.; Williams, C.; Mun, G.A. Effect of pH on the physico-mechanical properties and miscibility of methyl cellulose/poly(acrylic acid) blends. *Carbohydr. Polym.* **2014**, *101*, 415–422. [CrossRef]
259. Varshosaz, J.; Sajadi-Javan, Z.S.; Kouhi, M.; Mirian, M. Effect of bassorin (derived from gum tragacanth) and halloysite nanotubes on physicochemical properties and the osteoconductivity of methylcellulose-based injectable hydrogels. *Int. J. Biol. Macromol.* **2021**, *192*, 869–882. [CrossRef]
260. Blinova, A.A.; Blinov, A.V.; Baklanova, O.A.; Yasnaya, M.A.; Baklanov, I.S.; Siddiqui, A.A.; Golik, A.B.; Okolelova, A.I.; Gvozdenko, A.A.; Simonov, A.N.; et al. Study of Wound-Healing Ointment Composition based on Highly Dispersed Zinc Oxide Modified with Nanoscale Silver. *Int. J. Pharm. Phytopharm. Res.* **2021**, *11*, 134–142. [CrossRef]
261. Ghodrati Azadi, H.; Fathi, B.; Kazemi Mehrjerdi, H.; Maleki, M.; Shaterzadeh, H.; Abyazi, M.J. Macroscopic evaluation of wound healing activity of the Persian shallot, Allium hirtifolium in rat. *Iran. J. Vet. Sci. Technol.* **2011**, *3*, 31–38.
262. Volkova, N.; Yukhta, M.; Pavlovich, O.; Goltsev, A. Application of Cryopreserved Fibroblast Culture with Au Nanoparticles to Treat Burns. *Nanoscale Res. Lett.* **2016**, *11*, 22. [CrossRef] [PubMed]
263. Kim, M.H.; Park, H.; Shin, J.Y.; Park, W.H. Effect of vitamin derivatives on gelation rate and gel strength of methylcellulose. *Carbohydr. Polym.* **2018**, *196*, 414–421. [CrossRef] [PubMed]
264. Kim, M.H.; Park, H.; Nam, H.C.; Park, S.R.; Jung, J.Y.; Park, W.H. Injectable methylcellulose hydrogel containing silver oxide nanoparticles for burn wound healing. *Carbohydr. Polym.* **2018**, *181*, 579–586. [CrossRef] [PubMed]
265. Sriwiroch, W.; Chungsamarnyart, N.; Chantakru, S.; Pongket, P.; Saengprapaitip, K.; Pongchairerk, U.J.A.; Resources, N. The effect of *Pedilanthus tithymaloides* (L.) Poit crude extract on wound healing stimulation in mice. *Agric. Nat. Resour.* **2010**, *44*, 1121–1127.
266. Pislova, M.; Jackova, T.; Fajstavr, D.; Kolarova, K.; Svorcik, V. *Release of Silver Nanoparticles from Polysaccharide Films Based on Methylcellulose*; Tanger Ltd.: Slezska, Czech Republic, 2019; pp. 517–522.
267. Jung, H.S.; Kim, H.C.; Park, W.H. Robust methylcellulose hydrogels reinforced with chitin nanocrystals. *Carbohydr. Polym.* **2019**, *213*, 311–319. [CrossRef]
268. Denisova, E.V.; Suprunchuk, V.E.; Dronova, A.A. Biopolymeric materials containing brown algae polysaccharides. *Entomol. Appl. Sci. Lett.* **2017**, *4*, 19–23. [CrossRef]

269. Das, B.; Basu, A.; Maji, S.; Dutta, K.; Dewan, M.; Adhikary, A.; Maiti, T.K.; Chattopadhyay, D. Nanotailored hyaluronic acid modified methylcellulose as an injectable scaffold with enhanced physico-rheological and biological aspects. *Carbohydr. Polym.* **2020**, *237*, 116146. [CrossRef]
270. Nowald, C.; Penk, A.; Chiu, H.Y.; Bein, T.; Huster, D.; Lieleg, O. A Selective Mucin/Methylcellulose Hybrid Gel with Tailored Mechanical Properties. *Macromol. Biosci.* **2016**, *16*, 567–579. [CrossRef]
271. Huang, Y.; Guo, W.; Zhang, J.; Peng, X.; Li, G.; Zhang, L.-M.; Yang, L. Thermosensitive hydrogels based on methylcellulose derivatives for prevention of postoperative adhesion. *Cellulose* **2019**, *27*, 1555–1571. [CrossRef]
272. Laffleur, F.; Strasdat, B. Gelatin-based formulations for dermal application. *Eur. Polym. J.* **2019**, *118*, 542–550. [CrossRef]
273. Yeo, Y.H.; Park, W.H. Dual-crosslinked, self-healing and thermo-responsive methylcellulose/chitosan oligomer copolymer hydrogels. *Carbohydr. Polym.* **2021**, *258*, 117705. [CrossRef] [PubMed]
274. Wang, C.Y.; Liang, C.Y.; Wang, R.; Yao, X.L.; Guo, P.; Yuan, W.Z.; Liu, Y.; Song, Y.; Li, Z.H.; Xie, X.Y. The fabrication of a highly efficient self-healing hydrogel from natural biopolymers loaded with exosomes for the synergistic promotion of severe wound healing. *Biomater. Sci.* **2020**, *8*, 313–324. [CrossRef]
275. Mayol, L.; De Stefano, D.; De Falco, F.; Carnuccio, R.; Maiuri, M.C.; De Rosa, G. Effect of hyaluronic acid on the thermogelation and biocompatibility of its blends with methyl cellulose. *Carbohydr. Polym.* **2014**, *112*, 480–485. [CrossRef]
276. Jiang, J.; Han, X.; Xu, H. Injectable Methylcellulose and Hyaluronic Acid Hydrogel Containing Silver Nanoparticles for Their Effective Anti-Microbial and Wound Healing Activity in Nursing Care for Burn Injuries in Children. *J. Polym. Environ.* **2021**. under review. [CrossRef]
277. Kim, E.J.; Choi, J.S.; Kim, J.S.; Choi, Y.C.; Cho, Y.W. Injectable and Thermosensitive Soluble Extracellular Matrix and Methylcellulose Hydrogels for Stem Cell Delivery in Skin Wounds. *Biomacromolecules* **2016**, *17*, 4–11. [CrossRef]
278. Nagai, N.; Ogata, F.; Deguchi, S.; Ueno, A.; Kawasaki, N.; Ito, Y.J.B.; Bulletin, P. Combination ointment containing solid tranilast nanoparticles and dissolved sericin is efficacious for treating skin wound-healing deficits and redness in diabetic rats. *Biol. Pharm. Bull.* **2017**, *40*, 444–450. [CrossRef]
279. Park, C.H.; Jeong, L.; Cho, D.; Kwon, O.H.; Park, W.H. Effect of methylcellulose on the formation and drug release behavior of silk fibroin hydrogel. *Carbohydr. Polym.* **2013**, *98*, 1179–1185. [CrossRef]
280. Karavasili, C.; Tsongas, K.; Andreadis, I.I.; Andriotis, E.G.; Papachristou, E.T.; Papi, R.M.; Tzetzis, D.; Fatouros, D.G. Physico-mechanical and finite element analysis evaluation of 3D printable alginate-methylcellulose inks for wound healing applications. *Carbohydr. Polym.* **2020**, *247*, 116666. [CrossRef]
281. Rastin, H.; Ramezanpour, M.; Hassan, K.; Mazinani, A.; Tung, T.T.; Vreugde, S.; Losic, D. 3D bioprinting of a cell-laden antibacterial polysaccharide hydrogel composite. *Carbohydr. Polym.* **2021**, *264*, 117989. [CrossRef]
282. Mishra, R.K.; Ramasamy, K.; Lim, S.M.; Ismail, M.F.; Majeed, A.B. Antimicrobial and in vitro wound healing properties of novel clay based bionanocomposite films. *J. Mater. Sci. Mater. Med.* **2014**, *25*, 1925–1939. [CrossRef] [PubMed]
283. Niziol, M.; Paleczny, J.; Junka, A.; Shavandi, A.; Dawiec-Lisniewska, A.; Podstawczyk, D. 3D Printing of Thermoresponsive Hydrogel Laden with an Antimicrobial Agent towards Wound Healing Applications. *Bioengineering* **2021**, *8*, 79. [CrossRef] [PubMed]
284. Lan, B.Y.; Zhang, L.M.; Yang, L.Q.; Wu, J.F.; Li, N.; Pan, C.L.; Wang, X.Y.; Zeng, L.X.; Yan, L.; Yang, C.; et al. Sustained delivery of MMP-9 siRNA via thermosensitive hydrogel accelerates diabetic wound healing. *J. Nanobiotechnol.* **2021**, *19*, 130. [CrossRef] [PubMed]
285. Gyles, D.A.; Pereira, A.D.; Castro, L.D.; Santa Brigida, A.; Lamarao, M.L.N.; Barbosa, W.L.R.; Silva, J.O.C.; Ribeiro-Costa, R.M. Polyacrylamide-Metilcellulose Hydrogels Containing Aloe barbadensis Extract as Dressing for Treatment of Chronic Cutaneous Skin Lesions. *Polymers* **2020**, *12*, 690. [CrossRef]
286. Schuhladen, K.; Raghu, S.N.V.; Liverani, L.; Nescakova, Z.; Boccaccini, A.R. Production of a novel poly(e-caprolactone)-methylcellulose electrospun wound dressing by incorporating bioactive glass and Manuka honey. *J. Biomed. Mater. Res. Part B* **2021**, *109*, 180–192. [CrossRef]
287. Zhang, H.Y.; Xie, Y.W.; Tang, Y.; Ni, S.H.; Wang, B.; Chen, Z.; Liu, X.J. Development and characterization of thermo-sensitive films containing asiaticoside based on polyvinyl alcohol and Methylcellulose. *J. Drug Deliv. Sci. Technol.* **2015**, *30*, 133–145. [CrossRef]
288. Yadav, B.K.N.; Patel, G.C. Fabrication and characterization of coblended methyl cellulose with polyvinyl alcohol electrospun nanofibers as a carrier for drug delivery system. *Polym. Bull.* **2021**, *20*, 1–29. [CrossRef]
289. Sun, N.; Wang, T.; Yan, X. Self-assembled supermolecular hydrogel based on hydroxyethyl cellulose: Formation, in vitro release and bacteriostasis application. *Carbohydr. Polym.* **2017**, *172*, 49–59. [CrossRef]
290. Lin, P.J.; Liu, L.L.; He, G.H.; Zhang, T.; Yang, M.; Cai, J.Z.; Fan, L.H.; Tao, S.X. Preparation and properties of carboxymethyl chitosan/oxidized hydroxyethyl cellulose hydrogel. *Int. J. Biol. Macromol.* **2020**, *162*, 1692–1698. [CrossRef]
291. Noreen, A.; Zia, K.M.; Tabasum, S.; Khalid, S.; Shareef, R. A review on grafting of hydroxyethylcellulose for versatile applications. *Int. J. Biol. Macromol.* **2020**, *150*, 289–303. [CrossRef]
292. Yamane, L.T.; de Paula, E.; Jorge, M.P.; de Freitas-Blanco, V.S.; Montanari, I.; Figueira, G.M.; Anholeto, L.A.; de Oliveira, P.R.; Rodrigues, R.A. Acmella oleracea and Achyrocline satureioides as Sources of Natural Products in Topical Wound Care. *Evid.-Based Complement. Altern. Med.* **2016**, *2016*, 3606820. [CrossRef] [PubMed]
293. Rosic, R.; Kocbek, P.; Baumgartner, S.; Kristl, J. Electro-spun hydroxyethyl cellulose nanofibers: The relationship between structure and process. *J. Drug Deliv. Sci. Technol.* **2011**, *21*, 229–236. [CrossRef]

294. Raafat, A.I.; Ali, A.E.H. A novel Lawsonia inermis (Henna)/(hydroxyethylcellulose/polyvinylpyrrolidone) wound dressing hydrogel: Radiation synthesis, characterization and biological evaluation. *Polym. Bull.* **2019**, *76*, 4069–4086. [CrossRef]
295. Luo, P.F.; Liu, L.L.; Xu, W.Y.; Fan, L.H.; Nie, M. Preparation and characterization of aminated hyaluronic acid/oxidized hydroxyethyl cellulose hydrogel. *Carbohydr. Polym.* **2018**, *199*, 170–177. [CrossRef]
296. Ma, Y.L.; Zhou, G.; Ding, J.F.; Li, S.L.; Wang, G. Preparation and characterization of an agglomeration-cementing agent for dust suppression in open pit coal mining. *Cellulose* **2018**, *25*, 4011–4029. [CrossRef]
297. Ning, F.; Zhang, J.; Kang, M.X.; Ma, C.P.; Li, H.; Qiu, Z.M. Hydroxyethyl cellulose hydrogel modified with tannic acid as methylene blue adsorbent. *J. Appl. Polym. Sci.* **2021**, *138*, 49880. [CrossRef]
298. Kang, M.; Oderinde, O.; Han, X.; Fu, G.; Zhang, Z. Development of oxidized hydroxyethyl cellulose-based hydrogel enabling unique mechanical, transparent and photochromic properties for contact lenses. *Int. J. Biol. Macromol.* **2021**, *183*, 1162–1173. [CrossRef]
299. Zulkifli, F.H.; Shahitha, F.; Yusuff, M.M.; Hamidon, N.N.; Chahal, S. Cross-Linking Effect on Electrospun Hydroxyethyl Cellulose/Poly(Vinyl Alcohol) Nanofibrous Scaffolds. *Procedia Eng.* **2013**, *53*, 689–695. [CrossRef]
300. El Fawal, G.F.; Abu-Serie, M.M.; Hassan, M.A.; Elnouby, M.S. Hydroxyethyl cellulose hydrogel for wound dressing: Fabrication, characterization and in vitro evaluation. *Int. J. Biol. Macromol.* **2018**, *111*, 649–659. [CrossRef]
301. Taheri, H.; Hietala, M.; Oksman, K. One-step twin-screw extrusion process of cellulose fibers and hydroxyethyl cellulose to produce fibrillated cellulose biocomposite. *Cellulose* **2020**, *27*, 8105–8119. [CrossRef]
302. Fekete, T.; Borsa, J.; Takács, E.; Wojnárovits, L. Synthesis and characterization of superabsorbent hydrogels based on hydroxyethylcellulose and acrylic acid. *Carbohydr. Polym.* **2017**, *166*, 300–308. [CrossRef] [PubMed]
303. Zia, F.; Zia, K.M.; Aftab, W.; Tabasum, S.; Nazli, Z.-i.-H.; Mohammadi, M.; Zuber, M. Synthesis and characterization of graphene nanoplatelets-hydroxyethyl cellulose copolymer-based polyurethane bionanocomposite system. *Int. J. Biol. Macromol.* **2020**, *165*, 1889–1899. [CrossRef] [PubMed]
304. Zia, F.; Zia, K.M.; Nazli, Z.-i.-H.; Tabasum, S.; Khosa, M.K.; Zuber, M. Preparation of hydroxyethyl cellulose/halloysite nanotubes graft polylactic acid-based polyurethane bionanocomposites. *Int. J. Biol. Macromol.* **2020**, *153*, 591–599. [CrossRef]
305. Huang, S.; Shuyi, S.; Gan, H.; Linjun, W.; Lin, C.; Danyuan, X.; Zhou, H.; Lin, X.; Qin, Y. Facile fabrication and characterization of highly stretchable lignin-based hydroxyethyl cellulose self-healing hydrogel. *Carbohydr. Polym.* **2019**, *223*, 115080. [CrossRef]
306. Guo, M.; Yin, X.X.; Wang, P. C-13 NMR Characteristics and Thermal Properties of Hydroxyethyl Cellulose Grafting Poly(caprolactone) Copolymer. *Asian J. Chem.* **2013**, *25*, 4501–4505. [CrossRef]
307. Demina, T.S.; Birdibekova, A.V.; Svidchenko, E.A.; Ivanov, P.L.; Kuryanova, A.S.; Kurkin, T.S.; Khaibullin, Z.I.; Goncharuk, G.P.; Zharikova, T.M.; Bhuniya, S.; et al. Solid-State Synthesis of Water-Soluble Chitosan-g-Hydroxyethyl Cellulose Copolymers. *Polymers* **2020**, *12*, 611. [CrossRef]
308. Balaita, L.; Peptu, C.A.; Postolache, P.; Lisa, G.; Popa, M. Gelatin-hydroxyethyl cellulose magnetic microparticles as drug carriers: Preparation and characterization. *J. Optoelectron. Adv. Mater.* **2012**, *14*, 1023–1033.
309. Jamlang, J.; Marañon, M.T.; Rigor, G.J.; Tumolva, T. Developing a physically cross-linked hydroxyethyl cellulose hydrogel for wound dressing applications. In *Materials Science Forum*; Trans Tech Publications Ltd.: Kapellweg, Switzerland, 2019; pp. 3–12.
310. Arafa, A.A.; Nada, A.A.; Ibrahim, A.Y.; Zahran, M.K.; Hakeim, O.A. Greener therapeutic pH-sensing wound dressing based on Curcuma Longa and cellulose hydrogel. *Eur. Polym. J.* **2021**, *159*, 110744. [CrossRef]
311. Wang, C.W.; Niu, H.Y.; Ma, X.Y.; Hong, H.; Yuan, Y.; Liu, C.S. Bioinspired, Injectable, Quaternized Hydroxyethyl Cellulose Composite Hydrogel Coordinated by Mesocellular Silica Foam for Rapid, Noncompressible Hemostasis and Wound Healing. *ACS Appl. Mater. Interfaces* **2019**, *11*, 34595–34608. [CrossRef]
312. Zhang, Z.L.; Yang, R.D. Novel Nanocomposites Based on Hydroxyethyl Cellulose and Graphene Oxide. *Fiber Polym.* **2017**, *18*, 334–341. [CrossRef]
313. Zulkifli, F.H.; Hussain, F.S.J.; Zeyohannes, S.S.; Rasad, M.S.B.A.; Yusuff, M.M. A facile synthesis method of hydroxyethyl cellulose-silver nanoparticle scaffolds for skin tissue engineering applications. *Mater. Sci. Eng. C* **2017**, *79*, 151–160. [CrossRef] [PubMed]
314. Anumary, A.; Thanikaivelan, P.; Ashokkumar, M.; Kumar, R.; Sehgal, P.K.; Chandrasekaran, B. Synthesis and Characterization of Hybrid Biodegradable Films From Bovine Hide Collagen and Cellulose Derivatives for Biomedical Applications. *Soft Mater.* **2013**, *11*, 181–194. [CrossRef]
315. Zulkifli, F.H.; Jahir Hussain, F.S.; Abdull Rasad, M.S.B.; Mohd Yusoff, M. In vitro degradation study of novel HEC/PVA/collagen nanofibrous scaffold for skin tissue engineering applications. *Polym. Degrad. Stab.* **2014**, *110*, 473–481. [CrossRef]
316. Tohamy, K.M.; Soliman, I.E.; Mabrouk, M.; ElShebiney, S.; Beherei, H.H.; Aboelnasr, M.A.; Das, D.B. Novel polysaccharide hybrid scaffold loaded with hydroxyapatite: Fabrication, bioactivity, and in vivo study. *Mater. Sci. Eng. C* **2018**, *93*, 1–11. [CrossRef]
317. Kwon, S.S.; Kong, B.J.; Park, S.N. Physicochemical properties of pH-sensitive hydrogels based on hydroxyethyl cellulose-hyaluronic acid and for applications as transdermal delivery systems for skin lesions. *Eur. J. Pharm. Biopharm.* **2015**, *92*, 146–154. [CrossRef] [PubMed]
318. Li, H.Y.; Zhang, Z.W.; Godakanda, V.U.; Chiu, Y.J.; Angkawinitwong, U.; Patel, K.; Stapleton, P.G.; de Silva, R.M.; de Silva, K.M.N.; Zhu, L.M.; et al. The effect of collection substrate on electrospun ciprofloxacin-loaded poly (vinylpyrrolidone) and ethyl cellulose nanofibers as potential wound dressing materials. *Mater. Sci. Eng. C-Mater. Biol. Appl.* **2019**, *104*, 109917. [CrossRef]

319. Zhang, Y.; Sadgrove, M.P.; Mumper, R.J.; Jay, M. Transdermal Prodrug Delivery for Radionuclide Decorporation: Nonaqueous Gel Formulation Development and In Vitro and In Vivo Assessment. *Drug Dev. Res.* **2013**, *74*, 322–331. [CrossRef]
320. Lin, Y.M.; Li, M.; Xia, J.L.; Ding, H.Y.; Xu, L.N.; Yang, X.H.; Li, S.H. Synthesis of plant oil derived polyols and their effects on the properties of prepared ethyl cellulose composite films. *Cellulose* **2021**, *28*, 4211–4222. [CrossRef]
321. Ghorbani, M.; Ramezani, S.; Rashidi, M.R. Fabrication of honey-loaded ethylcellulose/gum tragacanth nanofibers as an effective antibacterial wound dressing. *Colloid Surf. A-Physicochem. Eng. Asp.* **2021**, *621*, 126615. [CrossRef]
322. Xu, J.Y.; Jia, H.G.; Yang, N.; Wang, Q.J.; Yang, G.X.; Zhang, M.Y.; Xu, S.P.; Zang, Y.; Ma, L.Q.; Jiang, P.F.; et al. High Efficiency Gas Permeability Membranes from Ethyl Cellulose Grafted with Ionic Liquids. *Polymers* **2019**, *11*, 1900. [CrossRef] [PubMed]
323. Mahnaj, T.; Ahmed, S.U.; Plakogiannis, F.M. Characterization of ethyl cellulose polymer. *Pharm. Dev. Technol.* **2013**, *18*, 982–989. [CrossRef] [PubMed]
324. Zhao, X.Y.; Yan, G.H.; Sun, Y.; Tang, X.; Zeng, X.H.; Lin, L.; Lei, T.Z. Preparation of Ethyl Cellulose Composite Film with Down Conversion Luminescence Properties by Doping Perovskite Quantum Dots. *ChemistrySelect* **2019**, *4*, 6516–6523. [CrossRef]
325. Djerafi, R.; Masmoudi, Y.; Crampon, C.; Meniai, A.; Badens, E. Supercritical anti-solvent precipitation of ethyl cellulose. *J. Supercrit. Fluids* **2015**, *105*, 92–98. [CrossRef]
326. Suksaeree, J.; Thuengernthong, A.; Pongpichayasiri, K.; Maneewattanapinyo, P.; Settharaksa, S.; Pichayakorn, W. Formulation and Evaluation of Matrix Type Transdermal Patch Containing Silver Nanoparticles. *J. Polym. Environ.* **2018**, *26*, 4369–4375. [CrossRef]
327. Su, X.C.; Yang, Z.; Tan, K.B.; Chen, J.F.; Huang, J.L.; Li, Q.B. Preparation and characterization of ethyl cellulose film modified with capsaicin. *Carbohydr. Polym.* **2020**, *241*, 116259. [CrossRef]
328. Li, H.Y.; Liu, K.L.; Sang, Q.Q.; Williams, G.R.; Wu, J.Z.; Wang, H.J.; Wu, J.R.; Zhu, L.M. A thermosensitive drug delivery system prepared by blend electrospinning. *Colloid Surf. B Biointerfaces* **2017**, *159*, 277–283. [CrossRef]
329. Girija, B.G.; Sailaja, R.R.N.; Biswas, S.; Deepthi, M.V. Mechanical and Thermal Properties of Eva Blended with Biodegradable Ethyl Cellulose. *J. Appl. Polym. Sci.* **2010**, *116*, 1044–1056. [CrossRef]
330. Chen, J.X.; Wu, D.F.; Pan, K.R. Effects of ethyl cellulose on the crystallization and mechanical properties of poly(beta-hydroxybutyrate). *Int. J. Biol. Macromol.* **2016**, *88*, 120–129. [CrossRef]
331. Li, X.; Jiang, F.T.; Ni, X.W.; Yan, W.L.; Fang, Y.P.; Corke, H.; Xiao, M. Preparation and characterization of konjac glucomannan and ethyl cellulose blend films. *Food Hydrocoll.* **2015**, *44*, 229–236. [CrossRef]
332. Borujeni, S.H.; Mirdamadian, S.Z.; Varshosaz, J.; Taheri, A. Three-dimensional (3D) printed tablets using ethyl cellulose and hydroxypropyl cellulose to achieve zero order sustained release profile. *Cellulose* **2020**, *27*, 1573–1589. [CrossRef]
333. Godakanda, V.U.; Li, H.Y.; Alquezar, L.; Zhao, L.X.; Zhu, L.M.; de Silva, R.; de Silva, K.M.N.; Williams, G.R. Tunable drug release from blend poly(vinyl pyrrolidone)-ethyl cellulose nanofibers. *Int. J. Pharm.* **2019**, *562*, 172–179. [CrossRef] [PubMed]
334. Yang, J.K.; Wang, K.; Yu, D.G.; Yang, Y.Y.; Bligh, S.W.A.; Williams, G.R. Electrospun Janus nanofibers loaded with a drug and inorganic nanoparticles as an effective antibacterial wound dressing. *Mater. Sci. Eng. C Mater. Biol. Appl.* **2020**, *111*, 110805. [CrossRef] [PubMed]
335. Iqbal, H.M.N.; Kyazze, G.; Locke, I.C.; Tron, T.; Keshavarz, T. Poly(3-hydroxybutyrate)-ethyl cellulose based bio-composites with novel characteristics for infection free wound healing application. *Int. J. Biol. Macromol.* **2015**, *81*, 552–559. [CrossRef]
336. Ahmadian, S.; Ghorbani, M.; Mahmoodzadeh, F. Silver sulfadiazine-loaded electrospun ethyl cellulose/polylactic acid/collagen nanofibrous mats with antibacterial properties for wound healing. *Int. J. Biol. Macromol.* **2020**, *162*, 1555–1565. [CrossRef] [PubMed]
337. Qiu, Y.; Wang, Q.Q.; Chen, Y.J.; Xia, S.F.; Huang, W.; Wei, Q.F. A Novel Multilayer Composite Membrane for Wound Healing in Mice Skin Defect Model. *Polymers* **2020**, *12*, 573. [CrossRef]
338. Kapileshwari, G.R.; Barve, A.R.; Kumar, L.; Bhide, P.J.; Joshi, M.; Shirodkar, R.K. Novel drug delivery system of luliconazole—Formulation and characterisation. *J. Drug Deliv. Sci. Technol.* **2020**, *55*, 101302. [CrossRef]
339. Ogawa, A.; Nakayama, S.; Uehara, M.; Mori, Y.; Takahashi, M.; Aiba, T.; Kurosaki, Y. Pharmaceutical properties of a low-substituted hydroxypropyl cellulose (L-HPC) hydrogel as a novel external dressing. *Int. J. Pharm.* **2014**, *477*, 546–552. [CrossRef]
340. Bhatt, N.; Gupta, P.M.; Naithani, S. Hydroxypropyl cellulose from alpha-cellulose isolated from Lantana camara with respect to DS and rheological behavior. *Carbohydr. Polym.* **2011**, *86*, 1519–1524. [CrossRef]
341. Lin, C.M.; Chang, Y.C.; Cheng, L.C.; Liu, C.H.; Chang, S.C.; Hsien, T.Y.; Wang, D.M.; Hsieh, H.J. Preparation of graphene-embedded hydroxypropyl cellulose/chitosan/polyethylene oxide nanofiber membranes as wound dressings with enhanced antibacterial properties. *Cellulose* **2020**, *27*, 2651–2667. [CrossRef]
342. Miao, L.; Hu, J.W.; Lu, M.G.; Tu, Y.Y.; Chen, X.; Li, Y.W.; Lin, S.D.; Li, F.; Hu, S.Y. Alkynyl-functionalization of hydroxypropyl cellulose and thermoresponsive hydrogel thereof prepared with P(NIPAAm-co-HEMAPCL). *Carbohydr. Polym.* **2016**, *137*, 433–440. [CrossRef]
343. Szabo, P.; Kallai-Szabo, B.; Kallai-Szabo, N.; Sebe, I.; Zelko, R. Preparation of hydroxypropyl cellulose microfibers by high-speed rotary spinning and prediction of the fiber-forming properties of hydroxypropyl cellulose gels by texture analysis. *Cellulose* **2014**, *21*, 4419–4427. [CrossRef]
344. Ishii, D.; Ueda, K.; Stroeve, P.; Nakaoki, T.; Hayashi, H. Transport properties of chemically crosslinked hydroxypropyl cellulose in solvated state. *Cell Chem. Technol.* **2016**, *50*, 755–760.
345. Ma, L.; Wang, L.L.; Wu, L.X.; Zhuo, D.X.; Weng, Z.X.; Ren, R.R. Cellulosic nanocomposite membranes from hydroxypropyl cellulose reinforced by cellulose nanocrystals. *Cellulose* **2014**, *21*, 4443–4454. [CrossRef]

346. El-Newehy, M.H.; El-Naggar, M.E.; Alotaiby, S.; El-Hamshary, H.; Moydeen, M.; Al-Deyab, S. Green Electrospining of Hydroxypropyl Cellulose Nanofibres for Drug Delivery Applications. *J. Nanosci. Nanotechnol.* **2018**, *18*, 805–814. [CrossRef]
347. Saraiva, D.V.; Chagas, R.; de Abreu, B.M.; Gouveia, C.N.; Silva, P.E.S.; Godinho, M.H.; Fernandes, S.N. Flexible and Structural Coloured Composite Films from Cellulose Nanocrystals/Hydroxypropyl Cellulose Lyotropic Suspensions. *Crystals* **2020**, *10*, 122. [CrossRef]
348. Bielska, D.; Karewicz, A.; Kaminski, K.; Kielkowicz, I.; Lachowicz, T.; Szczubialka, K.; Nowakowska, M. Self-organized thermo-responsive hydroxypropyl cellulose nanoparticles for curcumin delivery. *Eur. Polym. J.* **2013**, *49*, 2485–2494. [CrossRef]
349. El-Wakil, N.A.; Kassem, N.F.; Hassan, M.L. Hydroxypropyl cellulose/rice straw oxidized cellulose nanocrystals nanocomposites and their use in paper coating. *Ind. Crop. Prod.* **2016**, *93*, 186–192. [CrossRef]
350. Sudo, S. Dielectric Properties of the Free Water in Hydroxypropyl Cellulose. *J. Phys. Chem. B* **2011**, *115*, 2–6. [CrossRef] [PubMed]
351. Veerapur, R.S.; Gudasi, K.B.; Aminabhavi, T.M. Pervaporation dehydration of isopropanol using blend membranes of chitosan and hydroxypropyl cellulose. *J. Membr. Sci.* **2007**, *304*, 102–111. [CrossRef]
352. Onofrei, M.D.; Dobos, A.M.; Stoica, I.; Olaru, N.; Olaru, L.; Ioan, S. Lyotropic Liquid Crystal Phases in Cellulose Acetate Phthalate/Hydroxypropyl Cellulose Blends. *J. Polym. Environ.* **2014**, *22*, 99–111. [CrossRef]
353. Gan, S.C.; Bai, S.H.; Chen, C.; Zou, Y.L.; Sun, Y.J.; Zhao, J.H.; Rong, J.H. Hydroxypropyl cellulose enhanced ionic conductive double-network hydrogels. *Int. J. Biol. Macromol.* **2021**, *181*, 418–425. [CrossRef] [PubMed]
354. Lu, C.W.; Wang, C.P.; Zhang, D.H.; Wang, J.F.; Yong, Q.; Chu, F.X. Ultra-strong hydroxypropyl cellulose/polyvinyl alcohol composite hydrogel by combination of triple-network and mechanical training. *Int. J. Biol. Macromol.* **2021**, *184*, 200–208. [CrossRef] [PubMed]
355. Bjorn, C.; Noppa, L.; Salomonsson, E.N.; Johansson, A.L.; Nilsson, E.; Mahlapuu, M.; Hakansson, J. Efficacy and safety profile of the novel antimicrobial peptide PXL150 in a mouse model of infected burn wounds. *Int. J. Antimicrob. Agents* **2015**, *45*, 519–524. [CrossRef] [PubMed]
356. Wang, Y.W.; Shi, L.J.; Wu, H.P.; Li, Q.Y.; Hu, W.; Zhang, Z.B.; Huang, L.H.; Zhang, J.X.; Chen, D.J.; Deng, S.P.; et al. Graphene Oxide-IPDI-Ag/ZnO@Hydroxypropyl Cellulose Nanocomposite Films for Biological Wound-Dressing Applications. *ACS Omega* **2019**, *4*, 15373–15381. [CrossRef]
357. Wang, J.J.; Wei, J. Interpenetrating network hydrogels with high strength and transparency for potential use as external dressings. *Mater. Sci. Eng. C Mater. Biol. Appl.* **2017**, *80*, 460–467. [CrossRef]
358. Rojewska, A.; Karewicz, A.; Karnas, K.; Wolski, K.; Zajac, M.; Kaminski, K.; Szczubialka, K.; Zapotoczny, S.; Nowakowska, M. Pioglitazone-Loaded Nanostructured Hybrid Material for Skin Ulcer Treatment. *Materials* **2020**, *13*, 2050. [CrossRef]
359. Tolouei, S.; Hasheminia, S.J.; Narimani, M.; Khamesipour, A.; Shatalebi, M.A.; Hejazi, S.H. Leishmanicidal Activity of Films Containing Paromomycin and Gentamicin Sulfate both In Vitro and In Vivo. *Iran. J. Parasitol.* **2011**, *6*, 60–65.
360. Mohebian, Z.; Tajmohammadi, I.; Maroufi, L.Y.; Ramezani, S.; Ghorbani, M. A Novel Aloe Vera-Loaded Ethylcellulose/Hydroxypropyl Methylcellulose Nanofibrous Mat Designed for Wound Healing Application. *J. Polym. Environ.* **2021**, 11. [CrossRef]
361. Koneru, A.; Dharmalingam, K.; Anandalakshmi, R. Cellulose based nanocomposite hydrogel films consisting of sodium carboxymethylcellulose-grapefruit seed extract nanoparticles for potential wound healing applications. *Int. J. Biol. Macromol.* **2020**, *148*, 833–842. [CrossRef]
362. Dharmalingam, K.; Bordoloi, D.; Kunnumakkara, A.B.; Anandalakshmi, R. Preparation and characterization of cellulose-based nanocomposite hydrogel films containing CuO/Cu2O/Cu with antibacterial activity. *J. Appl. Polym. Sci.* **2020**, *137*, 49216. [CrossRef]
363. Dharmalingam, K.; Anandalakshmi, R. Fabrication, characterization and drug loading efficiency of citric acid crosslinked NaCMC-HPMC hydrogel films for wound healing drug delivery applications. *Int. J. Biol. Macromol.* **2019**, *134*, 815–829. [CrossRef] [PubMed]
364. Dharmalingam, K.; Bordoloi, D.; Kunnumakkara, A.B.; Anandalakshmi, R. Formation and characterization of zinc oxide complexes in composite hydrogel films for potential wound healing applications. *Polym. Compos.* **2020**, *41*, 2274–2287. [CrossRef]
365. Amanat, S.; Taymouri, S.; Varshosaz, J.; Minaiyan, M.; Talebi, A. Carboxymethyl cellulose-based wafer enriched with resveratrol-loaded nanoparticles for enhanced wound healing. *Drug Deliv. Transl. Res.* **2020**, *10*, 1241–1254. [CrossRef]
366. Rezvanian, M.; Tan, C.K.; Ng, S.F. Simvastatin-loaded lyophilized wafers as a potential dressing for chronic wounds. *Drug Dev. Ind. Pharm.* **2016**, *42*, 2055–2062. [CrossRef] [PubMed]

pharmaceuticals

Article

Ketogal Safety Profile in Human Primary Colonic Epithelial Cells and in Mice

Federica Sodano [1,2], Bice Avallone [3], Monica Tizzano [3], Chiara Fogliano [3], Barbara Rolando [1], Elena Gazzano [4], Chiara Riganti [5], Salvatore Magliocca [6], Mariarosaria Cuozzo [2], Stefania Albrizio [2], Antonio Calignano [2], Claudia Cristiano [2,*], Roberto Russo [2,†] and Maria Grazia Rimoli [2,†]

1 Department of Drug Science and Technology, University of Torino, 10125 Torino, Italy; federica.sodano@unito.it (F.S.); barbara.rolando@unito.it (B.R.)
2 Department of Pharmacy, University of Naples Federico II, 80131 Naples, Italy; mariarosaria.cuozzo@unina.it (M.C.); salbrizi@unina.it (S.A.); antonio.calignano@unina.it (A.C.); roberto.russo@unina.it (R.R.); rimoli@unina.it (M.G.R.)
3 Department of Biology, University of Naples Federico II, 80126 Naples, Italy; bice.avallone@unina.it (B.A.); monica.tizzano@unina.it (M.T.); chiara.fogliano@unina.it (C.F.)
4 Department of Life Sciences and Systems Biology, University of Torino, 10123 Torino, Italy; elena.gazzano@unito.it
5 Department of Oncology, University of Torino, 10126 Torino, Italy; chiara.riganti@unito.it
6 Department of Pharmaceutical and Pharmacological Sciences, University of Padova, 35131 Padova, Italy; magliocca.s@gmail.com
* Correspondence: claudia.cristiano@unina.it; Tel.: +39-0816-78411
† These authors contributed equally to this work.

Citation: Sodano, F.; Avallone, B.; Tizzano, M.; Fogliano, C.; Rolando, B.; Gazzano, E.; Riganti, C.; Magliocca, S.; Cuozzo, M.; Albrizio, S.; et al. Ketogal Safety Profile in Human Primary Colonic Epithelial Cells and in Mice. *Pharmaceuticals* **2021**, *14*, 1149. https://doi.org/10.3390/ph14111149

Academic Editor: Nuno Manuel Xavier

Received: 15 October 2021
Accepted: 9 November 2021
Published: 11 November 2021

Publisher's Note: MDPI stays neutral with regard to jurisdictional claims in published maps and institutional affiliations.

Abstract: In our previous studies, a ketorolac–galactose conjugate (ketogal) showed prolonged anti-inflammatory and analgesic activity, causing less gastric ulcerogenic effect and renal toxicity than its parent drug ketorolac. In order to demonstrate the safer profile of ketogal compared to ketorolac, histopathological changes in the small intestine and liver using three staining techniques before and after repeated oral administration in mice with ketorolac or an equimolecular dose of its galactosylated prodrug ketogal were assessed. Cytotoxicity and oxidative stress parameters were evaluated and compared in ketorolac- and ketogal-treated Human Primary Colonic Epithelial cells at different concentrations and incubation times. Evidence of mitochondrial oxidative stress was found after ketorolac treatment; this was attributable to altered mitochondrial membrane depolarization and oxidative stress parameters. No mitochondrial damage was observed after ketogal treatment. In ketorolac-treated mice, severe subepithelial vacuolation and erosion with inflammatory infiltrates and edematous area in the intestinal tissues were noted, as well as alterations in sinusoidal spaces and hepatocytes with foamy cytoplasm. In contrast, treatment with ketogal provided a significant improvement in the morphology of both organs. The prodrug clearly demonstrated a safer profile than its parent drug both in vitro and ex vivo, confirming that ketogal is a strategic alternative to ketorolac.

Keywords: ketorolac; ketogal; histological evaluation; colonic cytotoxicity; mitochondrial oxidative stress

1. Introduction

Non-steroidal anti-inflammatory drugs (NSAIDs) are the primary drugs of choice for the treatment of a variety of chronic and acute inflammatory conditions. In recent decades, the high incidence of chronic diseases, such as arthrosis, in the western world has led to an increase in the use of NSAIDs. Today, NSAIDs account for approximately 8% of total drug prescriptions [1].

However, NSAIDs are associated with several side effects [2], and much attention has been devoted to their toxic effects on the gastrointestinal (GI) tract, particularly the stomach, although a number of studies has demonstrated extensive damage to the intestine. In several cases, intestinal complications outnumbered gastric ones [3–6].

It is known that long-term use of NSAIDs can damage the small and large intestines in healthy subjects, causing bleeding and perforation, erosion, ulcers, and bowel obstruction [7,8]. This can lead to reduced intestinal absorption capacity and increased permeability [9,10]. In as many as 80% of users, chronic NSAID use can exacerbate pre-existing conditions, such as colitis and strictures, inflammatory bowel disease, irritable bowel syndrome, and diverticular disease. Proton-pump inhibitors (PPIs) are routinely co-prescribed with NSAIDs to prevent GI injury. However, while such co-administration leads to a reduction in gastric injury, it may worsen small bowel ulceration and bleeding [11,12]. This is because the intestinal damage induced by NSAIDs is more closely related to their enterohepatic circulation and the following repeated exposure of the intestinal epithelium to these drugs in the presence of bile [13]. In addition, chronic suppression of gastric acid by PPIs may trigger a bacterial overgrowth in the stomach and small intestine and consequently dysbiosis [13]. Dysbiosis, in turn, causes a significant increase in small intestinal permeability, which is considered the early event in the pathogenesis of NSAID enteropathy [13].

A variety of clinical studies has addressed NSAID-induced damage in the upper GI tract. This is particularly true for the small intestine because of its easy accessibility, which facilitates the diagnosis of damage. Upon endoscopic examination, acute bleeding and erosion of the gastroduodenal mucosa can be easily detected in most NSAIDs users [14]. However, NSAIDs can damage any area of the entire GI tract. No less than 40% of all NSAID-induced adverse events are in the lower GI tract, including the colon. These complications include increased permeability, mucosal inflammation, ulceration and hemorrhage, and strictures caused by fibrous diaphragms [15].

Hepatic injury is an additional side effect that has been poorly documented. The liver, because of its significant role in drug metabolism, often suffers from toxic side effects. In fact, hepatotoxicity is the most common reason for non-approval, restriction of indications, and sometimes withdrawal of drugs [16]. NSAIDs present a different spectrum of liver damage, ranging from asymptomatic and transient hypertransaminasemia to fulminant hepatic failure. Interestingly, NSAID-induced hepatotoxicity may be associated with six- to nine-fold greater liver damage in patients concomitantly taking other potentially hepatotoxic drugs (e.g., amoxicillin–clavulanic acid, proton pump inhibitors, phenobarbital, isoniazid) [17].

The main mechanism underlying the occurrence of the aforementioned NSAID lesions is the inhibition of cyclooxygenases (COX-1 and COX-2), which are the enzymes involved in the conversion of arachidonic acid into prostaglandins and other lipid mediators. They differ in their regulatory mechanisms and tissue distribution. In detail, COX-1 produces prostaglandins that regulate normal cellular processes, whereas COX-2 can be induced by inflammation or growth factors. Prostaglandins are responsible for restoring intestinal homeostasis. Indeed, they facilitate the regeneration of the intestinal epithelium [18]. NSAID-induced inhibition of prostaglandin secretion may lead to the collapse of the mucosal defense system and thus leave the intestine more exposed to a variety of insults. In general, NSAIDs selectively inhibit COX-2 as opposed to COX-1 for pain-relieving effects (anti-hyperalgesia).

The COX-independent topical effect of NSAIDs on the small bowel epithelium is another mechanism underlying NSAID injury. This activity plays an important role during the early stage of injury and is caused by mitochondrial impairment [19]. Somasundaram et al. have reported morphological changes of mitochondria in the initial process of small bowel ulceration after oral treatment with indomethacin [20]. Specifically, mitochondrial vacuolization 1 h after indomethacin treatment, mitochondrial swelling, and loss of cristae at 2 h were detected by electron microscopy [20,21]. These changes in mitochondria morphology could be attributable to the activity of NSAIDs to inhibit electron transport and uncouple the oxidative phosphorylation process. Indeed, in vivo treatments with certain NSAIDs caused the opening of the channel called the mitochondrial permeability transition pore (PTP) and thus an increase in permeability across the mitochondrial membrane [22–24]. In the case of diclofenac and ketorolac, the mitochondrial permeability transition occurred only under oxidative stress and in a calcium-dependent manner [25]. Despite the large number of studies, the exact biochemical mechanism by which NSAIDs inhibit, or uncouple, mitochondrial oxidative phosphorylation remains unclear.

Treatment with NSAIDs may also trigger a decrease in oxygen uptake [26]. This additional alteration, along with the NSAID-induced increase in mitochondrial permeability and changes in mitochondria morphology, appears to be mediated by the production of reactive oxygen species (ROS) [27]. ROS are likely produced by activation of xanthine oxidase and infiltration of neutrophils into the mucosa [27]. An in vitro study showed that an increase in xanthine oxidase activity with a concomitant decrease in xanthine dehydrogenase activity is associated with hyperproduction of superoxide anion, resulting in oxidative stress [28,29].

In this scenario, the exploitation of ketogal (ketorolac-d-galactos-6′-yl ester), the conjugate of a well-known drug, ketorolac, with a non-toxic molecule, galactose, may be an effective solution in terms of safety [30]. In order to investigate whether similar alterations to those described occurring in the small intestine also occur in the lower GI tract, we evaluated the cytotoxicity of ketorolac and ketogal against Human Primary Colonic Epithelial (CoEpi) cells. Another aim of the present study was to perform histopathological and morphological evaluations on the small intestine and liver, under the same experimental conditions used for the stomach and kidneys and described in our recent work [31].

2. Results

2.1. Stability of Ketogal and Ketorolac in PBS and CoEpi Cell Medium

Prior to the in vitro experiments, the chemical stability of ketogal and ketorolac was evaluated at 37 °C, both at pH 7.4 in PBS and in the cellular medium used for human CoEpi cells, as reported in Figure 1. These data allowed us to determine the extent of release of the parent drug, in our case ketorolac, from the galactosylated prodrug, ketogal, as a result of the susceptibility of the ketorolac–galactose conjugate to hydrolysis. Ketorolac was still stable after 24 h of incubation in both experimental conditions (data not shown). In addition, ketogal demonstrated good stability both under simulated physiological conditions and in cellular medium. Indeed, after 2 h of incubation, the percentage of ketogal remaining was about 78% in CoEpi cell medium and about 85% in PBS; ketorolac was the only hydrolysis product in all conditions (Figure 1) In detail, the half-life ($t_{\frac{1}{2}}$) of ketogal in CoEpi cell medium was found to be 6.4 h.

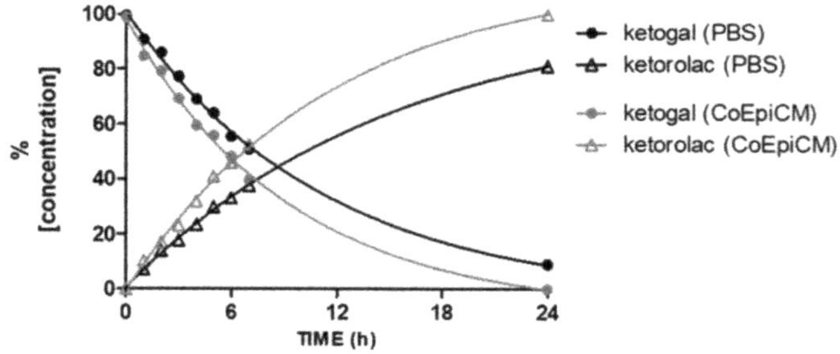

Figure 1. Stability of ketogal in PBS and CoEpi cell medium. Results are expressed as mean values (n = 3).

2.2. Human CoEpi Cell Viability

In order to predict the toxicity of ketogal in the colon, compared to ketorolac, human CoEpi cells were incubated at different times and at different concentrations of the parent drug and prodrug. The choice of three time points, 24, 48, and 72 h, was due to the remarkable and sustained anti-inflammatory activity of ketogal, as demonstrated in our past works [31,32]. The three concentrations were selected using a previous dose–response curve of CoEpi cells incubated with ketorolac and ketogal, where we observed no reduced viability in the range of 5–500 μM for either drug (data not shown). We therefore decided to start evaluating the lowest concentration of ketorolac that induced cytotoxicity, i.e., 1 mM after 72 h, and two higher concentrations (2.5 and 5 mM). The WST-1 assay was used to assess cell viability as it is related to mitochondrial activity and thus cellular metabolism (Figure 2A–C). As shown in Figure 2A, neither ketorolac nor ketogal displayed any significant toxicity after 24 h of incubation. In contrast, after 48 h of incubation (Figure 2B), both compounds exhibited a cytotoxic effect; starting at a concentration of 2.5 mM, the parent drug caused the death of about 30% of the cells, while the prodrug induced only a 10% reduction in cell viability under the same conditions. The cytotoxicity trend was dose-dependent. After 72 h of incubation (Figure 2C), significant ketorolac-induced cytotoxicity was observed at the lowest dose (1 mM), whereas ketogal did not reduce mitochondrial activity compared to untreated cells at this concentration. At concentrations of 2.5 and 5 mM, the number of viable cells was always significantly lower with ketorolac than with ketogal.

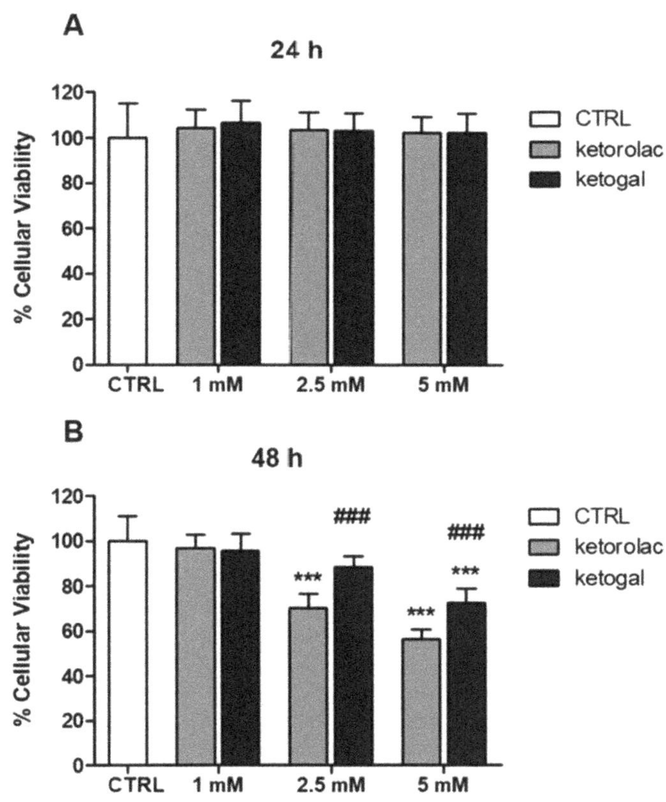

Figure 2. *Cont.*

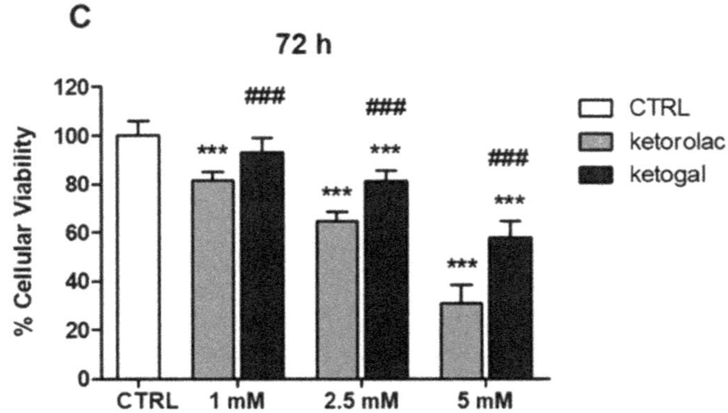

Figure 2. Effect of ketorolac and ketogal on the mitochondrial activity of human CoEpi cells. Cells were incubated for 24 (**A**), 48 (**B**), and 72 (**C**) h either in the absence (CTRL) or presence of 1, 2.5, or 5 mM ketorolac or ketogal. Data are presented as means ± SEM (n = 3). Significance of treated (ketorolac or ketogal) versus respective untreated cells: *** $p < 0.001$. Significance of ketogal vs. ketorolac: ### $p < 0.001$.

2.3. Quantitative Evaluation of Ketogal and Ketorolac in CoEpi Cells

In order to confirm that the presence of a carrier, D-galactose, improves not only the stability and toxicological profile, but also the drug uptake of ketorolac, the amounts of ketorolac and ketogal in cell lysates were evaluated (Figure 3A–C). The results are expressed in terms of ketorolac concentration. In the case of ketogal, the concentration of ketorolac refers to that released after enzymatic hydrolysis. As can be seen, at each time point (24, 48, and 72 h) and at each concentration (1, 2.5, 5 mM), the amount of ketorolac released from ketogal was always higher than that detected in cells incubated with ketorolac (Figure 3A–C). This is further confirmation that the prodrug approach, using galactose as the carrier, provides several advantages, including an increase in the effective concentration of the drug in the target cells.

2.4. Mitochondrial Depolarization and Oxidative Parameters

In order to understand whether the cytotoxic effects induced by ketorolac in human CoEpi cells were caused by the mitochondrial alterations that NSAIDs generally induce in the GI tract, mitochondrial depolarization was assessed by JC-1 staining. The results were expressed as a ratio of green (i.e., damaged)/red (i.e., not damaged) mitochondria, which is considered an index of mitochondrial depolarization and permeability transition. The incubation time chosen was 48 h at 2.5 mM concentration. In fact, as shown in Figure 2B, the first sign of cytotoxicity with both drugs occurred under these conditions. In line with previous results reporting mitochondrial damage elicited by NSAIDs [22], ketorolac induced significant mitochondrial depolarization. As shown in Figure 4A, mitochondrial depolarization in ketorolac-treated CoEpi cells was three times higher than that in untreated cells. On the other hand, ketogal revealed no significant differences compared to control samples. Therefore, in contrast to its parent drug, ketogal did not provoke any change in depolarization and permeability across the mitochondrial membrane. Mitochondrial oxidative stress parameters were then analyzed, including ROS and TBARS. As shown in Figure 4B, the amount of ROS detected in samples incubated with ketorolac was remarkably high. On the other hand, ketogal did not display any difference compared to the control. Thus, ketogal did not alter the level of ROS that the mitochondria of human CoEpi cells physiologically generate. The concentration of TBARS was higher in the mitochondria samples treated with ketorolac than in those treated with ketogal (Figure 4C). Indeed, the value of TBARS detected in ketogal samples was comparable to that of untreated ones.

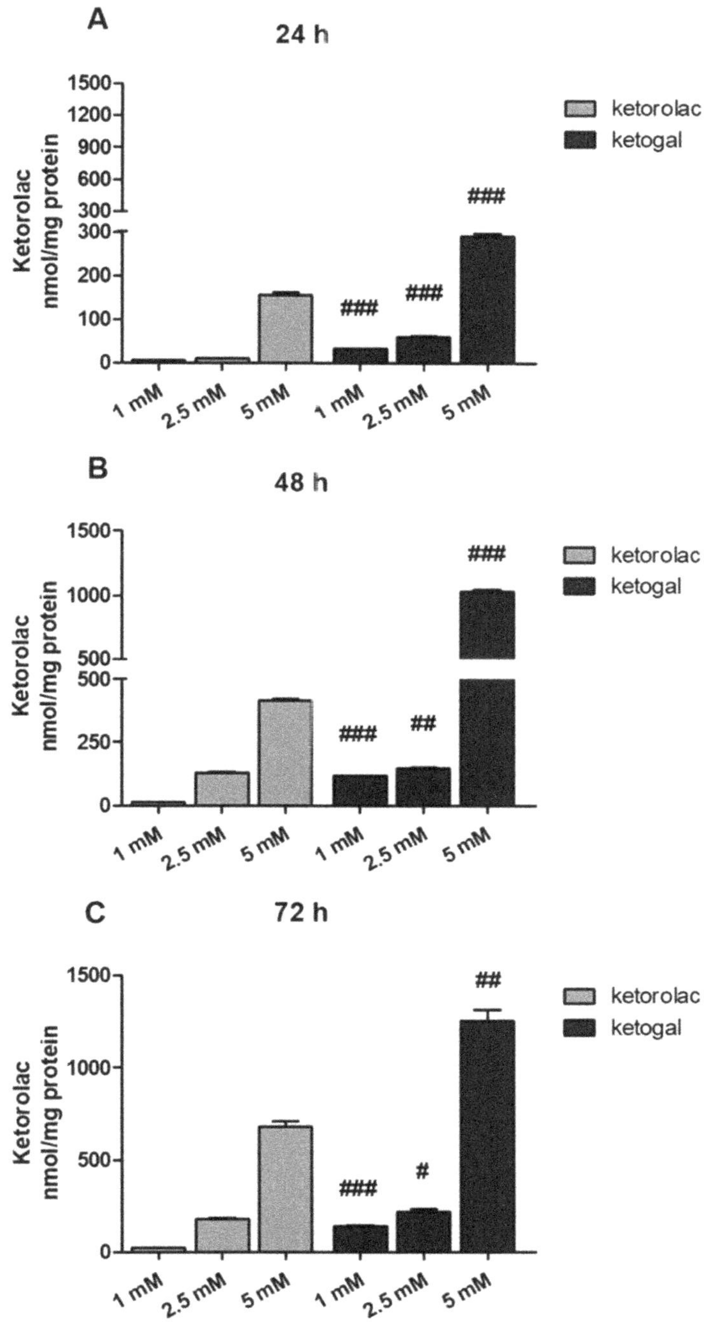

Figure 3. Amount of ketorolac in cell extracts of CoEpi cells incubated for 24 (**A**), 48 (**B**), or 72 (**C**) h in the presence of either 1, 2.5, or 5 mM ketorolac or ketogal. Data are presented as means ± SEM (n = 3). Significance of ketogal vs. ketorolac: # $p < 0.05$; ## $p < 0.01$; ### $p < 0.001$.

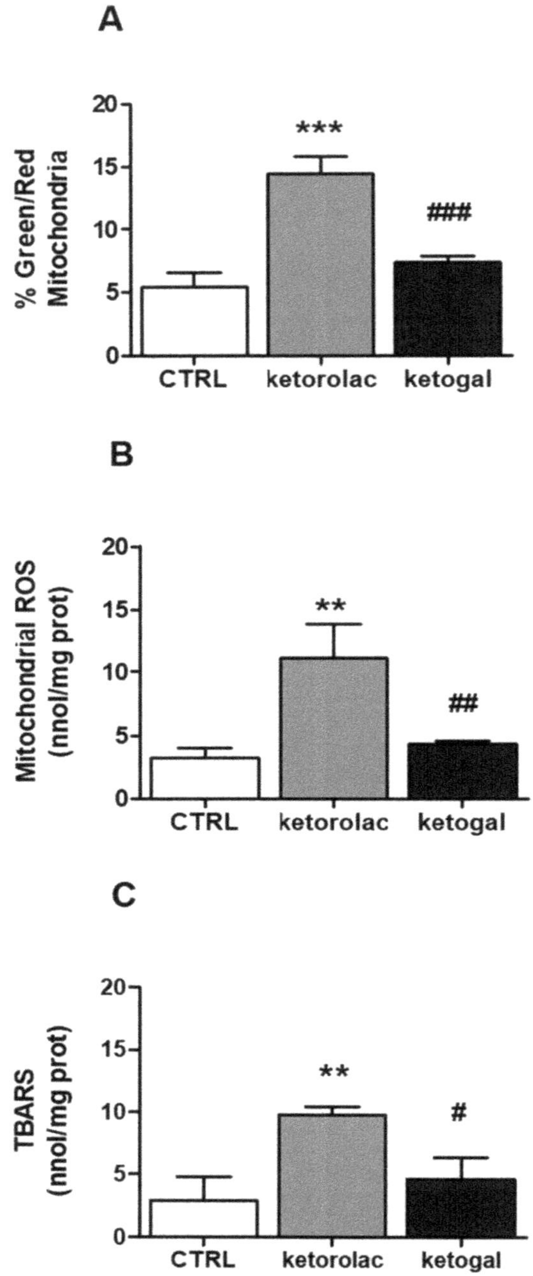

Figure 4. (**A**) Mitochondrial membrane potential was measured fluorometrically in triplicate using the JC-1 probe. Data are presented as means ± SEM (n = 3) of the percentage of green-fluorescent/red-fluorescent mitochondria. Significance of treated (ketorolac or ketogal) versus respective untreated cells: *** $p < 0.001$. Significance of ketogal vs. ketorolac: ### $p < 0.001$. (**B**) Mitochondrial ROS level was measured fluorometrically in triplicate using the MitoSOX Red probe. Data are presented as means ± SEM (n = 3). Significance of treated (ketorolac or ketogal) versus respective untreated cells:

** $p < 0.01$. Significance of ketogal vs. ketorolac: ## $p < 0.01$. (**C**) Mitochondrial TBARS levels were measured spectrophotometrically in triplicate using the TBARS assay kit. Data are presented as means ± SEM ($n = 3$). Significance of treated (ketorolac or ketogal) versus respective untreated cells: ** $p < 0.01$. Significance of ketogal vs. ketorolac: # $p < 0.05$.

2.5. Intracellular Glutathione Measurement

The higher amount of ROS and TBARS at the mitochondrial level observed in the ketorolac-treated samples was in parallel with a lower GSH/GSSG ratio in the whole cells. As shown in Figure 5, the GSH/GSSG ratio was higher for ketogal compared to ketorolac and, at the same time, comparable to that of the control samples. This confirms that ketogal undoubtedly has a safer profile than its parent drug.

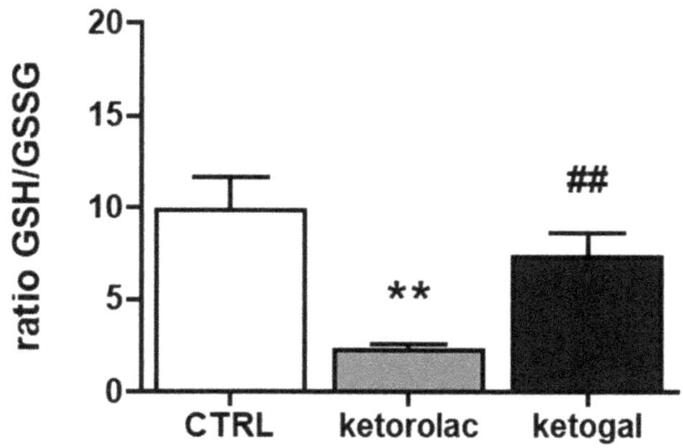

Figure 5. Glutathione measurement. GSH and GSSG were measured spectrophotometrically in triplicate in the cell lysates and expressed as nmol of glutathione/min/mg cellular protein. Data are presented as means ± SEM ($n = 3$) of GSH/GSSG ratio. Significance of treated (ketorolac or ketogal) versus respective untreated cells: ** $p < 0.01$. Significance of ketogal vs. ketorolac: ## $p < 0.01$.

2.6. Evaluation of Small Intestine and Liver Toxicity after Repeated Oral Administration of Ketorolac and Ketogal

In the intestine, hematoxylin–eosin-stained sections showed that the duodenum was arranged normally in the untreated and vehicle samples (grade 0): the mucosa, consisting of a single layer of epithelial cells, the lamina propria, and the muscularis mucosae; the sub-mucosa; the muscularis propria; and the serosa (Figure 6A,B). Enterocytes with uptake function, and goblet cells, which produce mucus, were well organized along the villi (Figure 6A′,B′). Mice treated with indomethacin, used as a positive control, and ketorolac showed compromised villi with subepithelial space and subepithelial vacuolation, erosion with loss of enterocytes, and goblet cells (grade 3) with presence of inflammatory infiltrates (Figure 6C,D). Deterioration of the villi was observed not only with alterations in their morphology but also with the halving of their lengths (Figure 7). In addition, edematous areas located in the mucosa and submucosa were observed (Figure 6D). Sections of the small intestine of ketogal-treated mice showed restoration of normal histology and the length of the villi, which appeared to be similar to that of untreated animals (Figure 6E). After treatment, neither inflammatory infiltrate nor erosion was evident (grade 2). On the other hand, PAS staining did not reveal substantial changes in mucus production in all observed specimens (Figure 6A′–E′).

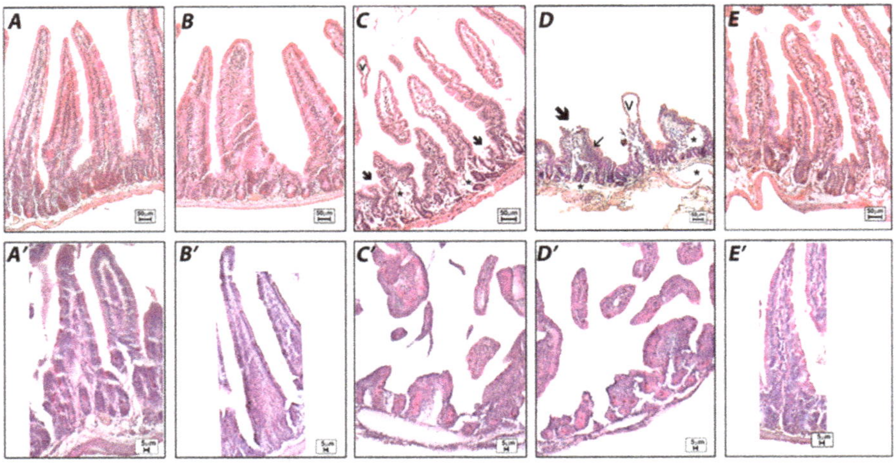

Figure 6. Histological observations of the different treatments on mouse small intestines. Hematoxylin–eosin (**A**–**E**) and PAS staining (**A′**–**E′**). Untreated (**A**) and vehicle (**B**) samples show normal small intestine morphology. Indomethacin (**C**) and ketorolac (**D**) samples demonstrate severe subepithelial vacuolations (v), eroded villi (arrow) with inflammatory infiltrates (small arrow), and edematous areas (*) with also a reduced villi length. The ketogal (**E**) sample demonstrates an improvement in intestine morphology. No significant changes in the production of mucus can be detected in any of the samples observed by PAS (6**A′**–**E′**).

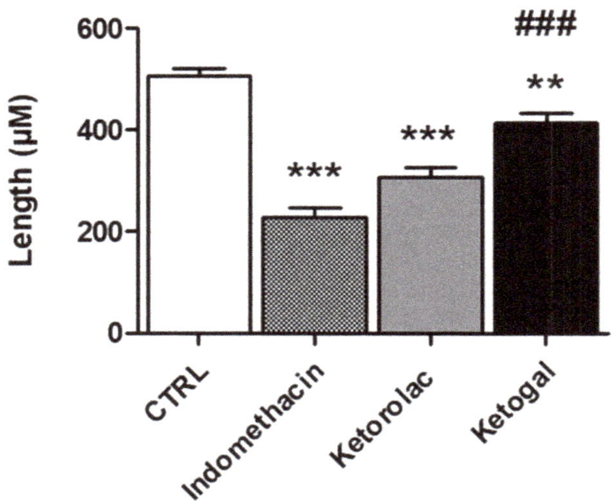

Figure 7. Effect of indomethacin, ketorolac, and ketogal on intestinal villi length (µm). Data are presented as means ± SEM (n = 3). Significance of indomethacin, ketorolac, or ketogal versus untreated: ** $p < 0.01$; *** $p < 0.001$. Significance of ketogal vs. ketorolac: ### $p < 0.001$.

Regarding the liver, with histological investigations of the untreated samples after hematoxylin–eosin staining, both the lobule and the portal triad showed typical tissue morphology with regular-sized hepatocytes and normal sinusoids (Figure 8A–A′). Histological evaluations of the vehicle samples displayed a regular shape, although the sinusoids were slightly dilated, which was likely due to fasting (Figure 8B–B′). Indomethacin-treated mice showed partial degeneration of liver structures, such as dilation of the sinusoids and macrovescicular hepatocytes, which appeared swollen or hydropic with peripheral displacement of the nuclei, especially around the central vein (Figure 8C–C′). However, treatment with ketorolac resulted in impaired sinusoidal spaces and hepatocytes with microvescicular steatosis characterized by foamy cytoplasm (Figure 8D–D′). Despite vacuolation of hepatocytes in some areas, a general improvement in tissue histology was noted in samples from ketogal-treated mice, with recovery of normal hepatocyte morphology (Figure 8E–E′).

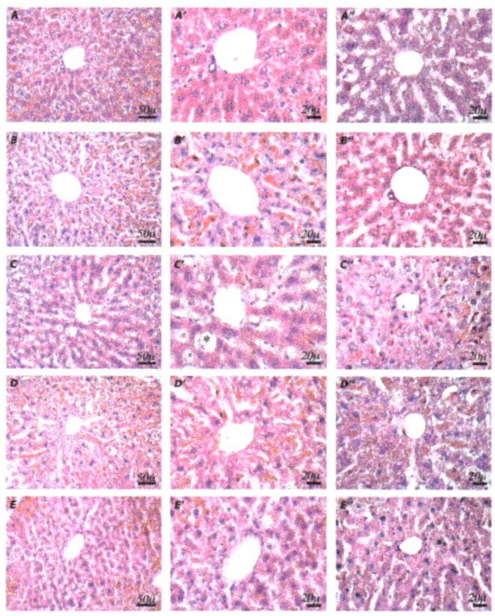

Figure 8. Histological observations of different treatments in mouse livers. Hematoxylin–eosin (**A–E′**) and PAS staining (**A″–E″**). Untreated (**A,A′**) and vehicle (**B,B′**) samples show normal organization of hepatocytes and sinusoids. Indomethacin (**C,C′**) samples demonstrate dilatations of the sinusoids and hydropic hepatocytes with displacement of nuclei (*) of several hepatocytes. In ketorolac (**D,D′**) samples, a foamy cytoplasm appearance is evident in the hepatocytes. Ketogal (**E,E′**) samples show a general improvement in hepatic tissue histology. PAS staining shows no substantial changes in the vehicle (**A″**), compared to the untreated (**B″**) samples. In indomethacin (**C″**) and ketorolac (**D″**) samples, a decrease in positivity is detectable. Ketogal (**E″**) samples show a slight improvement in terms of positivity.

PAS staining, which highlights mucopolysaccharides, showed a slight reduction in positivity in the vehicle compared to the untreated samples (Figure 8A″–B″), a significant decrease in positivity after indomethacin and ketorolac treatment (Figure 8C″–D″), and a slight improvement after ketogal treatment (Figure 8E″). Moreover, PAS staining was negative for the content of the numerous vesicles found in both indomethacin and ketorolac treatments.

Masson staining showed that there was a regular presence of collagen beneath the endothelium of the portal vein and central vein in both untreated and vehicle-treated samples (Figure 9A,B). An increase in collagen fibers was observed in the sections of the indomethacin-treated mice (Figure 9C). On the other hand, no sign of fibrosis was detected in samples treated with ketorolac or ketogal (Figure 9D,E).

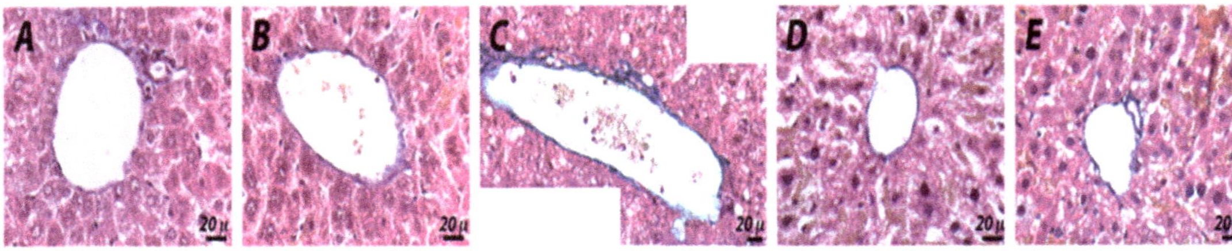

Figure 9. Histological observations of different treatments in mouse livers. Masson staining (**A–E**). In untreated and vehicle (**A**,**B**) samples, the regular presence of blue stained collagen fibers is observed. Indomethacin (**C**) samples show an increase in the collagen fibers below the endothelium of the portal and the central vein. In ketorolac (**D**) and ketogal (**E**) samples, no sign of fibrosis can be detected.

3. Discussion

Chronic treatment with ketorolac causes multiple side events despite its outstanding pharmacological profile. We have previously shown that esterification of ketorolac with a simple and natural carrier, such as D-galactose, improves its safety profile [31,32]. In particular, repeated administration of ketogal exerted significant pharmacological activity that was similar to that of its parent drug, ketorolac, but was less ulcerogenic and caused less renal toxicity [31].

In order to predict the effects on the lower GI tract, including the colon, we demonstrated the lower cytotoxicity of ketogal against human CoEpi cells. By increasing incubation times (in particular after 48 and 72 h) and drug dosing, the difference in toxicity between the parent drug and the prodrug became increasingly apparent. Moreover, at the same dosage, the amount of ketorolac actually delivered into human CoEpi cells was less than the amount of ketorolac resulting from hydrolysis of the prodrug. Therefore, ketogal also displayed improved uptake compared to ketorolac.

Mitochondrial oxidative stress parameters were measured, and changes in mitochondrial depolarization and permeability were explored with the aim of investigating whether ketorolac causes its toxic effects in human CoEpi cells through mitochondrial dysfunction, as previously reported [25–27], and comparing it with ketogal. Human CoEpi cells were chosen as an in vitro model for this study. This cell type is an excellent system for studying the impact of drugs on the human colon, such as NSAIDs. It has similar characteristics to the human primary cell monolayer system used by Bhatt et al. [33], and an equivalent ability to evaluate diclofenac induced-mitochondrial disorders in the small intestine. This model was not chosen by chance. In preliminary experiments, the toxicity of ketorolac and ketogal was previously modelled in vitro using transformed cell lines, including Caco-2 cells (data not shown). In accordance with Larregieu et al. [34], we found a miscorrelation between widely used Caco-2 colon carcinoma cell lines and CoEpi cells. This is due to the fact that the normal intestine, both the upper and lower tracts, differs from cancer cell lines, especially with regard to its metabolic activity. Cancer cells are known for their high metabolic rate and mitochondrial activity, which places limitations on the use of Caco-2 cells in studying the effects of NSAIDs on mitochondria [34]. For this reason, ketorolac- and ketogal-induced mitochondrial disorders were evaluated in a more physiological model of non-transformed colonic epithelial cells.

A significant increase in mitochondrial ROS levels was observed after incubation with ketorolac. This increase was paralleled to typical signs of oxidative damage, such as increased lipid peroxidation and a decrease in the intracellular GSH/GSSG ratio. The data were completely different for ketogal. At the same dose and incubation time, ketogal displayed no significant differences, compared to the control sample, in ROS and TBARS evaluations, or in the GSH/GSSG ratio measurements. Consequently, the lower viability of CoEpi cells, after 48 h of incubation with ketorolac and at 2.5 mM concentration (Figure 2B), was likely caused by the increased production of ROS of mitochondrial origin. According to the literature [35], reactive species, such as O_2^-, H_2O_2, and $OH^·$, generate perturbations at the mitochondrial level and also induce altered permeability. This was confirmed by a JC-1 assay. The depolarization of the mitochondrial membrane after incubation with ketorolac was three times higher than with ketogal and control samples. A change in mitochondrial depolarization leads to a change in pH and in the value of the potential gradient between the mitochondrial intramembranous space and the matrix [36]. After changing the value of the potential gradient, ATP cannot be produced due to the uncoupling between the electron transport chain and oxidative phosphorylation, leading to cell death. In the case of ketogal, the cells showed no altered mitochondrial polarization. We can speculate that the 10% reduction in cell viability observed after 48 h of incubation with 2.5 mM ketogal, compared to the 30% reduction caused by ketorolac, may be related to the different impact that the two drugs, ketorolac and ketogal, may have on mitochondrial homeostasis. Another possible mechanism by which ketogal caused 10% cell death in CoEpi cells may be due to ketorolac-induced damage resulting from hydrolysis of its galactosylated prodrug.

Once it was established that ketogal did not induce oxidative stress in the lower GI tract, consistent with most clinical studies of NSAIDs, we demonstrated the safer profile of ketogal compared with ketorolac in the other organs as well, particularly in the upper GI tract [3,7] and in the liver [16,17]. Consistently, we examined histopathological changes after repeated oral treatment in mice with ketorolac and ketogal, under the conditions previously adopted for the stomach and kidneys in our past work [31].

At the level of the small intestine, from a morphologic point of view, our data showed a significant alteration of the intestinal villi, as a result of the presence of edematous areas and erosion in samples treated with indomethacin and ketorolac. Damage to this type was not only related to a change in mucus production but also to the impairment of regular permeability. Indeed, it has been shown, especially in the case of indomethacin, that a change in the lipid content of the brush border membrane (BBM) can lead to an increase in permeability [27]. This increase in permeability due to NSAID exposure allows the entry of several aggressive luminal factors, such as bile, enzymes, and bacteria, and is considered an important event in the pathogenesis of NSAID-induced enteropathy and to cause ulcerogenic and erosive situations [6]. A considerable amount of data indicates that the modification of BBM, due to NSAID treatment, was mediated by oxidative stress [27]. Indeed, NSAIDs can damage the respiratory chain, resulting in mitochondrial dysfunction and free-radical formation, but can also contribute to neutrophil infiltration [19]. In agreement with these data, our results show significant alterations in both intestinal permeability and cell infiltration in mice treated with indomethacin and ketorolac. Intestinal villi were morphologically damaged, and their length was reduced. In contrast, edema, erosion and neutrophil infiltration were considerably lower after ketogal treatment, as was the length of the villi, which was comparable to that of untreated mice. With regard to mucus production, no significant differences were found in the tested samples. It is known that treatments with NSAIDs can cause disturbances in glycosylation, resulting in a loss of the glycocalyx layer. This could easily be explained by the widespread presence of erosion, caused by treatment with indomethacin and ketorolac, which impaired total mucus production. Completely different results were obtained with ketogal. In fact, an almost completely normal morphology was observed in mice treated with this prodrug.

Ketogal also appears to be safer for the liver than its parent drug. In particular, hepatocytes from mice treated with indomethacin and ketorolac displayed partial degeneration of the liver structure, including dilatation of the sinusoids and edema of several hepatocytes. Furthermore, NSAIDs are able to cause microvesiculation, which are often described as the beginning of hepatic steatosis [37,38]. Microvesicular steatosis can lead to excessive hepatic-fat deposition, increased ROS formation, mitochondrial dysfunction, and endoplasmic reticulum stress, inducing inflammation, cell death, and finally fibrosis [39]. In our experiments, repeated treatment with ketorolac and indomethacin induced evident microvesiculation. On the other hand, the absence of fibrosis and microvesiculation after chronic administration with ketogal demonstrated that only minor liver toxicity was present, despite its ability to accumulate in the liver, as reported in our previous work [32].

The results obtained for the small intestine, liver, and, previously, for the stomach and kidneys [31], demonstrate that ketogal undoubtedly has the potential to overcome most of the side effects associated with repeated administration of ketorolac. Indeed, its toxicological profile has consistently exceeded expectations in in vivo and ex vivo models [31,32]. The preservation of the pharmacological activity of the parent drug and its extended anti-inflammatory and analgesic properties are two additional strengths to indicate its safety.

4. Materials and Methods

4.1. Chemicals and Materials

Ketorolac and indomethacin were purchased from Sigma-Aldrich, Milan, Italy. Ketogal was synthesized as reported in Magliocca's work [40]. Plasticware for cultures was from Falcon (Becton Dickinson, Franklin Lakes, NJ, USA). Proteins in cell extracts were evaluated with the BCA kit from Sigma Chemical Co. (St. Louis, MO, USA). If not specified differently, the reagents were obtained by Sigma Chemical Co.

4.2. Stability of Ketogal and Ketorolac in Phosphate Buffered Saline (PBS) and Colonic Epithelial (CoEpi) Cell Medium

Ketorolac or ketogal (both at 10 mM) in DMSO were added to PBS (pH 7.4, 50 mM) or to CoEpi cell medium. The resulting solutions (100 μM) were maintained at 37 ± 0.5 °C. At given time intervals, PBS solutions were directly analyzed by RP-HPLC (described below); in the case of cell medium solutions, at fixed time intervals, 300 μL of reaction mixture were withdrawn and added to 300 μL of CH_3CN with 0.1% TFA. Samples were sonicated, vortexed, and then centrifuged for 10 min at 2150 g. The clear supernatants were filtered using 0.45 μm PTFE filters (Alltech) and analyzed via RP-HPLC (described below). The experiments were performed in triplicate. Pseudo-first-order half-times ($t_{\frac{1}{2}}$) for the chemical hydrolysis of ketogal to ketorolac were calculated from the linear slope of the plot of the logarithm of the remaining compound against time.

4.3. Cells

Human Primary Colonic Epithelial (CoEpi) cells (CliniSciences, Guidonia Montecello, Italy) were cultured on polylysine-coated plates in Colonic Epithelial Cell medium (CellBiologics, Chicago, IL, USA), supplemented with the Human CoEpi cell medium Supplement Kit (CellBiologics, Chicago, IL, USA) as per the manufacturer's instructions. Cells were maintained at 5% CO_2 and 37 °C and were used between passages 2 and 5.

4.3.1. Viability Assay

A total of 1×10^4 cells, seeded in a 96-well white plate, was incubated with the compounds for 24, 48, or 72 h. Cell viability was assessed with the Cell Proliferation Reagent WST-1 (Sigma Chemical Co., St. Louis, MO, USA). The tetrazolium salt WST-1 is cleaved to formazan by mitochondrial dehydrogenases: an increase in the number of viable cells results in an increase in enzyme activity, which leads to an increase in the amount of formazan. The formazan produced is directly proportional to the amount of mitochondrial dehydrogenase; therefore, the assay is a reliable method to measure the metabolic activity of cells [41]. Results were analyzed using a Synergy HT Multi-Mode Microplate Reader (BioTek Instruments, Winooski, VT, USA). The absorbance of the untreated cells was calculated as 100%, and the absorbance of all the other samples was expressed as percentage versus untreated cells.

4.3.2. Sample Preparation for the Quantitative Evaluation of Ketogal and Ketorolac in CoEpi Cells

CoEpi cells were incubated in the presence of either 1, 2.5, or 5 mM ketorolac or ketogal. After different incubation times (24, 48, or 72 h), cells were washed with PBS, collected, and lysed in 0.3 mL of ice-cold RIPA buffer (50 mM Tris, 10 mM EDTA, 1% *v*/*v* TritonX100; pH 7.5). Lysates were clarified by centrifugation at 15,000× *g*, for 10 min, and the pellets were discarded. The supernatants were transferred into fresh tubes for reverse-phase HPLC analysis. Each experimental point was performed in triplicate.

4.3.3. RP-HPLC Analysis

The reverse-phase HPLC procedure provided the separation and quantitation of ketogal and ketorolac. The amount of compounds in the cell extracts was expressed as nmol/mg protein. HPLC analysis was performed using an HP 1100 chromatograph system (Agilent Technologies, Palo Alto, CA, USA) equipped with a quaternary pump (model G1311A), a membrane degasser (G1379A), and a diode-array detector (DAD) (model G1315B) integrated into the HP1100 system. Data analyses were processed using a HP ChemStation system (Agilent Technologies). The analytical column was ZORBAX SB-Phenyl (250 × 4.6 mm, 5 µm; Agilent). The mobile phase consisted of 0.1% aqueous TFA and 0.1% TFA CH_3CN 60/40 *w*/*w*, at a flow rate = 1.0 mL/min. The injection volume was 20 µL (Rheodyne, Cotati, CA, USA). The column effluent was monitored at 310 nm, referenced against an 800 nm wavelength. The quantitation of ketogal and ketorolac was performed using calibration curves obtained by analyzing standard solutions of compounds in the previously described chromatographic conditions. The linearity of the calibration curves was determined in a concentration range of 0.5–100 µM ($r^2 > 0.999$).

4.3.4. Extraction of Mitochondria

Mitochondria were isolated as previously reported [42]. The pellet, enriched of mitochondria, was finally resuspended in 0.25 mL resuspension buffer (250 mM sucrose, 15 mM K_2HPO_4, 2 mM $MgCl_2$, 0.5 mM EDTA, 5% *w*/*v* BSA) and stored at −80 °C until use. Protein content was measured on a 50 µL aliquot.

4.3.5. Mitochondrial Depolarization

Cells were incubated in PBS with 5 µg/mL JC-1 (Biotium Inc., Freemont, CA, USA), for 30 min at 37 °C, then washed and rinsed with 300 µL PBS. As reported [43], polarized mitochondria emitted a red fluorescence, while depolarized and damaged mitochondria emitted a green fluorescence. The depolarization and damage to mitochondria was associated with an increase in green/red fluorescence intensity. Red and green fluorescence units were measured using a Synergy HT Multi-Mode microplate reader, with 550 nm excitation/600 nm emission for red fluorescence, and 485 nm excitation/535 nm emission for green fluorescence. Fluorescence units were expressed as percentage of green-fluorescent (i.e., depolarized) mitochondria/red-fluorescent (i.e., polarized) mitochondria.

4.3.6. Mitochondrial Reactive Oxygen Species (ROS)

Mitochondrial ROS were measured fluorometrically by incubating cells 10 min with 10 µM of MitoSOX Red (Invitrogen, Waltham, MA, USA), a superoxide-sensitive probe. Cell were washed and the red fluorescence, indicative of ROS generated within mitochondria, was read with a Synergy HT Multi-Mode microplate reader. Results were expressed as nmol/mg mitochondrial proteins.

4.3.7. Thiobarbituric Acid Reactive Substances (TBARS)

TBARS formed as a byproduct of lipid peroxidation and measured as a marker of lipid damage, were detected spectrophotometrically by measuring adduct TBARS-TBA, which absorbs at a wavelength of 535 nm [44], in mitochondrial extracts with the TBARS

Assay Kit (Cayman Chemicals, Ann Arbor, MI, USA), using a Synergy HT Multi-Mode microplate reader. Results were expressed as nmol/mg mitochondrial proteins.

4.3.8. Intracellular Glutathione Measurement

Intracellular glutathione was measured as previously described [45]. The kinetics of the reaction was monitored for 5 min at 415 nm, using a Synergy HT Multi-Mode Microplate Reader. The subtraction of GSSG from total GSH gave the value of GSH Results were expressed as nmol of glutathione/min/mg cellular protein, according to a titration curve prepared with serial dilutions (10 μM–0.1 nM) of 1:1 GSH + GSSG mix.

4.4. Animals

Male Swiss CD1 mice (30–35 g) of 10 weeks of age were purchased from Charles Rivers (Calco, Italy) and housed in cages in a room at 22 ± 1 °C on a 12/12 h light/dark cycle fed with a standard rodent chow diet and had ad libitum access to water. Each group of mice ($n = 6$) was treated for 5 days with either ketorolac (10 mg/kg/os), ketogal (16.3 mg/kg/os), or indomethacin (15 mg/kg/os), and a vehicle (carboxymethyl cellulose 0.5%). At the end of treatment, mice were fasted overnight for 12–14 h and then perfused for histological evaluations.

Ethics Statement

All experiments involving mice were carried out according to the ethical policies and procedures approved by the Institutional Committee on the Ethics of Animal Experiments (CSV) of the University of Naples "Federico II" and by the Ministry of Health (Approval no. 0084607, date 3 October 2014).

4.5. Histological Evaluations

Histological evaluations were performed on fasting mice deeply anesthetized with pentobarbitone (50 mg/kg, ip) and transcardially perfused with PBS, followed by 4% paraformaldehyde (PFA). Intestine and liver were collected and fixed in 4% buffered formaldehyde solution at 4 °C for 24 h. Samples were processed for wax embedding; serial sections (5 μm) were cut and stained with hematoxylin–eosin to highlight morphological alterations in the intestine and liver; periodic acid–Schiff (PAS) to underline mucin, glycogen, and glycoproteins; or Masson's Trichrome staining to assess the presence of fibrosis, in accordance with Pagliuca et al. [46]. Sections were oxidized in 0.5% periodic acid solution for 10 min, rinsed in double-distilled water, and stained with Schiff's reagent in the dark for 45 min. The reaction was blocked by repeated washing in 2.5% sodium bisulfite in 0.05 N HCl [47]. To perform Masson's trichrome, liver sections were stained in iron hematoxylin solution for 10 min and placed in Biebrich's scarlet fuchsin solution for 15 min, differentiated in phosphomolybdic acid solution, transferred directly to an aniline blue solution for 10 min, and further differentiated in 1% glacial acetic acid solution for 1 min. Observations of intestinal lesions were graded according to Park's Scale [48]. Intestinal villus length measurements were made on randomly selected digital image photographs of untreated and treated mice and analyzed using the software Image J 1.45.

4.6. Statistical Analyses

Data were analyzed using one-way ANOVA followed by post hoc Bonferroni's multiple comparison test to determine the significance of differences between groups. Results are reported as mean ± SEM of at least two-to-three independent experiments, performed in triplicate. Statistical significance was accepted at $p < 0.05$.

5. Conclusions

In conclusion, we can undoubtedly confirm that ketogal has a better safety and reliability profile than ketorolac. Interestingly, in the colon, ketogal was significantly less toxic than its parent drug. Notably, the prodrug induced only a 10% reduction in CoEpi cell viability, and this decrease was not due to any mitochondrial damage or alteration. Remarkably, after in vitro treatment with ketogal, mitochondrial oxidative stress parameters were comparable to control samples. Furthermore, in ketogal-treated mice, a significant improvement in the morphology of both the upper GI tract and the liver was observed, which appeared to be similar to that of untreated animals. With this information, ketogal could replace ketorolac and, at the same time, reduce the use of drugs, such as PPIs, that are normally prescribed to mitigate the side effects of NSAIDs. The use of a single drug could be the best alternative in terms of health and public spending.

Author Contributions: Conceptualization, F.S., R.R., C.C. and M.G.R.; Methodology, C.C. and C.R.; Validation, M.G.R., S.A., R.R. and A.C.; Formal Analysis, F.S., S.M., B.A., M.T., C.F. and E.G.; Investigation, B.R., M.C. and S.M.; Resources, R.R.; Data Curation, C.C. and F.S.; Writing—Original Draft Preparation, F.S. and C.C., Writing—Review and Editing, R.R., M.G.R., S.A. and A.C.; Supervision, R.R. and M.G.R.; Project Administration, C.C.; Funding Acquisition, R.R. All authors have read and agreed to the published version of the manuscript.

Funding: This research received no external funding.

Institutional Review Board Statement: The study was conducted according to the guidelines of the Declaration of Helsinki and approved by the Institutional Committee on the Ethics of Animal Experiments (CSV) of the University of Naples "Federico II" and by the Ministry of Health under protocol no. 0084607, date 3 October 2014.

Informed Consent Statement: Not applicable.

Data Availability Statement: The data presented in this study are available in article.

Acknowledgments: We thank Costanzo Costamagna at the Department of Oncology, University of Turin, for technical assistance, and Dale James Matthew Lawson for English language editing.

Conflicts of Interest: The authors declare no conflict of interest.

References

1. Jones, R. Nonsteroidal anti-inflammatory drug prescribing: Past, present, and future. *Am. J. Med.* **2001**, *110*, 4S–7S. [CrossRef]
2. Modi, C.M.; Mody, S.K.; Patel, H.B.; Dudhatra, G.B.; Avinash, K.; Madhavi, A. Toxicopathological overview of analgesic and anti-inflammatory drugs in animals. *J. Appl. Pharm. Sci.* **2012**, *2*, 149–157.
3. Bjarnason, I.; Hayllar, J.; Macpherson, A.J.; Russell, A.S. Side effects of nonsteroidal anti-inflammatory drugs on the small and large intestine in humans. *Gastroenterology* **1993**, *104*, 1832–1847. [CrossRef]
4. Whittle, B.J.R. Gastrointestinal effects of nonsteroidal anti-inflammatory drugs. *Fundam. Clin. Pharmacol.* **2003**, *17*, 301–313. [CrossRef] [PubMed]
5. Whittle, B.J.R. Mechanisms underlying intestinal injury induced by anti-inflammatory COX inhibitors. *Eur. J. Pharmacol.* **2004**, *500*, 427–439. [CrossRef]
6. Wallace, J.L. Mechanisms, prevention and clinical implications of nonsteroidal anti-inflammatory drug enteropathy. *World J. Gastroenterol.* **2013**, *19*, 1861–1876. [CrossRef]
7. Laudanno, O.M.; Cesolari, J.A.; Esnarriaga, J.; Flaherty, P.; Vada, J.; Guastalli, G.; San Miguel, P.; Bedini, O.A. In vivo selectivity of nonsteroidal anti-inflammatory drugs and gastrointestinal ulcers in rats. *Dig. Dis. Sci.* **2009**, *45*, 1359–1365. [CrossRef]
8. Sostres, C.; Gargallo, C.J.; Lanas, A. Nonsteroidal anti-inflammatory drugs and upper and lower gastrointestinal mucosal damage. *Arthritis Res. Ther.* **2013**, *15*, S3. [CrossRef]
9. Bjarnason, I.; Fehilly, B.; Smethurst, P.; Menzies, I.S.; Levi, A.J. Effects of nonsteroidal anti-inflammatory drugs on permeability of the small intestine in humans. *J. Rheumatol. Suppl.* **1992**, *19*, 83–84.
10. Choi, V.M.; Coates, J.E.; Chooi, J.; Thomson, A.B.; Russell, A.S. Small bowel permeability-a variable effect of NSAIDS. *Clin. Invest. Med.* **1995**, *18*, 357–361.
11. Freedberg, D.E.; Yang, Y.X.; Abrams, J.A. Proton Pump Inhibitors and Myocardial Infarction. *Gastroenterology* **2015**, *149*, 830–833. [CrossRef]
12. Lazarus, B.; Chen, Y.; Wilson, F.P.; Sang, Y.; Chanq, A.R.; Coresh, J.; Grams, M.E. Proton pump inhibitor use and the risk of chronic kidney disease. *JAMA Intern. Med.* **2016**, *176*, 238–246. [CrossRef]
13. Wallace, J.L.; Syer, S.; Denou, E.; De Palma, G.; Vong, L.; McKnight, W.; Jury, J.; Bolla, M.; Bercik, P.; Collins, S.M.; et al. Proton pump inhibitors exacerbate NSAID-induced small intestinal injury by inducing dysbiosis. *Gastroenterology* **2011**, *141*, 1314–1322. [CrossRef]
14. Ehsanullah, R.S.; Page, M.C.; Tildesley, G.; Wood, J.R. Prevention of gastroduodenal damage induced by non-steroidal anti-inflammatory drugs: Controlled trial of ranitidine. *BMJ* **1988**, *297*, 1017–1021. [CrossRef]

15. Lanas, A.; Ferrandez, A. NSAIDs and the colon. *Curr. Opin. Gastroenterol.* **2008**, *25*, 44–49. [CrossRef]
16. Donati, M.; Conforti, A.; Lenti, M.C.; Capuano, A.; Bortolami, O.; Motola, D.; Moretti, U.; Vannacci, A.; Rafaniello, C.; Vaccheri, A.; et al. Risk of acute and serious liver injury associated to nimesulide and other NSAIDs: Data from drug-induced liver injury case–control study in Italy. *Br. J. Clin. Pharmacol.* **2016**, *82*, 238–248. [CrossRef]
17. Bessone, F. Non-steroidal anti-inflammatory drugs: What is the actual risk of liver damage? *World J. Gastroenterol.* **2010**, *16*, 5651–5661. [CrossRef]
18. Takeuchi, K.; Tanaka, A.; Kato, S.; Amagase, K.; Satoh, H. Roles of COX inhibition in pathogenesis of NSAID-induced small intestinal damage. *Clin. Chim. Acta* **2010**, *411*, 459–466. [CrossRef]
19. Watanabe, T.; Tanigawa, T.; Nadatani, Y.; Otani, K.; Machida, H.; Okazaki, H.; Yamagami, H.; Watanabe, K.; Tominaga, K.; Arakawa, T. Mitochondrial disorders in NSAIDs-induced small bowel injury. *J. Clin. Biochem. Nutr.* **2011**, *48*, 117–121. [CrossRef]
20. Somasundaram, S.; Rafi, S.; Hayllar, J.; Sigthorsson, G.; Jacob, M.; Price, A.B.; Macpherson, A.; Mahmod, T.; Scott, D.; Wrigglesworth, J.M.; et al. Mitochondrial damage: A possible mechanism of the "topical" phase of NSAID induced injury to the rat intestine. *Gut* **1997**, *41*, 344–353. [CrossRef]
21. Somasundaram, S.; Hayllar, H.; Rafi, S.; Wrigglesworth, J.M.; Macpherson, A.J.S.; Bjarnason, I. The Biochemical Basis of Non-Steroidal Anti-Inflammatory Drug-Induced Damage to the Gastrointestinal Tract: A Review and a Hypothesis. *Scand. J. Gastroenterol.* **1995**, *30*, 289–299. [CrossRef]
22. Masubuchi, Y.; Nakayama, S.; Horie, T. Role of mitochondrial permeability transition in diclofenac-induced hepatocyte injury in rats. *Hepatology* **2002**, *35*, 544–551. [CrossRef]
23. Al-Nasser, J.A. Salicylate-induced kidney mitochondrial permeability transition is prevented by cyclosporin A. *Toxicol. Lett.* **1999**, *105*, 1–8. [CrossRef]
24. Trost, L.C.; Lemasters, J.J. Role of the mitochondrial permeability transition in salicylate toxicity to cultured rat hepatocytes: Implications for the pathogenesis of Reye' syndrome. *Toxicol. Appl. Pharmacol.* **1997**, *147*, 431–441. [CrossRef]
25. Lal, N.; Kumar, J.; Erdahl, W.E.; Pfeiffer, D.R.; Gadd, M.E.; Graff, G.; Yanni, J.M. Differential effects of non-steroidal anti-inflammatory drugs on mitochondrial dysfunction during oxidative stress. *Arch. Biochem. Biophys.* **2009**, *490*, 1–8. [CrossRef]
26. Jacob, M.; Bjarnason, I.; Simpson, R.J. In vitro evidence for mitochondrial effects of indomethacin. *Biochem. Soc. Trans.* **1998**, *26*, S315. [CrossRef]
27. Basivireddy, J.; Jacob, M.; Ramamoorthy, P.; Pulimood, A.B.; Balasubramanian, K.A. Indomethacin-induced free radical-mediated changes in the intestinal brush border membranes. *Biochem. Pharmacol.* **2003**, *65*, 683–695. [CrossRef]
28. Nalini, S.; Mathan, M.M.; Balasubramanian, K.A. Free radical-induced damage intestinal ischaemia reperfusion in normal and xanthine oxidase deficient rats. *Mol. Cell. Biochem.* **1993**, *124*, 59–66. [CrossRef]
29. Anup, R.; Aparna, V.; Pulimood, A.; Balasubramanian, K.A. Surgical stress and the small intestine: Role of oxygen free radicals. *Surgery* **1999**, *125*, 560–569. [CrossRef]
30. Melisi, D.; Curcio, A.; Luongo, E.; Morelli, E.; Rimoli, M.G. D-galactose as a vector for prodrug design. *Curr. Top. Med. Chem.* **2011**, *11*, 2288–2298. [CrossRef]
31. Russo, R.; De Caro, C.; Avallone, B.; Magliocca, S.; Nieddu, M.; Boatto, G.; Troiano, R.; Cuomo, R.; Cirillo, C.; Avagliano, C.; et al. Ketogal: A Derivative Ketorolac Molecule with Minor Ulcerogenic and Renal Toxicity. *Front. Pharmacol.* **2017**, *8*, 757. [CrossRef] [PubMed]
32. Curcio, A.; Sasso, O.; Melisi, D.; Nieddu, M.; La Rana, G.; Russo, R.; Gavini, E.; Boatto, G.; Abignente, E.; Calignano, A.; et al. Galactosyl prodrug of ketorolac: Synthesis, stability, and pharmacological and pharmacokinetic evaluations. *J. Med. Chem.* **2009**, *52*, 3794–3800. [CrossRef] [PubMed]
33. Bhatt, A.P.; Gunasekara, D.B.; Speer, J.; Reed, M.I.; Peña, A.N.; Midkiff, B.R.; Magness, S.T.; Bultman, S.J.; Allbritton, N.L.; Redinbo, M.R. Nonsteroidal Anti-Inflammatory Drug-Induced Leaky Gut Modeled Using Polarized Monolayers of Primary Human Intestinal Epithelial Cells. *ACS Infect. Dis.* **2018**, *4*, 46–52. [CrossRef] [PubMed]
34. Larregieu, C.A.; Benet, L.Z. Drug discovery and regulatory considerations for improving in silico and in vitro predictions that use Caco-2 as a surrogate for human intestinal permeability measurements. *AAPS J.* **2013**, *15*, 483–497. [CrossRef]
35. Omatsu, T.; Naito, Y.; Handa, O.; Hayashi, N.; Mizushima, K.; Qin, Y.; Hirata, I.; Adachi, S.; Okayama, T.; Kishimoto, E.; et al. Involvement of reactive oxygen species in indomethacin-induced apoptosis of small intestinal epithelial cells. *J. Gastroenterol.* **2009**, *44*, 30–34. [CrossRef]
36. Sherratt, H.S. Mitochondria: Structure and function. *Rev. Neurol.* **1991**, *147*, 417–430.
37. Lewis, J.H. NSAID-induced hepatotoxicity. *Clin. Liver Dis.* **1998**, *2*, 543–561. [CrossRef]
38. Amacher, D.E.; Chalasani, N. Drug-induced hepatic steatosis. *Semin. Liver Dis.* **2014**, *34*, 205–214.
39. Patel, V.; Sanyal, A.J. Drug-Induced Steatohepatitis. *Clin. Liver Dis.* **2013**, *17*, 533. [CrossRef]
40. Magliocca, S.; Sodano, F.; Nieddu, M.; Burrai, L.; Boatto, G.; Rimoli, M.G. New Galactosylated NSAIDS prodrugs in a green context: Synthesis and stability. *IJPSR* **2017**, *8*, 1575–1581.
41. Ishiyama, M.; Shiga, M.; Sasamoto, K.; Mizoguchi, M.; He, P.-G. A new sulfonated tetrazolium salt that produces a highly water-soluble formazan dye. *Chem. Pharm. Bull.* **1993**, *41*, 1118–1122. [CrossRef]
42. Campia, I.; Lussiana, C.; Pescarmona, G.; Ghigo, D.; Bosia, A.; Riganti, C. Geranylgeraniol prevents the cytotoxic effects of mevastatin in THP-1 cells, without decreasing the beneficial effects on cholesterol synthesis. *Br. J. Pharmacol.* **2009**, *158*, 1777–1786. [CrossRef]

43. Shao-Peng, L.; Hu, J.; Wei, J.; Ye, S.; Bu, J.; Xu, W.; Chen, S.; Wu, Y.; Wu, G.; Zhu, L.; et al. Silencing of circFoxO3 Protects HT22 Cells from Glutamate-Induced Oxidative Injury via Regulating the Mitochondrial Apoptosis Pathway. *Cell. Mol. Neurobiol.* **2020**, *40*, 1231–1242.
44. Dawn-Linsley, M.; Ekinci, F.J.; Ortiz, D.; Rogers, E.; Shea, T.B. Monitoring thiobarbituric acid-reactive substances (TBARs) as an assay for oxidative damage in neuronal cultures and central nervous system. *J. Neurosci. Methods* **2005**, *141*, 219–222. [CrossRef]
45. Gazzano, E.; Riganti, C.; Tomatis, M.; Turci, F.; Bosia, A.; Fubini, B.; Ghigo, D. Potential toxicity of nonregulated asbestiform minerals: Balangeroite from the western Alps. Part 3: Depletion of antioxidant defenses. *J. Toxicol. Environ. Health A* **2005**, *68*, 41–49. [CrossRef]
46. Pagliuca, C.; Cicatiello, A.G.; Colicchio, R.; Greco, A.; Cerciello, R.; Auletta, L.; Albanese, S.; Scaglione, E.; Pagliarulo, C.; Pastore, G.; et al. Novel Approach for Evaluation of Bacteroides fragilis Protective Role against Bartonella henselae Liver Damage in Immunocompromised Murine Model. *Front. Microbiol.* **2016**, *7*, 1750. [CrossRef]
47. Avallone, B.; Agnisola, C.; Cerciello, R.; Panzuto, R.; Simoniello, P.; Cretì, P.; Motta, C.M. Structural and functional changes in the zebrafish (Danio rerio) skeletal muscle after cadmium exposure. *Cell Biol. Toxicol.* **2015**, *31*, 273–283. [CrossRef]
48. Park, P.O.; Haglund, U.; Bulkley, G.B.; Fält, K. The sequence of development of intestinal tissue injury after strangulation ischemia and reperfusion. *Surgery* **1990**, *107*, 574–580.

pharmaceuticals

MDPI

Review

Inulin and Its Application in Drug Delivery

Franklin Afinjuomo, Sadikalmahdi Abdella, Souha H. Youssef, Yunmei Song and Sanjay Garg *

Pharmaceutical Innovation and Development Group, Clinical and Health Sciences, University of South Australia, Adelaide 5001, Australia; Franklin.Afinjuomo@unisa.edu.au (F.A.); sadikalmahdi.abdella@mymail.unisa.edu.au (S.A.); souha.youssef@mymail.unisa.edu.au (S.H.Y.); may.song@unisa.edu.au (Y.S.)

* Correspondence: sanjay.garg@unisa.edu.au; Tel.: +61-88-302-1575; Fax: +61-88-302-2389

Abstract: Inulin's unique and flexible structure, stabilization/protective effects, and organ targeting ability make it an excellent drug delivery carrier compared to other biodegradable polysaccharides. The three hydroxyl groups attached to each fructose unit serve as an anchor for chemical modification. This, in turn, helps in increasing bioavailability, improving cellular uptake, and achieving targeted, sustained, and controlled release of drugs and biomolecules. This review focuses on the various types of inulin drug delivery systems such as hydrogel, conjugates, nanoparticles, microparticles, micelles, liposomes, complexes, prodrugs, and solid dispersion. The preparation and applications of the different inulin drug delivery systems are further discussed. This work highlights the fact that modification of inulin allows the use of this polymer as multifunctional scaffolds for different drug delivery systems.

Keywords: inulin; drug delivery; nanoparticles; microparticles; hydrogels

Citation: Afinjuomo, F.; Abdella, S.; Youssef, S.H.; Song, Y.; Garg, S. Inulin and Its Application in Drug Delivery. *Pharmaceuticals* **2021**, *14*, 855. https://doi.org/10.3390/ph14090855

Academic Editors: Federica Sodano and Maria Grazia Rimoli

Received: 31 July 2021
Accepted: 24 August 2021
Published: 26 August 2021

1. Introduction

The use of natural polysaccharides for controlled drug delivery systems is drawing significant attention, as synthetics polymers used for this purpose are associated with drawbacks such as instability, expensive chemical modification [1] weak mechanical properties, poor receptor targeting properties [2], biodegradation, and excretion [3]. Natural polysaccharides possess properties that can address the multiple requirements for optimal drug delivery systems, including targeting and stealth properties [4]. In contrast to synthetic polymers, natural polysaccharides are non-toxic and biodegradable [5]. Inulin (INU) is a natural, highly abundant storage fructan polysaccharide that can be harvested from different plants such as chicory, dahlia, Jerusalem artichoke [6], vegetables, fruits, and grains, including garlic, leeks, bananas, onions, and wheat [7]. Genetically modified potatoes are also an important source of inulin [8] and enzymatic production [9]. The chemical structure of inulin is made up of linear chains of fructose units with a glucose unit linked to the reducing end [10]. The fructose units vary in length from 2–60 units. The variation in polydispersity and chain length can be traced to the plant species and time of harvesting [11,12]. The physicochemical properties of the polymer depend on the degree of polymerization (DP) and the presence of branching in the fructan [11]. Inulin has been used in drug delivery as a versatile polymer, and its versatility could be attributed partially to its flexible backbone. The flexibility of the backbone is, in turn, related to the absence of sugar rings in the polyethylene oxide-based backbone (Figure 1) [12]. In addition, the fructan is primarily made up of furanose groups. The molecular structure of inulin allows the polymers chain greater freedom to move as well as the ability to reassemble into a different range of structures [10,13]. This translates into structural flexibility and makes inulin an appealing and versatile material for use in different applications.

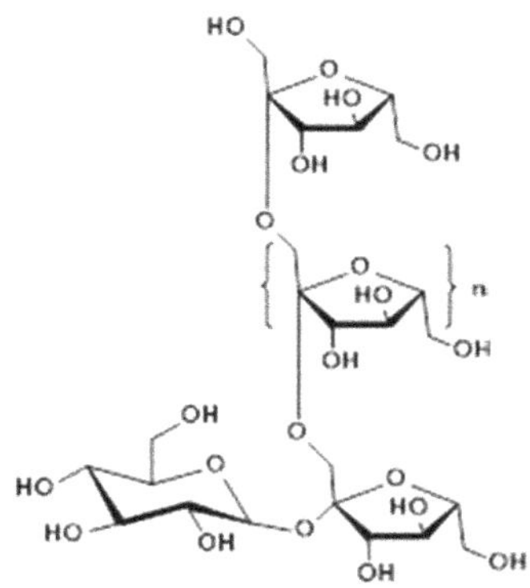

Figure 1. Inulin structure.

Another interesting feature of inulin's structure is the β (2−1) glycosidic bonds that halt the polymer from being digested by endogenous enzymes such as other typical carbohydrates. Instead, inulin is relatively chemically stable in the gastrointestinal tract (GIT) until it reaches the colon, where bacteria in the gut digest it, making it a good source of carbon for health-promoting probiotic bacteria [14]. The change in solubility with chain length makes inulin suitable for different applications in both the pharmaceutical and food industries [13]. For instance, inulin with long chain lengths is insoluble at body temperature, making them useful for particulate delivery systems, while soluble (lower chain length) inulin is ideal for kidney and renal function testing [15]. INU is considered a non-toxic and highly biodegradable polymer. It is commonly used as a food ingredient and was given the Generally Recognized As Safe (GRAS) status by the United States Food and Drug Administration (FDA) in 2002. INU has also been used extensively in food industries as a sugar and fat replacement [15–20], nutritional excipient [21,22], texture modifier [15,23,24], and for organoleptic improvement [25]. The glycosidic bonds that humans cannot digest but that can be broken down by Bifidobacteria in the gut make inulin a suitable dietary fiber with prebiotic properties [10,26–31]. Aside from being degraded by inulinase that is found only in the colon, this polysaccharide passes straight through the stomach and the small intestine intact making it a good candidate for colon targeting. Inulin's membrane stabilization ability is well established and has been used in pharmaceutical industries for several purposes, including to stabilize therapeutic proteins [32–36]. Modified inulin also improve the low dissolution of poorly soluble drugs [37–39], measurement of kidney function [40,41], modulate lipid production [42], glycemic control, as dietary fiber, prebiotic, treatment of constipation [43–45], vaccine adjuvant [46,47] and in cancer treatments [48–50] among others. Modified inulin has also been used for drug delivery [51] for targeting the colon [52–54] and pulmonary drug delivery [55]. Previous reviews by Barclay et al. [10] and Mensink et al. [13] explored the detailed physicochemical properties and pharmaceutical applications of inulin. Giri et al. also demonstrated the use of inulin derivatives for colon targeting [56]. The review by Jiwani et al. focused more on the delivery routes with emphasis on a few inulin deliveries forms [57]. The review by Akram et al. was not exhaustive and comprehensive because the review discussed few inulin delivery carriers [58]. Another good review from Wan and colleagues focused on inulin physiological functions and pharmaceutical uses only [59]. However, there is no detailed review focusing on the various types of drug delivery carriers such as liposomes, micelles, nanoparticles, microparticles, hydrogels, solid dispersion, and conjugates fabricated from the use of modified inulin. It is important to improve drug bioavailability, efficacy, reduce drug degradation as well as prevent serious adverse drug reactions. The improvement in the delivery of drugs helps minimize toxicities, improve patient compliance, and provide opportunities for pharmaceutical companies to market new products. Some of the delivery systems that have been successfully used to address delivery challenges include liposomes, micelles, hydrogels, nanoparticles, microparticles, microspheres, prodrugs, and solid dispersion. A review of these delivery systems will provide better insight into the use of inulin-based materials for researchers working in the area of pharmaceutical sciences. This gap was the reason and motivation behind the development of this inulin drug delivery review article. This review highlights the use of INU derivatives as multifunctional drug delivery scaffolds and vehicles. This review aims to fill that gap and draw attention to the huge potential of inulin as a multifunctional drug delivery scaffold (Figure 2).

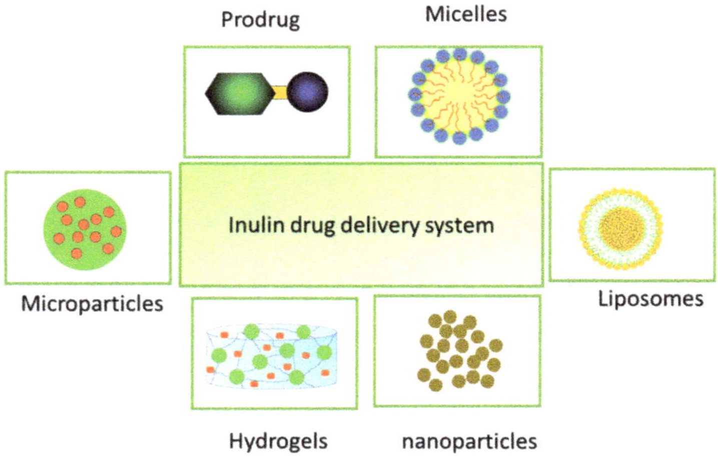

Figure 2. Inulin drug delivery carriers.

2. Inulin Drug Delivery Systems

The main focus of this review is to provide detailed insight into the various ways that modified inulin have been prepared and utilized for drug delivery applications. Inulin's unique properties, such as molecular flexibility, ability to stabilize proteins, and self-assembled structures, ease of chemical modification, and immune modulation, encourage researchers and formulation scientists to exploit it as a drug delivery scaffold and biomaterials. Inulin's ability to undergo chemical modification has been utilized to increase drug delivery to target cells such as cancers and genetic materials. The repeating fructose monomer has three hydroxyl groups (two secondary and one primary), which serve as a handle or platform to anchor other functional agents. Furthermore, Inulin stabilizes self-assembled proteins and lipid structures. For instance, it is noted that protein conformational changes and unfolding can result in loss of biological activity. To address this problem, there is a need to preserve the conformation by using protein stabilizers while formulating the products, which keeps the products safe during processing and for long-term storage [60]. One of the strategies commonly used for stabilizing proteins in vaccines and other pharmaceutical delivery systems involves adding polysaccharides. The abundant hydroxyl groups present in inulin allow the polysaccharide to replace interactions between water and proteins as they dry [61]. Additionally, the flexible molecular structure of the polysaccharide and the ability to form glassy polymeric arrangements with high glass transition temperatures facilitates adaption to the confirmation of the protein [62,63]. The reported mechanism of fructans providing membrane protection during drying includes inhibition of vesicle fusion and an increase in the lamellar spacing created by the polymer inserting deeply in between the layers [64]. Inulin demonstrated superior ability to stabilize proteins and lipid-based delivery systems when compared to many other polysaccharides due to the linear and flexible backbone [65,66].

For more than 60 years, low molecular weight, soluble inulin has been considered as the "gold standard" for assessing kidney function because it is predominantly eliminated by rapid renal excretion, making it a useful compound for measuring glomerular blood filtration rate [43,44]. The exceptional tropism of inulin towards the kidney makes this polysaccharide a potential carrier for kidney-targeted drug delivery. In addition to kidney targeting, many researchers have demonstrated the excellent colon targeting capacity of inulin. Colon targeting of drugs to the colon exploits the presence and variation of the bacteria population residing in various segments of the gastrointestinal tract. This is the basic concept of using polysaccharides degraded by colonic bacteria as a suitable carrier for colon targeting [67]. Inulin is particularly attractive for this purpose because it is not digested or absorbed in the upper GIT tract, whereas it is significantly hydrolyzed by inulinases produced by Bifidobacterium in the colon [65,66,68–75]. Inulin with a higher degree of polymerization (DP) is less soluble compared with those with lower DP. The difference in chain length of inulin can be used to tune solubility, which is then exploited for either particulate delivery systems (long-chain give insoluble inulin) or kidney and renal function testing (short-chain give soluble inulin). Inulin has been used in different drug delivery systems (Figure 2).

Different drug delivery systems:

- Hydrogel
- Micelles
- Liposomes
- Prodrugs/Conjugates
- Inulin complex/chelating
- Microparticles
- Nanoparticles
- Solid dispersion

2.1. Inulin Hydrogels

Hydrogel is a three-dimensional (3D) network obtained from hydrophilic polymers that can swell and hold a large amount of water or physiological fluid without dissolving [76–79]. Hydrogels maintain their structure due to the chemical or physical cross-linking of individual polymers. Among the delivery system and carriers available, interest in the use of hydrogel as biomaterials is growing rapidly due to their exceptional biocompatibility, biodegradation, good loading capacity, similarity to normal tissues in terms of physiological activity [80]. The fabrication of hydrogels can be carried out using either natural or synthetic polymers. Natural polysaccharides such as inulin, pectin, chitosan, hyaluronic acid, alginate, starch, gelatin, pullulan, dextran, guar gum are examples of polysaccharides that form hydrogels [81].

The functional groups present in the polymer structure, particularly the hydrophilic group such as hydroxyl, amine, and amide, confer on the hydrogel a good swelling capacity and ability to absorb water. In particular, the large number of hydroxyl groups in inulin provides a functional handle for chemical modification (Figure 1). Based on the method of fabrication, hydrogels can be either physically or chemically crosslinked hydrogels. The physical cross-linking of hydrogels involves weak forces such as ionic interactions, crystallization, and hydrophobic interactions between water-soluble polymer chains. The use of chemicals to crosslink the polymers results in chemical hydrogels [80]. Changes in external stimuli such as temperature, pH, or ion changes can provoke some form of physical crosslinking in hydrogel systems. However, some of these hydrogels are characterized by poor stability and mechanical properties. In contrast, hydrogels fabricated with the use of a crosslinker show remarkably better stability and mechanical properties. The disadvantage of these chemical crosslinkers is the cytotoxic issues, especially to the biological tissues.

By tuning some parameters during the hydrogel preparation, the swelling behavior, crosslinking density, [82] mesh size, [77] permeability, mechanical strength, and drug release can be modified to tailor drug delivery to target tissues. By modifying inulin, stable hydrogels with desired physicochemical properties can be fabricated for use in drug delivery. Different approaches and strategies such as free radical polymerization of aqueous solutions of methacrylic anhydride or glycidyl methacrylates modified inulin [66,71,83], radical copolymerization, grafting polymerization [84], Michael addition crosslinking [85], UV radiation, chemical crosslinkers [86], enzymatic [87] have been reported for fabrication of inulin hydrogels. The use of aromatic azo agent bis (methacryloylamino)azobenzene (BMAAB) as a crosslinker for methacrylated inulin resulted in a rigid hydrophobic region in the inulin hydrogel [72]. Fabrication of smart inulin hydrogels that are responsive to environmental stimuli was using a different approach. For instance, addition reaction pH-sensitive inulin hydrogel was synthesized via chemical crosslinking with thiol-ene conjugate addition [54]. Modification of inulin with divinyl sulfone (DV), followed by succinic anhydride (SA), before crosslinking the derivative with tri-thiolated reagent resulted in the formation of hydrogel with the pH-sensitive property. The succinate groups confer pH sensitivity properties on the hydrogel [54]. Another pH-sensitive hydrogel was fabricated by initially modifying inulin with methacrylic anhydride and subsequently with succinic anhydride to form methacrylated/succinilated derivative (INU-MA-SA) followed by UV radiation to obtain the pH-sensitive hydrogel [88]. The use of radical copolymerization of modified inulin derivative MA-inulin with acrylic acid (AAc) in phosphate buffer solution with (pH 6.5) using both APS and TMEDA as an initiator reagent resulted in the formation of pH-responsive hydrogel [89]. Similarly, the condensation reaction between modified inulin derivative with the carboxyl group (INU–SA) and the hydrazide groups of water-soluble poly(amino acid) compound α,β-polyaspartylhydrazide (PAHy) with EDC and NHSS as coupling reagent results in pH-sensitive hydrogels [74]. Afinjuomo and his colleagues [86] fabricated pH-sensitive hydrogels by crosslinking inulin with an acidic anhydride-pyromellitic dianhydride (PMDA). They used the crosslinker (PMDA) via esterification reaction to synthesize hydrogels with pH sensitivity. The pH-sensitive hydrogels were useful and particularly important when considering target a specific area in the GI tract due to variation in pH along the gastrointestinal tract (GIT). While bioactive agents can be better protected due to less swelling at low acidic pH until the arrival of the hydrogel at the desired site (Colon). The extensive enzymatic degradation allows the drug encapsulated in the hydrogel to be released in the colon. Most of the reported pH-sensitive inulin hydrogels utilized both the change in the pH along the GIT as well as the presence of inulinase, which is predominantly found in the colon to achieved targeted delivery of drugs to the colon via the oral route.

Afinjuomo et al. also reported the fabrication of novel injectable inulin hydrogels by using oxidized inulin and adipic acid dihydrazide as a crosslinker without the use of any catalyst or initiator [90]. The hydrogels are formed via Schiff-base reactions between oxidized inulin and adipic acid dihydrazide. In this case, the drug-loaded hydrogel shows enhanced in vitro release of 5FU in acidic pH compared to physiological pH. Inulin hydrogels are also formed via Michael addition crosslinking reaction between linear inulin and another crosslinking agent divinyl sulfone (DVS) in a sodium bis (2-ethylhexyl) sulfosuccinate (AOT) inverse microemulsion [91]. It is very clear from the above works that different chemical approaches have been devised for inulin hydrogel synthesis. However, the use of physical methods such as hydrogen bonding and supra-molecular chemistry needs further investigation. Cross-linking using enzymes and chemical synthesis in a water-friendly environment are also other possibilities that need attention. The fabrication of inulin hydrogels results in new biomaterials, which are subsequently exploited for the loading and encapsulation of different drug classes. Different classes of drugs such as NSAIDs, hormones, peptides, corticosteroids, anticancer, immunoglobulin G (IgG) [33], bovine serum albumin, and lysozyme. Inulin possesses great advantages and unique properties, especially for the delivery of drugs to the colon. This has been demonstrated in several drug molecules targeted to the colon. Loading of drugs into inulin hydrogel can be performed either by encapsulating the drug during crosslinking of the gel or permitting the drug to diffuse into the pores of the hydrogel after the crosslinking procedure [92]. Both methods have been reported for the loading of drugs into inulin hydrogel. Nonetheless, purification and removal of unreacted chemicals are very challenging with the encapsulation approach. The diffusion method involves adding the hydrogel into a drug solution and allowing it to swell to equilibrium.

The in vitro release studies from pH-sensitive inulin hydrogels, under conditions mimicking the GIT, formulated for targeted colon delivery, illustrated that inulin could effectively protect the drugs from degradation in the upper part of the GIT, particularly in the stomach and intestine, and allow the release of the drugs in the colon. The release of active drugs from inulin hydrogel is assumed to be complex, and various release mechanisms have been proposed. Previous works suggested that three different release mechanisms, namely diffusion, swelling, and chemically controlled for the release of active drugs from the hydrogel [93]. The diffusion mechanism could be the principal way by which active drugs are released, considering the larger hydrogel mesh size compared to most small drug molecules (Figure 3). In this case, hydrogel mesh size in the swollen state allows diffusion of smaller molecules. Based on this mechanism, a macromolecules protein and peptide will likely have poor diffusion. In swelling controlled mechanism, the inulin hydrogel transit from glass to rubbery state, and the drug diffusion occur at a faster rate than swelling of the hydrogel. The slow swelling may be the rate-controlling process for drug release. The hydrogel experiences a swelling-driven phase transition from a glassy state to a rubbery state. Chemical control releases involve conditions where hydrolytic or enzymatic cleavage of the polymer chain or bond controls the release of drugs. Most of the inulin hydrogels reported in the literature seem to follow the swelling-controlled release. The driving force for drug release from hydrogel is the influx of solvent molecules into the hydrogel. In this case, the diffusion of drugs occurs at a faster rate than the swelling of the hydrogel. Two phases are formed during swelling, namely the inner glassy phase and the outer rubbery phase. Various parameters such as type of monomer [94], the molecular mass of peptide/protein [65,74], polymer concentration, the degree of substitution, loading concentration of the proteins [65], polymers swelling, and the electric charge can modify the release of drug/peptide from inulin hydrogel. For instance, the release of glutathione (GSH) and oxytocin (OT) from hydrogel synthesized from inulin INU–SA crosslinked with α,β-polyaspartylhydrazide (PAHy) was reported to be inversely proportional to the equilibrium swelling [74,95], which was also in agreement with reports from Maris et al. [72].

Factors that influence drug release from the hydrogels include loading procedure, degree of substitution, crosslinking density, pH sensitivity, and the type of bond. Van den Mooter et al. [65] prepared inulin hydrogel and loaded BSA protein into MA-IN hydrogel via two different loading methods (incorporation of proteins during crosslinking and after crosslinking). The authors reported that the hydrogel with BSA added during crosslinking exhibits a slow release pattern due to physical entrapment of the protein in the hydrogel compared with BSA loading via swelling. The swelling method results in protein release within one hour, which can be attributed to the inability of larger BSA molecules to diffuse to the inner smaller pore and the movement of the drug to the outer surface during drying [65]. A higher degree of substitution delays the diffusion of BSA from the hydrogel when compared to lower DS (15.4 versus 8.1). Crosslinking density is directly related to the degree of substitution. There will be less swelling with increased cross-linking density. The release studies of ibuprofen [75] and diflunisal [88] under conditions mimicking the GIT show pH-dependent release attributed to the carboxylic pendant group.

Hydrogel from methacrylated inulin demonstrated a high swelling rate, which was significantly influenced by the degree of substitution, polymer concentration, and media ionic strength [70]. The mechanism of water transport in the swollen inulin gel is controlled by both inulin chain relaxation processes and diffusion (anomalous, i.e., diffusion and relaxation controlled) [70]. In this work, the media pH does not influence the swelling of non-ionic hydrogel. In contrast, pH-dependent inulin hydrogel formed between (INU–SA) and α,β-polyaspartylhydrazide (PAHy) exhibited lower swelling in acidic media. Due to the presence of ionic groups in the hydrogels, they showed greater swelling in water neutral pH compared to acidic pH. The carboxyl acid pendant in the hydrogel network confers and allows a pH-sensitive swelling. The reduced swelling of the gel in acidic media reduces the degradation of drugs in the stomach, making them ideal for drug delivery to the colon. The author reports protection of the encapsulated drug glutathione and oxytocin from the low acidic environment of the stomach [95].

The copolymeric compositions and crosslink density are other parameters that determine the degree of swelling of hydrogels. For example, in the formation of pH-responsive inulin hydrogels prepared from acrylic acid (AAc) and methacrylated inulin (MA-inulin) via radical copolymerization. The hydrogel demonstrated a sharp increase in swelling with an increase in pH from 2.2 to 7.4. At low pH, less swelling is observed due to the presence of the carboxylic acid pendant, but with further increase in pH, swelling increases due to the protonation of the carboxylic acid group [89]. Similarly, a pH-responsive hydrogel obtained from methacrylic anhydride modified inulin chains with succinic anhydride demonstrated a pH-sensitive swelling behavior due to the presence of the carboxylic acid pendant [88]. At low pH (1.0 and 4.0), the swelling ratio was low, but with an increase in pH (6.8 or 7.4), the swelling ratio was increased. The pH-responsive hydrogel synthesized from using PMDA crosslinker exhibited excellent and high swelling in water far higher than those previously reported for inulin pH-responsive [86]. This observation is justified by the electrostatic repulsion between the ionic group within the hydrogel network. For hydrogels with an ionic group such as carboxylic acid group/pendant, the ionization of the functional groups and dissociation of the hydrogen molecules within the polymer chain with change in pH play a key role in explaining why this hydrogel has reduced swelling in acidic media, which makes them ideal for drug delivery to the colon.

Hydrogel biomaterials must be cleared from the human body after the delivery of drug cargo. One pre-requisite for the use of hydrogels in humans is biodegradability. The degradation of inulin hydrogels can be controlled and manipulated by different factors, including the type and cleavage of the bond in the polymer network as well as how the polymers are connected. Inulin is an attractive polymer for colon targeting because of the site-specific degradation of the glycosidic linkages of inulin by colonic microflora. Of note, it is important to investigate the degradation of modified inulin derivatives used for hydrogel formation by subjecting them to inulinase or cecal contents.

The degradation of the modified inulin (INU-MA-SA) hydrogels after incubation with inulinase was investigated using High-Performance Liquid Chromatography (HPLC) and Thin-layer chromatography (TLC). The analysis of the result after 24 h showed the presence of a fructose spot on the TLC analysis, confirming the excellent biodegradation of inulin. About 53.3% of the hydrogel was degraded within a day, indicating that the crosslinking and functionalization of the inulin backbone did not prevent degradation of the inulin. In contrast, the radical crosslinking of N,N′-bis(methacryloylamino)azobenzene (B(MA)AB), and methacrylated inulin resulted in denser hydrophobic azo-inulin making accessibility to the inulinase enzyme very poor and difficult [96]. Bacteria strain present in the colon degraded dextran azo gel while inu-azo hydrogel showed less degradation. Adequate swelling is important for inulinase hydrolysis and degradation. Crosslinking agents can influence both the swelling and degradation behavior of inulin hydrogel. Excellent or sufficient swelling is necessary for the enzyme's degradation of hydrogel in the colon environment. Detection of fructose monomers as a degradation by-product from hydrogels prepared by Pitarresi et al. [54,97] after incubation with colonic enzymes inulinase is another confirmation of the excellent biodegradability of inulin hydrogels.

The degradation rate of inulin hydrogel decreases with an increase in the degree of substitution as more sites are required to be cleaved for degradation to occur. Increased crosslinking density of the gel makes them less degradable. It was noted that the amount of hydrogel degraded was directly proportional to the enzyme concentration. In general, by using modified inulin derivative, it is possible to tune or tailor the degradation kinetics of derived gels for colon targeting by using enzymes that are specifically localized in the colon.

Non-hydrogel Delivery (Beads, Vesicles, Pellets, tablets, and nano-hydrogel) systems.

The use of a non-hydrogel delivery system such as beads and vesicles [98] with inulin polymer as the starting material has been reported in the targeting of drugs to the colon. López-Molina et al. investigated the use of inulin vesicles obtained by solvent precipitation technique after modifying inulin with cinnamic acid chloride for the colon-specific delivery of methotrexate (MTX) [98]. The vesicles were biodegradable by inulinase. The release of the anticancer drug MTX was significantly increased after the breakdown of the inulin backbone by inulinase in cellular conditions [98]. Fabricated beads using natural polysaccharides inulin and alginate were successfully utilized to encapsulate three different probiotics strains [99]. The results from this work proved that the synbiotic formulations could prevent the premature release of the probiotics from the upper GIT and only permit release in the colon. The work demonstrated the potential of inulin to deliver live probiotics to the colon. Similarly, Bahadori and his colleagues [100] successfully synthesized nano-hydrogel loaded with 5-aminosalicylic acid (5-ASA) using negatively charged anionic Na-Alg nano-particles and coated with biodegradable quaternized inulin with 5-ASA encapsulated core-shell structured delivery system. The inulin component enhances the stability of the formulation in the upper GIT and only allows the release of 5ASA as inulin begins to degrade, resulting in the prolonged release of 5ASA. A new pH-sensitive phthalyl inulin (PI) tablet capable of protecting the probiotic bacteria from degradation from low gastric acidity (low pH) and only released the probiotics at the site of action was also developed [101]. The formulated tablets from phthalyl inulin (PI) show higher swelling in simulated intestinal fluid when compared to simulated gastric fluid. The difference in swelling was ascribed to the pH sensitivity of the inulin derivative. The formulated tablets with pH-responsive swelling are a promising approach for the oral delivery of probiotics. Hufnagel et al. [102] synthesized a novel pellet of inulin using hydrophobic and acetylated inulin derivatives. The inulin derivatives were subsequently loaded with 5-aminosalicylic acid. The pellets release about 80% of 5-aminosalicylic acid within 24 h under conditions mimicking the GIT intestinal environment. The developed formulations show promising results. The delivery platform can revolutionize the delivery of drugs during the treatment of Crohn and ulcerative colitis.

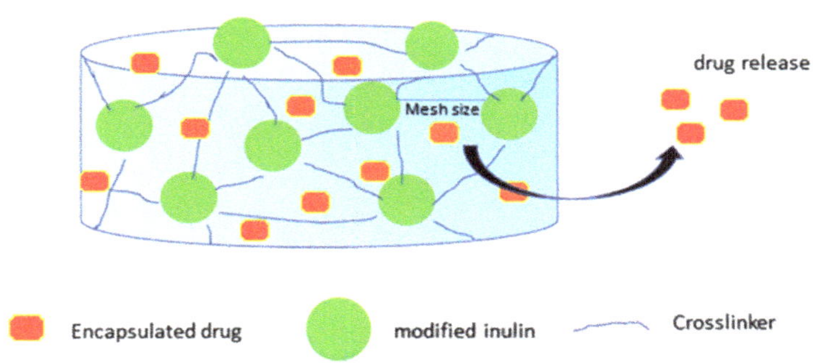

Figure 3. Hydrogel releasing of encapsulated drugs.

2.2. Inulin Micelles

Over the past two decades, the uses of polymeric micelles (PMs) as a versatile nanotechnology-based carrier system for poorly water-soluble drugs have gained tremendous attention [103–105]. The encapsulation of drugs with poor water solubility inside the hydrophobic core of PMs leads to significant improvement in solubility and bioavailability [106]. PMs are self-assembled core-shell nanostructures formed in an aqueous solution and consist of a hydrophobic core surrounded by a hydrophilic shell [106]. The accumulation of PMs into pathological tissues such as tumors and infracts with leaky vasculature is partly attributed to the small size ranging from 10–100 nm via enhanced permeability and retention (EPR) effect [107,108]. Interestingly, the small size of PM allows sterilization of micelle by filtration [109]. This system can be used for target drug delivery by modifying the outer surface using a specific ligand [110] (Figures 4 and 5). The type of micelle formed can be influenced by the type of intermolecular forces driving the formation of the micelle. Based on these, three different categories of PMs have been reported, namely amphiphilic micelles, polyion complex micelles, and metal complexation micelles [111]. When amphiphilic molecules are exposed to an aqueous medium, there is spontaneous self-assembly into supramolecular core/shell structures consisting of a hydrophobic core surrounded by a hydrophilic shell.

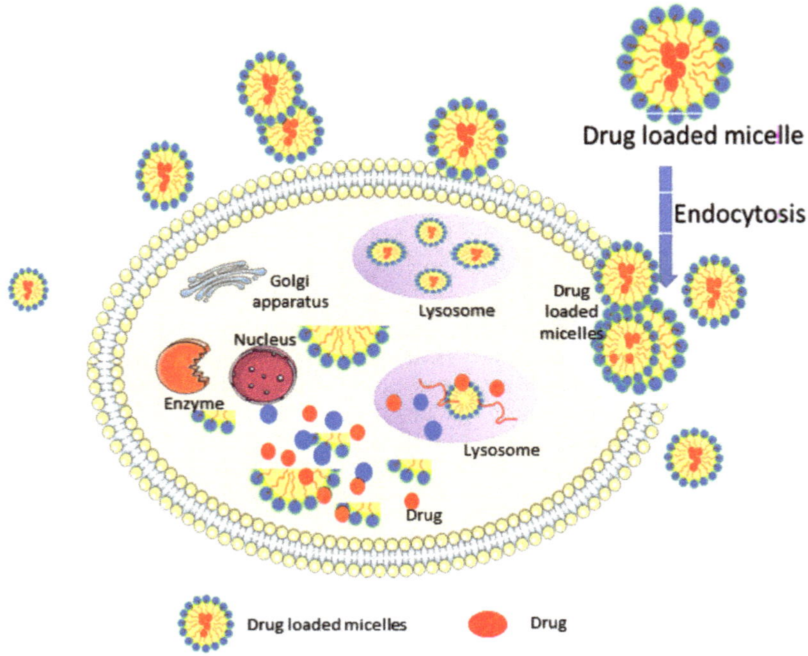

Figure 4. Polymeric micelle drug delivery concept.

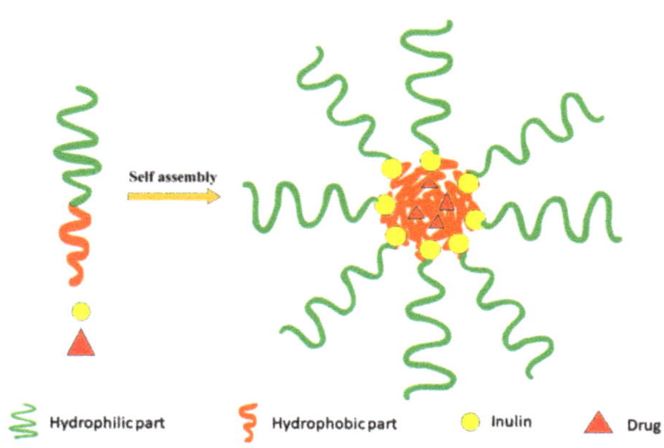

Figure 5. Inulin prodrug conjugate self-assembly into prodrug micelles.

Different shapes of micelles can be synthesized, for instance, shapes such as spheres, rods, vesicles, tubules, and lamellae can be formed depending on parameters such as length of hydrophobic/hydrophilic blocks and solution conditions [112,113]. Most polymeric micelles use polymers such as inulin as the hydrophobic domain. The excellent biocompatibility, biodegradability, and non-toxic nature of inulin make it an attractive polymer for the formulation of polymeric micelles. Inulin derivatives such as inulin-α-tocopherol succinate conjugates (INVITE), an amphiphilic polymer, form nano micelles that helps to deliver hydrophobic drugs such as curcumin via intravenous route for anticancer use Figure 6 [114]. The delivery of combination drugs such as curcumin and celecoxib with poor water solubility was further investigated with the use of INVITE micelles for the control of angiogenesis in cancer.

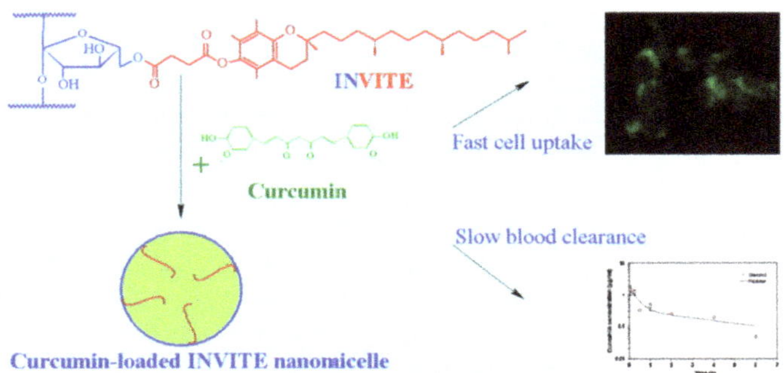

Figure 6. Inulin-D-α-Tocopherol succinate (INVITE) nanomicelles for intravenous administration of curcumin. Reproduced with permission [114]. Copyright 2015, American Chemical Society.

The partial replacement of the hydroxyl anchor on inulin with VITE results in the formation of hydrophobized INVITE with the ability to self-assemble into stable nanomicelles at low concentration (from 2.4×10^{-3} to 22.3×10^{-3} mM) [106]. The authors reported that the inulin micelles improved the solubility of poorly water-soluble and highly hydrophobic drugs, including curcumin and celecoxib (CLX) [106]. The release, cellular uptake of curcumin into the tumor cell was enhanced by the use of INVITE-based nanomicelles. Micelles can also be used for the delivery of different drugs. For instance, Tripodo, G et al. [115] designed modified inulin functionalized with vitamin E (INVITE) micelles for the efficient delivery of rifampicin for the treatment of infections caused by mycobacterium tuberculosis.

The modification of micelles surface with ligands can improve cellular uptake and increase the accumulation of the micelles in the target site. The potential use of polymeric INVITE micelles was further extended by decorating the surface of INVITE with biotin (BIO) to form INVITE-BIO micelles. This was then exploited for its long-circulating and targeting ability [116]. In addition, compared to INVITE micelles, the INVITE-BIO micelles demonstrated lower toxicity, long-circulating ability (48 h after IV administration), and improved drug accumulation in the tumor that over-express biotin through receptor-mediated endocytosis. In another study, inulin was modified with both hydrophilic chain poly-ethylene glycol (PEG) and hydrophobic ceramide to form a self-assembling amphiphilic inulin copolymer in water [117]. The micelle was then used to load and deliver doxorubicin, prolong blood circulation, and prevent recognition by the reticuloendothelial system due to the PEG component. The ceramide adds a hydrophobic effect and enhances antiproliferative signaling inside cells. The doxorubicin-loaded INU-ceramide-PEG2000 micelles demonstrated superior and preferential cytotoxic activity against cancer cells due to the enhanced uptake [117]. Similarly, Muley et al. designed a hydrophobically modified derivative of inulin (lauryl carbamate derivative of inulin). Inutec SP11 (INT) is another alternative to using a PEG block polymer in the delivery of paclitaxel via a micellar delivery system [118]. The resulting inulin amphiphile self-assembles in an aqueous environment with a low CMC of 27.8 μg/mL. The authors reported good hemocompatibility, low cytotoxicity, and enhanced antitumor ability in vitro and in vivo [118]. The formulation showed superior anticancer efficacy in mouse melanoma cells when compared with cremophor EL1 based Taxol formulation. In contrast to PEG, the INT micelles exhibited no production of antibodies or the accelerated blood clearance phenomenon, indicating that the INT micelles are safe carriers for IV administration of PTX [118].

In another study, a novel amphiphilic inulin copolymer INU-ED-RA was synthesized by partial functionalization of the inulin hydroxyl group with ethylenediamine (EDA) followed by coupling of retinoic acid (RA) to the INU-ED via amide bond using carboiimide chemistry [119]. The hydrophobic copolymer self-assembles in an aqueous medium with particle size less than 500 nm to form stable micelles. It has been used for the encapsulation of corticosteroids such as dexamethasone, triamcinolone, and triamcinolone acetonide [119]. The micellar delivery systems when freeze-dried resulted in physical stability for about three months and are compatible with ocular cells. In addition, the micelles possess mucoadhesive properties and enhanced permeation of the corticosteroid into the corneal and epithelial cells. The micelle facilitates effective internalization and uptake into an ocular cell. The drug-loaded micelles efficiently deliver the model drugs in high amounts compared to free drug suspension, demonstrating the potential of this novel system as a carrier for corticosteroids in the treatment of a degenerative vascular disorder.

On the other hand, smart lysosome-triggered delivery systems sensitive to pH environment that allow smart targeting of drug to tumor environment were synthesized using inulin. By using a four-step chemistry reaction, another pH-sensitive system, named INU-EDA-P, C-Doxo amphiphile was synthesized. Briefly, pentynoic acid was coupled to the modified inulin functionalized with amine via an amide bond, followed by attachment of the pH-sensitive citraconate pendants by reacting with citraconic anhydride. Lastly, dox was attached to the intermediate to form INU-EDA-P, C-Doxo using the ethyl-3-(3dimethylaminopropyl) carbodiimide (EDC)-mediated coupling reaction chemistry [120]. The final INU-EDA-P, C-Doxo prodrug conjugate self-assembles into prodrug micelles, which selectively release the linked DOX payload into the cancer cell cytoplasm. The inulin conjugates exhibited selective cytotoxicity against cancer cells. Interestingly, the conjugates self-assemble into organized nanosystems at low concentrations (0.33 mg mL^{-1}) due to volume difference during trafficking between tissue and cells, which ultimately results in the preferential release of DOX into the cancer site [120]. The citraconylamide spacer confers pH sensitivity, which triggers and promotes selective release and accumulation of the DOX in the tumor cells compared to normal and healthy cells. The presence of doxorubicin and pentynoic pendants in the conjugates contribute to the π-π structure and hydrophilic/lipophilic balance necessary for self-aggregation [120]. It was recommended that this drug delivery concept needs further investigation because of the selective release of drug cargo once inside the lysosome via pH-dependent stimuli, making it a potential lysosome-triggered delivery system [121]. Chehardoli et al. [122] reported the use of inulin-grafted stearate polymeric micelle-based carriers (In-g-St), fabricated from the reaction of stearoyl chloride and inulin for the precise delivery of betamethasone to the damaged joints area. The authors highlighted that the smaller micelles from inulin-grafted stearate allowed the controlled release of a poorly water-soluble drug such as betamethasone. Micelles formulated from inulin derivative conjugated to the peptide via an ester linkage were successfully used for the encapsulation of Ornidazole for colon targeted delivery [123]. The amphiphilic inulin derivative was capable of self-assembly in an aqueous environment into nanosized biomaterials, which enable the loading and encapsulation of ornidazole In this work, the author reported that the hydrophobic peptide arm supports and promote physical crosslinker, which resulted in self-assembly and formation of hydrophobic pocket and cores. However, despite the numerous advantages of polymeric micelles, there is a need for further research into stability and degradation.

2.3. Liposomes

Liposomes are drug delivery vesicles made of lipid that self-associate into bilayers with the ability to encapsulate the aqueous interior [124]. Liposomes offer several advantages, including circumventing the challenges of instability, cellular and tissue uptake obstacles, poor bioavailability, and poor in vivo biodistribution of drugs to the target site [125–127]. However, the setback to the use of the conventional type of liposomes as a drug delivery vehicle in clinical practice includes lack of specificity as well as recognition and degradation by the reticuloendothelial system (RES) due to their large particles size in the range of 500–5000 nm [128]. Novel approaches such as using smaller particles size (micro to nanosized scale) and surface modification with multifunctional carriers have been designed to improve the liposome drug delivery system [129–131]. Attaching targeting ligands to liposomes helped to specifically deliver the drug to target sites [132,133]. Inulin has been employed for sheath liposome strategy and as a component of the hydrophilic phase of the lamellar structure of liposomes. For instance, inulin has been used as a coating for liposomes and also for stabilization of labile therapeutic components in liposomes, polysomes, and lipoplexes [62–64]. They stabilized liposomes, providing opportunities for membrane interaction that results in increase acyl chain mobility and immobilization of the lipid head group regions due to their linear and flexible structure. Long-chain inulin conjugated to liposomes may give a dual advantage to the formulation (protect the drug from a harsh environment and increase the circulation time). Inulin derivative prepared by reacting oxidized inulin with ethylenediamine and diethylenetriaminepentaacetic acid and the liposomal encapsulation of this derivative with either $67Ga^{3+}$ or $111In^{3+}$ resulted in a stable complex in vivo [134]. The liposome-encapsulated inulin derivatives allowed loading and stability of heavy radioactive metal cations ($111In^{3+}$ or $67Ga^{3+}$) for in vivo administration. Moreover, the liposomes were utilized for investigating variation in tissue distribution among different drug targeting or delivery protocols. The advantage of this liposomal system is that it provides a non-invasive method of studying the transport and delivery of drugs [134]. Inulin's unique properties, such as macromolecular size, non-tissue distribution combined with the fact that it is not metabolized and only excreted in the urine, make the evaluation of intracellular drug delivery from the liposomal drug delivery system possible.

2.4. Prodrugs

The design of polymeric prodrug is another strategy for improving the therapeutic application of drugs [135]. Conjugation of drugs to polymer is to improve poor bioavailability low solubility, [136,137], short plasma half-life, and chemical instability of drugs [138]. Furthermore, it masks the bitter taste [139,140], reduces systemic metabolism [141,142], and minimizes drug resistance [143]. Pharmacokinetic and pharmacodynamic properties of drugs such as inadequate blood-brain barrier permeability [144] and lack of selectivity at the site of action can be improved using a product [145,146]. The use of inulin conjugate is aimed at designing sustained-release formulations by increasing plasma half-life [147,148], targeting drugs to a specific organ, for instance, colon targeting [149], improve cellular drug uptake via pinocytosis, and reducing toxicity. Inulin therapeutic benefit, blood compatibility, non-toxicity, and biodegradability make it attractive and a suitable drug carrier using the prodrug approach (Figure 7).

Modification of inulin by activating the hydroxyl group followed by nucleophilic attack and activation of carboxylic acid via EDC/NHS are two current strategies employed in the formation of the inulin prodrug. The two approaches result in esters and Schiff bases prodrug, respectively.

a. Prodrug with an ester linkage
b. Bioactive Schiff bases Prodrug

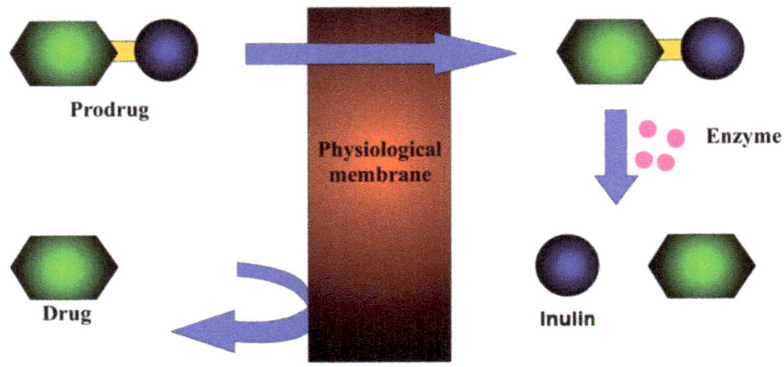

Figure 7. Diagram showing prodrug concept.

2.4.1. Prodrug with an Ester Linkage

Drug molecules with carboxyl and hydroxyl functional groups have poor bioavailability and limited diffusion across the lipid membrane due to ionization of the functional group under physiological conditions. The formation of inulin derivatives obtained by esterification reaction with other pro-moieties can give derivatives with the desired hydrophilicity, lipophilicity, and in vivo properties, as shown in Scheme 1. The rate of ester formation during the reaction can be increased by using either an acidic or basic catalyst such as 4-(dimethylamino) pyridine [150]. Metronidazole is the first-line drug for the treatment of anaerobic bacteria and protozoa infections, including amoebiasis. There is a need to target metronidazole to the colon for the treatment of conditions caused by the protozoa parasites. The parasite is also associated with complications such as hemorrhage and ulceration of the colon [151]. Vermeersun et al. reported the design of metronidazole inulin prodrug by succinylation of metronidazole to form metronidazole monosuccinate ester followed by coupling with inulin using 1,1′carbonyldiimidazole (CDI) chemistry [152] (Scheme 1). As expected and illustrated in Scheme 1, the use of dimethylamino-pyridine (DMAP), triethylamine (TEA) with the coupling agent CDI improves the yield and esterification degree of the reaction. The inulin prodrug showed adequate stability in the acid environment of the upper intestine, but the prodrug was swiftly cleaved to release the drug cargo on incubation with the cecal contents suggesting that the enzymatic cleavage and degradation of the prodrug is vital for its bioactivation leading to amoebicidal action of metronidazole. The author suggests that the systems could serve as colon-targeted delivery systems for metronidazole in the treatment of amoebiasis [152].

Hartzell et al. synthesized an ester-linked conjugate using 5-ASA and inulin to form 5-fASA-inulin a potential prodrug for colon-specific delivery of both 5-ASA and inulin to the gut [149]. The release of 5-fASA from the prodrug was a significant increase in the presence of gut bacteria. However, the 5-fASA released was not hydrolyzable to active agent 5-ASA by the gut bacteria. The prodrug delays the inulin fermentation in the colon, ensuring the beneficial effect of polymer persists for longer in the colon. The non-conversion of 5-fASA to the active agent 5-ASA makes this system unsuitable for the intended drug delivery approach. Therefore, it is necessary to investigate an alternative approach for the conjugation of 5-ASA to inulin to produce biodegradable inulin derivatives.

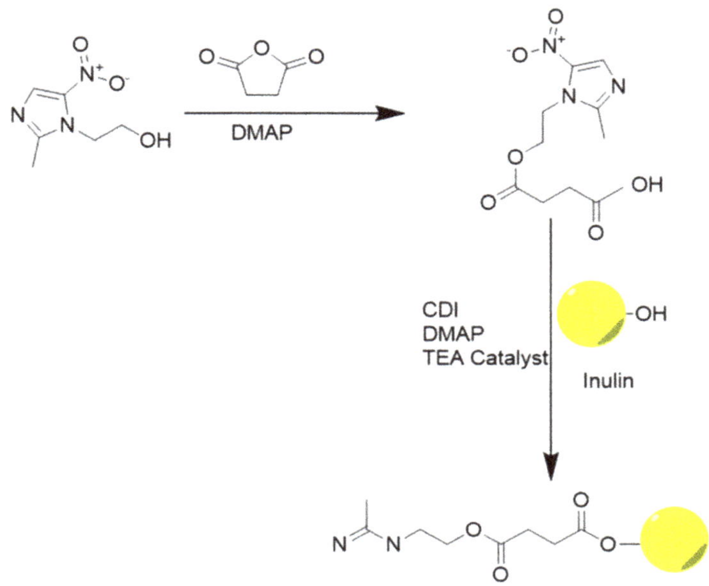

Scheme 1. Esterification of metronidazole to form a prodrug.

2.4.2. Bioactive Schiff Bases Prodrug

Amine-containing drugs often ionize under physiological pH, which facilitates faster diffusion, high first-pass effect, instability, and off-targeting of drugs into non-tumor sites [153,154], making the development of amine prodrug very challenging for formulation scientists. Developing a good amine prodrug was one way of overcoming this challenge. Schacht et al. [148], for instance, used the inulin prodrug of procainamide to improve its half-life and ensure a slow release of procainamide. In this work, the linkage of procainamide to inulin prevented the diffusion of the drugs to all parts of the body and ensured cellular uptake via pinocytosis. The conjugation involves oxidation of the hydroxyl group of inulin with sodium periodate, which resulted in the formation of dialdehyde followed by the coupling of procainamide via formation of the Schiff base and reduction with sodium cyanoborohydride. The procainamide moiety linked to the inulin backbone via hydrolytic imine conjugation bond was used to manipulate and tune the release of the drug. The in vitro release study showed that the drug was slowly liberated from the prodrug, and the rate was influenced by the ionic strength of the media [148]. The low yield of the prodrug was a limitation of this approach. The low yield can be attributed to the less reactive inulin hemiacetal intermediate from the oxidation process, which tends to reduce the effective attack of the intermediate by a nucleophilic agent. The problem of low yield was further addressed by using a different coupling method for the attachment of procainamide to inulin. In this case, inulin was activated by reacting with epichlorohydrin, and the derivative formed 3-chloro-2-hydropropyl was then reacted with procainamide in a neutral medium [147]. The author demonstrated that cellular uptake of the prodrug into yolk sac cells occurs via pinocytosis and the uptake process was dependent on the type of spacer and drug loading.

2.5. Inulin Chelating Agents/Complex/Polyplex

Magnetic resonance imaging (MRI) is one of the fastest and rapidly growing non-invasive diagnostic techniques that allow good and high-quality imaging of both healthy and diseased tissue/organ following the administration of a contrasting agent (CA) [155]. Currently, due to the high relaxation efficiency of Gd^{3+} -based contrasting agent, the vast majority of MRI scans are made after administering this CA [156,157]. However, Gd (III) toxicity in ionic form [157], long-term accumulation in the bone [158], and interference with protein and calcium-binding [159] limit its use for CA. In addition, this problem prevents the direct administration of this paramagnetic metal as contrasting agents (CA) for MRI. Hence, a complex is formed by attaching a ligand to Gd to overcome its toxicity. For this reason, one of the ways to address this limitation involves the formation of a low molecular weight Gd chelates/complex, formed by attaching ligands such as diethylenetriaminepenta-acetic acid (DTPA) and 1,4,7,10-tetra(carboxymethyl)-1,4,7,10-tetraazacyclodo-decane (DOTA). Unfortunately, the limited relaxivity (TR), rapid clearance, instability, short circulation time, toxicity, and precipitation of low molecular weight complex of Gd^{3+} limit their versatile application in MRI [157]. There was a significant compromise in the image resolution due to rapid elimination, making them unsuitable for angiography imaging.

As a result, there is a demand for novel contrast agents, which can address the abovementioned limitations. According to the relaxation theory, an increase in the rotational correlation time can be achieved by linking low molecular-weight Gd^{3+} chelates to sugar polysaccharides, which automatically translate to higher relaxation rates [160,161]. To achieve this, inulin polysaccharide has been explored as a platform for the attachment of chelates and complexing agents. In addition, macromolecular MR contrasting agents have increased the blood retention time and increased relaxivity due to the increased steric hindrance that translates into slower molecular tumbling. A lower concentration of agents is required for acquiring a satisfactory image during MRI testing because the macromolecular structure permits higher loading of the chelates and metal ions [162]. Compared with protein, which can provoke an immunogenic reaction, sugar macromolecular carriers are less likely to cause this problem [163].

Different derivatives and approaches have been used to synthesized inulin macromolecular MRI contrast agents. Corsi et al. reported the design of a new inulin conjugate (aminopropyl)inulin (API)linked via diethyl squarate to the DO3A ligand (1,4,7,10-tetraazacyclododecan-1,4,7-triacetic acid) [164]. The high molecular weight (12,000 and 22,000 Da) of the two synthesized macromolecular conjugates with DS of 0.7 and 1.5, respectively, prevented diffusion into extravascular space and improved the blood circulation time. Inulin flexibility facilitates the formation of a derivative with a higher degree of substitution that ensures rigid Gd^{3+} conjugates, which ultimately contributes to increasing rotational-correlation time [165]. The efficacy studies, according to the report by the authors, show that the final complex Gd37[(API-(DO3ASQ) exhibits an increase in relaxivity and was also five times more potent than commercial contrasting agents at 20 MHz and 37 °C. Furthermore, the high relaxivity (20.3 mm^{-1} s^{-1}) for this inulin complex gives credibility to its use as a promising contrast agent.

To increase the rotational correlation time, Lebduskova et al. designed a novel inulin conjugate and a Gd (III) chelate by covalently attaching carboxylic-phosphorus acid derivative of diethylenetriamine with inulin [166]. The novel agent formed in this work allowed the replacement of the central carboxylic pendant arm of 1,1,4,7,7-diethylenetriaminepentaacetic acid (H5DTPA) with a phosphinic acid functional group to form an intermediate, which was then subsequently linked to O-aminopropylinulin. The efficacy of the novel macromolecular contrast agent was evaluated at 20 MHz and 37 °C and the resulting NMRD profile showed that the inulin derivative was four times more potent than the commercially available contrast agents. This is mainly due to an increased rotational correlation time (866 ps at 298 K). The kinetic stability of the novel macromolecules was similar to the clinically approved MRI contrasting agent [Gd (DTPA)(H_2O)]2^-. Moreover, the inulin macromolecular structure allows the attachment of about 24 Gd^3+ ions per molecule, meaning lower concentration can be used for MRI enhancement. Toxicity will also be reduced as a result of using low concentration, which would help to increase MRI enhancement. The author reported that the final conjugate molecular weight of about 23,110 partly contributes to the favorable residence lifetime.

However, the poor kinetic stability of the previous contrast agents based on inulin macromolecules prompted Granato and his co-workers to design a stable Gd^3+ complex by conjugating modified inulin derivative O-(aminopropyl)inulin with DOTA-NCSA[1,4,7,10-tetraazacyclododecane-1,4,7,10-tetraacetic acid mono(p-isothiocyanatoanilide)] [167]. This Gd–inulin complex displayed superior kinetic stability compared to the conjugates prepared previously using a carboxylic-phosphorus acid derivative of diethylenetriamine [166] and API-DO3ASQ conjugates [164]. The relaxivity at 20MHz and 37 °C for these contrasting agents was higher than the model compound (21.7 vs 5 s^{-1} mM^{-1}), and this was attributed to reduced molecular tumbling originating from attachment to the inulin carrier.

The functionalization of the conjugates with targeting moiety improves the specificity of the CA by promoting vascular confinement to compromised tissue only. The rotational correlation times at 37 °C for the GdLx–API conjugate is about 17 times higher compared to GdLx. The authors reported that both GdLx and GdLx–API conjugate did not interact with human serum albumin. The high molecular weight of the CA macromolecule contributes to an increasing plasma circulation time, making it suitable for cardiovascular angiographic imaging. Inulin serves as an excellent carrier due to its flexibility [162,164,166,167] that allows the formation of inulin derivatives with high DS. The variable molecular weight of inulin 1500 and 5000 Da (depending on the degree of polymerization (10–30) support the formation of the derivative with a high molecular weight macromolecule suitable for use as a contrasting agent. Other properties of inulin, such as low viscosity, inertness, and not being metabolized in the blood, make it ideal for CA application [167]. The use of different tethers can influence the rotational correlation time of the CA. For instance, DO3A ligand, H5DTPA, and DOTA-NCSA when attached to inulin have rotational correlation time of 735, 866, and 1460 ps, respectively, at 37 °C, and the difference can be attributed to the use of a different tether to the same inulin backbone, which ultimately affects the mobility and rigidity of the final product. The excellent biodegradability of inulin coupled with its macromolecular size limits plasma clearance and promotes vascular confinement to cancer or diseased tissue. This makes inulin complexes ideal and more efficient contrast-enhancing agents than low molecular weight complexes. Nonetheless, the slow clearance of inulin complexes from the body may result in Gd^{3+} ions accumulation and toxicities [157]. From the studies above, it is clear that Gd- inulin complex is efficient, but there is a need to prove the safety of the Gd- inulin complex to realize its full potential in CA application. Short interfering RNAs (siRNAs) have been used in silencing the expression of genes that are involved in the progression of diseases such as cancer, cardiovascular, neurodegenerative, metabolic diseases, and viral illnesses [168]. The effective cellular uptake is obstructed due to their inherent negative charge and poor diffusion across the membrane. In addition, within the endosomes, there is the sequestration of the gene, which reduced the number of active drugs. Even though cationic polymers such as PEI exhibits high siRNA condensation and transfection ability, their full potential use is limited due to their non-biodegradability and cytotoxicity. All of these points further establish that there is a need for an improved delivery system.

Different strategies have been employed to overcome degradation, cellular uptake, and off-target problem. The use of cationic polymer is a very common and effective way for gene delivery because the polyplexes formed between the cationic polymer and nucleic acids result in reduced degradation and endosomal escape while improving cellular uptake. The use of functionalized and modified inulin represents one approach to overcome the toxicity obstacle without the need for extensive and multiple-step chemical modification. Different functional groups such as ethylendiammine (EDA), diethylediamine (DETA), spermine (SPM) have been conjugated to the inulin backbone [169]. Polycation synthesized by conjugation of inulin with diethylenediamine possesses positive charges at physiological pH, which allow its use as a carrier for siRNA [170]. The Inu-DETA copolymer can bind and release the siRNA in JHH6 cells with less cytotoxicity. Further modification of the INU-DETA imparted improved properties in the biomaterial. The introduction of imidazole (IMI) to the already synthesized INU-DETA copolymer results in the formation of INU-IMI-DETA, which exhibits pH-dependent response as well as strong cationic characteristics [171]. This new derivative demonstrated better cellular uptake due to self-aggregation into Inulin Complex Nanoaggregates (ICONs) triggered activated by physiological salt concentration and pH elevation [171]. ICON shows a superior ability to condense a higher amount of siRNA molecules because the protonation of the imidazole group will prevent escape from the endosomal/lysosomal compartment. The functionalization of the INU-IMI-DETA for active targeting are potential strategies that need further exploitation for future gene therapy delivery. The search for new biocompatible and safe inulin derivatives for gene delivery led to the use of the atom transfer radical polymerization (ATRP) approach to graft poly (DMAEMA) onto the inulin backbone [172].

The inulin cationic copolymer formed a strong complex with RNA/DNA via electrostatic interaction. The author reported that a low hemolytic rate and high transfection efficiency highlight that the modified inulin copolymer (DMAEMA)(PDIN) might be an excellent biocompatible candidate for gene delivery systems. The transfection efficiency of this copolymer was three times higher than that of the popular lipofectamine 2000 transfection reagent. The possibility of using modified inulin derivatives for the delivery of iron as well as iron supplements in the diet was demonstrated by Pitarresi and his colleague's findings [173]. The complex formed between iron, carboxylated, and thiolated inulin derivative (INU-SA and INU-SA-Cys) was exploited for the oral delivery of iron. The authors reported that the release profile of iron from the copolymer delivery system in a condition mimicking the intestinal tract was excellent, and the derivative also demonstrated good mucoadhesion properties [173]. In addition to the release, the modified inulin was broken down by inulinase despite modification.

2.6. Microparticles

Inulin microparticles have been used to deliver drugs, proteins, vaccines antigen, and other agents by adsorbing, attaching, or entrapping. Depending on the desired physiochemical properties, inulin microparticles can be prepared by either chemical or physical methods. The desired particle size and chemical properties of the drug/active agents determine the technique to be used. Inulin biodegradability is important when used as microparticles because the drug cargo entrapped in the matrix is released when the particle degrades. Furthermore, another good reason for using inulin can be attributed to its FDA-approved status, and the hydroxyl group in the structure provide a chemical handle and platform for modification. The degradation rate and drug release can be tuned by varying some parameters during the synthesis of inulin microparticles. In addition, the ability of inulin particles to be functionalized with targeting ligand makes it an excellent and ideal carrier. Modification of inulin results in derivatives with improved properties without changing the fundamental skeleton of the inulin backbone. Inulin microparticles have been reported as a carrier and adjuvant for vaccines. For instance, apart from stabilizing vaccines, delta inulin microparticles were reported to exhibit potent immunomodulatory properties making them suitable adjuvant [174]. Incorporating antigen into inulin microparticles resulted in targeted delivery of the antigen to the APC as well as adjuvant effects (dual function) [175,176]. By using special incubation conditions, seven different isoforms of inulin were fabricated by Silva and co-workers [46], but only the γ-inulin and higher isoform, insoluble semicrystalline particulate forms at 37 °C, possessed highly potent immunomodulatory properties [46,174]. Inulin semicrystalline particles, especially the delta inulin, showed promising results both as adjuvant and delivery systems for the vaccine [177]. The image characterization of these semi-crystalline forms of inulin using SEM and AFM morphology studies showed that they are regular discoid structures made up of piles of lamellar sheets with a diameter of about 1–2 μm, which facilitates their uptake by monocytes [177]. This unique size of Delta inulin (γ-inulin) microparticles allows efficient internalization by antigen-presenting cells (APCs). In addition, the excellent tropisms of these particles towards monocytes and dendritic cells are partly attributed to the surface molecular organization of the particles. In contrast, the soluble inulin does not possess immunomodulatory properties.

Another distinctive advantage of γ-inulin (delta inulin) microparticles is the ability to stimulate both the humoral and cellular immune responses without activating NFKb or inflammatory response [178]. These microparticles share a different mechanism when compared to other polysaccharide adjutants. Delta inulin activates the complement via the ACP, unlike other polysaccharides such as dextran, mannan, chitin, and glutan, which activate NFKb and inflammatory response [179–181]. Delta inulin remains a strong candidate for human vaccine adjuvant development because it is highly efficacious, potent, safe, and well-tolerated both in human, animal young foals, and pregnant mares [182,183].

In recent times, the use of soluble inulin as an adjuvant was reported. However, the soluble inulin particles in this work did not demonstrate potent immunomodulatory properties similar to the delta inulin polymorph [184]. Water-soluble inulin with particle size of 1.5 ± 0.12 μm was exploited as a drug delivery carrier for ovalbumin and as a vaccine adjuvant [184]. The ovalbumin-loaded inulin microparticles (SIMs) were synthesized by the emulsion–precipitation technique. Characterization of the microparticles by SEM shows that the particles are spherical in shape and the average diameter is about 1.5 μm promoted the uptake of the antigen into the APC's. Mice immunized with SIMs show significantly higher antibody titers and enhance cellular uptake of ovalbumin antigen by APC compared to alum plus ova or ovalbumin alone. The burst release of antigen was minimal (20% in less than 30 min) compared to about 50% release in 5 min with inulin microparticles encapsulated with BABCH. The microparticles allow the release of ovalbumin in a controlled manner, with about 70% of the ovalbumin released over 16 h. Unfortunately, the SIM shows poor Th1 type immune response, and this may require the use of agents such as CpG, MPLA, which can help increase the immune response to address this limitation. Another strategy used for the fabrication of microparticles was the emulsion/solvent evaporation method. Gallovic et al. evaluated the potential of soluble inulin microparticles with a diameter greater than 250 nm produced from acetylated inulin by emulsion/solvent evaporation method as carrier and vaccine adjuvant for ovalbumin (OVA) antigen [185].

This work exploited the change in pH along the phagocytic pathway to trigger the release of loaded antigen and immunostimulatory hydrolytic by-products from acid-sensitive acetylated inulin microparticles. The result shows that the inulin microparticles were able to induce a variable and significant amount of TNF-α production compared to unmodified inulin due to the cellular uptake of the Ace-IN microparticles by APC. Following immunization in mice, there was higher antibody production in mice administered with ovalbumin encapsulated inulin microparticles compared to pure soluble ovalbumin. This work demonstrated that inulin microparticles can efficiently deliver ovalbumin to APC as well as act as an adjuvant. The degradation of the acetylated inulin can be tuned and manipulated to the desired parameter. For example, varying the reaction time during the synthesis of acetylated inulin (from 20 min, 1 day to 3 days) results in the formation of 0.4, 20.4, and 26.7% of cyclic acetal coverages, which provide a chemistry platform to tune the degradation of the acetylated inulin. The ability to control the degradation of acetylated inulin allows the control of the immunostimulatory and TNF-α production [185]. The large size of this microparticle still allows passive targeting to the APC.

The acid-labile modified inulin makes it pH-sensitive, thereby preventing burst release until it reaches acidic phagolysosomes of APCs and also, the production of neutral degradative by-product gives these derivatives some advantage as a suitable carrier for antigen. It is worthwhile to point out that acetylated inulin particles were recently exploited as both adjuvant and antigen delivery vehicles against infectious disease and cancer [186]. The modified acetylated inulin inability to activate cells with TLR-4 antagonists and mice lacking TLR signaling support the assumption that they are toll-like receptor-4 (TLR-4) agonists. The encapsulation of acetylated inulin particles with antigen confers the ability to imitate pathogens, further translating into better antigen delivery (about 47 times more potent) compared to soluble antigen alone.

It is also possible to use engineered inulin microparticles to target monocytes and macrophages, which can be useful in the treatment of inflammatory disorders, tuberculosis, AIDS, and other infectious diseases. Afinjuomo et al. [187,188] used Delta inulin particles that were engineered via conjugation chemistry to efficiently deliver two different antitubercular drugs, namely pyrazinoic acid and isoniazid to host macrophages. They showed that the synthesized inulin conjugates could be used for intracellular delivery to the tubercle bacilli reservoir (Figure 8). The modified inulin microparticle shows a diameter of about 1–2 μm from the SEM results, which help contributes to the excellent uptake and tropism towards monocytes and dendritic cells.

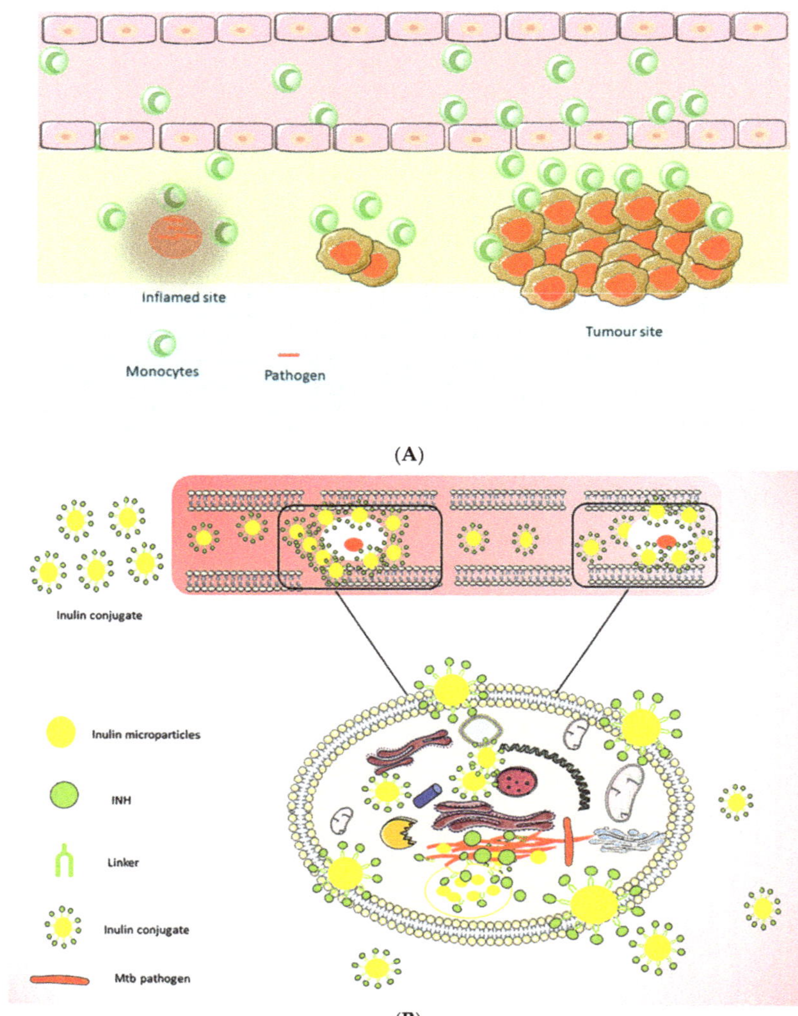

Figure 8. (**A**) schematic diagram showing monocytes targeted delivery to inflamed and tumor site. (**B**) Isoniazid intracellular targeted delivery to infected monocytes with inulin microparticles. Reprinted from [187].

Overall, the result demonstrates that the inulin delivery system can target the monocytes/macrophages harboring the intracellular Mtb pathogens (Figure 8B) [187]. Apart from infectious disease, the use of delta inulin was further exploited to deliver and promote uptake of anticancer drug doxorubicin to the monocytes [189]. This delivery system was also utilized to target doxorubicin to the lymphoid organs. The authors demonstrated that pH-dependent release for the conjugate as well as cellular uptake.

An alternative approach was reported for synthesizing acetylated inulin microparticles. By using the spray drying technique, gallic acid (GA) was encapsulated in modified acetylated inulin starch [190]. The impact of acetylation of inulin on the encapsulation efficiency and release behavior of gallic acid in water was investigated. The introduction of acetyl groups into the starch molecules led to a greater GA encapsulating capacity while the same introduction of an acetyl group into inulin resulted in decreased GA encapsulation efficiency. Due to the structural difference between inulin and starch, the acetylation of starch results in structural alteration and stronger hydrogen bonding. In contrast, modification of inulin with acetyl group reduces the water solubility. The in vitro release profile demonstrates that inulin acetate was superior and better than starch acetate as a carrier for the delivery of gallic acid. The GA release profile was similar for both unmodified and acetyl starch. However, inulin acetate particles release GA at a lower rate compared to native unmodified inulin. A change in the physicochemical properties of the new derivative (decreased hydrophilic, solubility, and swelling ability) of modified inulin might be a reason for the change in release rate.

Another fascinating strategy used in the formation of microsphere from inulin derivatives is the solvent precipitation method. Acetylated and succinylated inulin were synthesized by reacting inulin with both acetic and succinic anhydride. Using solvent precipitation method, microsphere of inulin acetate and inulin acetate succinate with both drug and no drug were prepared [191]. Cationic compounds such as chlorhexidine and chymotrypsin were used as model drugs. The microspheres diameter varies from 0.5 to 4 μm for inulin acetate and inulin/chymotrypsin microspheres from 90 to 130 μm for inulin acetate/chlorhexidine microspheres. The in vitro drug release results show that the microspheres allow the extended release of the incorporated drugs. Another feasible approach in the formation of microsphere from inulin involves the coacervation technique. Water-soluble drug E, E-bis (aminobenzylidene) cycloheptanone [(E, E)-BABCH] was encapsulated using inulin and modified inulin acetate as starting material with and without 1,12-dodecanedicarboxylic acid [192]. The microspheres diameter was about 0.5 to 5 microns and the influence of variable parameters such as drug mass, speed, stirring, and centrifugation time on the encapsulation efficiency of the modified inulin was investigated. The in vitro release profile for unmodified inulin and inulin acetate shows burst release of about 58–62% within 5 min. However, the microsphere formed with 1,12-dodecanedicarboxylic acid reduce the burst release to about 32% within 15 min and allows controlled release of the remaining drug over 2.5 days. Inulin microparticles find application in colon targeting. Inulin microparticles were prepared from acetylated inulin for colonic delivery of indomethacin by electrospraying technique. The acetylation of inulin before microsphere formation improves the formation of inulin microspheres without trading off the colon targeting property of inulin [193]. The in vitro release of indomethacin under simulated GI conditions similar to the GIT shows that these microparticles were stable to acid degradation of the stomach environment, and about 60% of loaded indomethacin was only released in the simulated colonic fluid in the presence of inulinase after 18 h in the colonic fluid. The author concludes that indomethacin was release from the drug-loaded microparticles due to the degradation of the inulin glycosidic linkage by the inulinase microflora of the colon.

2.7. Inulin Nanoparticles

Nanoparticles can target cancer cells passively via leaky endothelium (EPR) or actively by binding to specific receptors using ligands on the surface of the nanoparticles

(Figures 9 and 10) [194]. The size of nanoparticles controls the interaction of the nanoparticles with cells, trafficking as well as internalization. This translates to the delivery efficiency of the system.

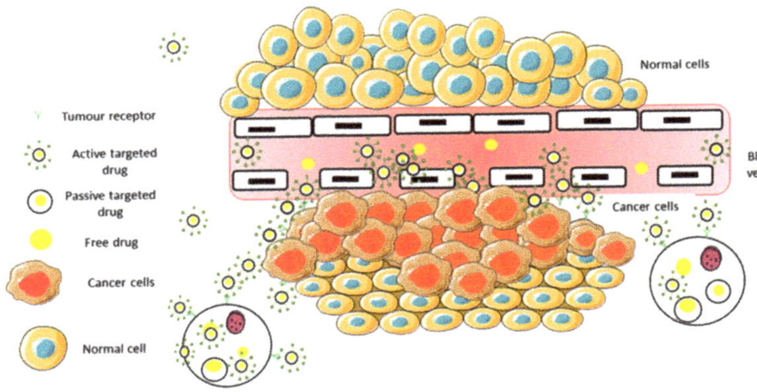

Figure 9. Active binding of the ligand attached to the nanocarrier surface to receptors commonly overexpressed by the cancer cells.

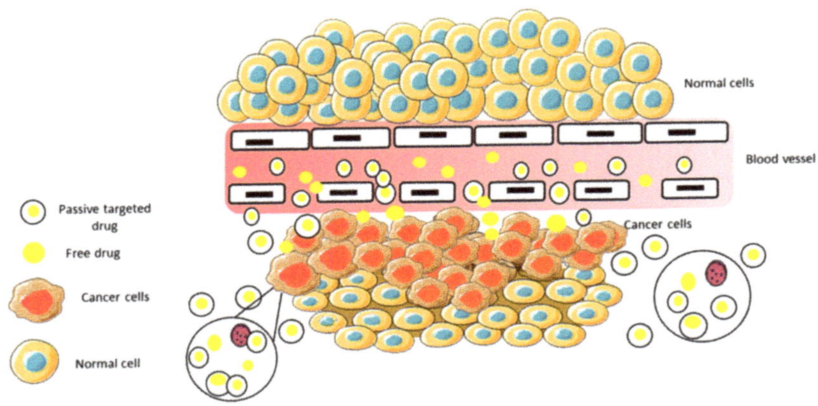

Figure 10. Passive targeting of cancer cells using nanoparticles via leaky endothelium.

2.7.1. Active Targeting

Self-Assembling for Active Targeting

The heterogeneous tumor's extracellular matrix in cancer cells limits and prevents the accumulation of nanoparticles inside cancer cells. To overcome this challenge, Zhang et al. [195] developed RGD-peptide conjugated inulin-ibuprofen nanoparticles for the smart delivery of epirubicin. The capacity of the modified inulin–ibuprofen conjugate to undergo self-assembly into nanoparticles with inulin forming the hydrophilic shell and the ibuprofen forming the hydrophobic core was utilized to enhance the active targeting of the inulin-based nano-carrier in the delivery of epirubicin (Scheme 2). The chance of premature drug release outside the target site can be minimized by attaching targeting ligand such as RGD peptide to the surface of the nanoparticles. TEM analysis showed that both EPB loaded nanoparticles and RGD conjugated nanoparticles are round with a diameter of about 135 nm and 138 nm, respectively. The in vitro release studies revealed a pH-dependent pattern with slow-release at neutral pH 7.4 and significant release of EPB at the acidic environment (pH 5.0) of cancer and lysosomes. The conjugation of

RGD-peptide to the EPB loaded nanoparticles improves the cellular uptake and cytotoxicity of the nanoparticles compared with EPB loaded nanoparticles. The result suggests that this approach may be useful in the smart delivery of drugs to solid tumors.

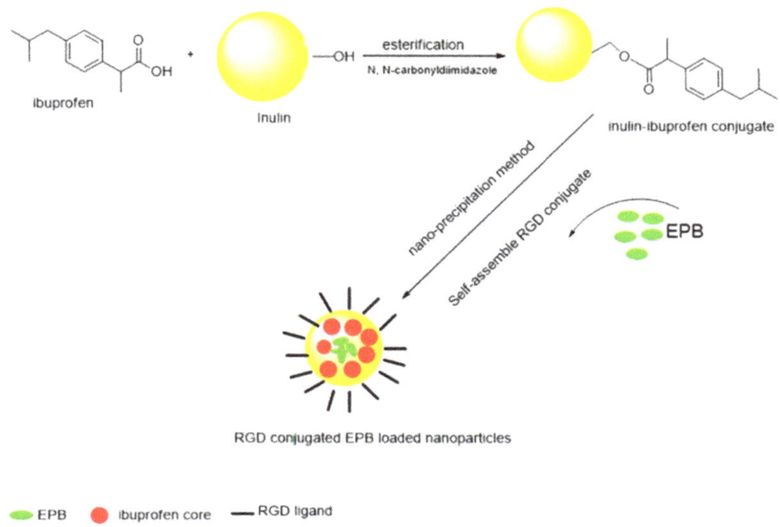

Scheme 2. Formation of EPB loaded nanoparticles by self-assembly of inulin-ibuprofen [195].

The use of folic acid as a ligand can also offer the specific targeting capability for the inulin derivative towards tumor cells. This strategy was exploited by Licciardi et al. [196]. Modified inulin derivative acted as a coating agent for gold nanoparticles loaded with DOX and functionalized with folic acid ligand to achieve targeted and selective delivery to MCF7 cancer cells. In this study, Inulin-2-aminoethyl-carbamate (INU-EDA) was conjugated with folic acid using 1-Ethyl-3-(3-dimethylaminopropyl) carbodiimide (EDC) and N-hydroxy succinimide (NHS) as coupling agents to form inu-FA derivative. This derivative was subsequently used to functionalize and coat the surface of 40 nm gold nanospheres. The interaction between the amino pendant group in the inulin-FA derivative and the gold helped to maintain and stabilize the nanoparticles. The characterization of the nanoparticles using DLS and UV showed that DOX was successfully adsorbed onto the surface of the gold nanoparticles due to electrostatic interaction and possibly the formation of the gold complex [AuCl(NH2R)] [197]. The authors reported a hydrodynamic diameter of 40 nm from the DLS result. This small size improve the targeting behavior leading to increased anticancer activity.

This approach of using folate ligand results in increased uptake of nanoparticles by MCF7 cell line overexpressing folate compared to folate receptor-deficient cell line such as human bronchial epithelial cells via receptor-mediated endocytosis. In addition to the cellular uptake, the targeting moiety folic acid ligand enhances the accumulation of the nanoparticles into the MCF7 cells.

2.7.2. Passive Targeting

Gold Nanoparticles (GNPs)

To address the aggregation problem that often arises when citrate is used as a capping ligand [198]. Licciardi and colleagues [196] used inulin derivative (INU-EDA) and thiolated polyethyleneglycol (PEG-SH) as a coating agent and pre-stabilizing agent, respectively, for the fabrication of 40 nm gold nanospheres [199]. Characterization of the gold nanoparticles using TEM shows no change in particles size or evidence of aggregation after the modification. In addition, UV, DLS, and zeta potential analysis validate the stability of the DOX nanosystem in different media. The HPLC analysis results revealed that about 12% of DOX was loaded, and encapsulation efficiency was about 60%. The high encapsulation efficiency was attributed to physical interactions of inulin backbone INU-EDA and DOX and also to the hydrophobic interactions of DOX with the gold core. Using a co-culture cell model it was demonstrated that the nanosystem allowed the selective accumulation of the nanoparticles in malignant cancer cells. The authors suggested that the use of inulin derivative INU-EDA improved loading of DOX and stabilization of the Au@PEG-INU NPs in a buffer for 14 days. Interestingly, the nanosystem allowed slow, sustained, and selective release of DOX into cancer cells. Further potential application of this system in light-controlled drug release may be feasible because INU-EDA can be used for coating different sizes and shapes of gold nanoparticles.

Inulin Based Magnetic Nanoparticles

SUPERPARAMAGNETIC NANOPARTICLES (SPIONs)

One major challenge limiting the use of iron-oxide particles is the rapid elimination from the blood by the reticuloendothelial system (RES) [200,201]. Modification of the iron oxide particle is one of the effective strategies that can be employed to reduce RES clearance. The surface modification of iron oxide particles can improve stability, biocompatibility and allow multifunctional use of the particles as a platform for drug delivery [202] and imaging agent [203]. The magnetic nanoparticle can be used to address limitations such as invasive long-term treatment, treatment failure, tumor resistance, severe adverse effects, and non-selective accumulation to target cells [204,205]. SPION based drug delivery exploits a dual-targeting strategy (leaky structure of tumor vasculature-EPR and magnetic targeting to increase targeting efficiency [206]. SPION is superior to other drug delivery carriers as it allows imaging and drug delivery simultaneously. In addition, it offers better targeting efficiency. However, its use is limited by the formation of acidic by-products, which could result in inflammation and poor biodegradability. Suitable inulin derivatives have been used to overcome the drawbacks above. Inulin biocompatibility and biodegradability make inulin-coated superparamagnetic iron oxide nanoparticles a suitable platform for the delivery of drug and imaging agents. In addition, inulin improves stability and prolongs the blood circulation of SPION. Scialabba and co-workers synthesized amphiphilic inulin derivatives used for coating SPIONs [207]. This derivative, once placed in aqueous media, increases the loading of DOX significantly during the self-assembling process [207]. Characterization of the IC-SPIONs using SEM and DLS analysis confirmed the stability of the modified inulin with measurement of 21 ± 8 mV for zeta potential and spherical morphology a diameter of about 50 nm. The hydrophobic interaction between the aromatic ring of DOX and the squalenoyl appendage moieties leads to the formation of smaller nanoparticles, and this allows the loading of higher DOX ($11.6 \pm 0.5\%$ (*w/w*)). In addition, the squalenoyl portion adds hydrophobicity to the inulin backbone, which assists in stabilizing the SPION coating via van der Waals forces. The release studies in two different media simulating the different human body compartments revealed a slow release pattern with about 12% of doxorubicin released after two days, irrespective of the pH of the media. The nanoparticle size (~55 nm) allowed the efficient uptake and internalization of the particles into the cytosol compartment area. Compared with free doxorubicin, the inulin-coated SPION showed three-fold anticancer activity under the control of an external magnetic field. The inulin coated SPION allow preferential accumulation and release of DOX with the external magnetic field (Figure 11).

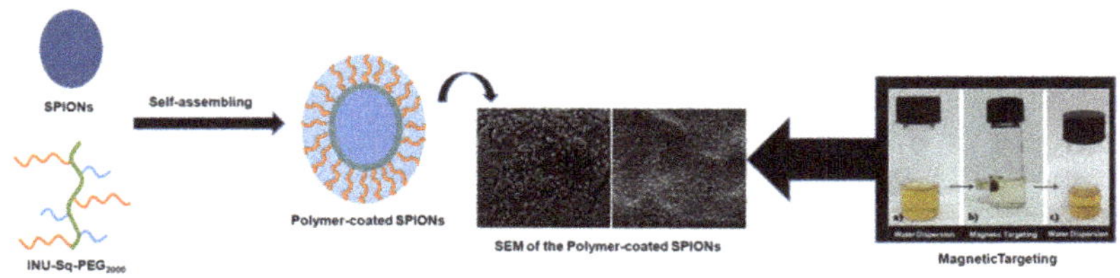

Figure 11. Inulin-based polymer coated SPIONs loaded with DOX. Reproduced with permission from [207] Copyright 2014, Elsevier.

In a similar study, the versatility of inulin derivatives was exploited to obtain carboxymethyl inulin (CMI). It was subsequently used for coating iron oxide for use as a magnetic fluid hyperthermia agent [208]. This work involves coating the surface of iron oxide particles decorated with amine silane with carboxylmethyl inulin using amide covalent linkage. Physicochemical characterization using different techniques such as FTIR, SEM, DLS, and SQUID magnetometry analysis showed evidence of new band appearance, hydrodynamic diameter 70 ± 10 nm, and polydisperse particles. The inulin coated iron nanoparticles were not only less toxic to five different cancer cells despite their higher concentration of 1.5 mg/mL, but also a significant amount was uptaken by Caco-2 cells. Moreover, the amide attachment of CMI through the APS layer of the iron oxide gives better stability to the nanosystem when stored in water and other media. The fact that no change in hydrodynamic diameter with change in temperature (25 °C and 60 °C) was observed makes this nanosystem a potential magnetic fluid hyperthermia agent.

Functionalized Superparamagnetic Nanoparticles

To increase the active targeting capacity of inulin nanoparticles towards specific cancers cell that over-express folic acid on the surface, Scialabba et al. developed an amphiphilic inulin derivative using folic acid as a ligand, which was subsequently used as a coating material for SPIONs [209]. Folic acid is over-expressed in various human cancers such as colon, breast, lung, and ovary and prostate cancer. This encourages the engineering of an efficient drug delivery platform capable of selectively binding to the target cell. In this work, a new amphiphilic graft copolymer INU-LA-PEG-FA copolymer, which can self-assemble in aqueous media, was synthesized by a reaction that involves two steps. First, lipoic acid (LA) was attached to the inulin backbone, followed by conjugation with folic acid via PEG spacer by reacting with NH2-PEG-FA. The final copolymer INU-LA-PEG-FA copolymer was then employed as a coating and stabilizing agent for SPION, and the surface is enriched with folic acid to improve the selective targeting of cancer cells. The change in the size from 14 to 131 nm as well as a surface charge from ~−30 to −5 Mv promote and helped the successful adsorption of the copolymer to the surface of the SPION. The cell viability and erythrolysis studies showed that FA-SPIONs were not toxic to different cancers cells (HCT116, MCF7). In addition, the in vitro and in vivo MRI experiments revealed that FA-SPIONs were efficient as a contrasting agent for MRI analysis combined with superior targeting ability attributed to the surface ligand folic acid. The use of PEG spacer prolonged the half-life of the drug delivery system by reducing their removal by the reticuloendothelial system.

Magnetoplexes

The development of new and tunable RNA and DNA delivery platforms for the treatment of gene-related disorders is drawing attention after the approval of Glybera by the European Medicines Agency (EMA) [210]. Nevertheless, low transfection efficiency, unloading of the siRNAs to the target site, poor siRNAs uptake, and degradation limit the potential use of RNA and DNA by most routes [211,212]. For this reason, the use of non-viral vectors for the delivery of RNA and DNA is becoming popular due to the low cost of production, safety, less immunotoxicity, and low pathogenicity when compared to the viral approaches. Apart from the drug and ligand conjugation, the hydroxyl group of inulin serves as an anchor in the formation of polycations derivative that can easily complex with the negatively charged siRNA and DNA. Inulin derivatives have been used in coating magnetic nanoparticles in the delivery of therapeutic genes. In one approach, inulin derivative was used for dual purposes (coating SPION and complexation with siRNA). Licciardi et al. enhanced siRNA delivery by developing SPION stabilized with inulin-ethylenediamine derivative obtained by enhanced microwave synthesis (EMS) [213]. The inulin coated SPIONs were less toxic to 16HBE and HCT116 cancer cell lines, and at the same time, their transfection efficiency was significantly increased. The cellular uptake and accumulation of the siRNA in the cells were also enhanced in the presence of an external magnet [213].

Self-Assembled Inulin Nanoparticles

Zhang and co-workers developed an amphiphilic inulin derivative by the esterification reaction between the functional hydroxyl group of inulin and the carboxylic acid groups of ibuprofen using carbodiimide chemistry. The inulin was subsequently self-assembled into nanoparticles by the nanoprecipitation method [214]. The amphiphilic inulin derivative self-assemble into nanoparticles with a hydrophobic core and hydrophilic shells in an aqueous environment. By chemically attaching hydrophobic moiety such as ibuprofen to inulin hydrophilic backbone allows the formation of self-assembled nanoparticles (SNP). The authors used a pyrene fluorescence probe to confirm the formation of nanoparticles by investigating critical aggregation behavior. The physicochemical characterization of the nanoparticles was conducted using ^{1}H NMR, TEM, and DLS, and it was confirmed that ibuprofen was attached to the inulin backbone, with a hydrodynamic diameter of 141 nm for the drug-loaded nanoparticles with spherical shape nanoparticles. In a similar study, methylprednisolone was enveloped into the inner hydrophobic core of the nanoparticles, and the in vitro release in phosphate buffer solution by dialysis method showed a sustained release pattern with about 94.9% of the drug released in 4 days. The cell viability studies showed that both empty nanoparticles and drug-loaded methylprednisolone nanoparticles were less toxic to the Rat Schwann cell line RSC-96. It should be emphasized that this drug-loaded nanoparticle may be utilized in the smart and targeted delivery of drugs such as methylprednisolone for spinal cord injury.

Nanostructured Carbon-Nanosheets

The quest for a multifaceted approach to the improvement and treatment of cancer was the compelling reason and motivation for the development of a nanosystem with RGO-induced hyperthermia and a smart chemotherapy delivery system (Figure 12). Inulin-doxorubicin prodrug conjugate (INU-EDA-P, C-Doxo) with proven pH sensitivity and self-assembling properties was functionalized with PEG-biotin using Huisgen cycloaddition click chemistry [215]. The use of Huisgen cycloaddition click chemistry to functionalize inulin prodrug (INU- EDA-P, C) is a classic example of using molecular designs and modification reactions to form novel polymeric materials with advanced function.

In this case, the functionalizing of the inulin derivative with biotin allowed preferential selectivity of the delivery system towards cancer cells. The available alkyne groups in the inulin prodrug (INU- EDA-P, C) support conjugation of the prodrug with biotin, which is the targeting agent via the PEG spacer arm. The addition of the reduced graphene oxide (RGO) and biotinylated Inulin-doxorubicin conjugate results in a Π-Π-staked nanosystem with a synergic anticancer effect. The stability, self-assembling formation of the stable π-staked system was confirmed using UV, AFM, and DSC analysis. It was easy to control the amount of DOX loaded, which is an important factor that can influence the efficacy of nanoparticles in vivo by varying the ratio of conjugate/RGO. The biotinylated system shows both higher efficacy and potency compared with the non-targeted system, which is attributed to preferential and enhanced uptake as well as efficient drug release at acidic tumor pH. The application of an 810 nm laser diode cold laser beam resulted not only in the release of DOX but also physically burnt/damaged the cells. This was evidenced by a decrease in cell viability and blocked cell proliferation. The result demonstrated the dual ability of this nanosystem to function as a thermoablation agent with a hyperthermia-triggered drug release mechanism. This nanosystem could significantly broaden the use of inulin nanoparticles in both drug delivery and biomedical fields.

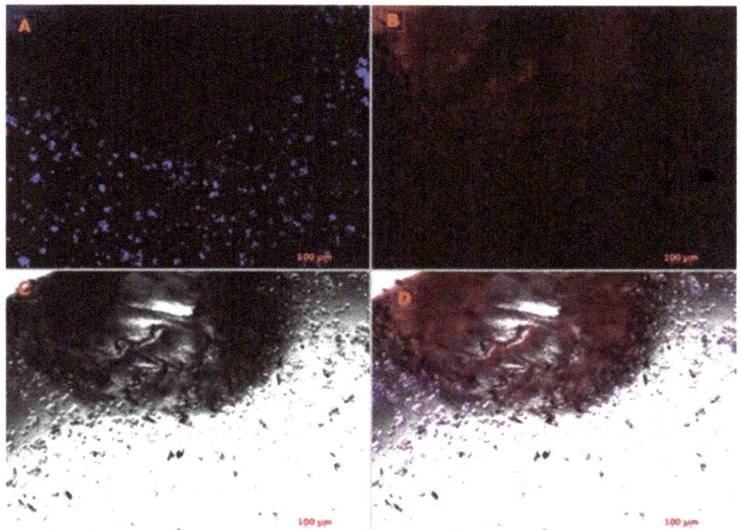

Figure 12. Thermoablation of RGO/CJ-PEGBT treated MCF-7 cells obtained with an 810 nm laser fitted at 7×10^{-2} W mm^{-3}: blu nuclei (**A**), red doxorubicin (**B**), bright field (**C**), merging of all fluorescence (**D**); magnification 5×. Reproduced with permission from [215]. Copyright 2015, American Chemical Society.

2.8. Solid Dispersion

Increasing both the dissolution rate and bioavailability of drugs with poor water solubility (Biopharmaceutics Classification System class II and IV) remains a challenging task for formulation scientists. One way of dealing with this problem involves the use of solid dispersions (SDs) (Figure 13). Drugs with low and poor water solubility may be dispersed in solid dispersion by using a hydrophilic carrier, which helps to increase the surface area of the drug available for dissolution, improved drug wetting, reduce particles sizes of drugs and improve the solubility of drugs under physiologically relevant conditions [216].

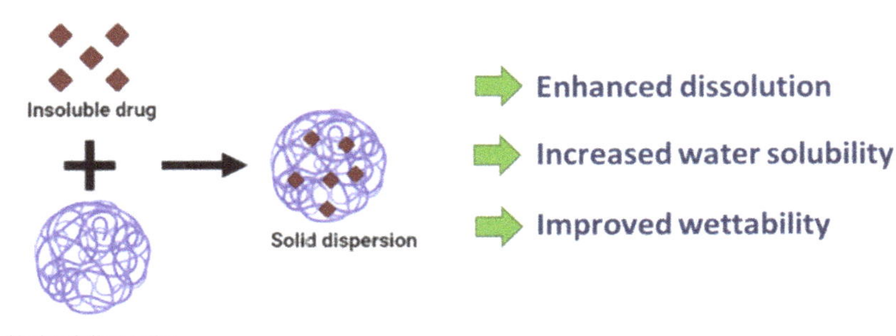

Figure 13. The formation of solid dispersion and the benefits.

Hydrophilic carriers such as inulin can be utilized as solid dispersion carriers to improve and enhance the solubility of drugs. The use of inulin solid dispersion technology to boost the release and bioavailability of poorly soluble drugs such as nifedipine (218), diazepam [217–219], lipophilic drugs [218], (fenofibrate, ritonavir, and efavirenz) cyclosporine A [220], curcumin, irbesartan [221], and itraconazole have been reported. The high glass transition temperature of inulin makes it a suitable carrier in solid dispersion formulation because this allows stability of the final formulation. Inulin glassy SD improved the stability of labile THC D9-tetrahydrocannabinol [222], diazepam [218,223], and fenofibrate [218] due to the high glass transition temperature of inulin. The dissolution of the carrier results in faster hydration of the poorly water-soluble drugs. One of the limitations with a sugar glass solid dispersion system is low drug release due to crystallization after exceeding certain drug load percentages with the use of sugars such as sucrose or trehalose [219]. However, compared to other simple sugars such as sucrose or trehalose as a carrier for solid dispersion, inulin was a better candidate because it exhibits slower dissolution and a higher crystallization threshold [219]. The superiority of inulin over polyvinylpyrrolidone (PVP) as a carrier for solid dispersion for different drugs with poor water solubilities such as diazepam, ritonavir, and efavirenz was demonstrated [218]. Inulin lauryl carbamate (Inutec SPI) is a surfactant formed by covalently attaching lipophilic alkyl side chains to the hydrophilic inulin backbone. The use of Inutec SPI as SD permits the loading of a high amount of drug without crystallization during dissolution. The inulin derivative showed superior dissolution behavior and improved stability of the final formulation due to its surface-active nature. The use of Inutec SPI solid dispersion outperformed pure itraconazole because solid dispersion carriers of itraconazole yield better result outcomes [224]. The dissolution behavior, releases, and bioavailability for a class IV drug TMC240 (HIV protease inhibitor characterized by poor solubility and permeating properties) were significantly improved by using Inutec SP1 solid dispersion technology [225]. Apart from Inutec SPI, the use of derivatives with different moieties attached to the inulin backbone as solid dispersion for dissolution enhancement is possible. The dissolution and bioavailability of irbesartan were increased by using poly (acrylic acid) grafted inulin copolymer [221]. The inulin backbone was grafted with hydrophilic moieties, which leads to a higher dissolution rate and bioavailability of water-insoluble irbesartan. The incorporation of poorly soluble HTC into the inulin solid dispersion matrix not only increases the stability but also the bioavailability [226]. The stable dry powder obtained with glassy inulin polymer allowed the development of a more convenient sublingual dosage formulation with immediate release within 3 min compared to about half an hour for a tablet.

There is a high probability of undesired recrystallization with higher drug loading with solid dispersion technology, which can be minimized with the addition of surfactant [37]. Interestingly, despite the high drug loading, this problem did not arise with solid dispersions of diazepam [219] and cyclosporine A [220]. Inulin solid dispersion of cyclosporine A (CSA) prepared by spray freeze-drying surpass formulation with a physical mix of both pure CSA and spray freeze-dried inulin even at higher drug loading [220]. The solid dispersion results in the formation of large porous particles deemed suitable for the pulmonary route due to deep lung deposition and high emitted dose. The use of inulin in the SD increased the specific surface area and wettability of CSA, making it a suitable carrier for the design of pulmonary formulation useful for the treatment of lung transplant patients. The superior dissolution enhancement properties of inulin as solid dispersion can be ascribed to different mechanisms such as the formation of flower-like platelets of inulin around the drug [221,227], surface-active nature of the inulin derivative [218] as well as higher surface area and improves wetting of drugs agents [220]. Another important factor that can influence drug dissolution and release of solid dispersion is the chain length of inulin used in the formation of the solid dispersion. Drug dissolution from solid dispersion may be hindered by using inulin with higher molecular weight. The higher molecular weight inulin does not give better solid dispersion because of the slower dissolution of the inulin matrix. Lower molecular weight inulin due to the oligomeric nature performed better than higher molecular weight, as demonstrated with solid dispersion of tacrolimus with 1.8 kDa and 4 kDa inulin [228]. The smaller chain 1.8 kDa has higher solubility in an aqueous environment compared to the longer 4 kDa inulin chain [219]. It is possible to use solid inulin dispersion as a suitable carrier for incorporating higher rovement in the dissolution and physical stability of formulations.

The comprehensive summary of the different inulin drug delivery systems is provided below in Table 1.

Table 1. Showing a summary of the different delivery systems, drugs delivered, method of fabrication and key outcomes.

Delivery System	Drug Delivered	Method	Key Outcomes
Hydrogels and Beads, Vesicles, Pellets, tablets, and nano-hydrogel	NSAIDs, hormones, peptides, corticosteroids, anticancer, immunoglobulin G (IgG), bovine serum albumin, and lysozyme, MTX, 5-aminosalicylic acid, probiotics strains	radical polymerization radical copolymerization Michael addition crosslinking UV radiation, chemical crosslinkers solvent precipitation	Allow targeted and localized delivery of drugs to the colon
Micelles	curcumin and celecoxib rifampicin doxorubicin	Self-assembly micelles using modified inulin Amphiphilic micelles	Delivers high payload to cancer cells Use of biotin can also improve targeting
liposomes	Quercetin	Shealth Coating	Stabilizes liposome Coat liposome Liposome uptake marker
prodrugs	Metronidazole Procainamide 5-ASA	Conjugation chemistry ester and Schiff base prodrug	Improve stability Increase plasma half-life
Complexes/chelating agent	Gd–inulin complex iron supplements	Chelation and complex formation	Better contrasting agents
Microparticles	Vaccine adjuvants Isoniazid, pyrazinoic acid Doxorubicin Gallic acid Chlorhexidine Indomethacin	Conjugation chemistry coacervation technique Spray drying	Better cellular uptake Targeted delivery to infected cells
Nanoparticles Gold nanoparticles Supermagnetic nanoparticles (SPION) Self-assembly Nanostructured carbon sheet	Ibuprofen Doxorubicin Methylprednisolone	Modification of inulin via chemistry	Attaching ligand such as RGD, biotin improves cellular uptake Better targeting to cancer cells
Solid dispersion	Nifedipine Diazepam, fenofibrates Ritonavir, itraconazole, curcumin, Irbesartan	Inulin as hydrophilic polymer	Result in high loading without recrystallization Improve dissolution and physical stability

3. Toxicological/Clinical Studies Performed Related to Inulin or Inulin-Containing Products

There is limited data on clinical trials of drug formulations containing inulin. However, several reports on the use of inulin microparticles, specifically Advax as vaccine adjuvant in different vaccines, indicate that delta inulin was well tolerated and safe [178,183]. There is no toxic reaction or concern with the safety profile. The use of these inulin microparticles in vaccines tested in humans was found to be safe. This excellent safety profile makes it a good and potential candidate for vaccine adjuvant. Recently, inulin microparticles were utilized as an adjuvant for SARS-CoV-2 vaccine (COVID Vaccine), which is currently moving into human clinical trials.

4. Conclusions and Expert Opinion

This review highlighted the application of inulin as a drug carrier. Inulin possesses good biocompatibility, biodegradability, molecular flexibility, membrane-stabilizing effect, immune modulation effect, and easy chemical modification due to the hydroxyl group. This makes inulin an excellent candidate as a carrier for drug delivery systems and biomaterials. The scope of its use can be increased via modification of its chemical structure. The hydroxyl group serves as an anchor and active site for the modification of inulin. This further allows the fabrications of different inulin drug delivery systems such as hydrogel, micelles, prodrugs, liposomes, solid dispersion, microparticles, and nanoparticles. The exceptional tropism of inulin towards the kidney allows targeted delivery of drugs to the kidney. The colon targeted delivery is possible since inulin is only hydrolyzed by inulinases produced by Bifidobacterium in the colon. The hydroxyl group also confers in inulin good swelling in water when utilized for fabrication of hydrogels. The microparticles and nanoparticle size range permit efficient uptake by the macrophages and monocytes. By functionalization of the existing inulin derivatives with targeting ligands such as biotin, folic acid, and RGD-peptide, the targeting efficiency can be improved. The use of solid dispersion constructed with inulin allows the improvement of the solubility and bioavailability of drugs with poor water solubility. It is worth mentioning that Inulin's potential application for light-controlled, lysosome-triggered, and pH-targeted delivery needs further investigation.

Author Contributions: The authors contributed to this review. F.A.; writing—original draft preparation and editing, S.A.; writing—review and editing, S.H.Y.; writing—review and editing, Y.S.; writing—review and editing and S.G.; supervision, review, and editing. All authors have read and agreed to the published version of the manuscript.

Funding: This research received no external funding.

Institutional Review Board Statement: Not applicable.

Informed Consent Statement: Not applicable.

Data Availability Statement: Not applicable.

Acknowledgments: The authors would also like to thank Candace Day for her help with drawing.

Conflicts of Interest: The authors declare no conflict of interest.

References

1. Hines, D.J.; Kaplan, D.L. Poly(lactic-co-glycolic) Acid-Controlled-Release Systems: Experimental and Modeling Insights. *Crit. Rev. Ther. Drug Carr. Syst.* **2013**, *30*, 257–276. [CrossRef]
2. Lim, T.Y.; Poh, C.K.; Wang, W. Poly (lactic-co-glycolic acid) as a controlled release delivery device. *J. Mater. Sci. Mater. Electron.* **2009**, *20*, 1669–1675. [CrossRef]
3. Tardif, H.; Arnold, C.; Hayes, C.; Eagar, K. Establishment of the Australasian Electronic Persistent Pain Outcomes Collaboration. *Pain Med.* **2016**, *18*, 201–1018. [CrossRef]
4. Liu, Z.; Jiao, Y.; Wang, Y.; Zhou, C.; Zhang, Z. Polysaccharides-based nanoparticles as drug delivery systems. *Adv. Drug Deliv. Rev.* **2008**, *60*, 1650–1662. [CrossRef]
5. Torres, F.G.; Troncoso, O.P.; Pisani, A.; Gatto, F.; Bardi, G. Natural Polysaccharide Nanomaterials: An Overview of Their Im-munological Properties. *Int. J. Mol. Sci.* **2019**, *20*, 5092. [CrossRef]
6. Kaur, N.; Gupta, A.K. Applications of inulin and oligofructose in health and nutrition. *J. Biosci.* **2002**, *27*, 703–714. [CrossRef] [PubMed]
7. Mensink, M.A.; Frijlink, H.W.; Maarschalk, K.V.D.V.; Hinrichs, W.L. Inulin, a flexible oligosaccharide I: Review of its physicochemical characteristics. *Carbohydr. Polym.* **2015**, *130*, 405–419. [CrossRef]
8. Van Arkel, J.; Sevenier, R.; Hakkert, J.C.; Bouwmeester, H.J.; Koops, A.J.; van der Meer, I.M. Tailor-made fructan synthesis in plants: A review. *Carbohydr. Polym.* **2013**, *93*, 48–56. [CrossRef] [PubMed]

9. Mutanda, T.; Mokoena, M.P.; Olaniran, A.O.; Wilhelmi, B.; Whiteley, C.G. Microbial enzymatic production and applications of short-chain fructooligosaccharides and inulooligosaccharides: Recent advances and current perspectives. *J. Ind. Microbiol. Biotechnol.* **2014**, *41*, 893–906. [CrossRef]
10. Barclay, T.; Ginic-Markovic, M.; Cooper, P.; Petrovsky, N. Inulin—A versatile polysaccharide with multiple pharmaceutical and food chemical uses. *J. Excip. Food Chem.* **2010**, *1*, 27–50.
11. Saengthongpinit, W.; Sajjaanantakul, T. Influence of harvest time and storage temperature on characteristics of inulin from Jerusalem artichoke (Helianthus tuberosus L.) tubers. *Postharvest Biol. Technol.* **2005**, *37*, 93–100. [CrossRef]
12. Koch, K.; Andersson, R.; Rydberg, I.; Åman, P. Influence of harvest date on inulin chain length distribution and sugar profile for six chicory (Cichorium intybus L) cultivars. *J. Sci. Food Agric.* **1999**, *79*, 1503–1506. [CrossRef]
13. Mensink, M.A.; Frijlink, H.W.; Maarschalk, K.V.D.V.; Hinrichs, W.L. Inulin, a flexible oligosaccharide. II: Review of its pharmaceutical applications. *Carbohydr. Polym.* **2015**, *134*, 418–428. [CrossRef] [PubMed]
14. Leyva-Porras, C.; López-Pablos, A.; Alavrez, C.; Perez-Urizar, J.; Saavedra, Z. *Physical Properties of Inulin and Technological Applications*; Ramawat, K.G., Mérillon, J.M., Eds.; Springer: New York, NY, USA, 2015; pp. 959–984.
15. Keenan, D.F.; Resconi, V.C.; Kerry, J.P.; Hamill, R.M. Modelling the influence of inulin as a fat substitute in comminuted meat products on their physico-chemical characteristics and eating quality using a mixture design approach. *Meat Sci.* **2014**, *96*, 1384–1394. [CrossRef]
16. Rodriguez Furlan, L.T.; Padilla, A.P.; Campderros, M.E. Development of reduced fat minced meats using inulin and bovine plasma proteins as fat replacers. *Meat Sci.* **2014**, *96*, 762–768. [CrossRef] [PubMed]
17. Kocer, D.; Hicsasmaz, Z.; Bayindirli, A.; Katnas, S. Bubble and pore formation of the high-ratio cake formulation with polydextrose as a sugar- and fat-replacer. *J. Food Eng.* **2007**, *78*, 953–964. [CrossRef]
18. Rodríguez-García, J.; Salvador, A.; Hernando, I. Replacing Fat and Sugar with Inulin in Cakes: Bubble Size Distribution, Physical and Sensory Properties. *Food Bioprocess Technol.* **2014**, *7*, 964–974. [CrossRef]
19. Mittal, S.; Bajwa, U. Effect of fat and sugar substitution on the quality characteristics of low calorie milk drinks. *J. Food Sci. Technol.* **2011**, *49*, 704–712. [CrossRef]
20. Pintor, A.; Severiano-Pérez, P.; Totosaus, A. Optimization of fat-reduced ice cream formulation employing inulin as fat replacer via response surface methodology. *Food Sci. Technol. Int.* **2013**, *20*, 489–500. [CrossRef]
21. Van Loo, J.; Coussement, P.; De Leenheer, L.; Hoebregs, H.; Smits, G. On the presence of Inulin and Oligofructose as natural in-gredients in the western diet. *Crit. Rev. Food Sci. Nutr.* **1995**, *35*, 525–552. [CrossRef]
22. Coussement, P.A.A. Inulin and Oligofructose: Safe Intakes and Legal Status. *J. Nutr.* **1999**, *129*, 1412S–1417S. [CrossRef] [PubMed]
23. Rodriguez Furlan, L.T.; Perez Padilla, A.; Campderros, M.E. Improvement of gluten-free bread properties by the incorporation of bovine plasma proteins and different saccharides into the matrix. *Food Chem.* **2015**, *170*, 257–264. [CrossRef] [PubMed]
24. Rezaei, R.; Khomeiri, M.; Aalami, M.; Kashaninejad, M. Effect of inulin on the physicochemical properties, flow behavior and probiotic survival of frozen yogurt. *J. Food Sci. Technol.* **2014**, *51*, 2809–2814. [CrossRef]
25. Laguna, L.; Primo-Martín, C.; Salvador, A.; Sanz, T. Inulin and Erythritol as Sucrose Replacers in Short-dough Cookies: Sensory, Fracture, and Acoustic Properties. *J. Food Sci.* **2013**, *78*, S777–S784. [CrossRef] [PubMed]
26. Kelly, G. Inulin-type prebiotics—A review: Part 1. *Altern. Med. Rev. J. Clin. Ther.* **2008**, *13*, 315–329.
27. Matusek, A.; Merész, P.; Le, T.K.D.; Örsi, F. Effect of temperature and pH on the degradation of fructo-oligosaccharides. *Eur. Food Res. Technol.* **2009**, *228*, 355–365. [CrossRef]
28. Roberfroid, M.B. Introducing inulin-type fructans. *Br. J. Nutr.* **2005**, *93*, S13–S25. [CrossRef]
29. Shoaib, M.; Shehzad, A.; Omar, M.; Rakha, A.; Raza, H.; Sharif, H.R.; Shakeel, A.; Ansari, A.; Niazi, S. Inulin: Properties, health benefits and food applications. *Carbohydr. Polym.* **2016**, *147*, 444–454. [CrossRef]
30. Chaito, C.; Judprasong, K.; Puwastien, P. Inulin content of fortified food products in Thailand. *Food Chem.* **2016**, *193*, 102–105. [CrossRef]
31. Karimi, R.; Azizi, M.H.; Ghasemlou, M.; Vaziri, M. Application of inulin in cheese as prebiotic, fat replacer and texturizer: A review. *Carbohydr. Polym.* **2015**, *119*, 85–100. [CrossRef]
32. Grasmeijer, N.; Stankovic, M.; de Waard, H.; Frijlink, H.W.; Hinrichs, W.L.J. Unraveling protein stabilization mechanisms: Vitrifica-tion and water replacement in a glass transition temperature controlled system. *Biochim. Biophys. Acta (BBA) Proteins Proteom.* **2013**, *1834*, 763–769. [CrossRef] [PubMed]
33. Haj-Ahmad, R.R.; Elkordy, A.A.; Chaw, C.S.; Moore, A. Compare and contrast the effects of surfactants (Pluronic®F-127 and Cremophor®EL) and sugars (β-cyclodextrin and inulin) on properties of spray dried and crystallised lysozyme. *Eur. J. Pharm. Sci.* **2013**, *49*, 519–534. [CrossRef]
34. Saluja, V.; Amorij, J.-P.; Kapteyn, J.; de Boer, A.; Frijlink, H.; Hinrichs, W. A comparison between spray drying and spray freeze drying to produce an influenza subunit vaccine powder for inhalation. *J. Control. Release* **2010**, *144*, 127–133. [CrossRef] [PubMed]
35. Hinrichs, W.; Prinsen, M.; Frijlink, H.W. Inulin glasses for the stabilization of therapeutic proteins. *Int. J. Pharm.* **2001**, *215*, 163–174. [CrossRef]
36. Wahjudi, M.; Murugappan, S.; van Merkerk, R.; Eissens, A.C.; Visser, M.R.; Hinrichs, W.L.; Quax, W.J. Development of a dry, stable and inhalable acyl–homoserine–lactone–acylase powder formulation for the treatment of pulmonary Pseudomonas aeruginosa infections. *Eur. J. Pharm. Sci.* **2013**, *48*, 637–643. [CrossRef]

37. Srinarong, P.; Faber, J.; Visser, M.; Hinrichs, W.; Frijlink, H.W. Strongly enhanced dissolution rate of fenofibrate solid dispersion tablets by incorporation of superdisintegrants. *Eur. J. Pharm. Biopharm.* **2009**, *73*, 154–161. [CrossRef]
38. Broesder, A.; Berends, J.M.E.; Scheepers, S.M.; Nguyen, D.N.; Frijlink, H.W.; Hinrichs, W.L.J. Ileo-Colon Targeting of the Poorly Water-Soluble Drug Celecoxib Using a pH-Dependent Coating in Combination with Self-Emulsifying Drug Delivery or Solid Dispersion Systems. *Pharmaceutics* **2021**, *13*, 731. [CrossRef] [PubMed]
39. Swarup, P.; Agrawal, G.P. Solid Dispersion: A Mechanistic and Realistic Approach on Antihypertensive Drug as a Drug Carrier System. *ASSAY Drug Dev. Technol.* **2021**, *19*, 282–289. [CrossRef] [PubMed]
40. Traynor, J.; Mactier, R.; Geddes, C.C.; Fox, J.G. How to measure renal function in clinical practice. *BMJ* **2006**, *333*, 733–737. [CrossRef]
41. Orlando, R.; Floreani, M.; Padrini, R.; Palatini, P. Determination of inulin clearance by bolus intravenous injection in healthy subjects and ascitic patients: Equivalence of systemic and renal clearances as glomerular filtration markers. *Br. J. Clin. Pharmacol.* **1998**, *46*, 605–609. [CrossRef] [PubMed]
42. Davidson, M.H.; Maki, K.C. Effects of Dietary Inulin on Serum Lipids. *J. Nutr.* **1999**, *129*, 1474S–1477S. [CrossRef]
43. Hond, E.D.; Geypens, B.; Ghoos, Y. Effect of high performance chicory inulin on constipation. *Nutr. Res.* **2000**, *20*, 731–736. [CrossRef]
44. Marteau, P.; Jacobs, H.; Cazaubiel, M.; Signoret, C.; Prevel, J.-M.; Housez, B. Effects of chicory inulin in constipated elderly people: A double-blind controlled trial. *Int. J. Food Sci. Nutr.* **2010**, *62*, 164–170. [CrossRef] [PubMed]
45. Leenen, C.H.M.; Dieleman, L.A. Inulin and Oligofructose in Chronic Inflammatory Bowel Disease. *J. Nutr.* **2007**, *137*, 2572S–2575S. [CrossRef] [PubMed]
46. Silva, D.G.; Cooper, P.D.; Petrovsky, N. Inulin-derived adjuvants efficiently promote both Th1 and Th2 immune responses. *Immunol. Cell Biol.* **2004**, *82*, 611–616. [CrossRef]
47. Cooper, P.D.; Steele, E.J. The adjuvanticity of gamma inulin. *Immunol. Cell Biol.* **1988**, *66*, 345–352. [CrossRef] [PubMed]
48. Korbelik, M.; Cooper, P.D. Potentiation of photodynamic therapy of cancer by complement: The effect of γ-inulin. *Br. J. Cancer* **2007**, *96*, 67–72. [CrossRef] [PubMed]
49. Weir, C.; Oksa, A.; Millar, J.; Alexander, M.; Kynoch, N.; Walton-Weitz, Z.; MacKenzie-Wood, P.; Tam, F.; Richards, H.; Naylor, R.; et al. The Safety of an Adjuvanted Autologous Cancer Vaccine Platform in Canine Cancer Patients. *Vet. Sci.* **2018**, *5*, 87. [CrossRef] [PubMed]
50. Cooper, P.D.; Carter, M. The anti-melanoma activity of inulin in mice. *Mol. Immunol.* **1986**, *23*, 903–908. [CrossRef]
51. Anda-Flores, D.; Carvajal-Millan, E.; Campa-Mada, A.; Lizardi-Mendoza, J.; Rascon-Chu, A.; Tanori-Cordova, J.; Martínez-López, A.L. Polysaccharide-Based Nanoparticles for Colon-Targeted Drug Delivery Systems. *Polysaccharides* **2021**, *2*, 626–647.
52. Antunes, J.C.; Seabra, C.L.; Domingues, J.M.; Teixeira, M.O.; Nunes, C.; Costa-Lima, S.A.; Homem, N.C.; Reis, S.; Amorim, M.T.P.; Felgueiras, H.P. Drug Targeting of Inflammatory Bowel Diseases by Biomolecules. *Nanomaterials* **2021**, *11*, 2035. [CrossRef]
53. Sun, Q.; Arif, M.; Chi, Z.; Li, G.; Liu, C.-G. Macrophages-targeting mannosylated nanoparticles based on inulin for the treatment of inflammatory bowel disease (IBD). *Int. J. Biol. Macromol.* **2021**, *169*, 206–215. [CrossRef]
54. Pitarresi, G.; Tripodo, G.; Calabrese, R.; Craparo, E.F.; Licciardi, M.; Giammona, G. Hydrogels for Potential Colon Drug Release by Thiol-ene Conjugate Addition of a New Inulin Derivative. *Macromol. Biosci.* **2008**, *8*, 891–902. [CrossRef]
55. Zijlstra, G.S.; Ponsioen, B.J.; Hummel, S.A.; Sanders, N.; Hinrichs, W.L.; de Boer, A.H.; Frijlink, H.W. Formulation and process development of (recombinant human) deoxyribonuclease I as a powder for inhalation. *Pharm. Dev. Technol.* **2009**, *14*, 358–368. [CrossRef]
56. Giri, S.; Dutta, P.; Giri, T.K. Inulin-based carriers for colon drug targeting. *J. Drug Deliv. Sci. Technol.* **2021**, *64*, 102595. [CrossRef]
57. Imran, S.; Gillis, R.B.; Kok, S.M.; Harding, S.E.; Adams, G.G. Application and use of Inulin as a tool for therapeutic drug delivery. *Biotechnol. Genet. Eng. Rev.* **2012**, *28*, 33–46. [CrossRef] [PubMed]
58. Akram, W.; Joshi, R.; Garud, N. Inulin: A promising carrier for controlled and targeted drug delivery system. *J. Drug Deliv. Ther.* **2019**, *9*, 437–441. [CrossRef]
59. Wan, X.; Guo, H.; Liang, Y.; Zhou, C.; Liu, Z.; Li, K.; Niu, F.; Zhai, X.; Wang, L. The physiological functions and pharmaceutical applications of inulin: A review. *Carbohydr. Polym.* **2020**, *246*, 116589. [CrossRef] [PubMed]
60. Wang, W. Instability, stabilization, and formulation of liquid protein pharmaceuticals. *Int. J. Pharm.* **1999**, *185*, 129–188. [CrossRef]
61. Crowe, J.H.; Crowe, L.M.; Carpenter, J.F.; Rudolph, A.; Wistrom, C.A.; Spargo, B.; Anchordoguy, T. Interactions of sugars with membranes. *Biochim. Biophys. Acta (BBA) Rev. Biomembr.* **1988**, *947*, 367–384. [CrossRef]
62. Buitink, J.; Leprince, O. Glass formation in plant anhydrobiotes: Survival in the dry state. *Cryobiology* **2004**, *48*, 215–528. [CrossRef] [PubMed]
63. Crowe, J.H.; Carpenter, J.F.; Crowe, L.M. The Role of Vitrification in Anhydrobiosis. *Annu. Rev. Physiol.* **1998**, *60*, 73–103. [CrossRef] [PubMed]
64. Vereyken, I.J.; Chupin, V.; Islamov, A.; Kuklin, A.; Hincha, D.K.; de Kruijff, B. The Effect of Fructan on the Phospholipid Organization in the Dry State. *Biophys. J.* **2003**, *85*, 3058–3065. [CrossRef]
65. Van den Mooter, G.; Vervoort, L.; Kinget, R. Characterization of methacrylated inulin hydrogels designed for colon targeting: In vitro release of BSA. *Pharm. Res.* **2003**, *20*, 303–307. [CrossRef]
66. Hinrichs, W.L.; Sanders, N.N.; De Smedt, S.C.; Demeester, J.; Frijlink, H.W. Inulin is a promising cryo- and lyoprotectant for PEGylated lipoplexes. *J. Control. Release* **2005**, *103*, 465–479. [CrossRef] [PubMed]

67. Lopez-Molina, D.; Navarro-Martinez, M.D.; Rojas Melgarejo, F.; Hiner, A.N.; Chazarra, S.; Rodriguez-Lopez, J.N. Molecular prop-erties and prebiotic effect of inulin obtained from artichoke (Cynara scolymus L.). *Phytochemistry* **2005**, *66*, 1476–1484. [CrossRef]
68. Flamm, G.; Glinsmann, W.; Kritchevsky, D.; Prosky, L.; Roberfroid, M. Inulin and Oligofructose as Dietary Fiber: A Review of the Evidence. *Crit. Rev. Food Sci. Nutr.* **2001**, *41*, 353–362. [CrossRef]
69. Roberfroid, M.; Delzenne, N. Dietary Fructans. *Annu. Rev. Nutr.* **1998**, *18*, 117–143. [CrossRef]
70. Vervoort, L.; Mooter, G.V.D.; Augustijns, P.; Kinget, R. Inulin hydrogels. I. Dynamic and equilibrium swelling properties. *Int. J. Pharm.* **1998**, *172*, 127–135. [CrossRef]
71. Vervoort, L.; Vinckier, I.; Moldenaers, P.; Mooter, G.V.D.; Augustijns, P.; Kinget, R. Inulin hydrogels as carriers for colonic drug targeting. Rheological characterization of the hydrogel formation and the hydrogel network. *J. Pharm. Sci.* **1999**, *88*, 209–214. [CrossRef]
72. Maris, B.; Verheyden, L.; Van Reeth, K.; Samyn, C.; Augustijns, P.; Kinget, R.; Mooter, G.V.D. Synthesis and characterisation of inulin-azo hydrogels designed for colon targeting. *Int. J. Pharm.* **2001**, *213*, 143–152. [CrossRef]
73. Zabot, G.L.; Silva, E.K.; Azevedo, V.M.; Meireles, M.A.A. Replacing modified starch by inulin as prebiotic encapsulant matrix of lipophilic bioactive compounds. *Food Res. Int.* **2016**, *85*, 26–35. [CrossRef] [PubMed]
74. Mandracchia, D.; Denora, N.; Franco, M.; Pitarresi, G.; Giammona, G.; Trapani, G. New Biodegradable Hydrogels Based on Inulin and alpha, beta-Polyaspartylhydrazide Designed for Colonic Drug Delivery: In Vitro Release of Glutathione and Oxytocin. *J. Biomater. Sci. Polym. Ed.* **2011**, *22*, 313–328. [CrossRef]
75. Tripodo, G.; Pitarresi, G.; Palumbo, F.S.; Craparo, E.F.; Giammona, G. UV-photocrosslinking of inulin derivatives to produce hy-drogels for drug delivery application. *Macromol. Biosci.* **2005**, *5*, 1074–1084. [CrossRef] [PubMed]
76. Sharpe, L.A.; Daily, A.M.; Horava, S.D.; Peppas, N.A. Therapeutic applications of hydrogels in oral drug delivery. *Expert Opin. Drug Deliv.* **2014**, *11*, 901–915. [CrossRef] [PubMed]
77. Peppas, N.A.; Bures, P.; Leobandung, W.; Ichikawa, H. Hydrogels in pharmaceutical formulations. *Eur. J. Pharm. Biopharm.* **2000**, *50*, 27–46. [CrossRef]
78. Caló, E.; Khutoryanskiy, V.V. Biomedical applications of hydrogels: A review of patents and commercial products. *Eur. Polym. J.* **2015**, *65*, 252–267. [CrossRef]
79. Buwalda, S.J.; Boere, K.W.; Dijkstra, P.J.; Feijen, J.; Vermonden, T.; Hennink, W.E. Hydrogels in a historical perspective: From simple networks to smart materials. *J. Control. Release* **2014**, *190*, 254–273. [CrossRef] [PubMed]
80. Tian, B.; Hua, S.; Tian, Y.; Liu, J. Chemical and physical chitosan hydrogels as prospective carriers for drug delivery: A review. *J. Mater. Chem. B* **2020**, *8*, 10050–10064. [CrossRef]
81. Hamedi, H.; Moradi, S.; Hudson, S.M.; Tonelli, A.E. Chitosan based hydrogels and their applications for drug delivery in wound dressings: A review. *Carbohydr. Polym.* **2018**, *199*, 445–460. [CrossRef]
82. Weber, L.M.; Lopez, C.G.; Anseth, K.S. Effects of PEG hydrogel crosslinking density on protein diffusion and encapsulated islet survival and function. *J. Biomed. Mater. Res. Part. A* **2008**, *90*, 720–729. [CrossRef] [PubMed]
83. Vervoort, L.; Mooter, G.V.D.; Augustijns, P.; Busson, R.; Toppet, S.; Kinget, R. Inulin hydrogels as carriers for colonic drug targeting: I. Synthesis and characterization of methacrylated inulin and hydrogel formation. *Pharm. Res.* **1997**, *14*, 1730–1737. [CrossRef]
84. Spizzirri, U.G.; Altimari, I.; Puoci, F.; Parisi, O.I.; Iemma, F.; Picci, N. Innovative antioxidant thermo-responsive hydrogels by radical grafting of catechin on inulin chain. *Carbohydr. Polym.* **2011**, *84*, 517–523. [CrossRef]
85. Palumbo, F.S.; Fiorica, C.; Di Stefano, M.; Pitarresi, G.; Gulino, A.; Agnello, S.; Giammona, G. In situ forming hydrogels of hyaluronic acid and inulin derivatives for cartilage regeneration. *Carbohydr. Polym.* **2015**, *122*, 408–416. [CrossRef]
86. Afinjuomo, F.; Barclay, T.; Song, Y.; Parikh, A.; Petrovsky, N.; Garg, S. Synthesis and characterization of a novel inulin hydrogel crosslinked with pyromellitic dianhydride. *React. Funct. Polym.* **2019**, *134*, 104–111. [CrossRef]
87. Ferreira, L.; Carvalho, R.; Gil, M.H.; Dordick, J.S. Enzymatic synthesis of inulin-containing hydrogels. *Biomacromolecules* **2002**, *3*, 333–341. [CrossRef] [PubMed]
88. Castelli, F.; Sarpietro, M.G.; Micieli, D.; Ottimo, S.; Pitarresi, G.; Tripodo, G.; Carlisi, B.; Giammona, G. Differential scanning calorimetry study on drug release from an inulin-based hydrogel and its interaction with a biomembrane model: pH and loading effect. *Eur. J. Pharm. Sci.* **2008**, *35*, 76–85. [CrossRef] [PubMed]
89. Chiu, H.-C.; Hsu, Y.-H.; Lin, P.-J. Synthesis of pH-sensitive inulin hydrogels and characterization of their swelling properties. *J. Biomed. Mater. Res.* **2002**, *61*, 146–152. [CrossRef]
90. Afinjuomo, F.; Fouladian, P.; Parikh, A.; Barclay, T.G.; Song, Y.; Garg, S. Preparation and Characterization of Oxidized Inulin Hy-drogel for Controlled Drug Delivery. *Pharmaceutics* **2019**, *11*, 356. [CrossRef]
91. Sahiner, N.; Sagbas, S.; Yoshida, H.; Lyon, L.A. Synthesis and Properties of Inulin Based Microgels. *Colloid Interface Sci. Commun.* **2014**, *2*, 15–18. [CrossRef]
92. Kim, S.W.; Bae, Y.H.; Okano, T. Hydrogels: Swelling, Drug Loading, and Release. *Pharm. Res.* **1992**, *09*, 283–290. [CrossRef]
93. Bhattarai, N.; Gunn, J.; Zhang, M. Chitosan-based hydrogels for controlled, localized drug delivery. *Adv. Drug Deliv. Rev.* **2010**, *62*, 83–99. [CrossRef]
94. Wang, J.; Zhao, X.; Zhou, C.; Wang, C.; Zheng, Y.; Ye, K.; Li, C.; Zhou, G. Effects of gellan gum and inulin on mixed-gel properties and molecular structure of gelatin. *Food Sci. Nutr.* **2021**, *9*, 1336–1346. [CrossRef]

95. Martinez, A.W.; Caves, J.M.; Ravi, S.; Li, W.S.; Chaikof, E.L. Effects of crosslinking on the mechanical properties, drug release and cytocompatibility of protein polymers. *Acta Biomater.* **2014**, *10*, 26–33. [CrossRef] [PubMed]
96. Stubbe, B.; Maris, B.; Mooter, G.V.D.; De Smedt, S.C.; Demeester, J. The in vitro evaluation of 'azo containing polysaccharide gels' for colon delivery. *J. Control. Release* **2001**, *75*, 103–114. [CrossRef]
97. Pitarresi, G.; Triolo, D.; Giorgi, M.; Fiorica, C.; Calascibetta, F.; Giammona, G. Inulin-Based Hydrogel for Oral Delivery of Flutamide: Preparation, Characterization, and in vivo Release Studies. *Macromol. Biosci.* **2012**, *12*, 770–778. [CrossRef] [PubMed]
98. López-Molina, D.; Chazarra, S.; How, C.W.; Pruidze, N.; Navarro-Perán, E.; Garcia-Canovas, F.; Garcia-Ruiz, P.A.; Rojas-Melgarejo, F.; Rodríguez-López, J.N. Cinnamate of inulin as a vehicle for delivery of colonic drugs. *Int. J. Pharm.* **2015**, *479*, 96–102. [CrossRef]
99. Atia, A.; Gomaa, A.; Fernandez, B.; Subirade, M.; Fliss, I. Study and Understanding Behavior of Alginate-Inulin Synbiotics Beads for Protection and Delivery of Antimicrobial-Producing Probiotics in Colonic Simulated Conditions. *Probiotics Antimicrob. Proteins* **2018**, *10*, 157–167. [CrossRef]
100. Bahadori, F.; Akinan, B.S.; Akyıl, S.; Eroğlu, M.S. Synthesis and engineering of sodium alginate/inulin core-shell nano-hydrogels for controlled-release oral delivery of 5-ASA. *Org. Commun.* **2019**, *12*, 132–142. [CrossRef]
101. Kim, W.S.; Cho, C.S.; Hong, L.; Han, G.G.; Kil, B.J.; Kang, S.K.; Kim, D.D.; Choi, Y.J.; Huh, C.S. Oral Delivery of Probiotics Using pH-Sensitive Phthalyl Inulin Tablets. *J. Microbiol. Biotechnol.* **2019**, *29*, 200–208. [CrossRef]
102. Hufnagel, B.; Muellner, V.; Hlatky, K.; Tallian, C.; Vielnascher, R.; Guebitz, G.M.; Wirth, M.; Gabor, F. Chemically modified inulin for intestinal drug delivery—A new dual bioactivity concept for inflammatory bowel disease treatment. *Carbohydr. Polym.* **2021**, *252*, 117091. [CrossRef]
103. Cho, H.; Lai, T.C.; Tomoda, K.; Kwon, G.S. Polymeric Micelles for Multi-Drug Delivery in Cancer. *AAPS PharmSciTech* **2014**, *16*, 10–20. [CrossRef] [PubMed]
104. Jones, M.-C.; Leroux, J.-C. Polymeric micelles—A new generation of colloidal drug carriers. *Eur. J. Pharm. Biopharm.* **1999**, *48*, 101–111. [CrossRef]
105. Lukyanov, A.N.; Torchilin, V.P. Micelles from lipid derivatives of water-soluble polymers as delivery systems for poorly soluble drugs. *Adv. Drug Deliv. Rev.* **2004**, *56*, 1273–1289. [CrossRef]
106. Mandracchia, D.; Tripodo, G.; Trapani, A.; Ruggieri, S.; Annese, T.; Chlapanidas, T.; Trapani, G.; Ribatti, D. Inulin based micelles loaded with curcumin or celecoxib with effective anti-angiogenic activity. *Eur. J. Pharm. Sci.* **2016**, *93*, 141–146. [CrossRef]
107. Jain, R.K. Delivery of molecular and cellular medicine to solid tumors. *Adv. Drug Deliv. Rev.* **1997**, *26*, 71–90. [CrossRef]
108. Torchilin, V.P. Structure and design of polymeric surfactant-based drug delivery systems. *J. Control. Release* **2001**, *73*, 137–172. [CrossRef]
109. Mu, L.; Elbayoumi, T.; Torchilin, V. Mixed micelles made of poly(ethylene glycol)–phosphatidylethanolamine conjugate and d-α-tocopheryl polyethylene glycol 1000 succinate as pharmaceutical nanocarriers for camptothecin. *Int. J. Pharm.* **2005**, *306*, 142–149. [CrossRef] [PubMed]
110. Shahin, M.; Ahmed, S.; Kaur, K.; Lavasanifar, A. Decoration of polymeric micelles with cancer-specific peptide ligands for active targeting of paclitaxel. *Biomaterials* **2011**, *32*, 5123–5133. [CrossRef] [PubMed]
111. Gaucher, G.; Dufresne, M.-H.; Sant, V.; Kang, N.; Maysinger, D.; Leroux, J.-C. Block copolymer micelles: Preparation, characterization and application in drug delivery. *J. Control. Release* **2005**, *109*, 169–188. [CrossRef]
112. Yu, Y.; Zhang, L.; Eisenberg, A. Morphogenic Effect of Solvent on Crew-Cut Aggregates of Apmphiphilic Diblock Copolymers. *Macromolecules* **1998**, *31*, 1144–1154. [CrossRef]
113. Choucair, A.; Eisenberg, A. Control of amphiphilic block copolymer morphologies using solution conditions. *Eur. Phys. J. E* **2003**, *10*, 37–44. [CrossRef]
114. Tripodo, G.; Pasut, G.; Trapani, A.; Mero, A.; Lasorsa, F.M.; Chlapanidas, T.; Trapani, G.; Mandracchia, D. Inulin-d-α-Tocopherol Succinate (INVITE) Na-nomicelles as a Platform for Effective Intravenous Administration of Curcumin. *Biomacromolecules* **2015**, *16*, 550–557. [CrossRef] [PubMed]
115. Tripodo, G.; Perteghella, S.; Grisoli, P.; Trapani, A.; Torre, M.L.; Mandracchia, D. Drug delivery of rifampicin by natural micelles based on inulin: Physicochemical properties, antibacterial activity and human macrophages uptake. *Eur. J. Pharm. Biopharm.* **2019**, *136*, 250–258. [CrossRef] [PubMed]
116. Mandracchia, D.; Rosato, A.; Trapani, A.; Chlapanidas, T.; Montagner, I.M.; Perteghella, S.; Di Franco, C.; Torre, M.L.; Trapani, G.; Tripodo, G. Design, synthesis and evaluation of biotin decorated inulin-based polymeric micelles as long-circulating nanocarriers for targeted drug delivery. *Nanomed. Nanotechnol. Biol. Med.* **2017**, *13*, 1245–1254. [CrossRef]
117. Licciardi, M.; Scialabba, C.; Sardo, C.; Cavallaro, G.; Giammona, G. Amphiphilic inulin graft co-polymers as self-assembling micelles for doxorubicin delivery. *J. Mater. Chem. B* **2014**, *2*, 4262–4271. [CrossRef]
118. Muley, P.; Kumar, S.; El Kourati, F.; Kesharwani, S.S.; Tummala, H. Hydrophobically modified inulin as an amphiphilic carbohydrate polymer for micellar delivery of paclitaxel for intravenous route. *Int. J. Pharm.* **2016**, *500*, 32–41. [CrossRef]
119. Di Prima, G.; Saladino, S.; Bongiovì, F.; Adamo, G.; Ghersi, G.; Pitarresi, G.; Giammona, G. Novel inulin-based mucoadhesive micelles loaded with corticosteroids as potential transcorneal permeation enhancers. *Eur. J. Pharm. Biopharm.* **2017**, *117*, 385–399. [CrossRef] [PubMed]
120. Mauro, N.; Campora, S.; Scialabba, C.; Adamo, G.; Licciardi, M.; Ghersi, G.; Giammona, G. Self-organized environment-sensitive inu-lin-doxorubicin conjugate with a selective cytotoxic effect towards cancer cells. *Rsc. Adv.* **2015**, *5*, 32421–32430. [CrossRef]

121. Giammona, G.; Mauro, N.; Scialabba, C. Inulin for Cancer Therapy: Present and Perspectives. *Int. J. Pharm. Res. Rev.* **2016**, *5*, 63–69.
122. Chehardoli, G.; Norouzian, P.; Firozian, F. Inulin-Grafted Stearate (In-g-St) as the Effective Self-Assembling Polymeric Micelle: Synthesis and Evaluation for the Delivery of Betamethasone. *J. Nanomater.* **2020**, *2020*, 1–8. [CrossRef]
123. Shivhare, K.; Garg, C.; Priyam, A.; Gupta, A.; Sharma, A.K.; Kumar, P. Enzyme sensitive smart inulin-dehydropeptide conjugate self-assembles into nanostructures useful for targeted delivery of ornidazole. *Int. J. Biol. Macromol.* **2018**, *106*, 775–783. [CrossRef]
124. Monteiro, N.; Martins, A.; Reis, R.L.; Neves, N.M. Liposomes in tissue engineering and regenerative medicine. *J. R. Soc. Interface* **2014**, *11*, 20140459. [CrossRef]
125. Koning, G.A.; Storm, G. Targeted drug delivery systems for the intracellular delivery of macromolecular drugs. *Drug Discov. Today* **2003**, *8*, 482–483. [CrossRef]
126. Metselaar, J.M.; Storm, G. Liposomes in the treatment of inflammatory disorders. *Expert Opin. Drug Deliv.* **2005**, *2*, 465–476. [CrossRef] [PubMed]
127. Sercombe, L.; Veerati, T.; Moheimani, F.; Wu, S.; Sood, A.K.; Hua, S. Advances and Challenges of Liposome Assisted Drug Delivery. *Front. Pharmacol.* **2015**, *6*, 286. [CrossRef]
128. Mufamadi, M.S.; Pillay, V.; Choonara, Y.E.; Du Toit, L.C.; Modi, G.; Naidoo, D.; Ndesendo, V.M.K. A Review on Composite Liposomal Technologies for Specialized Drug Delivery. *J. Drug Deliv.* **2011**, *2011*, 1–19. [CrossRef] [PubMed]
129. Hua, S.; Wu, S.Y. The use of lipid-based nanocarriers for targeted pain therapies. *Front. Pharmacol.* **2013**, *4*, 143. [CrossRef] [PubMed]
130. Sapra, P.; Allen, T. Ligand-targeted liposomal anticancer drugs. *Prog. Lipid Res.* **2003**, *42*, 439–462. [CrossRef]
131. Kundu, S.K.; Sharma, A.R.; Lee, S.-S.; Sharma, G.; Doss, C.G.P.; Yagihara, S.; Kim, D.-Y.; Nam, J.-S.; Chakraborty, C. Recent Trends of Polymer Mediated Liposomal Gene Delivery System. *BioMed Res. Int.* **2014**, *2014*, 1–15. [CrossRef]
132. Nobs, L.; Buchegger, F.; Gurny, R.; Allemann, E. Current methods for attaching targeting ligands to liposomes and nano-particles. *J. Pharm. Sci.* **2004**, *93*, 1980–1992. [CrossRef]
133. Zylberberg, C.; Gaskill, K.; Pasley, S.; Matosevic, S. Engineering liposomal nanoparticles for targeted gene therapy. *Gene. Ther.* **2017**, *24*, 441–452. [CrossRef]
134. Essien, H.; Lai, J.Y.; Hwang, K.J. Synthesis of diethylenetriaminepentaacetic acid conjugated inulin and utility for cellular uptake of liposomes. *J. Med. Chem.* **1988**, *31*, 898–901. [CrossRef] [PubMed]
135. Rautio, J.; Kumpulainen, H.; Heimbach, T.; Oliyai, R.; Oh, D.; Järvinen, T.; Savolainen, J. Prodrugs: Design and clinical applications. *Nat. Rev. Drug Discov.* **2008**, *7*, 255–270. [CrossRef] [PubMed]
136. Stella, V.J.; Nti-Addae, K.W. Prodrug strategies to overcome poor water solubility. *Adv. Drug Deliv. Rev.* **2007**, *59*, 677–694. [CrossRef] [PubMed]
137. Muller, C.E. Prodrug approaches for enhancing the bioavailability of drugs with low solubility. *Chem. Biodivers.* **2009**, *6*, 2071–2083. [CrossRef]
138. Peterson, L.W.; McKenna, C.E. Prodrug approaches to improving the oral absorption of antiviral nucleotide analogues. *Expert Opin. Drug Deliv.* **2009**, *6*, 405–420. [CrossRef] [PubMed]
139. Karaman, R. Prodrugs for masking bitter taste of antibacterial drugs—a computational approach. *J. Mol. Model.* **2013**, *19*, 2399–2412. [CrossRef] [PubMed]
140. Shin, S.C.; Choi, J.S. Enhanced bioavailability of atenolol by transdermal administration of the ethylene-vinyl acetate matrix in rabbits. *Eur. J. Pharm. Biopharm.* **2003**, *56*, 439–443. [CrossRef]
141. D'Souza, M.; Venkataramanan, R.; D'Mello, A.; Niphadkar, P. An alternative prodrug approach for reducing presystemic metabolism of drugs. *Int. J. Pharm.* **1986**, *31*, 165–167. [CrossRef]
142. Aungst, B.J.; Matz, N. Prodrugs to reduce presystemic metabolism. In *Solvent Systems and Their Selection in Pharmaceutics and Biopharmaceutics*; Augustijns, J., Brewste, M.E., Eds.; Springer: Amsterdam, The Netherlands, 2007; pp. 339–355.
143. Min, Y.; Mao, C.; Chen, S.; Ma, G.; Wang, J.; Liu, Y. Combating the Drug Resistance of Cisplatin Using a Platinum Prodrug Based Delivery System. *Angew. Chem. Int. Ed.* **2012**, *51*, 6742–6747. [CrossRef]
144. Placzek, A.T.; Ferrara, S.J.; Hartley, M.D.; Sanford-Crane, H.S.; Meinig, J.M.; Scanlan, T.S. Sobetirome prodrug esters with en-hanced blood-brain barrier permeability. *Bioorg. Med. Chem.* **2016**, *24*, 5842–5854. [CrossRef]
145. Rautio, J.; Laine, K.; Gynther, M.; Savolainen, J. Prodrug Approaches for CNS Delivery. *AAPS J.* **2008**, *10*, 92–102. [CrossRef] [PubMed]
146. Giang, I.; Boland, E.L.; Poon, G.M.K. Prodrug Applications for Targeted Cancer Therapy. *AAPS J.* **2014**, *16*, 899–913. [CrossRef] [PubMed]
147. Remon, J.; Duncan, R.; Schacht, E. Polymer-drug combinations: Pinocytic uptake of modified polysaccharides containing procainamide moieties by rat visceral yolk sacs cultured -itin vitro. *J. Control. Release* **1984**, *1*, 47–56. [CrossRef]
148. Schacht, E.; Buys, L.; Vermeersch, J.; Remon, J.P. Polymer-drug combinations: Synthesis and characterization of modified polysaccharides containing procainamide moieties. *J. Control. Release* **1984**, *1*, 33–46. [CrossRef]
149. Hartzell, A.L.; Maldonado-Gómez, M.X.; Yang, J.; Hutkins, R.W.; Rose, D.J. In vitro digestion and fermentation of 5-formyl-aminosailcylate-inulin: A potential prodrug of 5-aminosalicylic acid. *Bioact. Carbohydr. Diet. Fibre* **2013**, *2*, 8–14. [CrossRef]
150. Stevens, C.V.; Meriggi, A.; Booten, K. Chemical Modification of Inulin, a Valuable Renewable Resource, and Its Industrial Applications. *Biomacromolecules* **2001**, *2*, 1–16. [CrossRef]

151. Mura, C.; Valenti, D.; Floris, C.; Sanna, R.; De Luca, M.A.; Fadda, A.M.; Loy, G. Metronidazole prodrugs: Synthesis, physico-chemical properties, stability, and ex vivo release studies. *Eur. J. Med. Chem.* **2011**, *46*, 4142–4150. [CrossRef]
152. Vermeersch, J.; Vandoorne, F.; Permentier, D.; Schacht, E. Macromolecular Prodrugs of Metronidazole. Esterification of Hydroxyl Containing Polymers With Metronidazole Monosuccinate. *Bull. Des. Sociétés Chim. Belg.* **2010**, *94*, 591–596. [CrossRef]
153. Wang, W.; Jiang, J.; Ballard, E.C.; Wang, B. Prodrug approaches to the improved delivery of peptide drugs. *Curr. Pharm. Des.* **1999**, *5*, 265–287.
154. Simplício, A.L.; Clancy, J.M.; Gilmer, J.F. Prodrugs for Amines. *Molecules* **2008**, *13*, 519–547. [CrossRef]
155. Barsanti, C.; Lenzarini, F.; Kusmic, C. Diagnostic and prognostic utility of non-invasive imaging in diabetes management. *World J. Diabetes* **2015**, *6*, 792–806. [CrossRef]
156. Caravan, P.; Ellison, J.J.; McMurry, T.J.; Lauffer, R.B. Gadolinium(III) Chelates as MRI Contrast Agents: Structure, Dynamics, and Applications. *Chem. Rev.* **1999**, *99*, 2293–2352. [CrossRef] [PubMed]
157. Mohs, A.M.; Lu, Z.R. Gadolinium(III)-based blood-pool contrast agents for magnetic resonance imaging: Status and clinical po-tential. *Expert Opin Drug Deliv.* **2007**, *4*, 149–164. [CrossRef]
158. Weinmann, H.J.; Brasch, R.C.; Press, W.R.; Wesbey, G.E. Characteristics of gadolinium-DTPA complex: A potential NMR contrast agent. *Am. J. Roentgenol.* **1984**, *142*, 619–624. [CrossRef] [PubMed]
159. Biagi, B.A.; Enyeart, J.J. Gadolinium blocks low- and high-threshold calcium currents in pituitary cells. *Am. J. Physiol. Physiol.* **1990**, *259*, C515–C520. [CrossRef]
160. Aime, S.; Botta, M.; Fasano, M.; Terreno, E. Lanthanide(III) chelates for NMR biomedical applications. *Chem. Soc. Rev.* **1998**, *27*, 19–29. [CrossRef]
161. Peters, J.; Huskens, J.; Raber, D. Lanthanide induced shifts and relaxation rate enhancements. *Prog. Nucl. Magn. Reson. Spectrosc.* **1996**, *28*, 283–350. [CrossRef]
162. Huang, C.-H.; Tsourkas, A. Gd-based macromolecules and nanoparticles as magnetic resonance contrast agents for mo-lecular imaging. *Curr. Top. Med. Chem.* **2013**, *13*, 411–421. [CrossRef]
163. Baxter, A.B.; Lazarus, S.C.; Brasch, R.C. In Vitro Histamine Release Induced by Magnetic Resonance Imaging and Iodinated Contrast Media. *Investig. Radiol.* **1993**, *28*, 308–312. [CrossRef]
164. Corsi, D.M.; Elst, L.V.; Muller, R.N.; van Bekkum, H.; Peters, J.A. Inulin as a carrier for contrast agents in magnetic resonance imaging. *Chemistry* **2001**, *7*, 64–71. [CrossRef]
165. Rebizak, R.; Schaefer, M.; Dellacherie, É. Macromolecular contrast agents for magnetic resonance imaging: Influence of polymer content in ligand on the paramagnetic properties. *Eur. J. Pharm. Sci.* **1999**, *7*, 243–248. [CrossRef]
166. Lebduskova, P.; Kotek, J.; Hermann, P.; Vander, E.L.; Muller, R.N.; Lukes, I.; Peters, J.A. A gadolinium(III) complex of a carbox-ylic-phosphorus acid derivative of diethylenetriamine covalently bound to inulin, a potential macromolecular MRI contrast agent. *Bioconjug. Chem.* **2004**, *15*, 881–889. [CrossRef] [PubMed]
167. Granato, L.; Laurent, S.; Elst, L.V.; Djanashvili, K.; Peters, J.A.; Muller, R.N. The Gd3+ complex of 1,4,7,10-tetraazacyclododecane-1,4,7,10-tetraacetic acid mono(p-isothiocyanatoanilide) conjugated to inulin: A potential stable macromolecular contrast agent for MRI. *Contrast Media Mol. Imaging* **2011**, *6*, 482–491. [CrossRef] [PubMed]
168. Sioud, M. Promises and Challenges in Developing RNAi as a Research Tool and Therapy. *Pept. Microarrays* **2011**, *703*, 173–187. [CrossRef]
169. Cavallaro, G.; Sardo, C.; Scialabba, C.; Licciardi, M.; Giammona, G.; Lamberti, G. Smart inulin-based polycationic nanodevices for siRNA delivery. *Curr. Drug Deliv.* **2016**, *13*, 1. [CrossRef]
170. Sardo, C.; Farra, R.; Licciardi, M.; Dapas, B.; Scialabba, C.; Giammona, G.; Grassi, M.; Grassi, G.; Cavallaro, G. Development of a simple, biocompatible and cost-effective Inulin-Diethylenetriamine based siRNA delivery system. *Eur. J. Pharm. Sci.* **2015**, *75*, 60–71. [CrossRef]
171. Sardo, C.; Craparo, E.F.; Porsio, B.; Giammona, G.; Cavallaro, G. Improvements in Rational Design Strategies of Inulin De-rivative Polycation for siRNA Delivery. *Biomacromolecules* **2016**, *17*, 2352. [CrossRef] [PubMed]
172. Wang, J.; Fu, X.; Bao, Y. Cationization of Inulin via Atom Transfer Radical Polymerization for Gene Delivery. *Chem. J. Chin. Univ.* **2014**, *35*, 2124–2130.
173. Pitarresi, G.; Tripodo, G.; Cavallaro, G.; Palumbo, F.S.; Giammona, G. Inulin–iron complexes: A potential treatment of iron deficiency anaemia. *Eur. J. Pharm. Biopharm.* **2008**, *68*, 267–276. [CrossRef]
174. Cooper, P.D.; Barclay, T.; Ginic-Markovic, M.; Petrovsky, N. The polysaccharide inulin is characterized by an extensive series of periodic isoforms with varying biological actions. *Glycobiology* **2013**, *23*, 1164–1174. [CrossRef]
175. Dolter, K.E.; Evans, C.F.; Ellefsen, B.; Song, J.; Boente-Carrera, M.; Vittorino, R.; Rosenberg, T.J.; Hannaman, D.; Vasan, S. Immunogenicity, safety, biodistribution and persistence of ADVAX, a prophylactic DNA vaccine for HIV-1, delivered by in vivo electroporation. *Vaccine* **2011**, *29*, 795–803. [CrossRef]
176. Honda-Okubo, Y.; Saade, F.; Petrovsky, N. Advax, a polysaccharide adjuvant derived from delta inulin, provides im-proved influenza vaccine protection through broad-based enhancement of adaptive immune responses. *Vaccine* **2012**, *30*, 5373–5381. [CrossRef]
177. Cooper, P.D.; Petrovsky, N. Delta inulin: A novel, immunologically active, stable packing structure comprising beta-D-[2 -> 1] poly(fructo-furanosyl) alpha-D-glucose polymers. *Glycobiology* **2011**, *21*, 595–606. [CrossRef]

178. Petrovsky, N.; Cooper, P.D. Advax™, a novel microcrystalline polysaccharide particle engineered from delta inulin, provides robust adjuvant potency together with tolerability and safety. *Vaccine* **2015**, *33*, 5920–5926. [CrossRef]
179. Petrovsky, N.; Cooper, P.D. Carbohydrate-based immune adjuvants. *Expert Rev. Vaccines* **2011**, *10*, 523–537. [CrossRef]
180. Nishiyama, A.; Tsuji, S.; Yamashita, M.; Henriksen, R.A.; Myrvik, Q.N.; Shibata, Y. Phagocytosis of N-acetyl-d-glucosamine particles, a Th1 adjuvant, by RAW 264.7 cells results in MAPK activation and TNF-α, but not IL-10, production. *Cell. Immunol.* **2006**, *239*, 103–112. [CrossRef] [PubMed]
181. Huang, H.; Ostroff, G.R.; Lee, C.K.; Wang, J.P.; Specht, C.A.; Levitz, S.M. Distinct patterns of dendritic cell cytokine release stimulated by fungal beta-glucans and toll-like receptor agonists. *Infect. Immun.* **2009**, *77*, 1774–1781. [CrossRef] [PubMed]
182. Bielefeldt-Ohmann, H.; Prow, A.N.; Wang, W.; Tan, C.S.; Coyle, M.; Douma, A.; Hobson-Peters, J.; Kidd, L.; Hall, A.R.; Petrovsky, N. Safety and immunogenicity of a delta inulin-adjuvanted inactivated Japanese encephalitis virus vaccine in pregnant mares and foals. *Veter. Res.* **2014**, *45*, 130. [CrossRef] [PubMed]
183. Gordon, D.L.; Sajkov, D.; Woodman, R.J.; Honda-Okubo, Y.; Cox, M.M.; Heinzel, S.; Petrovsky, N. Randomized clinical trial of immu-nogenicity and safety of a recombinant H1N1/2009 pandemic influenza vaccine containing Advax polysaccharide adjuvant. *Vaccine* **2012**, *30*, 5407–5416. [CrossRef] [PubMed]
184. Kumar, S.; Tummala, H. Development of Soluble Inulin Microparticles as a Potent and Safe Vaccine Adjuvant and Delivery System. *Mol. Pharm.* **2013**, *10*, 1845–1853. [CrossRef] [PubMed]
185. Gallovic, M.D.; Montjoy, D.G.; Collier, M.A.; Do, C.; Wyslouzil, B.E.; Bachelder, E.M.; Ainslie, K.M. Chemically modified inulin micro-particles serving dual function as a protein antigen delivery vehicle and immunostimulatory adjuvant. *Biomater. Sci.* **2016**, *4*, 483–493. [CrossRef] [PubMed]
186. Kumar, S.; Kesharwani, S.S.; Kuppast, B.; Rajput, M.; Bakkari, M.A.; Tummala, H. Discovery of inulin acetate as a novel immune-active polymer and vaccine adjuvant: Synthesis, material characterization, and biological evaluation as a toll-like receptor-4 agonist. *J. Mater. Chem. B* **2016**, *4*, 7950–7960. [CrossRef]
187. Afinjuomo, F.; Barclay, T.G.; Parikh, A.; Chung, R.; Song, Y.M.; Nagalingam, G.; Triccas, J.; Wang, L.; Liu, L.; Hayball, J.D.; et al. Synthesis and Characterization of pH-Sensitive Inulin Conjugate of Isoniazid for Monocyte-Targeted Delivery. *Pharmaceutics* **2019**, *11*, 555. [CrossRef] [PubMed]
188. Afinjuomo, F.; Barclay, T.G.; Parikh, A.; Song, Y.; Chung, R.; Wang, L.; Liu, L.; Hayball, J.D.; Petrovsky, N.; Garg, S. Design and Characterization of Inulin Con-jugate for Improved Intracellular and Targeted Delivery of Pyrazinoic Acid to Monocytes. *Pharmaceutics* **2019**, *11*, 243. [CrossRef] [PubMed]
189. Wang, L.; Song, Y.; Parikh, A.; Joyce, P.; Chung, R.; Liu, L.; Afinjuomo, F.; Hayball, J.D.; Petrovsky, N.; Barclay, T.G.; et al. Doxorubicin-Loaded Delta Inulin Conjugates for Controlled and Targeted Drug Delivery: Development, Characterization, and In Vitro Evaluation. *Pharmaceutics* **2019**, *11*, 581. [CrossRef]
190. Robert, P.; García, P.; Reyes, N.; Chávez, J.; Santos, J. Acetylated starch and inulin as encapsulating agents of gallic acid and their release behaviour in a hydrophilic system. *Food Chem.* **2012**, *134*, 1–8. [CrossRef]
191. Wu, X.Y.; Lee, P.I. Preparation and characterization of inulin ester microspheres as drug carriers. *J. Appl. Polym. Sci.* **2000**, *77*, 833–840. [CrossRef]
192. Poulain, N.; Dez, I.; Perrio, C.; Lasne, M.-C.; Prud'Homme, M.-P.; Nakache, E. Microspheres based on inulin for the controlled release of serine protease inhibitors: Preparation, characterization and in vitro release. *J. Control. Release* **2003**, *92*, 27–38. [CrossRef]
193. Jain, A.K.; Sood, V.; Bora, M.; Vasita, R.; Katti, D.S. Electrosprayed inulin microparticles for microbiota triggered targeting of colon. *Carbohydr. Polym.* **2014**, *112*, 225–234. [CrossRef]
194. Samuli, H.; Catherine, P.; Jean-Pierre, B. Passive and Active Tumour Targeting with Nanocarriers. *Curr. Drug Dis. Cover. Technol.* **2011**, *8*, 188–196.
195. Zhang, L.; Li, G.; Gao, M.; Liu, X.; Ji, B.; Hua, R.; Zhou, Y.; Yang, Y. RGD-peptide conjugated inulin-ibuprofen nanoparticles for targeted delivery of Epirubicin. *Colloids Surf. B Biointerfaces* **2016**, *144*, 81–89. [CrossRef] [PubMed]
196. Licciardi, M.; Volsi, A.L.; Mauro, N.; Scialabba, C.; Cavallaro, G.; Giammona, G. Preparation and Characterization of Inulin Coated Gold Nanoparticles for Selective Delivery of Doxorubicin to Breast Cancer Cells. *J. Nanomater.* **2016**, *2016*, 1–12. [CrossRef]
197. Kumar, A.; Mandal, S.; Selvakannan, P.R.; Pasricha, R.; Mandale, A.B.; Sastry, M. Investigation into the Interaction between Surface-Bound Alkylamines and Gold Nanoparticles. *Langmuir* **2003**, *19*, 6277–6282. [CrossRef] [PubMed]
198. Polte, J.; Ahner, T.T.; Delissen, F.; Sokolov, S.; Emmerling, F.; Thünemann, A.F.; Kraehnert, R. Mechanism of Gold Nanoparticle Formation in the Classical Citrate Synthesis Method Derived from Coupled In Situ XANES and SAXS Evaluation. *J. Am. Chem. Soc.* **2010**, *132*, 1296–1301. [CrossRef] [PubMed]
199. Volsi, A.L.; De Aberasturi, D.J.; Henriksen-Lacey, M.; Giammona, G.; Licciardi, M.; Liz-Marzán, L.M. Inulin coated plasmonic gold nanoparticles as a tumor-selective tool for cancer therapy. *J. Mater. Chem. B* **2016**, *4*, 1150–1155. [CrossRef]
200. Liu, L.; Hitchens, T.K.; Ye, Q.; Wu, Y.; Barbe, B.; Prior, D.E.; Li, W.F.; Yeh, F.-C.; Foley, L.M.; Bain, D.J.; et al. Decreased reticuloendothelial system clearance and increased blood half-life and immune cell labeling for nano- and micron-sized superparamagnetic iron-oxide particles upon pre-treatment with Intralipid. *Biochim. Biophys. Acta BBA Gen. Subj.* **2013**, *1830*, 3447–3453. [CrossRef]
201. Chouly, C.; Pouliquen, D.; Lucet, I.; Jeune, J.J.; Jallet, P. Development of superparamagnetic nanoparticles for MRI: Effect of particle size, charge and surface nature on biodistribution. *J. Microencapsul.* **1996**, *13*, 245–255. [CrossRef]
202. Durán, J.; Arias, J.L.; Gallardo, V.; Delgado, A. Magnetic Colloids as Drug Vehicles. *J. Pharm. Sci.* **2008**, *97*, 2948–2983. [CrossRef]

203. Corot, C.; Robert, P.; Idée, J.-M.; Port, M. Recent advances in iron oxide nanocrystal technology for medical imaging. *Adv. Drug Deliv. Rev.* **2006**, *58*, 1471–1504. [CrossRef]
204. Kawasaki, E.S.; Player, A. Nanotechnology, nanomedicine, and the development of new, effective therapies for cancer. *Nanomed. Nanotechnol. Biol. Med.* **2005**, *1*, 101–109. [CrossRef]
205. Liang, X.-J.; Chen, C.; Zhao, Y.; Wang, P.C. Circumventing Tumor Resistance to Chemotherapy by Nanotechnology. *CRISPR Cas Methods* **2010**, *596*, 467–488.
206. Laurent, S.; Saei, A.; Behzadi, S.; Panahifar, A.; Mahmoudi, M. Superparamagnetic iron oxide nanoparticles for delivery of therapeutic agents: Opportunities and challenges. *Expert Opin. Drug Deliv.* **2014**, *11*, 1449–1470. [CrossRef]
207. Scialabba, C.; Licciardi, M.; Mauro, N.; Rocco, F.; Ceruti, M.; Giammona, G. Inulin-based polymer coated SPIONs as potential drug delivery systems for targeted cancer therapy. *Eur. J. Pharm. Biopharm.* **2014**, *88*, 695–705. [CrossRef]
208. Santiago-Rodríguez, L.; Lafontaine, M.M.; Castro, C.; Méndez-Vega, J.; Latorre-Esteves, M.; Juan, E.J.; Mora, E.; Torres-Lugo, M.; Rinaldi, C. Synthesis, stability, cellular uptake, and blood circulation time of carboxymethyl-inulin coated magnetic nanoparticles. *J. Mater. Chem. B* **2013**, *1*, 2807–2817. [CrossRef]
209. Scialabba, C.; Puleio, R.; Peddis, D.; Varvaro, G.; Calandra, P.; Cassata, G.; Cicero, L.; Licciardi, M.; Giammona, G. Folate targeted coated SPIONs as efficient tool for MRI. *Nano Res.* **2017**, *10*, 3212–3227. [CrossRef]
210. Riley, M.K.; Vermerris, W. Recent Advances in Nanomaterials for Gene Delivery—A Review. *Nanomaterials* **2017**, *7*, 94. [CrossRef] [PubMed]
211. Ramamoorth, M.; Narvekar, A. Non Viral Vectors in Gene Therapy- An Overview. *J. Clin. Diagn. Res. JCDR* **2015**, *9*, GE01–GE6. [CrossRef] [PubMed]
212. Lam, J.K.W.; Chow, M.Y.T.; Zhang, Y.; Leung, S.W.S. siRNA Versus miRNA as Therapeutics for Gene Silencing. *Mol. Ther Nucleic Acids* **2015**, *4*, e252. [CrossRef]
213. Licciardi, M.; Volsi, A.L.; Sardo, C.; Mauro, N.; Cavallaro, G.; Giammona, G. Inulin-Ethylenediamine Coated SPIONs Magnetoplexes: A Promising Tool for Improving siRNA Delivery. *Pharm. Res.* **2015**, *32*, 3674–3687. [CrossRef]
214. Zhang, L.; Li, Y.; Wang, C.; Li, G.; Zhao, Y.; Yang, Y. Synthesis of methylprednisolone loaded ibuprofen modified inulin based nanoparticles and their application for drug delivery. *Mater. Sci. Eng. C* **2014**, *42*, 111–115. [CrossRef] [PubMed]
215. Mauro, N.; Scialabba, C.; Cavallaro, G.; Licciardi, M.; Giammona, G. Biotin-Containing Reduced Graphene Oxide-Based Nanosystem as a Multieffect Anticancer Agent: Combining Hyperthermia with Targeted Chemotherapy. *Biomacromolecules* **2015**, *16*, 2766–2775. [CrossRef]
216. Leuner, C.; Dressman, J. Improving drug solubility for oral delivery using solid dispersions. *Eur. J. Pharm. Biopharm.* **2000**, *50*, 47–60. [CrossRef]
217. Srinarong, P.; Kouwen, S.; Visser, M.R.; Hinrichs, W.; Frijlink, H.W. Effect of drug-carrier interaction on the dissolution behavior of solid dispersion tablets. *Pharm. Dev. Technol.* **2009**, *15*, 460–468. [CrossRef] [PubMed]
218. Srinarong, P.; Hamalainen, S.; Visser, M.R.; Hinrichs, W.L.; Ketolainen, J.; Frijlink, H.W. Surface-active derivative of inulin (Inutec(R) SP1) is a superior carrier for solid dispersions with a high drug load. *J. Pharm. Sci.* **2011**, *100*, 2333–2342. [CrossRef] [PubMed]
219. Van Drooge, D.J.; Hinrichs, W.L.; Frijlink, H.W. Anomalous dissolution behaviour of tablets prepared from sugar glass-based solid dispersions. *J. Control. Release* **2004**, *97*, 441–452. [CrossRef] [PubMed]
220. Zijlstra, G.S.; Rijkeboer, M.; Jan van Drooge, D.; Sutter, M.; Jiskoot, W.; van de Weert, M.; Hinrichs, W.L.; Frijlink, H.W. Characterization of a cyclo-sporine solid dispersion for inhalation. *AAPS J.* **2007**, *9*, E190–E199. [CrossRef] [PubMed]
221. Fares, M.M.; Salem, M.t.S.; Khanfar, M. Inulin and poly(acrylic acid) grafted inulin for dissolution enhancement and pre-liminary controlled release of poorly water-soluble Irbesartan drug. *Int. J. Pharm.* **2011**, *410*, 206–211. [CrossRef]
222. Van Drooge, D.-J.; Hinrichs, W.L.J.; Dickhoff, B.H.J.; Elli, M.N.A.; Visser, M.R.; Zijlstra, G.S.; Frijlink, H.W. Spray freeze drying to produce a stable Δ9-tetrahydrocannabinol containing inulin-based solid dispersion powder suitable for inhalation. *Eur. J. Pharm. Sci.* **2005**, *26*, 231–240. [CrossRef]
223. Fares, M.M.; Salem, M.T.S. Dissolution enhancement of curcumin via curcumin–prebiotic inulin nanoparticles. *Drug Dev. Ind. Pharm.* **2015**, *41*, 1785–1792. [CrossRef] [PubMed]
224. Van den Mooter, G.; Weuts, I.; De Ridder, T.; Blaton, N. Evaluation of Inutec SP1 as a new carrier in the formulation of solid dispersions for poorly soluble drugs. *Int. J. Pharm.* **2006**, *316*, 1–6. [CrossRef] [PubMed]
225. Visser, M.R.; Baert, L.; Klooster, G.V.T.; Schueller, L.; Geldof, M.; Vanwelkenhuysen, I.; De Kock, H.; De Meyer, S.; Frijlink, H.W.; Rosier, J.; et al. Inulin solid dispersion technology to improve the absorption of the BCS Class IV drug TMC. *Eur. J. Pharm. Biopharm.* **2010**, *74*, 233–238. [CrossRef] [PubMed]
226. Van Drooge, D.J.; Hinrichs, W.L.J.; Wegman, K.A.M.; Visser, M.R.; Eissens, A.C.; Frijlink, H.W. Solid dispersions based on inulin for the stabilisation and formulation of Δ9-tetrahydrocannabinol. *Eur. J. Pharm. Sci.* **2004**, *21*, 511–518. [CrossRef] [PubMed]
227. Kiumarsi, M.; Majchrzak, D.; Jäger, H.; Song, J.; Lieleg, O.; Shahbazi, M. Comparative study of instrumental properties and sensory profiling of low-calorie chocolate containing hydrophobically modified inulin. Part II: Proton mobility, topological, tribological and dynamic sensory properties. *Food Hydrocol.* **2021**, *110*, 106144. [CrossRef]
228. Srinarong, P.; Pham, B.T.; Holen, M.; Van Der Plas, A.; Schellekens, R.C.; Hinrichs, W.L.; Frijlink, H.W. Preparation and physicochemical evaluation of a new tacrolimus tablet formulation for sublingual administration. *Drug Dev. Ind. Pharm.* **2011**, *38*, 490–500. [CrossRef]

MDPI
St. Alban-Anlage 66
4052 Basel
Switzerland
www.mdpi.com

Pharmaceuticals Editorial Office
E-mail: pharmaceuticals@mdpi.com
www.mdpi.com/journal/pharmaceuticals

www.ingramcontent.com/pod-product-compliance
Lightning Source LLC
Chambersburg PA
CBHW061609220926
51850CB00005B/461
9783725804665